INTRODUCTION TO MATRICES AND LINEAR TRANSFORMATIONS

A SERIES OF BOOKS IN THE MATHEMATICAL SCIENCES

EDITOR: Victor Klee

THIRD EDITION

INTRODUCTION TO MATRICES AND LINEAR TRANSFORMATIONS

DANIEL T. FINKBEINER II

Kenyon College

W. H. FREEMAN AND COMPANY

San Francisco

Library of Congress Cataloging in Publication Data

Finkbeiner, Daniel Talbot, 1919–
   Introduction to matrices and linear transformations.
   (A Series of books in the mathematical sciences)
   Includes index.
   1. Matrices. 2. Algebras, Linear. I. Title.
QA188.F56 1978   512.9'43   78-18257
ISBN 0-7167-0084-0

AMS 1970 subject classification: 15-01

INTRODUCTION TO MATRICES AND LINEAR TRANSFORMATIONS
THIRD EDITION

9 8 7 6 5 4 3 2 1

PREFACE

Persons familiar with earlier editions of this book will observe a number of changes and improvements in this new version. Although the previous approach and organization have been retained, virtually all the exposition has been rewritten, with more illustrative examples, new exercises, and an expanded solutions section. Students learn mathematics by practicing it, and practice can be stimulated by having detailed solutions available as guidance. Students should be encouraged to read *all* the exercises and to note especially those that extend the ideas of the text.

A change of notation has been adopted for the new edition in that linear mappings are treated as *left*-hand operators on vectors, in conformity with customary function notation.

More significantly, this revision begins concretely to allow students time to develop understanding before general concepts are introduced. The book begins with the familiar problem of solving a system of linear equations. This is used to introduce the concept of a vector and to motivate the ideas of vector and matrix algebra. Throughout the book Gaussian elimination is used as a unifying computational technique.

Metric notions of Euclidean space are introduced at an early stage to establish a familiar geometric setting in which the concepts of linear algebra can be interpreted. Euclidean spaces are reexamined more generally in Chapter 8.

The book can be used flexibly for a variety of courses in linear algebra. In view of the differences in pace and emphasis of mathematics instruction at various institutions, the following time estimates should be regarded only as rough guidelines, reflecting experience at Kenyon College. The book contains enough material for a year of linear algebra, with Chapters 1–5 constituting a suitable first-semester course. (If time permits, Chapter 10 can be included as a significant application of special interest to students of business and economics. Alternatively, Chapter 8 can be covered to deepen the earlier exposure to inner product spaces.) Chapters 6–9 comprise a suitable second-semester course, with Chapters 10 and 11 as optional material if time permits.

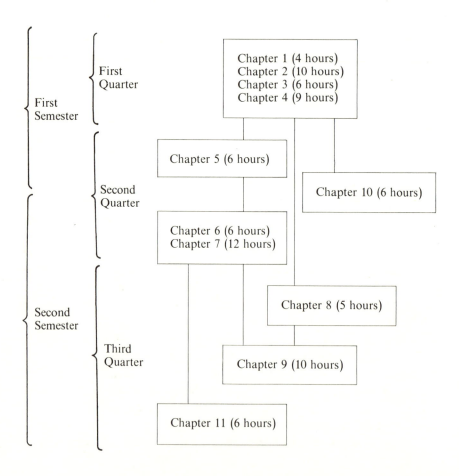

Courses in linear algebra currently are elected by students ranging from freshmen to seniors and with widely different professional goals. Consequently, such courses must be paced carefully, with time allowed for review and assimilation of ideas. As an average rule an instructor should allow at least three class hours for every two sections of text material, allotting some class time for questions and discussions of exercises.

Chapter interdependencies and an illustrative arrangement of courses are shown in the diagram, but each instructor should adjust this schedule to the needs of the class.

I am indebted to many persons—in particular, the Kenyon students who pointed out errors in an earlier draft of this material, to Wendell Lindstrom who offered thoughtful suggestions for improvement, and to Hope Weir who typed the manuscript with rare expertise. The errors and flaws that remain are my responsibility, and I shall welcome the help of readers in finding them and in sending me their comments and criticisms. Special gratitude is expressed to Charles Rice, Richard Hoppe, Gerald Chaplin, George Fowler, Steve Alex, Danny Vaughn, James Carhart, and the late Thomas L. Bogardus, Jr., whose contributions to this work were indirect but most essential.

May 1978   DANIEL T. FINKBEINER II

CONTENTS

INTRODUCTION TO MATRICES AND LINEAR TRANSFORMATIONS

CHAPTER 1
LINEAR
EQUATIONS

Linear algebra is concerned primarily with mathematical systems of a particular type (called *vector spaces*), functions of a particular type (called *linear mappings*), and the algebraic representation of such functions by matrices. If you have completed a course in calculus, you are already familiar with some examples of vector spaces, such as the real number system $\mathbb{R}$ and the Euclidean plane. You also have studied functions from $\mathbb{R}$ to $\mathbb{R}$, so at least superficially the study of linear algebra appears to be a natural extension and generalization of your previous studies. But you should be forewarned that the degree of generalization is substantial and the methods of linear algebra are significantly different from those of calculus.

A glance at the Table of Contents will reveal many terms and topics that might be unfamiliar to you at this stage in your mathematical development. Therefore, as you study this material you will need to pay close attention to the definitions and theorems, assimilating each idea as it arises, gradually building your mathematical vocabulary and your ability to utilize new concepts and techniques. You are urged to make a practice of reading all the exercises and noting the results they contain, whether or not you solve them in detail.

The contents of this book are a blend of formal theory and computational techniques related to that theory. We begin with the problem, familiar from secondary school algebra, of solving a system of linear equations, thereby introducing the idea of a vector space informally. Vector spaces are not defined formally until Section 3 of Chapter 2. At that point, and from time to time thereafter, you are urged to study Appendix A.1, where algebraic systems are explained briefly but generally. You might not need that much generality to understand the concept of a vector space, but firm familiarity with the notion of an algebraic system will greatly accelerate your ability to feel comfortable with the ideas of linear algebra.

Individuals acquire mathematical sophistication and maturity at different rates, and you should not expect to achieve instant success in assimilating some of the more subtle concepts of this course. With patience, persistence, and plenty of practice with specific examples and exercises, you can anticipate steady progress in developing your capacity for abstract thought and careful reasoning. Moreover, you will greatly enhance your insight into the nature of mathematics and your appreciation of its power and beauty.

1.1   SYSTEMS OF LINEAR EQUATIONS

The central focus of this book is the concept of *linearity*. Persons who have studied mathematics through a first course in calculus already are familiar with examples of linearity in elementary algebra, coordinate geometry, and calculus, but they probably are not yet aware of the extent to which linear methods pervade mathematical theory and application. Such awareness will develop gradually throughout this book as we explore the properties and significance of linearity in various mathematical settings.

We begin with the familiar example of a line L in the real coordinate plane, which can be described algebraically by a *linear equation* in two variables:

$$L: ax + by = d.$$

A point (x_0, y_0) of the plane lies on the line L if and only if the real number $ax_0 + by_0$ has the value d. The formal expression

$$ax + by$$

is called a *linear combination* of x and y.

By analogy a linear combination of three variables has the form

$$ax + by + cz,$$

where a, b, and c are constants. Any equation of the form

$$ax + by + cz = d$$

is called a linear equation in three variables. If you have studied the geometry of three-dimensional space, you will recall that the graph of a linear equation in three variables is a *plane*, rather than a line. This is a significant observation: *the word linear refers to the algebraic form of an equation rather than to the geometric object that is its graph.* The two meanings coincide only for the case of two variables—that is, for the coordinate plane. In general, a linear equation in n variables has the form

$$c_1 x_1 + c_2 x_2 + c_3 x_3 + \cdots + c_n x_n = d,$$

where at least one $c_i \neq 0$. For $n > 3$ the graph of this equation in n-dimensional space is called a *hyperplane*.

Applications of mathematics to science and social science frequently lead to the need to solve a *system* of several linear equations in several variables, the coefficients being real numbers:

$$
\begin{aligned}
a_{11}x_1 + a_{12}x_2 + \cdots + a_{1n}x_n &= d_1, \\
a_{21}x_1 + a_{22}x_2 + \cdots + a_{2n}x_n &= d_2, \\
&\ \vdots \\
a_{m1}x_1 + a_{m2}x_2 + \cdots + a_{mn}x_n &= d_m.
\end{aligned}
$$

(1.1)

The number m of equations might be less than, equal to, or greater than the number n of variables. *A solution* of the System 1.1 is an ordered n-tuple $(c_1, \ldots, c_n)$ of real numbers having the property that the substitution

$$
\begin{aligned}
x_1 &= c_1, \\
x_2 &= c_2, \\
&\ \vdots \\
x_n &= c_n,
\end{aligned}
$$

simultaneously satisfies each of the m equations of the system. *The solution* of (1.1) is the set of *all* solutions, and *to solve* the system means to describe the set of all solutions. As we shall see, this set can be finite or infinite.

This problem is considered in algebra courses in secondary school for the case $m = 2 = n$, and sometimes for other small values of m and n. But a large scale linear model in contemporary economics might require the solution of a system of perhaps 83 equations in 115 unknowns. Hence we need to find very efficient procedures for solving (1.1), regardless of the values of m and n, in a finite number of computational steps. Any fixed set of instructions that is guaranteed to solve a particular type of problem in a finite number of steps is called an *algorithm*. Many algorithms exist for solving systems of linear equations, but one of the oldest methods, introduced by Gauss, is also one of the most efficient. Gaussian elimination, and various algorithms related to it, operate on the principle of exchanging the given system (1.1) for another system (1.1A) that has precisely the same set of solutions but one that is easier to solve. Then (1.1A) is exchanged for still another system (1.1B) that has the same solutions as (1.1) but is éven easier than (1.1A) to solve. By increasing the ease of solution at each step, after m or fewer exchanges we obtain a system with the same solutions as (1.1) and in an algebraic form that easily produces the solution. For convenience, we say that two systems of linear equations are *equivalent* if and only if each solution of each system is also a solution of the other.

We first illustrate this idea with a specific example. Soon we shall be able to verify that the following two systems are equivalent, and for the moment we shall assume that they are.

$$6x_1 + 2x_2 - x_3 + 5x_4 = -8, \qquad x_1 \quad - x_3 + x_4 = -3,$$
$$3x_1 + 2x_2 + x_3 + 3x_4 = -1, \quad \text{and} \quad x_2 + 2x_3 \quad = \quad 4,$$
$$4x_1 + x_2 \quad - x_3 + 3x_4 = -6, \qquad \qquad x_3 - x_4 = \quad 2.$$

Obviously, we would prefer to solve the second system. To do so we let x_4 be any number, say c. Then

$$x_4 = \quad c,$$
$$x_3 = \quad 2 + x_4 = 2 + c,$$
$$x_2 = \quad 4 - 2x_3 = 4 - 2(2 + c) = -2c,$$
$$x_1 = -3 + x_3 - x_4 = -3 + (2 + c) - c = -1,$$

and we conclude that for any number c the ordered quadruple

$$\begin{pmatrix} -1 + 0c \\ 0 - 2c \\ 2 + c \\ 0 + c \end{pmatrix}$$

is a solution of the second system and hence of the first. Furthermore, it is easy to see that any solution of the second system must be of that form, and therefore we have produced the complete solution of the first system. There are infinitely many solutions because each value of c produces a different solution. When a system has infinitely many solutions, a complete description of all solutions involves one, two, or more arbitrary constants.

The second system is easy to solve because of its special algebraic form: one of the variables (x_1) appears with nonzero coefficient in the first equation but in no subsequent equation, another variable (x_2) appears with nonzero coefficient in the second equation but in no subsequent equation, and so on. A system of this nature is said to be in *echelon form*. To solve a system that already is in echelon form we first consider the last equation; we solve for the first variable of that equation in terms of the constant term and the subsequent variables. Each subsequent variable may be assigned an arbitrary value. In this case

$$x_4 = c,$$

$$x_3 = 2 + x_4 = 2 + c.$$

Then we consider the next to last equation; we solve for the first variable of that equation, assigning an arbitrary value to any subsequent variable whose value is not already assigned. For this example,

$$x_2 = 4 - 2x_3 = 4 - 2(2 + c) = -2c.$$

Continuing in the same way with each preceding equation, we eventually obtain the complete solution of the system.

What we need, therefore, is a process that leads from a given system of linear equations to an equivalent system that is in echelon form. And that is precisely the process that Gaussian elimination provides, as we now shall see. Beginning with a system in the form (1.1), we can assume that x_1 has a nonzero coefficient in at least one of the m equations. Furthermore, because the solution of a system does not depend on the order in which the equations are written, we can assume further that $a_{11} \neq 0$. Thus we can solve the first equation for x_1 in terms of the other variables:

$$x_1 = a_{11}^{-1}(d_1 - a_{12}x_2 - a_{13}x_3 - \cdots - a_{1n}x_n).$$

We then replace x_1 by this expression in each of the *other* equations. The resulting equations then contain variables x_2 through x_m, and after collecting the coefficients of each of these variables we obtain the equivalent system

(1.1A)
$$
\begin{aligned}
a_{11}x_1 + a_{12}x_2 + a_{13}x_3 + \cdots + a_{1n}x_n &= d_1, \\
b_{22}x_2 + b_{23}x_3 + \cdots + b_{2n}x_n &= e_2, \\
&\ \ \vdots \\
b_{m2}x_2 + b_{m3}x_3 + \cdots + b_{mn}x_n &= e_m.
\end{aligned}
$$

At this stage we need not be concerned with explicit formulas for the new coefficient b_{ij} and the new constants e_i, where $i \geq 2$ and $j \geq 2$. Such formulas result immediately from a bit of routine algebra, and we record the results here for future reference.

$$
b_{ij} = a_{ij} - a_{i1}a_{11}^{-1}a_{1j},
$$
$$
e_i = d_i - a_{i1}a_{11}^{-1}d_1.
$$

The system (1.1A) is said to be obtained from (1.1) by means of a *pivot operation* on the nonzero entry a_{11}.

The second stage of Gaussian elimination leaves the first equation of (1.1A) untouched but repeats the pivot process on the reduced system of $m - 1$ equations in $n - 1$ variables:

$$
\begin{aligned}
b_{22}x_2 + b_{23}x_3 + \cdots + b_{2n}x_n &= e_2, \\
b_{32}x_2 + b_{33}x_3 + \cdots + b_{3n}x_n &= e_3, \\
&\ \ \vdots \\
b_{m2}x_2 + b_{m3}x_3 + \cdots + b_{mn}x_n &= e_m.
\end{aligned}
$$

Conceivably each coefficient b_{i2} is zero; if so, we look at the coefficients b_{i3}, in order, and continue in this way until we find the first *nonzero* coefficient, say b_{rs}. Again because we can write these equations in any order without changing the solutions, we can assume that $r = 2$. Then we pivot on b_{2s}; that is, we solve for x_s as

$$
x_s = b_{2s}^{-1}(e_2 - b_{2,s+1}x_{s+1} - \cdots - b_{2n}x_n),
$$

and substitute this expression for x_s into each of the last $m - 2$ equations.

Together with the original first equation the new system, equivalent to (1.1) and to (1.1A), is of this form:

$$a_{11}x_1 + \cdots + a_{1s}x_s + a_{1,s+1}x_{s+1} + \cdots + a_{1n}x_n = d_1,$$

(1.1B)
$$b_{2s}x_s + b_{2,s+1}x_{s+1} + \cdots + b_{2n}x_n = e_2,$$

$$c_{3,s+1}x_{s+1} + \cdots + c_{3n}x_n = f_3,$$

$$.$$
$$.$$
$$.$$

$$c_{m,s+1}x_{s+1} + \cdots + c_{mn}x_n = f_m.$$

Then the pivot process is repeated again on the last $m - 2$ equations of (1.1B), leaving the first two equations untouched. Continuing in this manner, we eventually obtain a system that is equivalent to (1.1) and is in echelon form.

To illustrate the method of Gaussian elimination we return to our previous example of three equations in four unknowns. The first equation is

$$6x_1 + 2x_2 - x_3 + 5x_4 = -8.$$

We pivot on the coefficient 6 by solving for x_1,

$$x_1 = \tfrac{1}{6}(-8 - 2x_2 + x_3 - 5x_4),$$

substituting this expression in the last two equations, and collecting like terms. The result, which you should verify on scratch paper, is the equivalent system,

$$6x_1 + 2x_2 - x_3 + 5x_4 = -8,$$
$$x_2 + \tfrac{3}{2}x_3 + \tfrac{1}{2}x_4 = 3,$$
$$-\tfrac{1}{3}x_2 - \tfrac{1}{3}x_3 - \tfrac{1}{3}x_4 = -\tfrac{2}{3}.$$

Now we pivot on the coefficient 1 by solving the second equation for x_2,

$$x_2 = 3 - \tfrac{3}{2}x_3 - \tfrac{1}{2}x_4,$$

substituting this expression for x_2 in the third equation, and collecting like terms. Again you should verify that the result is

$$6x_1 + 2x_2 - x_3 + 5x_4 = -8,$$
$$x_2 + \tfrac{3}{2}x_3 + \tfrac{1}{2}x_4 = 3,$$
$$\tfrac{1}{6}x_3 - \tfrac{1}{6}x_4 = \tfrac{1}{3}.$$

Although this new system is in echelon form, we can improve its appearance by multiplying each side of the second equation by 2 and each side of the third equation by 6, obtaining an equivalent system in echelon form:

$$6x_1 + 2x_2 - x_3 + 5x_4 = -8,$$
$$2x_2 + 3x_3 + x_4 = 6,$$
$$x_3 - x_4 = 2.$$

The last equation contains two variables. We assign arbitrary values to all but one, say $x_4 = c$. Then $x_3 = 2 + c$. Using these values for x_3 and x_4 in the second equation, we have $x_2 = -2c$, and then from the first equation we obtain $x_1 = -1$, which agrees with our previous solution.

Suppose we now replace the second equation of this system with a new equation, obtained by adding the two left-hand members and the two right-hand members of the second and third equations,

$$2x_2 + 4x_3 = 8,$$

or equivalently

$$x_2 + 2x_3 = 4.$$

The resulting system is then

$$6x_1 + 2x_2 - x_3 + 5x_4 = -8,$$
$$x_2 + 2x_3 = 4,$$
$$x_3 - x_4 = 2,$$

and it is equivalent to the preceding system. Now we replace the first equation by the equation obtained by subtracting the third equation from the first equation,

$$6x_1 + 2x_2 - 2x_3 + 6x_4 = -10,$$

and then immediately replace that equation by the equation obtained by twice subtracting the second equation from it,

$$6x_1 - 6x_3 + 6x_4 = -18,$$

or in simpler form

$$x_1 - x_3 + x_4 = -3.$$

Then the new system, also in echelon form, is

$$x_1 \quad - \ x_3 + \ x_4 = -3,$$
$$x_2 + 2x_3 \qquad = \quad 4,$$
$$x_3 - \ x_4 = \quad 2.$$

Note that this is precisely the system that we solved when this example was originally introduced.

Let us summarize what we have observed:

(1) A system of m linear equations in n variables is easily solved if that system is in echelon form.
(2) Gaussian elimination is a systematic procedure for replacing a given system of linear equations by an equivalent system that is in echelon form.
(3) Two equivalent systems of linear equations can both be in echelon form and still not be identical; that is, different methods of reducing a system of linear equations to echelon form can produce different (but equivalent) systems of equations in echelon form.

In the next section we shall use these observations to simplify and to formalize Gaussian elimination as a practical computational method for solving systems of linear equations. In Section 1.3 we shall analyze the various types of solutions that can occur; these types are illustrated in the following exercises.

EXERCISES 1.1

1. Use the method of Gaussian elimination to solve each of the following systems of linear equations.

(i)
$$x_1 + \ x_2 + \ x_3 = \quad 3,$$
$$2x_1 \qquad + \ x_3 = \quad 4,$$
$$2x_2 + \ x_3 = \quad 2.$$

(ii)
$$x_1 + 2x_2 + \ x_3 = -1,$$
$$6x_1 + \ x_2 + \ x_3 = -4,$$
$$2x_1 - 3x_2 - \ x_3 = \quad 0,$$
$$-x_1 - 7x_2 - 2x_3 = \quad 7,$$
$$x_1 - \ x_2 \qquad = \quad 1.$$

(iii) $2x_1 + x_2 + 5x_3 = 4,$
$3x_1 - 2x_2 + 2x_3 = 2,$
$5x_1 - 8x_2 - 4x_3 = 1.$

(iv) $x_1 - x_2 + x_3 - x_4 + x_5 = 1,$
$2x_1 - x_2 + 3x_3 + 4x_5 = 2,$
$3x_1 - 2x_2 + 2x_3 + x_4 + x_5 = 1,$
$x_1 + x_3 + 2x_4 + x_5 = 0.$

(v) $x_1 - 2x_2 + 3x_3 = 1,$
$2x_1 - 3x_2 + 5x_3 = 4,$
$3x_1 - 2x_2 + 5x_3 = 11.$

(vi) $2x_1 + 2x_2 - 3x_3 + 4x_4 = 1,$
$x_1 - 2x_2 + x_3 - x_4 = 2,$
$4x_1 - 2x_2 - x_3 + 2x_4 = -1.$

2. In the following system of linear equations the symbol b represents a number whose value is unspecified.

$$x_1 + 3x_2 + 2x_3 = 3,$$

$$-3x_1 + x_2 + 4x_3 = 1,$$

$$5x_1 + 7x_2 + 2x_3 = b.$$

(i) Use Gaussian elimination to find an equivalent system that is in echelon form.

(ii) What value must b have in order that the system have a solution?

(iii) If b is assigned the value determined in (ii), does the system have more than one solution? Write the complete solution.

3. Consider the system (1.1) of m linear equations in n variables.

(i) Let (1.1C) denote the system obtained by replacing the first equation of (1.1) by

$$ka_{11}x_1 + ka_{12}x_2 + \cdots + ka_{1n}x_n = kd_1,$$

where k is any nonzero constant. Explain why (1.1) and (1.1C) are equivalent. Also explain why (1.1) and (1.1C) are not necessarily equivalent if $k = 0$.

(ii) Let (1.1D) denote the system obtained by replacing the first equation of (1.1) by

$$(a_{11} + ka_{21})x_1 + \cdots + (a_{1n} + ka_{2n}) = (d_1 + kd_2).$$

Explain why (1.1) and (1.1D) are equivalent.

(iii) Let (1.1E) denote the system obtained by interchanging the positions of the first two equations of (1.1). Explain why (1.1) and (1.1E) are equivalent.

4. A system of two linear equations in two unknowns,

$$ax + by = e,$$

$$cx + dy = f,$$

can be interpreted geometrically as two lines in the real coordinate plane. The solution of the system consists of all points that lie simultaneously on both lines. By considering the possible points of intersection of two lines, show that this linear system can have no solutions, exactly one solution, or infinitely many solutions. Are these the only possibilities?

5. As a special case of the system (1.1), suppose that $d_1 = d_2 = \cdots = d_m = 0$; let the ordered n-tuples $U = (u_1, \ldots, u_n)$ and $V = (v_1, \ldots, v_n)$ denote two solutions.

(i) Show that $(u_1 + v_1, \ldots, u_n + v_n)$ is a solution.

(ii) Show that $(bu_1, \ldots, bu_n)$ is a solution for any constant b.

(iii) Deduce that for any constants b and c,

$$(bu_1 + cv_1, \ldots, bu_n + cv_n)$$

is a solution. (This last n-tuple can also be denoted by $bU + cV$, and it is therefore referred to as a *linear combination* of the solutions U and V.)

1.2 MATRIX REPRESENTATION OF A LINEAR SYSTEM

After solving a few systems of linear equations by hand, we recognize that a lot of unnecessary writing is involved, even for small values of m and n. However, if we agree to arrange the work so that the symbols x_j for the n variables always appear in the natural order, we can dispense with writing the symbols for those variables because the required computations involve only the constants in each equation. Thus we can represent the essential information of the linear system (1.1) in skeleton form by the array of numbers

(1.2)

$$\begin{pmatrix} a_{11} & a_{12} & \cdots & a_{1n} & d_1 \\ a_{21} & a_{22} & \cdots & a_{2n} & d_2 \\ \cdot & \cdot & & \cdot & \cdot \\ \cdot & \cdot & & \cdot & \cdot \\ \cdot & \cdot & & \cdot & \cdot \\ a_{m1} & a_{m2} & \cdots & a_{mn} & d_m \end{pmatrix}$$

This rectangular array has m horizontal *rows* and $n + 1$ vertical *columns*. The first row displays the coefficients of the variables and the constant term of the first equation; the second row displays the corresponding information of the second equation, and so on. The first column displays all of the coefficients of x_1 in the order in which the equations of the system are written; the second column displays the coefficients of x_2, and so on. The last column displays the constant terms on the right-hand sides of the equations, and the vertical line replaces the equality signs. Such an array is called a *matrix*.

> **Definition 1.1** A rectangular array A of real numbers arranged in r rows and s columns is called a real r-by-s *matrix*. The number in row i and column j is denoted a_{ij}; the first subscript is the *row index* and the second subscript is the *column index*. A one-by-n matrix is called a *row vector*, and an m-by-one matrix is called a *column vector*.

From the preceding section we recall that our strategy for solving a linear system was to replace the original system by an equivalent system in echelon form. This was accomplished by a succession of m or fewer exchanges, each exchange producing a new system equivalent to the original system but closer to echelon form than the previous system. In this section we shall develop the matrix analogue of that strategy.

We have previously observed (Exercise 1.1-3) that each of the following operations on a linear system produces a new system that is equivalent to the one on which the operation was performed:

(M) *Multiply* each side of any equation of the system by the same *nonzero* constant.

(R) *Replace* any equation of the system by the equation obtained by adding to each side of the given equation a multiple c of the corresponding side of some other equation of the system.

(P) *Permute* the equations of the system; that is, rewrite the same equations in different order.

Because each row of the matrix (1.2) represents the corresponding equation in the system (1.1), these operations on the equations of a linear system give rise to three types of *elementary row operations* on a matrix:

$M_i(c)$: *Multiply* each entry in row i by the nonzero constant c,

$R_{i, i+cj}$: *Replace* row i by the sum of row i and c times row j, $j \neq i$,

P_{ij}: *Permute* (interchange) row i and row j, $j \neq i$.

A matrix that results from applying any one of these operations to the rows of a given matrix will represent a new linear system that is equivalent to the system represented by the given matrix. Hence we are led to adopt the following terminology.

Definition 1.2   Let A and B denote two r-by-s matrices. B is said to be *row equivalent* to A if and only if B can be derived from A by applying successively a finite number of elementary row operations.

By the same analogy we arrive at a definition of row echelon form for matrices.

Definition 1.3   An r-by-s matrix is said to be in *row echelon form* if and only if

(a) for some $k \leq r$, each of the first k rows contains a nonzero entry, and each entry of the last $r - k$ rows is zero, and

(b) the first nonzero entry in each nonzero row is 1, and it occurs in a column to the right of the first nonzero entry in any preceding row.

The matrix E, shown below, is in row echelon form with $r = 4$, $s = 5$, and $k = 3$:

$$E = \begin{pmatrix} 1 & 2 & 3 & 4 & 0 \\ 0 & 0 & 1 & 3 & 2 \\ 0 & 0 & 0 & 1 & 1 \\ 0 & 0 & 0 & 0 & 0 \end{pmatrix}.$$

Perhaps you noticed that our definition of row echelon form for matrices is not quite an exact analogue of our description of a linear system in echelon form, because previously we did not require the first nonzero coefficient in each equation to be 1. But by dividing each side of each equation by the first nonzero coefficient of that equation, we obtain an equivalent system in echelon form and having 1 as the first nonzero coefficient of each equation. The reason for standardizing echelon forms in this way will be apparent later in our study.

Next we demonstrate the use of matrices and elementary row operations in solving a system of linear equations, returning to the example introduced in Section 1.1. The matrix that represents the original system is

$$A = \begin{pmatrix} 6 & 2 & -1 & 5 & | & -8 \\ 3 & 2 & 1 & 3 & | & -1 \\ 4 & 1 & -1 & 3 & | & -6 \end{pmatrix}.$$

Using R_i to denote row i of the *current* matrix and an arrow to denote "is replaced by," we perform the following elementary row operations in succession to produce zero in the first column of the second and third rows.

$R_1 \rightarrow \frac{1}{6}R_1$:   multiply the first row by $\frac{1}{6}$.

$R_2 \rightarrow R_2 - 3R_1$: replace the second row by the sum of R_2 and $-3R_1$.

$R_3 \rightarrow R_3 - 4R_1$: replace the third row by the sum of R_3 and $-4R_1$.

Be sure that you understand the notation. Each row operation produces a new current matrix, and the rows of that new matrix are used in the next operation. Thus the second operation, $R_2 \rightarrow R_2 - 3R_1$, instructs you to replace the second row of the current matrix

$$\begin{pmatrix} 1 & \frac{1}{3} & -\frac{1}{6} & \frac{5}{6} & | & -\frac{4}{3} \\ 3 & 2 & 1 & 3 & | & -1 \\ 4 & 1 & -1 & 3 & | & -6 \end{pmatrix}$$

by the sum of the second row and -3 times the first row of that matrix. If you perform these calculations on scratch paper you will verify that the resulting matrix is

$$A_1 = \begin{pmatrix} 1 & \frac{1}{3} & -\frac{1}{6} & \frac{5}{6} & | & -\frac{4}{3} \\ 0 & 1 & \frac{3}{2} & \frac{1}{2} & | & 3 \\ 0 & -\frac{1}{3} & -\frac{1}{3} & -\frac{1}{3} & | & -\frac{2}{3} \end{pmatrix}.$$

Next we apply row operations on A_1 to produce 0 in the second column of the third row, which can be accomplished as follows:

$R_3 \rightarrow R_3 + \frac{1}{3}R_2$: replace the third row by the sum of R_3 and $\frac{1}{3}R_2$.

Again you should verify that the result is

$$A_2 = \begin{pmatrix} 1 & \frac{1}{3} & -\frac{1}{6} & \frac{5}{6} & | & -\frac{4}{3} \\ 0 & 1 & \frac{3}{2} & \frac{1}{2} & | & 3 \\ 0 & 0 & \frac{1}{6} & -\frac{1}{6} & | & \frac{1}{3} \end{pmatrix}.$$

Finally multiply R_3 of A_2 by 6 to produce a matrix that is row equivalent to A and is in row echelon form.

$$A_3 = \begin{pmatrix} 1 & \frac{1}{3} & -\frac{1}{6} & \frac{5}{6} & | & -\frac{4}{3} \\ 0 & 1 & \frac{3}{2} & \frac{1}{2} & | & 3 \\ 0 & 0 & 1 & -1 & | & 2 \end{pmatrix}.$$

The solution of the linear system represented by A_3 is easily obtained from A_3. The last row represents the equation

$$x_3 - x_4 = 2.$$

Let $x_4 = c$, an arbitrary number. Then $x_3 = 2 + c$. The second row represents the equation

$$x_2 + \frac{3}{2}x_3 + \frac{1}{2}x_4 = 3,$$

so

$$x_2 = 3 - \frac{3}{2}(2 + c) - \frac{1}{2}c = -2c.$$

Similarly,

$$x_1 = -\frac{4}{3} - \frac{1}{3}(-2c) + \frac{1}{6}(2 + c) - \frac{5}{6}c = -1.$$

Hence the solution can be written as the column vector

$$\begin{pmatrix} -1 + 0c \\ 0 - 2c \\ 2 + 1c \\ 0 + 1c \end{pmatrix} = \begin{pmatrix} -1 \\ 0 \\ 2 \\ 0 \end{pmatrix} + \begin{pmatrix} 0c \\ -2c \\ 1c \\ 1c \end{pmatrix} = \begin{pmatrix} -1 \\ 0 \\ 2 \\ 0 \end{pmatrix} + c \begin{pmatrix} 0 \\ -2 \\ 1 \\ 1 \end{pmatrix}.$$

If you are not familiar with the method of adding numerical vectors (component-by-component) and the method of multiplying a vector by a number (component-by-component), you may either accept these calculations tentatively, pending our study of vector algebra in the next chapter, or read Section 2.1 at this time.

It is important to notice that the matrix A_1 represents the system obtained in the previous section by a pivot operation on the coefficient 6 in the first equation. And A_2 represents the system obtained previously by a pivot on the coefficient 1 in the second equation of the system represented by A_1. Hence we see that the process of Gaussian elimination was carried out in this example by a systematic use of elementary row operations, rather than by referring to the formulas for pivot operations, cited in Section 1.1.

For emphasis we now describe the Gaussian procedure for reducing a matrix to row echelon form by means of elementary row operations. Given an r-by-s matrix

$$A = \begin{pmatrix} a_{11} & a_{12} & \cdots & a_{1s} \\ a_{21} & a_{22} & \cdots & a_{2s} \\ & & & \\ & & & \\ & & & \\ a_{r1} & a_{r2} & \cdots & a_{rs} \end{pmatrix},$$

we examine in order the entries of the first column to locate the first nonzero entry in the first column. If all entries in the first column are 0 we move to the next column and continue the search. If each entry of the matrix is 0, the matrix is already in row echelon form. Otherwise, we thus locate the first nonzero entry, say a_{pq}. If $p \neq 1$ we then interchange R_1 and R_p. Then we have a matrix of the form

$$B = \begin{pmatrix} 0 & \cdots & 0 & b_{1q} & \cdots & b_{1s} \\ 0 & \cdots & 0 & b_{2q} & \cdots & b_{2s} \\ & & & & & \\ & & & & & \\ & & & & & \\ 0 & \cdots & 0 & b_{rq} & \cdots & b_{rs} \end{pmatrix},$$

where $b_{1q} = a_{pq} \neq 0$. The matrix B is simply the matrix A with R_1 and R_p interchanged. We now divide R_1 by b_{1q}, and for each $i > 1$ we add to R_i the required multiple of R_1 to produce 0 in row i and column q. The resulting matrix is

$$C = \begin{pmatrix} 0 & \cdots & 0 & 1 & * & \cdots & * \\ 0 & \cdots & 0 & 0 & * & \cdots & * \\ & & & & & & \\ & & & & & & \\ & & & & & & \\ 0 & \cdots & 0 & 0 & * & \cdots & * \end{pmatrix},$$

where the exact values for the entries marked with an asterisk are not now of concern. We now repeat this entire procedure for the matrix composed of the last $r - 1$ rows of C. After r or fewer such repetitions, we obtain a matrix that is row equivalent to A and is in row echelon form.

It is easy to see that the required multiple of R_1 to be added to R_i (after dividing R_1 by b_{1q}) is $-b_{iq}$. Hence in C the entry in row $i > 1$ and column $j > q$ is

$$c_{ij} = b_{ij} - b_{iq} b_{1q}^{-1} b_{1j},$$

which for $q = 1$ confirms the formula stated in Section 1.1 for the process of Gaussian elimination.

To recapitulate, in this section we have seen how the work of solving a system of linear equations can be lessened somewhat by representing the system in matrix form. Furthermore, we have formally expressed the process of Gaussian elimination in terms of elementary row operations on matrices. Stated concisely, a linear system can be solved by first reducing the corresponding matrix to row echelon form and then solving the linear system represented by that matrix.

Actually, additional row operations can be applied to a matrix in row echelon form to simplify further the procedure for solving a linear system. We illustrate with the matrix E that appears immediately following Definition 1.3. The idea is to use the leading 1 in each nonzero row to produce 0 in that column in each preceding row. Starting with the leading 1 in R_3 of E we perform the operations

$$R_2 \to R_2 - 3R_3,$$

$$R_1 \to R_1 - 4R_3.$$

The result is

$$E_1 = \begin{pmatrix} 1 & 2 & 3 & 0 & -4 \\ 0 & 0 & 1 & 0 & -1 \\ 0 & 0 & 0 & 1 & 1 \\ 0 & 0 & 0 & 0 & 0 \end{pmatrix}.$$

Now moving to the leading 1 in R_2 of E_1 we let

$$R_1 \to R_1 - 3R_2$$

to obtain

$$E_2 = \begin{pmatrix} 1 & 2 & 0 & 0 & -1 \\ 0 & 0 & 1 & 0 & -1 \\ 0 & 0 & 0 & 1 & 1 \\ 0 & 0 & 0 & 0 & 0 \end{pmatrix}.$$

The matrix E_2 is in row echelon form and it satisfies the additional property

   (c) the first nonzero entry in each nonzero row is the only nonzero entry in that column.

A matrix that satisfies (c) in addition to (a) and (b) of Definition 1.3 is said to be in *reduced echelon form.* It can be proved that for each matrix A there is one and only one matrix E in *reduced* echelon form that is row equivalent to E. We shall return to this idea in Sections 4.3 and 6.1.

   If we regard E as representing a system of four linear equations in four unknowns, the solution is immediately obvious from E_2. The third row of E_2 tells us that $x_4 = 1$, the second row requires that $x_3 = -1$, and the first row yields $x_2 = c$ (arbitrary) and $x_1 = -1 - 2c$. Hence the solution, written in column vector notation, is

$$\begin{pmatrix} -1 - 2c \\ 0 + 1c \\ -1 + 0c \\ 1 + 0c \end{pmatrix} = \begin{pmatrix} -1 \\ 0 \\ -1 \\ 1 \end{pmatrix} + c \begin{pmatrix} -2 \\ 1 \\ 0 \\ 0 \end{pmatrix}.$$

   The reduced echelon form of a matrix results from a modified form of the Gaussian method of solving linear systems; the modified method is called *Gauss-Jordan elimination.*

EXERCISES 1.2

   1. For each of the following matrices use elementary row operations to find a matrix in row echelon form that is row equivalent to the given matrix.

(i) $A = \begin{pmatrix} 1 & 2 & 3 \\ 0 & 1 & 2 \\ 2 & 3 & 0 \end{pmatrix}.$

(ii) $B = \begin{pmatrix} 1 & 3 & 2 \\ 5 & 7 & 2 \\ -3 & 1 & 4 \end{pmatrix}.$

(iii) $C = \begin{pmatrix} -1 & 5 & 6 \\ 1 & 2 & 1 \\ -1 & -3 & -2 \end{pmatrix}.$

(iv) $D = \begin{pmatrix} 1 & 1 & 1 & 0 \\ -1 & 1 & 2 & 1 \\ 1 & 1 & 4 & 4 \end{pmatrix}.$

2. Determine the reduced echelon forms of the matrices in Exercise 1.

3. Solve each of the following systems of linear equations by first reducing a corresponding matrix to echelon form or to reduced echelon form.

(i) $\begin{aligned} x_1 - x_2 + 2x_3 &= 3, \\ 3x_1 - 4x_2 + 5x_3 &= 9, \\ x_1 + x_2 + x_3 &= 6. \end{aligned}$

(ii) $\begin{aligned} x_1 - x_2 \phantom{{}+ 2x_3} &= 2, \\ -x_1 + x_2 + 2x_3 &= -1, \\ x_1 - x_2 + 4x_3 &= 4. \end{aligned}$

(iii) $\begin{aligned} x_1 + 2x_2 + 4x_3 &= 7, \\ x_1 \phantom{{}+ 2x_2} + 2x_3 &= -2, \\ 2x_1 + 3x_2 + 7x_3 &= 9. \end{aligned}$

4. Show that each of the three elementary row operations can be "undone" by an elementary row operation. (For example, if we apply P_{ik} to B to obtain B_1, and then apply P_{ik} to B_1, the result is B. The corresponding question for $M_i(c)$ also is answered readily, and the question for $R_{i,\,i+cj}$ is only slightly more difficult.)

1.3   SOLUTIONS OF A LINEAR SYSTEM

Up to this point our attention has been focused mainly on the mechanics of solving a system of m linear equations in n variables. Now we want to describe the solutions of such a system. Any one solution is an ordered n-tuple of numbers, which we can write as a column vector

$$\begin{pmatrix} c_1 \\ c_2 \\ \cdot \\ \cdot \\ \cdot \\ c_n \end{pmatrix},$$

having the property that the substitution $x_i = c_i$, $i = 1, \ldots, n$, satisfies each of the m equations of the system. The complete solution S of the system is the set of all individual solutions, and we seek to describe the set S. Can S be the void set; that is, can a linear system fail to have any solution? If the system has one solution, can it have more than one? How many more? And so on.

We begin by distinguishing two types of linear systems. The linear system (1.1) is said to be *homogeneous* if and only if $d_1 = d_2 = \cdots = d_m = 0$. If any d_i is nonzero, the system is *nonhomogeneous*. We first consider the homogeneous case:

$$a_{11}x_1 + a_{12}x_2 + \cdots + a_{1n}x_n = 0,$$
$$a_{21}x_1 + a_{22}x_2 + \cdots + a_{2n}x_n = 0,$$

(1.3)

$$a_{m1}x_1 + a_{m2}x_2 + \cdots + a_{mn}x_n = 0.$$

There exists at least one solution; namely, when each $x_i = 0$. Suppose that there are two solutions,

$$U = \begin{pmatrix} u_1 \\ u_2 \\ \cdot \\ \cdot \\ \cdot \\ u_n \end{pmatrix} \quad \text{and} \quad V = \begin{pmatrix} v_1 \\ v_2 \\ \cdot \\ \cdot \\ \cdot \\ v_n \end{pmatrix}.$$

Then for each $i = 1, \ldots, m$,

$$a_{i1}u_1 + a_{i2}u_2 + \cdots + a_{in}u_n = 0,$$

$$a_{i1}v_1 + a_{i2}v_2 + \cdots + a_{in}v_n = 0.$$

Adding these equations yields

$$a_{i1}(u_1 + v_1) + a_{i2}(u_2 + v_2) + \cdots + a_{in}(u_n + v_n) = 0.$$

Since column vectors of the same length can be added, component by component, we deduce the following significant fact.

(A) *If each of two column vectors U and V is a solution of (1.3), then so is their sum, $U + V$.*

Furthermore for any fixed number k, and for each $i = 1, \ldots, m$,

$$k(a_{i1}u_1 + a_{i2}u_2 + \cdots + a_{in}u_n) = 0,$$

$$a_{i1}(ku_1) + a_{i2}(ku_2) + \cdots + a_{in}(ku_n) = 0.$$

Thus

(B) *If a column vector U is a solution of (1.3) and if k is any number, then kU is also a solution.*

By combining Statements A and B we can conclude:

(C) *If U and V are solutions of* (1.3) *and if b and c are any numbers, then bU + cV is also a solution.*

Statement C can be rephrased as follows: any linear combination of solutions of a homogeneous system is also a solution of the system. So now we have a concise description of the set S of all solutions of a homogeneous system: *the zero n-tuple is a solution, and any linear combination of solutions is a solution.* Conceivably the zero solution is the only solution. But suppose there exists a solution B in which at least one component is nonzero. Then each real number k produces a solution kB. We conclude that every homogeneous system has either exactly one solution or infinitely many solutions.

For the case $n = 3$ let us look at the system geometrically. The graph in three-dimensional space of a homogeneous linear equation in three variables,

$$ax + by + cz = 0,$$

is a plane through the origin. A system of m such equations defines m planes, each passing through the origin. Those planes have at least the origin O as a point of intersection. If there is another point P common to all of those planes, then each point on the line through O and P lies on each of those planes. If a point Q not on the line through O and P lies on each plane, then every point on the plane determined by the three points O, P, and Q also lies on each of the m planes of the system. Hence the solution set S of a homogeneous system of m equations in three variables is either a single point (the origin), or an entire line through the origin, or an entire plane through the origin.

Now let us turn to the general, nonhomogeneous system. For $m = n = 2$ this problem was described geometrically in Exercise 1.1-4. You should review that exercise at this time before reading on.

One of the possibilities in the nonhomogeneous case is that no solution exists. This circumstance will be revealed when the matrix of the system is brought into echelon form; at least one of the rows will be of the form

$$(0 \quad 0 \quad \cdots \quad 0 \quad d), \text{ where } d \neq 0.$$

This corresponds to the equation

$$0x_1 + 0x_2 + \cdots + 0x_n = d \neq 0,$$

which is a contradiction for all choices of $x_1, \ldots, x_n$.

Now assume that at least one solution W of the nonhomogeneous system exists:

$$W = \begin{pmatrix} w_1 \\ \cdot \\ \cdot \\ \cdot \\ w_n \end{pmatrix}.$$

And let U denote any solution of the associated homogeneous system obtained by replacing each d_i in the nonhomogeneous system by 0. Then for each i we have

$$a_{i1} w_1 + a_{i2} w_2 + \cdots + a_{in} w_n = d_i ,$$

$$a_{i1} u_1 + a_{i2} u_2 + \cdots + a_{in} u_n = 0.$$

It follows that

$$a_{i1}(w_1 + u_1) + a_{i2}(w_2 + u_2) + \cdots + a_{in}(w_n + u_n) = d_i .$$

That is, $W + U$ is a solution of the nonhomogeneous system. Conversely let Y be any solution of the nonhomogeneous system, and let W be a known solution, as before. Then for each i,

$$a_{i1} y + a_{i2} y + \cdots + a_{in} y_n = d_i ,$$

$$a_{i1} w_1 + a_{i2} w_2 + \cdots + a_{in} y_n = d_i ,$$

$$a_{i1}(y_1 - w_1) + a_{i2}(y_2 - w_2) + \cdots + a_{in}(y_n - w_n) = 0.$$

Hence $Y - W$ is a solution of the associated homogeneous system. That is, $Y = W + U$ for some solution U of the homogeneous system. Hence if the nonhomogeneous system has any solution W, the complete solution can be obtained by adding W to each vector in the set of all solutions of the associated homogeneous equation.

Geometrically, for $n = 3$ a single linear equation,

$$ax + by + cz = d,$$

represents a plane in three-dimensional space. That plane passes through the origin if and only if $d = 0$. Several such equations represent several planes, which might have no points in common. The equations of the associated homogeneous system represent planes that pass through the origin, and each is parallel to the corresponding plane of the nonhomogeneous system. The set of all solutions of the homogeneous system is a point, or a line, or a plane, each containing the origin. If the planes of the nonhomogeneous system do

intersect at a point W, then the complete set of solutions of the nonhomogeneous system is the point W, or a line through W, or a plane through
W, according to the nature of the complete set of solutions of the homogeneous system.

These observations demonstrate that vectors and matrices provide a
natural and convenient language in which to study systems of linear equations. In the next three chapters we shall develop the necessary algebraic and
geometric groundwork for further study of systems of linear equations and
many other aspects of mathematics in which linearity plays a crucial role.
Frequently in that development the answers to quite different questions will
depend critically on properties of the solution set of a system of linear
equations. After we have developed the basic tools of linear mathematics we
shall reexamine some of the ideas of this chapter.

EXERCISES 1.3

1. Describe geometrically the set of all solutions to each of the following
systems of linear equations.

   (i) The system in Exercise 1.2-3(i).

   (ii) The system in Exercise 1.2-3(ii).

   (iii) The system in Exercise 1.2-3(iii).

   (iv)

$$x_1 - 2x_2 + 2x_3 = 0,$$

$$4x_1 - 8x_2 + 8x_3 = 0,$$

$$-3x_1 + 6x_2 - 6x_3 = 0.$$

2. Let A denote the matrix that corresponds to a homogeneous system
of three equations in three variables. Suppose that A is row equivalent to a
matrix in the form

$$\begin{pmatrix} 1 & a & b & 0 \\ 0 & 1 & c & 0 \\ 0 & 0 & 0 & 0 \end{pmatrix}.$$

Describe geometrically the nature of the set of all solutions of that system.

3. Consider the general system of two linear equations in two variables,

$$ax + by = e,$$

$$cx + dy = f.$$

Assume that a is nonzero and at least one of b and d is nonzero, but otherwise make no other assumptions about the constants.

(i) Under what conditions on the constants will the system have no solution (that is, the solution set is void)?

(ii) Under what conditions on the constants will the system have exactly one solution (that is, the solution set contains exactly one element)?

(iii) Under what conditions on the constants will the solution set be infinite?

4. Determine a relationship among the constants a, b, and c which is a necessary and sufficient condition that the following system has a solution:

$$x_1 + x_2 + x_3 + x_4 = a,$$
$$5x_2 + 2x_3 + 4x_4 = b,$$
$$3x_1 - 2x_2 + x_3 - x_4 = c.$$

If that condition is satisfied, will there necessarily be more than one solution? Explain.

CHAPTER 2
LINEAR SPACES

2.1 VECTOR ALGEBRA IN $\mathbb{R}^n$

The purpose of this chapter is to develop a general algebraic structure for the study of linearity and to interpret that structure in geometric terms that are familiar from our past experience with the real coordinate plane and three-dimensional space. That basic algebraic structure is called a *linear space*, or more commonly a *vector space*.

In the preceding chapter we used the word *vector* to refer to an ordered n-tuple of real numbers, written either as a row or a column. Later we shall see that there are many other mathematical entities that justifiably can be called vectors, so it is important to realize that, although any real n-tuple can be regarded as a vector, not every object that we will call a vector can be represented as an ordered n-tuple.

The system of all real numbers will be denoted by $\mathbb{R}^2$, and the set of all ordered n-tuples of real numbers by $\mathbb{R}^n$. An element of $\mathbb{R}$ (a real number) will be denoted by a lowercase letter, such as a, b, x, y. An element of $\mathbb{R}^n$ (a real n-tuple) will be denoted by a capital letter, such as A, B, X, Y. For example, an element of $\mathbb{R}^4$ might be denoted by A, and if we need to exhibit the numerical components of A, we will write either

$$A = (a_1, a_2, a_3, a_4) \qquad \text{or} \qquad A = \begin{pmatrix} a_1 \\ a_2 \\ a_3 \\ a_4 \end{pmatrix}.$$

Proper use of notation is important in any mathematical study, particularly in subjects like linear algebra that involve several types of mathematical objects. You are strongly urged both to pay close attention to the notational conventions that we shall introduce as need arises and to use notation properly in your own work.

Vector algebra in $\mathbb{R}^n$ is concerned with two types of objects: ordered n-tuples of real numbers (called *vectors*) and real numbers (called *scalars*). Two n-tuples are equal if and only if for each j the entry in position j of one vector equals the entry in position j of the other vector. Two algebraic operations involve vectors:

(1) *Vector addition.* The sum of two n-tuples is an n-tuple. If we have $A = (a_1, a_2, \ldots, a_n)$ and $B = (b_1, b_2, \ldots, b_n)$, then we.define

$$A \oplus B = (a_1 + b_1, a_2 + b_2, \ldots, a_n + b_n).$$

(2) *Scalar multiple of a vector.* The product of a scalar and an n-tuple is an n-tuple. If k is in $\mathbb{R}$ and if $A = (a_1, \ldots, a_n)$, then we define

$$k \odot A = (ka_1, ka_2, \ldots, ka_n).$$

The special symbols $\oplus$ and $\odot$ are used for the two arithmetic operations on vectors to point out that vector operations are different from scalar operations. But because we use lowercase letters for scalars and capital letters for vectors, no confusion should arise by writing $A + B$ for vector addition and kA for the scalar multiple of a vector.

It should be noted that scalar multiple and vector sum are precisely the operations needed to form linear combinations of vectors, and every linear combination of vectors in $\mathbb{R}^n$ is a vector in $\mathbb{R}^n$.

The two definitions of vector operations are consistent with the physical interpretation of a vector as a quantity that has both magnitude and direction, such as force or velocity. Any such quantity can be represented by an arrow. If a rectangular coordinate system is placed with the origin at the tail of the arrow, then the tip of the arrow determines uniquely a point in that coordinate system and therefore an ordered set of real numbers that specify

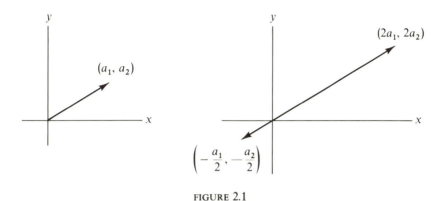

FIGURE 2.1

the coordinates of that point. Figure 2.1 illustrates scalar multiples of a vector in $\mathbb{R}^2$, whereas Figure 2.2 illustrates vector addition in $\mathbb{R}^3$. (For $n > 3$ sketches in $\mathbb{R}^n$ are somewhat more difficult to draw.)

Using elementary analytic geometry, you can easily verify that when vectors A and B extend from the origin O along two different lines, the vector $A + B$ extends from O to the fourth vertex of the parallelogram with A, O, and B as vertices. This is an expression of the physical observation that if two forces, F_1 and F_2, act on a point O, the resultant force F has direction and magnitude equal to that of the diagonal from O of the parallelogram

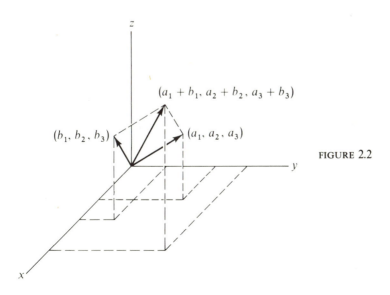

FIGURE 2.2

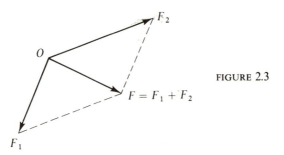

FIGURE 2.3

having the given forces as adjacent sides. This is called the *parallelogram principle* for vector addition, illustrated in Figure 2.3.

We now list some immediate consequences of the definitions of the two algebraic operations for ordered *n*-tuples. Each of these properties can be verified by direct computation, using the corresponding algebraic properties of real numbers.

(1) Vector sum is *commutative*:
$$A + B = B + A \text{ for all } A, B \text{ in } \mathbb{R}^n.$$

(2) Vector sum is *associative*:
$$(A + B) + C = A + (B + C) \text{ for all } A, B, C \text{ in } \mathbb{R}^n.$$

(3) There exists in $\mathbb{R}^n$ a *zero vector*, denoted Z:
$$A + Z = A = Z + A \text{ for all } A \text{ in } \mathbb{R}^n.$$

(4) Each vector A in $\mathbb{R}^n$ has an *additive inverse* in $\mathbb{R}^n$, denoted $-A$:
$$A + (-A) = Z = (-A) + A.$$

(5) For all scalars a, b in $\mathbb{R}$ and each vector A in $\mathbb{R}^n$
$$(a + b)A = aA + bA.$$

(6) For each scalar a in $\mathbb{R}$ and all vectors A, B in $\mathbb{R}^n$
$$a(A + B) = aA + aB.$$

(7) For all scalars a, b in $\mathbb{R}$ and each vector A in $\mathbb{R}^n$
$$(ab)A = a(bA).$$

(8) For each vector A in $\mathbb{R}^n$
$$1A = A.$$

Observe that Properties 1–4 concern vector addition, whereas Properties 5–8 concern scalar multiples of vectors. Properties 2–4 describe a type of algebraic structure that is called a *group*—a set of elements on which there is defined a closed binary operation that satisfies Properties 2–4. Properties 1–4 describe a *commutative group*. Properties 5 and 6 describe modified forms of distributivity, and Property 7 describes a modified associative law. Property 8, which seems almost trivial, further relates the algebra of scalars to the algebra of vectors.

In Section 2.3 we shall describe the term *linear space* by means of a formal definition in which Properties 1–8 appear as *postulates* rather than as provable properties of familiar objects (in the present case, real numbers and real *n*-tuples). At that time we shall adopt a more inclusive interpretation of the terms *vector* and *scalar*. Then we shall prove the following additional properties, which are easily verified for real *n*-tuples.

(9) For each A in $\mathbb{R}^n$, $0A = Z$ (the zero *n*-tuple).
(10) For each A in $\mathbb{R}^n$, $(-1)A = -A$ (the additive inverse of A).
(11) For each a in $\mathbb{R}$, $aZ = Z$ (the zero *n*-tuple).

We now use the parallelogram principle to describe informally some lines in $\mathbb{R}^n$. Observe that a real *n*-tuple A can be interpreted either as a point in space or as an arrow from the origin to that point, whichever is more convenient. By the parallelogram principle the point $A + B$ can be reached from the origin by traveling along the arrow A to point A and then traveling from A along a line segment that projects in the same direction as the arrow B and is as long as the arrow B. More generally, for any real number t, the vector $A + tB$ corresponds to a point X on the line L that passes through the point A and is parallel to arrow B. See Figure 2.4.

If X denotes an arbitrary point on L, then the equation

$$X = A + tB, t \text{ in } \mathbb{R}$$

is a *parametric vector equation* for that line. In terms of components, this vector equation has the form

$$\begin{pmatrix} x_1 \\ x_2 \\ \cdot \\ \cdot \\ \cdot \\ x_n \end{pmatrix} = \begin{pmatrix} a_1 \\ a_2 \\ \cdot \\ \cdot \\ \cdot \\ a_n \end{pmatrix} + t \begin{pmatrix} b_1 \\ b_2 \\ \cdot \\ \cdot \\ \cdot \\ b_n \end{pmatrix} = \begin{pmatrix} a_1 + tb_1 \\ a_2 + tb_2 \\ \cdot \\ \cdot \\ \cdot \\ a_n + tb_n \end{pmatrix}.$$

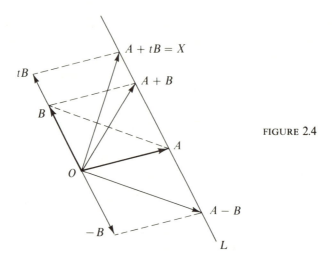

FIGURE 2.4

Hence for each $i = 1, 2, \ldots, n$ we have

$$x_i = a_i + tb_i.$$

If each component of B is nonzero, then

$$t = \frac{x_i - a_i}{b_i} \text{ for } i = 1, \ldots, n,$$

and we can eliminate the parameter t to write *Cartesian equations* of the line L through A and parallel to the segment OB:

$$\frac{x_1 - a_1}{b_1} = \frac{x_2 - a_2}{b_2} = \cdots = \frac{x_n - a_n}{b_n}.$$

The numbers $(b_1, b_2, \ldots, b_n)$ are called *direction numbers* of L.

Again referring to Figure 2.4, we observe that the vector $A - B$ is obtained by translating the directed line segment *from B to A* parallel to itself, carrying its initial point to the origin. That segment is the "other" diagonal of the parallelogram with O, A, and B as three vertices.

To illustrate further this way of thinking about points and lines in $\mathbb{R}^n$, let us describe the line M passing through A and B. Because the directed line segment from B to A is parallel to the vector $A - B$, we can reach any point X on M by traveling from O to B and then traveling an appropriate distance along M. Thus a parametric vector equation for M has the form

$$X = B + t(A - B),$$

$$X = tA + (1 - t)B.$$

An astute reader might object to these physical and geometric inter-pretations of real n-tuples, because they rely on concepts that have not yet been defined for $\mathbb{R}^n$. For example, such terms as "rectangular coordinate system," "magnitude," "direction," "parallel," and "distance" have intui-tive meaning for us in $\mathbb{R}^2$ and $\mathbb{R}^3$ but perhaps not in $\mathbb{R}^n$. Thus the objection is valid, and we defer further investigation of the geometry of $\mathbb{R}^n$ until the next section, where the notions of distance and angle in $\mathbb{R}^n$ are formally introduced.

EXERCISES 2.1

1. In $\mathbb{R}^2$ let $A = (2, -1)$ and $B = (3, 3)$. Using arrows from the origin to represent vectors, sketch and label on one set of axes $A, B, 2A, -B, 2A - B$, $B + 2A$.

2. With A and B as in the preceding exercise locate and label the points defined by

$$tA + (1 - t)B$$

for the following values of t: $t = -1, 0, \frac{1}{2}, \frac{7}{8}, 1, 2$. What locus is generated by letting t assume all real values?

3. In $\mathbb{R}^3$ let $A = (1, 1, 3)$ and $B = (2, -1, 1)$. Describe geometrically the locus generated by each of the following:

(i) tA, as t varies over $\mathbb{R}$.

(ii) $B + tA$, as t varies over $\mathbb{R}$.

(iii) $B + tA$, as t varies over the interval $t \geq 0$.

(iv) $tB + (1 - t)A$, as t varies over the interval $0 \leq t \leq 1$.

4. Let $A = (2, -1, 6)$ and $B = (4, 1, 3)$, and let L denote the line through the point A and parallel to the vector B.

(i) Write a parametric vector equation for L.

(ii) Eliminate the parameter to obtain Cartesian equations for L.

5. With A and B as in the preceding exercise, let M denote the line through the points A and B.

(i) Write a parametric vector equation for M.

(ii) Eliminate the parameter to obtain Cartesian equations for M.

6. By eliminating the parameter t from the next to last equation in this section, write Cartesian equations for the line M through two arbitrary points A and B in $\mathbb{R}^n$. Assume that $a_i \neq b_i$ for every i.

7. Verify Properties 2 and 6 for $\mathbb{R}^n$.

8. Prove Properties 9–11 directly from Properties 1–8, *without* assuming that A is an ordered n-tuple of real numbers. For example, to prove Property 9 start with the special case of Property 5 in which $a = 1$ and $b = 0$; then apply successively Properties 8, 4, and 3.

2.2 EUCLIDEAN n-SPACE

In the previous section we defined vector addition and scalar multiple of a vector as the two basic algebraic operations for $\mathbb{R}^n$, and we listed some algebraic properties of $\mathbb{R}^n$ that are consequences of those definitions. Those two operations on vectors are needed to ensure that each linear combination of vectors in $\mathbb{R}^n$ is also a vector in $\mathbb{R}^n$, a basic requirement of any linear space. However, linearity alone provides no metric information for $\mathbb{R}^n$. How do we compute the familiar quantities—length, distance, and angle—that permeate Euclidean geometry?

Let us begin with the length of a vector. Each n-tuple $A = (a_1, a_2, \ldots, a_n)$ is to be assigned a *length*, which we shall denote by $\|A\|$. Guided by experience, we want $\|A\|$ to be a *positive real number*, except for the special case of the zero n-tuple Z for which $\|Z\| = 0$. It is reasonable to expect that $\|A\|$ can be computed from the components a_i of A, and that the length of kA is $|k|$ times the length of A. Because we hope to extend to $\mathbb{R}^n$ the geometric properties of the Euclidean plane, we also want the familiar theorems of Euclidean geometry to remain valid in $\mathbb{R}^n$; for example, the Pythagorean Theorem, the Law of Cosines, and the theorem that states that the length of one side of a triangle never exceeds the sum of the lengths of the other two sides. Because any triangle OAP can be regarded as half of the parallelogram $OAPB$ (Figure 2.5), the last-mentioned theorem can be expressed as

$$\|A + B\| \leq \|A\| + \|B\|,$$

which is called the *triangle inequality*.

In seeking a suitable definition for length in $\mathbb{R}^n$ it is natural to try to extend the familiar definitions for $n = 1, 2, 3$.

In $\mathbb{R}$, $A = (a_1)$ and $\|A\| = |a_1| = \sqrt{a_1^2}$.

In $\mathbb{R}^2$, $A = (a_1, a_2)$ and $\|A\| = \sqrt{a_1^2 + a_2^2}$.

In $\mathbb{R}^3$, $A = (a_1, a_2, a_3)$ and $\|A\| = \sqrt{a_1^2 + a_2^2 + a_3^2}$.

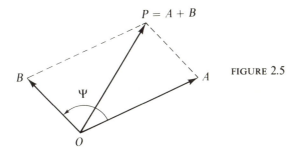

FIGURE 2.5

In $\mathbb{R}^n$ we shall define the *length* $\|A\|$ of the vector $A = (a_1, a_2, \ldots, a_n)$ by the formula

(2.1) $$\|A\| = \sqrt{a_1^2 + a_2^2 + \cdots + a_n^2}.$$

Referring again to Figure 2.4, we define the *distance* from point B to point A to be the length of the vector $A - B$:

(2.2) $$d(A, B) = \|A - B\| = \sqrt{(a_1 - b_1)^2 + \cdots + (a_n - b_n)^2}.$$

Thus

$$\begin{aligned}
\|A - B\|^2 &= (a_1 - b_1)^2 + (a_2 - b_2)^2 + \cdots + (a_n - b_n)^2 \\
&= (a_1^2 + a_2^2 + \cdots + a_n^2) + (b_1^2 + b_2^2 + \cdots + b_n^2) \\
&\quad - 2(a_1 b_1 + a_2 b_2 + \cdots + a_n b_n) \\
&= \|A\|^2 + \|B\|^2 - 2(a_1 b_1 + a_2 b_2 + \cdots + a_n b_n).
\end{aligned}$$

Because $\|A\|$, $\|B\|$, and $\|A - B\|$ are the lengths of three sides of a triangle, this formula reminds us of the Law of Cosines,

$$\|A - B\|^2 = \|A\|^2 + \|B\|^2 - 2\|A\|\,\|B\| \cos \Psi(A, B).$$

Indeed if the Law of Cosines is to be valid in $\mathbb{R}^n$, we are compelled to define the angle $\Psi(A, B)$ between vector A and vector B in such a way that

(2.3) $$\cos \Psi(A, B) = \frac{a_1 b_1 + a_2 b_2 + \cdots + a_n b_n}{(a_1 a_1 + \cdots + a_n a_n)^{1/2}(b_1 b_1 + \cdots + b_n b_n)^{1/2}}.$$

If you look carefully at formulas 2.1, 2.2, and 2.3, you should observe that the central metric concepts—length, distance, and angle—can all be expressed in terms of the real-valued function of $2n$ variables,

$$f(x_1, \ldots, x_n; y_1, \ldots, y_n) = x_1 y_1 + x_2 y_2 + \cdots + x_n y_n.$$

Definition 2.1   Given any two real n-tuples $X = (x_1, \ldots, x_n)$ and $Y = (y_1, \ldots, y_n)$, the *scalar product* of X and Y is the real number $X \cdot Y$ that is defined by the rule

$$X \cdot Y = x_1 y_1 + x_2 y_2 + \cdots + x_n y_n.$$

The scalar product is also called the *dot product* or the *inner product*.

The following properties of the scalar product are easily verified for $\mathbb{R}^n$.

(1) $(dA + eB) \cdot C = d(A \cdot C) + e(B \cdot C),$
    $A \cdot (dB + eC) = d(A \cdot B) + e(A \cdot C).$
(2) $A \cdot B = B \cdot A.$
(3) $A \cdot A > 0$ if $A \neq Z.$

The second statement in (1) asserts that the scalar product of A with a linear combination of two vectors equals that same linear combination of the scalar products of A with each of the two vectors; that is, the scalar product is a *linear function* of its second factor. The first statement of (1) asserts that the scalar product is also a linear function of its first factor. Combining these two assertions, we say that the scalar product is a real *bilinear* function. Property 2 states that the scalar product is commutative (or *symmetric*). Property 3 guarantees that the inner product of a nonzero vector with itself is positive; a bilinear function having that property is said to be *positive definite*. In short, the scalar product in $\mathbb{R}^n$ is a real, bilinear, symmetric, and positive definite function, defined for all pairs of vectors in $\mathbb{R}^n$.

Because distance and angle can be expressed in terms of the scalar product, we could have begun our discussion of metric concepts in $\mathbb{R}^n$ by defining the scalar product. Then length, distance, and angle would be defined in terms of the scalar product, as follows:

$$\|A\| = (A \cdot A)^{1/2},$$

$$d(A, B) = \|A - B\|,$$

$$\cos \Psi(A, B) = \frac{A \cdot B}{\|A\| \, \|B\|}.$$

This is the approach that we shall use later on when we study metric notions in other linear spaces. For now we shall continue to investigate *Euclidean n-space $\mathscr{E}^n$*, by which we mean the vector space $\mathbb{R}^n$ for which the metric concepts are defined as above in terms of the scalar product specified in Definition 2.1.

We now state and prove a theorem of central importance in $\mathscr{E}^n$, called the *Schwarz inequality*.

Theorem 2.1

$$|A \cdot B| \leq \|A\| \|B\| \text{ for all } A, B \text{ in } \mathbb{R}^n.$$

Proof   If either $A = Z$ or $B = Z$, the statement $0 \leq 0$ follows immediately. Otherwise, let A and B be nonzero vectors, and let t be any real number. Then using Properties 1, 2, and 3 we have

$$0 \leq \|A + tB\|^2 = (A + tB) \cdot (A + tB)$$
$$= A \cdot A + t(A \cdot B) + t(B \cdot A) + t^2(B \cdot B)$$
$$= A \cdot A + 2t(A \cdot B) + t^2(B \cdot B).$$

The last expression to the right of the equality sign is a quadratic function of the real variable t. Its graph is a parabola that opens upward, and the inequality tells us that the graph never goes below the t-axis. That means that the quadratic equation

$$(B \cdot B)t^2 + 2(A \cdot B)t + (A \cdot A) = 0$$

does *not* have two distinct real roots. Hence the discriminant of the quadratic formula cannot be positive:

$$[2(A \cdot B)]^2 - 4(B \cdot B)(A \cdot A) \leq 0,$$
$$|A \cdot B| \leq (A \cdot A)^{1/2}(B \cdot B)^{1/2} = \|A\| \|B\|.$$

If you try to prove the Schwarz inequality by working with n-tuples you will appreciate the simplicity of the preceding proof. That proof also has the virtue of using only the properties that characterize an inner product rather than the specific algebraic form of the dot product.

Now we shall apply what we have learned to obtain further geometric information about $\mathscr{E}^n$.

Theorem 2.2   Length has the following properties:
   (a) $\|A\| > 0$ if $A \neq Z$, and $\|Z\| = 0$,
   (b) $\|cA\| = |c| \|A\|$ for all c in $\mathbb{R}$ and all A in $\mathbb{R}^n$,
   (c) $\|A + B\| \leq \|A\| + \|B\|$ for all A, B in $\mathbb{R}^n$.

Proof   Statement (a) follows from Property 3 of the scalar product and the definition of length in terms of the scalar product. Similarly, (b) follows from the bilinearity of the scalar product. To prove (c) we compute

$$\|A + B\|^2 = (A + B) \cdot (A + B) = A \cdot A + 2A \cdot B + B \cdot B$$
$$= \|A\|^2 + 2A \cdot B + \|B\|^2$$
$$\leq \|A\|^2 + 2\|A\|\,\|B\| + \|B\|^2 = (\|A\| + \|B\|)^2.$$

The Schwarz inequality was used at the next to last step.

Theorem 2.3   Distance has the following properties:

(a) $d(A, B) > 0$ if $A \neq B$, and $d(A, A) = 0$,

(b) $d(A, B) = d(B, A)$ for all A, B in $\mathbb{R}^n$,

(c) $d(A, B) \leq d(A, C) + d(C, B)$ for all A, B, C in $\mathbb{R}^n$.

Proof   Exercise.

Any nonvoid set on which is defined a function d having the properties asserted in Theorem 2.3 is called a *metric space*. Hence $\mathcal{E}^n$ is a metric space; however not every metric space is Euclidean.

Theorem 2.4   Two nonzero vectors A and B in $\mathbb{R}^n$ are orthogonal (perpendicular) if and only if $A \cdot B = 0$.

Proof   The definition of cos $\Psi(A, B)$ provides an immediate proof.

Thus the scalar product provides a simple and direct computational criterion for orthogonality. Because the scalar product of any vector with the zero vector is zero, we define the zero vector to be orthogonal to every vector, including itself. As another example of orthogonality we now state and prove the Pythagorean Theorem and its converse. Be sure that you understand why Theorem 2.5, as stated, deserves to be called the Pythagorean Theorem. Other familiar theorems of Euclidean geometry are included in the exercises.

Theorem 2.5 $\|A + B\|^2 = \|A\|^2 + \|B\|^2$ if and only if $A \cdot B = 0$.

Proof $\|A + B\|^2 = (A + B) \cdot (A + B) = A \cdot A + 2A \cdot B + B \cdot B$
$$= \|A\|^2 + \|B\|^2 + 2A \cdot B.$$

Hence the equality asserted in the theorem holds if and only if $A \cdot B = 0$. What could be simpler!

As a final illustration of these methods we consider the problem of finding a vector X in $\mathscr{E}^n$ that is orthogonal to each of m specified vectors A_1, ..., A_m. If we denote X by $(x_1, \ldots, x_n)$ and A_i by $(a_{i1}, \ldots, a_{in})$, then X is simultaneously orthogonal to the given vectors A_i if and only if

$$A_i \cdot X = 0 \qquad \text{for } i = 1, \ldots, m.$$

Written out in detail, this is a homogeneous system of m linear equations in n variables:

$$a_{11}x_1 + a_{12}x_2 + \cdots + a_{1n}x_n = 0,$$
$$a_{21}x_1 + a_{22}x_2 + \cdots + a_{2n}x_n = 0,$$
$$\cdot$$
$$\cdot$$
$$\cdot$$
$$a_{m1}x_1 + a_{m2}x_2 + \cdots + a_{mn}x_n = 0.$$

The methods of Chapter 1 enable us to find the set of all solutions of this system, and each of those solution vectors is orthogonal to every A_i.

EXERCISES 2.2

1. In $\mathbb{R}$ the length of the "vector" (a) is given by $|a|$. One might try to extend this notion to define length in $\mathbb{R}^2$ by the formula

$$\|(a_1, a_2)\| = |a_1| + |a_2|.$$

Which of the three desirable properties of length (stated in Theorem 2.2) does this definition satisfy and which does it not satisfy?

2. Verify that the scalar product of Definition 2.1 satisfies Properties 1–3 as claimed in the text.

3. Prove the following properties of the scalar product:
   (i) If $A = Z$ then $A \cdot B = 0$ for every B in $\mathbb{R}^n$.
   (ii) If $A \cdot B = 0$ for every B in $\mathbb{R}^n$, then $A = Z$.
   (iii) If $A \cdot B = A \cdot C$ for every A in $\mathbb{R}^n$, then $B = C$.

4. Prove the following statements. A sketch will help you to follow the reasoning. Let A and B be nonzero vectors.
   (i) $\|A\|^{-1}A$ is a vector of unit length.

(ii) The length of the perpendicular projection of B onto A is $\|B\| \cos \Psi(A, B)$, which can also be written as

$$\frac{A \cdot B}{\|A\|}.$$

(iii) The component of B along A is $\dfrac{A \cdot B}{A \cdot A} A$.

(iv) The component of B perpendicular to A is $B - \dfrac{A \cdot B}{A \cdot A} A$.

5. In $\mathbb{R}^3$ let $A = (3, -1, 2)$ and $B = (-1, 5, 4)$.

(i) Show that A and B are orthogonal.

(ii) Determine a vector of unit length that is orthogonal to both A and B.

6. Each of the following sets of points in $\mathscr{E}^3$ describes the vertices of a triangle. Classify each triangle as acute, right, or obtuse. Is any of these triangles isosceles?

(i) $A = (1, 1, 1)$, $B = (3, -2, 2)$, $C = (4, 2, 1)$.

(ii) $A = (1, 1, 1)$, $B = (0, 4, -1)$, $C = (-5, 3, 0)$.

(iii) $A = (2, 3, -1)$, $B = (1, 5, 1)$, $C = (4, 2, 1)$.

7. Prove Theorem 2.3.

8. Prove each of the following statements in $\mathscr{E}^n$. Also rephrase each statement as a theorem in the usual language of Euclidean geometry.

(i) $(A + B) \cdot (A + B) + (A - B) \cdot (A - B) = 2A \cdot A + 2B \cdot B$.

(ii) $(A - B) \cdot (A - B) = (A - C) \cdot (A - C) + (B - C) \cdot (B - C)$ if and only if $(A - C) \cdot (B - C) = 0$.

9. Express each of the following theorems of Euclidean geometry in the notation of $\mathscr{E}^3$, and then prove the theorem.

(i) The diagonals of a parallelogram are perpendicular if and only if the parallelogram is a rhombus (equilateral).

(ii) The midpoints of the sides of any quadrilateral (not necessarily planar) are the vertices of a plane parallelogram.

10. Prove in $\mathscr{E}^n$: $\|A + tB\| \geq \|A\|$ for every t in $\mathbb{R}$ if and only if $A \cdot B = 0$. Interpret geometrically.

2.3   VECTOR SPACES

In this section we shall use the information developed for $\mathbb{R}^n$ in Section 2.1 to define a more general algebraic system, called a *vector space*, which underlies the study of matrices and linear transformations. As we have seen, vector algebra in $\mathbb{R}^n$ involves two types of quantities—real numbers (*scalars*) and ordered n-tuples of real numbers (*vectors*). It so happens that much of the mathematical theory that applies to $\mathbb{R}^n$ applies equally well to other algebraic systems in which there are scalars (not necessarily real numbers) and vectors (not necessarily real n-tuples).

For example, consider the set $\mathscr{P}$ of all *polynomial functions* with rational coefficients. Each member p of $\mathscr{P}$ has the form

$$p(x) = a_0 + a_1 x + a_2 x^2 + \cdots + a_n x^n + \cdots,$$

where each a_i is a rational number and where *only a finite number of the coefficients are nonzero.* If at least one a_i is nonzero, the *degree* of p is the largest exponent among those terms that have a nonzero coefficient. The polynomial that has each $a_i = 0$ is called the *zero* polynomial, and it is said to have *no degree.* Thus every polynomial *except* the zero polynomial is assigned a degree. The nonzero constant polynomial $p(x) = a_0 \neq 0$ has degree zero, but the zero polynomial is not assigned a degree. This distinction is important to preserve even though it sometimes requires an unnatural form of language. Let q be another polynomial in $\mathscr{P}$.

$$q(x) = b_0 + b_1 x + \cdots + b_m x^m + \cdots.$$

Then the sum $p + q$ is defined to be the polynomial obtained by adding the coefficients of like powers in p and q:

$$(p + q)(x) = (a_0 + b_0) + (a_1 + b_1)x + \cdots + (a_r + b_r)x^r + \cdots.$$

Because the sum of rational numbers is rational, $p + q$ is in $\mathscr{P}$ whenever p is in $\mathscr{P}$ and q is in $\mathscr{P}$. Observe that this rule for adding rational polynomials is analogous to the rule for adding real n-tuples. A polynomial here is regarded as having infinitely many coefficients, but all except a finite number of those coefficients are zero. That is, a rational polynomial can be regarded as an ordered, countably infinite-tuple of rational numbers that ends with an infinite string of zeros.

The scalar multiple kp of a rational number k and the polynomial p is defined by

$$(kp)(x) = ka_0 + ka_1 x + \cdots + ka_n x^n + \cdots.$$

Because the product of two rationals is rational, kp is in $\mathscr{P}$ whenever k is rational and p is in $\mathscr{P}$. It is now a simple matter to verify that the eight properties of vector algebra listed in Section 2.1 apply to the system in which the vectors are polynomials with rational coefficients and the scalars are rational numbers.

As another example, consider the set $\mathscr{D}$ of all real valued functions that have a derivative at each point of the interval $0 < x < 1$, with operations defined, as usual, by

$$(f + g)(x) = f(x) + g(x),$$

$$(kf)(x) = kf(x),$$

for all f and g in $\mathscr{D}$ and all k in $\mathbb{R}$. Because the sum and scalar multiples of functions that are differentiable at b are also differentiable at b, the set $\mathscr{D}$ is closed under the two basic operations of a linear space. Again it is easy to verify that Properties 1–8 of Section 2.1 are satisfied, where we regard the functions in $\mathscr{D}$ as vectors and the real numbers as scalars.

These examples reveal the essential nature of a vector space: there are two types of objects (a set $\mathscr{V}$ of vectors and a set $\mathscr{F}$ of scalars); there are two operations on vectors (sum and scalar multiple), there are two operations on scalars (sum and product), and these four operations satisfy certain rules. The set $\mathscr{F}$ of scalars is assumed to form a type of number system that is called a *field*. If you aren't familiar with the definition of a field, see Appendix A.1. For much of our study the particular choice of the scalar field will play a subordinate role. For the present you may think of $\mathscr{F}$ as the field of real numbers. The sum of two vectors must be a vector, and vector sum must satisfy Properties 1–4. The scalar multiple of a vector must be a vector, and scalar multiple must satisfy Properties 5–8. The resulting algebraic system $\mathscr{V}(\mathscr{F})$ is called a vector space $\mathscr{V}$ over a field $\mathscr{F}$.

Before describing a linear space formally we return to a matter of notation. As discussed in Section 2.1, we shall use lowercase Latin letters to denote elements of $\mathscr{F}$, whether or not $\mathscr{F}$ is the field of real numbers. But because we wish to consider vector spaces other than $\mathbb{R}^n$, we shall agree to use lowercase Greek letters to denote the elements of $\mathscr{V}$. When $\mathscr{V}$ is known to be $\mathbb{R}^n$, we may use either lowercase Greek letters or uppercase Latin letters to denote vectors. If you are unfamiliar with the names and forms of Greek letters, you are urged to practice with those listed at the top of page 41 before proceeding.

Definition 2.2 A *vector space over a field* is an algebraic system $\mathscr{V}(\mathscr{F})$ that consists of a set $\mathscr{F}$ of objects called *scalars*, a set $\mathscr{V}$ of objects

Name	Form
alpha	α
beta	β
gamma	γ
delta	δ
epsilon	ε
zeta	ζ
eta	η
theta	θ
lambda	λ
mu	μ
nu	ν
xi	ξ
rho	ρ
sigma	σ
tau	τ
phi	Φ
psi	Ψ

called *vectors*, and four binary operations, $+$, $\cdot$, $\oplus$, $\odot$, that satisfy the following postulates:

(a) $\mathscr{F}$ is a field relative to the operations $+$ and $\cdot$;

(b) For all α, β in $\mathscr{V}$, $\alpha \oplus \beta$ is in $\mathscr{V}$, and

  (1) $\alpha \oplus \beta = \beta \oplus \alpha$,

  (2) $(\alpha \oplus \beta) \oplus \gamma = \alpha \oplus (\beta \oplus \gamma)$ for all γ in $\mathscr{V}$,

  (3) there is a vector θ in $\mathscr{V}$ such that $\gamma \oplus \theta = \gamma$ for all γ in $\mathscr{V}$,

  (4) for each γ in $\mathscr{V}$ there is a vector $-\gamma$ in $\mathscr{V}$ such that $-\gamma \oplus \gamma = \theta$;

(c) For all a, b in $\mathscr{F}$ and all α, β in $\mathscr{V}$, $a \odot \alpha$ is in $\mathscr{V}$, and

  (5) $(a + b) \odot \alpha = (a \odot \alpha) \oplus (b \odot \alpha)$,

  (6) $a \odot (\alpha \oplus \beta) = (a \odot \alpha) \oplus (a \odot \beta)$,

  (7) $(a \cdot b) \odot \alpha = a \odot (b \odot \alpha)$,

  (8) $1 \odot \alpha = \alpha$.

The vector θ is called the *zero vector*, and the vector $-\alpha$ is called the *negative* of α.

We observe immediately that using Latin letters for scalars and Greek letters for vectors removes the need to use symbols that distinguish vector sum and scalar sum or that distinguish scalar multiple of a vector from the product of two scalars. For example, Postulates 6 and 7 can be written without ambiguity as

$$a(\alpha + \beta) = a\alpha + a\beta,$$

$$(ab)\alpha = a(b\alpha).$$

When a choice of the scalar field $\mathscr{F}$ is clear from the context or is immaterial to the discussion, we frequently write $\mathscr{V}$ instead of $\mathscr{V}(\mathscr{F})$. A vector space over $\mathbb{R}$ is called a *real vector space*, and a vector space $\mathscr{V}(\mathbb{C})$ over the complex numbers is called a *complex vector space*.

Further Examples of Vector Spaces

(A) Let $\mathscr{F}$ be any field, and let $\mathscr{P}_2$ be the set of all polynomials having coefficients in $\mathscr{F}$ and either having degree not exceeding 2 or having no degree. Thus an element of $\mathscr{P}_2$ is a function of the form

$$p(x) = a_0 + a_1 x + a_2 x^2,$$

where each a_i may be any element of $\mathscr{F}$. If q is in $\mathscr{P}_2$, where

$$q(x) = b_0 + b_1 x + b_2 x^2,$$

then $p + q$ is in $\mathscr{P}_2$, because

$$(p + q)(x) = (a_0 + b_0) + (a_1 + b_1)x + (a_2 + b_2)x^2.$$

Also cp is in $\mathscr{P}_2$ for each c in $\mathscr{F}$, because

$$(cp)(x) = (ca_0) + (ca_1)x + (ca_2)x^2.$$

The zero polynomial is in $\mathscr{P}_2$. Also the polynomial r, where

$$r(x) = (-a_0) + (-a_1)x + (-a_2)x^2,$$

is in $\mathscr{P}_2$ and is the negative of p. Hence Postulates 3 and 4 are satisfied in $\mathscr{P}_2$. But Postulates 1, 2, and 5–8 are automatically satisfied by the rules of polynomial algebra, so $\mathscr{P}_2(\mathscr{F})$ is a vector space.

(B) Let $\mathscr{V}(\mathscr{F})$ be any vector space over $\mathscr{F}$. By Postulate 3 there must exist at least one vector in $\mathscr{V}$, the zero vector θ. Let $[\theta]$ denote the set consisting of the single vector θ. Because θ is its own negative, Postulates 3 and 4 are satisfied in $[\theta]$, and the other six properties hold in $[\theta]$ because they hold in $\mathscr{V}(\mathscr{F})$. $[\theta]$ is called the *zero space*.

(C) Let $\mathscr{F}$ be any field, and let $\mathscr{V} = \mathscr{F}$. Thus in this example the set of vectors is the same as the set of scalars. Addition of vectors is defined as addition in $\mathscr{F}$, and scalar multiple of a vector is defined as multiplication in $\mathscr{F}$. Postulates 1–8 are easily seen to be satisfied. Hence any field $\mathscr{F}$ can be regarded as a vector space over $\mathscr{F}$ itself.

(D) As a variation of the previous example, we recall that any complex number α can be written in the form

$$\alpha = a + ib, \text{ where } a \text{ and } b \text{ are in } \mathbb{R} \text{ and } i^2 = -1.$$

If $\beta = c + id$, then $\alpha + \beta = (a + c) + i(b + d)$. And if k is in $\mathbb{R}$, then $k\alpha = ka + ikb$. The zero complex number is $(0 + i0)$, and the negative of α is $-a + i(-b)$; it follows that the field $\mathbb{C}$ of complex numbers is a vector space $\mathbb{C}(\mathbb{R})$ over $\mathbb{R}$. But $\mathbb{C}$ is also a vector space $\mathbb{C}(\mathbb{C})$ over itself. Later we shall see how the two spaces $\mathbb{C}(\mathbb{R})$ and $\mathbb{C}(\mathbb{C})$ are distinguished from each other.

In Section 2.1 we claimed that Properties 9–11 of $\mathbb{R}^n$ must hold in any vector space. We now prove that in full detail.

Theorem 2.6 Let $\mathscr{V}(\mathscr{F})$ be any vector space, let θ denote the zero vector and $-\alpha$ denote the negative of α. Then for all α in $\mathscr{V}$ and all a in $\mathscr{F}$,

(a) $0\alpha = \theta$,

(b) $(-1)\alpha = -\alpha$,

(c) $a\theta = \theta$.

Proof (a) $\alpha = 1\alpha = (1 + 0)\alpha = 1\alpha + 0\alpha = \alpha + 0\alpha$ by Postulates 8 and 5. If we add $-\alpha$ to each side, we obtain

$$-\alpha + \alpha = -\alpha + (\alpha + 0\alpha)$$
$$\theta = (-\alpha + \alpha) + 0\alpha = \theta + 0\alpha = 0\alpha + \theta = 0\alpha$$

by Postulates 4, 2, 1, and 3.

(b) $\alpha + (-1)\alpha = 1\alpha + (-1)\alpha = [1 + (-1)]\alpha = 0\alpha = \theta$ by

Postulates 8 and 5 and Theorem 2.6(a). By adding $-\alpha$ to each side we obtain

$$-\alpha + [\alpha + (-1)\alpha] = -\alpha + \theta,$$
$$[-\alpha + \alpha] + (-1)\alpha = -\alpha,$$
$$\theta + (-1)\alpha = -\alpha$$
$$(-1)\alpha = -\alpha$$

by Postulates 2, 3, 1, and 4.

(c) $a\theta = a(0\alpha) = (a0)\alpha = 0\alpha = \theta$

by Theorem 2.6(a) and Postulate 7.

EXERCISES 2.3

1. Let $\mathscr{V}$ be the set of all ordered triples of real numbers that satisfy certain conditions, specified below. In each case determine whether $\mathscr{V}(\mathbb{R})$ is a vector space relative to the usual addition and scalar multiple of triples. Justify your answers.

(i) $a_1 = 0$; a_2 and a_3 arbitrary.

(ii) $a_1 = -a_3$; a_2 arbitrary.

(iii) a_1, a_2 arbitrary; $a_3 = 1 + a_1 - a_2$.

(iv) a_1, a_2 arbitrary; $a_3 = a_1 - a_2$.

(v) a_1, a_2 arbitrary; $a_3 = a_1 a_2$.

(vi) $a_1 = a_2 = -a_3$.

2. Let $\mathscr{V}$ be the set of all real-valued functions that satisfy certain conditions, specified below. In each case determine whether $\mathscr{V}(\mathbb{R})$ is a vector space relative to the usual addition and scalar multiple of functions. Justify your answers.

(i) All functions that attain a relative maximum at $x = 0$.

(ii) All functions that attain either a relative maximum or a relative minimum at $x = 0$.

(iii) All functions that are continuous at $x = 0$.

(iv) All *even* functions: $f(x) = f(-x)$ for all x.

(v) All *odd* functions: $f(x) = -f(-x)$ for all x.

3. A polynomial with rational coefficients can be regarded as an infinite sequence $\{a_i\}$, $i = 1, 2, \ldots$, of rational numbers in which each term of the

sequence beyond a certain point is zero; that is, for some integer N, $a_i = 0$ whenever $i > N$. Let $\mathscr{V}$ denote the set of *all* infinite sequences of rational numbers, and let $\mathbb{Q}$ denote the field of rational numbers. Is $\mathscr{V}(\mathbb{Q})$ a vector space relative to the usual term-by-term addition and scalar multiple of sequences? Justify your answer.

4. Let $\mathscr{V}$ be the set of all real polynomials of degree two. Cite at least two specific reasons why $\mathscr{V}(\mathbb{R})$ fails to satisfy the definition of a vector space.

5. Let α be any vector of a vector space $\mathscr{V}(\mathscr{F})$, and let $\mathscr{W}$ denote the set of all scalar multiples of α.

(i) Verify that $\mathscr{W}(\mathscr{F})$ is a vector space.

(ii) In the special case in which $\mathscr{V}(\mathscr{F})$ is $\mathbb{R}^3$, describe $\mathscr{W}(\mathscr{F})$ geometrically; you will need to distinguish two cases, $\alpha = 0$ and $\alpha \neq 0$.

6. Prove that the set of all functions f that satisfy the differential equation

$$f''(x) + p(x)f'(x) + q(x)f(x) = 0$$

forms a real vector space. (You do not need to know the specific functions that are solutions to conclude that they form a vector space.)

7. Each real polynomial that does not have degree greater than two is of the form

$$p(x) = a_0 + a_1 x + a_2 x^2,$$

and therefore it determines uniquely an ordered triple (a_0, a_1, a_2) of real numbers. This suggests that the vector space $\mathscr{P}_2(\mathbb{R})$ might be related in some way to the vector space $\mathbb{R}^3$. Explore this idea and write a brief description of your investigation.

8. Let $\mathbb{Q}$ denoted the field of rational numbers, and let $\mathscr{V}$ be the set of all real numbers of the form $a + b\sqrt{2}$, where a and b are in $\mathbb{Q}$. Prove in detail that $\mathscr{V}(\mathbb{Q})$ is a vector space.

2.4 SUBSPACES

In the study of any algebraic system one normally is interested in discovering ways in which the whole system can be constructed from simple subsystems. For example, the coordinate line $\mathbb{R}$ is a vector space, and the coordinate plane $\mathbb{R}^2$ can be constructed from any two coordinate lines that have a common zero point but no other common point. From Exercise 2.3–5 it follows that each line through the origin of $\mathbb{R}^3$ is itself a vector space, a *subspace* of $\mathbb{R}^3$. In this section we explore the concept of subspace for any vector space.

Definition 2.3   Let $\mathscr{S}$ be any set of vectors in $\mathscr{V}(\mathscr{F})$. Then $\mathscr{S}$ is called a *subspace* of $\mathscr{V}(\mathscr{F})$ if and only if $\mathscr{S}(\mathscr{F})$ is a vector space.

It is understood, of course, that the operations of vector sum and scalar multiple for vectors in $\mathscr{S}$ are the same as for vectors in $\mathscr{V}$. This implies that some of Properties 1–8 hold automatically for vectors in $\mathscr{S}$ because those properties hold universally for vectors in $\mathscr{V}$. If we reexamine Definition 2.2, we can discover precisely which postulates we need to check to determine whether or not a subset $\mathscr{S}$ of $\mathscr{V}$ is a subspace of $\mathscr{V}$. Because the scalars used for $\mathscr{S}$ are those used for $\mathscr{V}$, Statement (a) is satisfied. In Statement (b) we must determine whether $\alpha + \beta$ is in $\mathscr{S}$ whenever α and β are in $\mathscr{S}$. Also we must verify that θ is in $\mathscr{S}$, and that $-\alpha$ and $c\alpha$ are in $\mathscr{S}$ whenever α is in $\mathscr{S}$ and c is in $\mathscr{F}$. Postulates 1, 2, and 5–8 are universally valid in $\mathscr{V}$ and hence in $\mathscr{S}$. The following theorem establishes a concise, convenient test for determining whether or not a subset $\mathscr{S}$ of $\mathscr{V}$ forms a subspace of $\mathscr{V}$.

Theorem 2.7   Let $\mathscr{S}$ be a nonvoid subset of vectors of $\mathscr{V}(\mathscr{F})$. Then $\mathscr{S}$ is a subspace if and only if each linear combination of vectors of $\mathscr{S}$ is a vector in $\mathscr{S}$.

Proof   The condition clearly is necessary. Conversely, assume that $a\alpha + b\beta$ is in $\mathscr{S}$ for all α, β in $\mathscr{S}$ and all a, b in $\mathscr{F}$. Choosing $a = 1 = b$, we see that $\alpha + \beta$ is in $\mathscr{S}$ whenever α and β are in $\mathscr{S}$. Choosing $b = 0$ and a arbitrary, we see that $\mathscr{S}$ is closed with respect to scalar multiples. To see that $-\alpha$ is in $\mathscr{S}$, choose $a = -1$ and $b = 0$. Then it follows that $\alpha + (-\alpha) = \theta$ is in $\mathscr{S}$. The other parts of Definition 2.2 are automatically satisfied for $\mathscr{S}$, so $\mathscr{S}$ is a vector space over $\mathscr{F}$.

We remark that it is sometimes more convenient to verify separately that $\mathscr{S}$ is closed with respect to vector sum and with respect to scalar multiples of a vector:

$$\alpha + \beta \text{ is in } \mathscr{S} \text{ whenever } \alpha \text{ and } \beta \text{ are in } \mathscr{S},$$

$$c\alpha \text{ is in } \mathscr{S} \text{ whenever } \alpha \text{ is in } \mathscr{S} \text{ and } c \text{ is in } \mathscr{F}.$$

These two conditions together are easily seen to be equivalent to the condition that $\mathscr{S}$ be closed with respect to the formation of linear combinations.

To facilitate reference to members of a set and to its subsets, we shall henceforth use the following notation:

A set will be denoted by a capital letter either in script or block form:

$$\alpha \in \mathcal{S} \text{ means that } \alpha \text{ is a member of } \mathcal{S},$$

$$\mathcal{T} \subseteq \mathcal{S} \text{ means that } \mathcal{T} \text{ is a subset of } \mathcal{S}.$$

The *void* or *empty* set is denoted by Φ, and the notation

$$\mathcal{Y} = \{x \in \mathcal{X} \,|\, P(x)\}$$

defines the set $\mathcal{Y}$ as "the set of all members x of the set $\mathcal{X}$ which have the property $P(x)$." For example,

$$\{\alpha \in \mathcal{E}^3 \,|\, \|\alpha\| < 1\}$$

describes the set of all points in Euclidean three-space that lie in the interior of the sphere of radius one with center at the origin.

Examples of Subspaces

(A) In any vector space let $\mathcal{S}$ have θ as its only member. Then $\mathcal{S}$ is a subspace, called the *zero subspace*.

(B) In any vector space let $\alpha \neq \theta$, and let

$$\mathcal{S} = \{c\alpha \,|\, c \in \mathcal{F}\}.$$

$\mathcal{S}$ is the set of all scalar multiples of a single vector, and it is a subspace (Exercise 2.3–5). From our experience in $\mathbb{R}^n$ it is natural to refer to $\mathcal{S}$ as a line through the origin.

(C) In any vector space let α, β be any vectors, and let

$$\mathcal{S} = \{a\alpha + b\beta \,|\, a, b \in \mathcal{F}\}.$$

$\mathcal{S}$ is the set of all linear combinations of vectors α and β. If $\alpha = \beta = \theta$, $\mathcal{S}$ is the zero space. If $\alpha \neq \theta$ and $\beta = k\alpha$ for some $k \in \mathcal{F}$, then $\mathcal{S}$ is a line through the origin along the vector α. If $\alpha \neq \theta$ and $\beta \neq k\alpha$ for all $k \in \mathcal{F}$, then using $\mathbb{R}^n$ as a model, we think of α and β as vectors that extend from the origin in two different directions. The set $\mathcal{S}$ of all linear combinations of α and β is then a plane through the origin.

(D) In the space $\mathcal{P}_2(\mathbb{Q})$ of all polynomials with rational coefficients and not of degree exceeding 2, the following are a few of the subspaces: the zero subspace, the space $\mathscr{P}_0(\mathbb{Q})$ of all rational constant functions, the spaces $\mathscr{P}_1(\mathbb{Q})$ and $\mathscr{P}_2(\mathbb{Q})$, and the space of all polynomials in $\mathscr{P}_2(\mathbb{Q})$ with zero as constant term:

$$f(x) = a_1 x + a_2 x^2.$$

On the other hand, the set of all first-degree polynomials is *not* a subspace because the sum of two first-degree polynomials is not necessarily of first degree. Nor is the zero polynomial a member of that set.

(E) In the space $\mathscr{C}(a, b)$ of all real functions that are continuous on the interval $a < x < b$, let $\mathscr{D}(a, b)$ be the set of all functions that have a derivative at each point of the interval $a < x < b$. From elementary calculus we know that any differentiable function is continuous, so $\mathscr{D}(a, b)$ is a subset of $\mathscr{C}(a, b)$. Also $\mathscr{D}(a, b)$ is a subspace of $\mathscr{C}(a, b)$ because any linear combination of differentiable functions is differentiable.

We now provide a general method of generating subspaces of any vector space.

Definition 2.4   Let S be any nonvoid subset of vectors in $\mathscr{V}(\mathscr{F})$. The set $[S]$ of all *linear combinations* of vectors in S is the collection of all *finite* sums of the form

$$a_1 \alpha_1 + a_2 \alpha_2 + \cdots + a_m \alpha_m,$$

where $a_i \in \mathscr{F}$, $\alpha_i \in S$, and $m = 1, 2, 3, \ldots$.

Theorem 2.8   The set $[S]$ of all linear combinations of a nonvoid subset S of vectors of $\mathscr{V}(\mathscr{F})$ is a subspace of $\mathscr{V}(\mathscr{F})$.
Proof   Any linear combination of two linear combinations of vectors of S is itself a linear combination of vectors of S.

If S is void, we arbitrarily define $[S]$ to be the zero subspace, which contains the single vector θ. Although this convention might seem peculiar now, in Section 2.6 we shall see that it is quite convenient. Thus for any set S of vectors, $[S]$ denotes the subspace of all linear combinations formed from vectors of S, and it is called the *subspace generated by S* or the *subspace spanned by S.*

Any subspace, of course, is a vector space, and frequently we can simplify the investigation of a question about $\mathscr{V}$ by considering the same question on suitably chosen subspaces of $\mathscr{V}$ and by knowing how $\mathscr{V}$ is related to those subspaces. In particular, we wish to know how various subspaces are related to each other.

Definition 2.5 Let $\mathscr{S}$ and $\mathscr{T}$ be subspaces of $\mathscr{V}$.

    (a) The *intersection* of $\mathscr{S}$ and $\mathscr{T}$, denoted $\mathscr{S} \cap \mathscr{T}$, is the set of all vectors common to $\mathscr{S}$ and $\mathscr{T}$:

$$\mathscr{S} \cap \mathscr{T} = \{\alpha \in \mathscr{V} \mid \alpha \in \mathscr{S} \text{ and } \alpha \in \mathscr{T}\}.$$

    (b) The *sum* of $\mathscr{S}$ and $\mathscr{T}$, denoted $\mathscr{S} + \mathscr{T}$, is the set of all vectors obtained by summing each vector of $\mathscr{S}$ with each vector of $\mathscr{T}$:

$$S + \mathscr{T} = \{\sigma + \tau \mid \sigma \in \mathscr{S} \text{ and } \tau \in \mathscr{T}\}.$$

Theorem 2.9 If $\mathscr{S}$ and $\mathscr{T}$ are subspaces of $\mathscr{V}$, then $\mathscr{S} \cap \mathscr{T}$ and $\mathscr{S} + \mathscr{T}$ are also subspaces of $\mathscr{V}$.

Proof Let $\alpha, \beta \in \mathscr{S} \cap \mathscr{T}$ and $a, b \in \mathscr{F}$. Then $\alpha, \beta \in \mathscr{S}$ and $\alpha, \beta \in \mathscr{T}$. Since $\mathscr{S}$ and $\mathscr{T}$ are subspaces, $a\alpha + b\beta \in \mathscr{S}$ and $a\alpha + b\beta \in \mathscr{T}$. Hence $a\alpha + b\beta \in \mathscr{S} \cap \mathscr{T}$. Since $\mathscr{S} \cap \mathscr{T}$ is closed under the formation of linear combinations, it is a subspace of $\mathscr{V}$. Now let $\alpha_1, \alpha_2 \in \mathscr{S} + \mathscr{T}$. Then $\alpha_i = \sigma_i + \tau_i$ for some $\sigma_i \in \mathscr{S}$ and some $\tau_i \in \mathscr{T}$, for $i = 1, 2$. Then

$$a\alpha_1 + b\alpha_2 = a(\sigma_1 + \tau_1) + b(\sigma_2 + \tau_2)$$
$$= (a\sigma_1 + b\sigma_2) + (a\tau_1 + b\tau_2).$$

But because $\mathscr{S}$ is a subspace, $a\sigma_1 + b\sigma_2 \in \mathscr{S}$. Similarly $a\tau_1 + b\tau_2 \in \mathscr{T}$. Hence $a\alpha_1 + b\alpha_2 \in \mathscr{S} + \mathscr{T}$, which shows by Theorem 2.7 that $\mathscr{S} + \mathscr{T}$ is a subspace of $\mathscr{V}$.

Note in particular that $\mathscr{S} + \mathscr{T}$ contains all the vectors of $\mathscr{S}$ and all the vectors of $\mathscr{T}$ and hence contains all the vectors in the set union $\mathscr{S} \cup \mathscr{T}$ of $\mathscr{S}$ and $\mathscr{T}$. But the set $\mathscr{S} \cup \mathscr{T}$ is not necessarily closed relative to the formation of linear combinations and hence is not necessarily a subspace. Indeed $\mathscr{S} + \mathscr{T} = [\mathscr{S} \cup \mathscr{T}]$, the subspace generated by the set union of $\mathscr{S}$ and $\mathscr{T}$. On the other hand, the set intersection of two subspaces is a subspace, the largest subspace that is simultaneously a subspace of $\mathscr{S}$ and of $\mathscr{T}$.

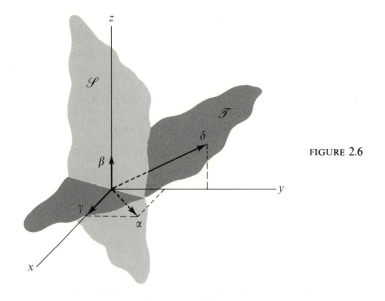

FIGURE 2.6

Example   In $\mathbb{R}^3$ let

$$\alpha = (1, 1, 0),$$
$$\beta = (0, 0, 1),$$
$$\gamma = (1, 0, 0),$$
$$\delta = (0, 2, 1),$$
$$\mathscr{S} = [\alpha, \beta],$$
$$\mathscr{T} = [\gamma, \delta].$$

The space $[\alpha]$ spanned by α is the line $y = x$ in the $z = 0$ plane—that is, the set of all real multiples of α, and similarly for the other vectors. Then $\mathscr{S}$ is the vertical plane $y = x$, because a point of $\mathscr{S}$ is any vector of the form $a\alpha + b\beta = (a, a, b)$ for any real numbers a, b. See Figure 2.6.

Similarly, $\mathscr{T}$ is the set of all vectors of the form $c\gamma + d\delta = (c, 2d, d)$, a sloping plane $y = 2z$ containing the x-axis. These two planes intersect in $\mathscr{S} \cap \mathscr{T}$, the line $x = y = 2z$ along the vector $(2, 2, 1) = 2\alpha + \beta = 2\gamma + \delta$. It is easily verified that $\mathscr{S} + \mathscr{T} = \mathbb{R}^3$, because any vector (x, y, z) of $\mathbb{R}^3$ can be written as $\sigma + \tau$, where

$$\sigma = y\alpha + z\beta = (y, y, z) \in \mathscr{S} \text{ and } \tau = (x - y)\gamma = (x - y, 0, 0) \in \mathscr{T}.$$

We observe two related facts about this representation of (x, y, z). First, the vector δ was not used, nor needed, because δ can be expressed as a linear combination of α, β, and γ:

$$\delta = (0, 2, 1) = (2, 2, 0) + (0, 0, 1) - (2, 0, 0) = 2\alpha + \beta - 2\gamma.$$

Second, the representation of a vector as a sum $\sigma + \tau$ is not unique, because $\mathscr{S} \cap \mathscr{T} = [(2, 2, 1)]$. We have

$$
\begin{aligned}
(x, y, z) &= (y, y, z) + (x - y, 0, 0) \\
&= [(y, y, z) + k(2, 2, 1)] + [(x - y, 0, 0) - k(2, 2, 1)] \\
&= (y + 2k, y + 2k, z) + (x - y - 2k, -2k, -k) \in \mathscr{S} + \mathscr{T}
\end{aligned}
$$

for any real number k.

As we shall now see, the representation of a vector space $\mathscr{V}$ as sum of two subspaces $\mathscr{M}$ and $\mathscr{N}$ produces a unique representation for each vector in $\mathscr{V}$ when and only when $\mathscr{M} \cap \mathscr{N}$ is as small as possible, namely the zero subspace $[\theta]$.

Definition 2.6   Let $\mathscr{M}$ and $\mathscr{N}$ be subspaces of a vector space $\mathscr{V}$. $\mathscr{V}$ is said to be the *direct sum* of $\mathscr{M}$ and $\mathscr{N}$, written

$$\mathscr{V} = \mathscr{M} \oplus \mathscr{N},$$

if and only if $\mathscr{M} + \mathscr{N} = \mathscr{V}$ and $\mathscr{M} \cap \mathscr{N} = [\theta]$.

Theorem 2.10 $\mathscr{V} = \mathscr{M} \oplus \mathscr{N}$ if and only if every vector $\xi \in \mathscr{V}$ has a unique representation

$$\xi = \mu + v,$$

for some $\mu \in \mathscr{M}$ and some $v \in \mathscr{N}$.

Proof   Assume that $\mathscr{V} = \mathscr{M} \oplus \mathscr{N}$. Then because $\mathscr{V} = \mathscr{M} + \mathscr{N}$, each $\xi \in \mathscr{V}$ can be expressed as $\xi = \mu + v$ for suitable $\mu \in \mathscr{M}$ and $v \in \mathscr{N}$, and we need only prove uniqueness. Suppose also that $\xi = \mu_1 + v_1$. Then

$$\mu_1 + v_1 = \mu + v,$$

$$\mu_1 - \mu = v - v_1.$$

But $\mu_1 - \mu \in \mathscr{M}$ and $v - v_1 \in \mathscr{N}$, and by hypothesis $\mathscr{M}$ and $\mathscr{N}$ have only the zero vector in common. Hence $\mu_1 = \mu$ and $v_1 = v$, so the representation is unique. To prove the converse, assume that each $\xi \in \mathscr{V}$

is expressed uniquely as the sum of some $\mu \in \mathcal{M}$ and $v \in \mathcal{N}$. Clearly, $\mathcal{M} + \mathcal{N} = \mathcal{V}$. Let $\alpha \in \mathcal{M} \cap \mathcal{N}$. Then

$$\alpha = \alpha + \theta, \text{ where } \alpha \in \mathcal{M} \text{ and } \theta \in \mathcal{N}$$

$$= \theta + \alpha, \text{ where } \theta \in \mathcal{M} \text{ and } \alpha \in \mathcal{N}.$$

Because any such representation of α is unique, we have $\alpha = \theta$. Hence $\mathcal{M} \cap \mathcal{N} = [\theta]$, and therefore $\mathcal{V} = \mathcal{M} \oplus \mathcal{N}$.

For example, each subspace of $\mathbb{R}^3$ (other than $[\theta]$ and $\mathbb{R}^3$ itself) is either a line through the origin or a plane through the origin. Any two planes through the origin contain a common line through the origin; any two lines through the origin generate a plane rather than $\mathbb{R}^3$. Hence to assert that $\mathbb{R}^3$ is the direct sum of two proper subspaces $\mathcal{M}$ and $\mathcal{N}$ means, for example, that $\mathcal{M}$ is a plane through the origin and $\mathcal{N}$ is a line that intersects that plane only at the origin.

EXERCISES 2.4

1. If $\mathscr{C}[0, 1]$ is the space of all real-valued functions that are continuous on the interval $0 \leq x \leq 1$, which of the following subsets are *not* subspaces? Cite a specific reason for each such case.

    (i) All polynomial functions.

    (ii) All polynomials of degree greater than two.

    (iii) All continuous functions that satisfy $f(0) = 0$.

    (iv) All continuous functions that satisfy $f(0) = 1$.

    (v) All continuous functions that satisfy $f(x) \leq 0$.

    (vi) All continuous functions that satisfy $|f(x)| \leq 0$.

    (vii) All constant functions.

2. Which of the following subsets of $\mathbb{R}^n$ are *not* subspaces? Cite a specific reason in each case, using Theorem 2.7.

    (i) $\{(a_1, \ldots, a_n) \,|\, a_1 = a_2^2\}$.

    (ii) $\{(a_1, \ldots, a_n) \,|\, a_1 < a_2\}$.

    (iii) $\{(a_1, \ldots, a_n) \,|\, a_1 + \cdots + a_n = 0\}$.

    (iv) $\{(a_1, \ldots, a_n) \,|\, a_1 + \cdots + a_n = 1\}$.

    (v) $\{(a_1, \ldots, a_n) \,|\, a_1 = 0 \text{ or } a_2 = 0\}$.

3. In $\mathbb{R}^3$ let $\alpha = (2, 1, 0)$, $\beta = (1, 0, -1)$, $\gamma = (-1, 1, 1)$, $\delta = (0, -1, 1)$. Let $\mathscr{S} = [\alpha, \beta]$ and $\mathscr{T} = [\gamma, \delta]$. Describe $\mathscr{S}$, $\mathscr{T}$, $\mathscr{S} \cap \mathscr{T}$, $\mathscr{S} \cup \mathscr{T}$, and $\mathscr{S} + \mathscr{T}$.

4. Referring to Exercise 3, determine a vector ξ such that

$$\mathbb{R}^3 = \mathscr{S} \oplus [\xi].$$

Explain geometrically how ξ can be chosen to satisfy that condition.

5. Let $\mathscr{S}$ and $\mathscr{T}$ be subspaces of $\mathscr{V}$ such that $\mathscr{S}$ is a subset of $\mathscr{T}$. Is $\mathscr{S}$ necessarily a subspace of $\mathscr{T}$? Prove your answer.

6. Let $\mathscr{S}$, $\mathscr{T}$, and $\mathscr{U}$ be subspaces of $\mathscr{V}$.

(i) Prove that $(\mathscr{S} \cap \mathscr{U}) + (\mathscr{T} \cap \mathscr{U}) \subseteq (\mathscr{S} + \mathscr{T}) \cap \mathscr{U}$.

(ii) Show by an example in $\mathbb{R}^2$ that equality need not hold in (i).

(iii) Prove that equality must hold in (i) if $\mathscr{S} \subseteq \mathscr{U}$.

7. Let $\mathscr{S}$ be a subspace of a vector space $\mathscr{V}$ over $\mathscr{F}$. For each vector $\xi \in \mathscr{V}$, the $\mathscr{S}$-*coset* of ξ is defined to be the set

$$(\xi + \mathscr{S}) = \{\xi + \alpha \,|\, \alpha \in \mathscr{S}\}.$$

(Geometrically in $\mathbb{R}^3$ if $\mathscr{S}$ is a plane through the origin, $(\xi + \mathscr{S})$ is a plane that is parallel to $\mathscr{S}$ and passes through the tip of the vector ξ.)

(i) Show that $(\xi + \mathscr{S})$ is a subspace if and only if $\xi \in \mathscr{S}$.

(ii) Show that $(\xi + \mathscr{S}) = (\eta + \mathscr{S})$ if and only if $\xi - \eta \in \mathscr{S}$.

A new vector space over $\mathscr{F}$ can be formed from the collection $\mathscr{C}$ of all $\mathscr{S}$-cosets by defining the sum of two cosets and the scalar multiple of a coset:

$$(\xi_1 + \mathscr{S}) \oplus (\xi_2 + \mathscr{S}) = (\xi_1 + \xi_2 + \mathscr{S}),$$

$$a \odot (\xi_1 + \mathscr{S}) = (a\xi_1 + \mathscr{S}).$$

(iii) Show that these operations are unambiguously defined; that is, if $\eta_1 \in (\xi_1 + \mathscr{S})$ and $\eta_2 \in (\xi_2 + \mathscr{S})$, then

$$(\xi_1 + \mathscr{S}) \oplus (\xi_2 + \mathscr{S}) = (\eta_1 + \eta_2 + \mathscr{S}),$$

$$a \odot (\xi_1 + \mathscr{S}) = (a\eta_1 + \mathscr{S}).$$

(iv) Show that the set $\mathscr{C}$ of $\mathscr{S}$-cosets is a vector space over $\mathscr{F}$. This space is called the *quotient space of $\mathscr{V}$ modulo $\mathscr{S}$* and is denoted $\mathscr{V}/\mathscr{S}$.

8. In the space $\mathscr{V}(\mathbb{Q})$ of all sequences of rational numbers (Exercise 2.3-3), let $\mathscr{S}$ denote the set of all sequences in which all but a finite number of terms are zero. Is $\mathscr{S}(\mathbb{Q})$ a subspace of $\mathscr{V}(\mathbb{Q})$? Justify your answer.

9. In the space $\mathscr{V}(\mathbb{R})$ of all functions from $\mathbb{R}$ to $\mathbb{R}$ let

$$\mathscr{E} = \{f \in \mathscr{V} \mid f(-x) = f(x) \text{ for all } x \text{ in } \mathbb{R}\},$$

$$\mathscr{O} = \{f \in \mathscr{V} \mid f(-x) = -f(x) \text{ for all } x \text{ in } \mathbb{R}\}.$$

($\mathscr{E}$ is the set of all *even* functions and $\mathscr{O}$ is the set of all *odd* functions.) Prove that $\mathscr{E}$ and $\mathscr{O}$ are subspaces of $\mathscr{V}$ and that

$$\mathscr{V} = \mathscr{E} \oplus \mathscr{O}.$$

This shows that any function on $\mathbb{R}$ can be expressed uniquely as the sum of an even function and an odd function.

2.5   LINEAR INDEPENDENCE

We come now to one of the most important concepts in the study of vector spaces, that of *linear independence*. Recall from the previous section the example, illustrated by Figure 2.6, in which four numerical vectors in $\mathbb{R}^3$ were used to describe various subspaces. In that example we observed that the vector δ could be expressed as a linear combination of the other three vectors and therefore was redundant, in the presence of α, β, and γ, in generating $\mathbb{R}^3$. The set $\{\alpha, \beta, \gamma, \delta\}$ is therefore said to be *linearly dependent*.

Definition 2.7   A set S of vectors of a vector space $\mathscr{V}(\mathscr{F})$ is said to be *linearly dependent* if and only if there exist a finite set of distinct vectors $\{\gamma_1, \ldots, \gamma_k\}$ in S and a finite set of scalars $\{c_1, \ldots, c_k\}$ in $\mathscr{F}$ such that *at least one of the scalars is nonzero* and such that

$$c_1\gamma_1 + c_2\gamma_2 + \cdots + c_k\gamma_k = 0.$$

Otherwise S is said to be *linearly independent*.

The zero vector can always be written as a linear combination of given vectors simply by using the scalar 0 for each coefficient. That particular linear combination is called *trivial*. The definition states that some *nontrivial* linear combination of vectors of a linearly dependent set yields the zero vector, but that no nontrivial linear combination of vectors of a linearly independent set yields the zero vector. If the zero vector is a nontrivial linear combination of vectors, we have

$$c_1 \gamma_1 + c_2 \gamma_2 + \cdots + c_m \gamma_m = 0,$$

where not all scalars c_i are zero. Let c_j be the first nonzero scalar. We obtain

$$\gamma_j = -\frac{1}{c_j}(c_{j+1} \gamma_{j+1} + \cdots + c_m \gamma_m).$$

In this case the vector γ_j *depends* linearly on the other vectors in the sense that γ_j is a linear combination of the others. Clearly any set of vectors is either linearly independent or linearly dependent, any set that contains the zero vector is dependent, and any set that consists of a single nonzero vector is independent. Also, the void set is linearly independent, because it is not linearly dependent.

Theorem 2.11   Any subset of a linearly independent set is linearly independent, and any set that contains a linearly dependent set is linearly dependent.
Proof   Exercise.

To construct linearly independent sets we can proceed inductively. Beginning with a linearly independent set S (for example, the void set), we can extend S to a larger linearly independent set by adjoining any vector $\xi \in \mathscr{V}$ that is not in the subspace $[S]$. Such a vector will exist unless $[S] = \mathscr{V}$.

Theorem 2.12   Let $S \subseteq \mathscr{V}$ be linearly independent. Then for any vector ξ, $S \cup \{\xi\}$ is linearly independent if and only if $\xi \notin [S]$.
Proof   Let S and ξ be as described. First suppose that $\xi \in [S]$; then for suitable scalars a_i and vectors $\alpha_i \in S$,

$$\xi = a_1 \alpha_1 + a_2 \alpha_2 + \cdots + a_k \alpha_k.$$

If $a_i = 0$ for each i, then $\xi = 0$ and $S \cup \{\xi\}$ is linearly dependent. If some $a_i \neq 0$, again $S \cup \{\xi\}$ is linearly dependent. Conversely, suppose that $\xi \notin [S]$, and consider any equation of the form

$$b_1 \alpha_1 + b_2 \alpha_2 + \cdots + b_m \alpha_m + b_{m+1} \xi = 0,$$

where each $\alpha_i \in S$. If $b_{m+1} \neq 0$, then ξ is a linear combination of vectors of S, and so $\xi \in [S]$, contrary to our assumption. Hence $b_{m+1} = 0$, and because S is linearly independent, $b_i = 0$ for all $i \leq m$. Then $S \cup \{\xi\}$ is linearly independent.

This theorem provides a useful geometric interpretation of linearly independent sets in $\mathbb{R}^n$. The void set is linearly independent and spans the zero subspace. If we adjoin any nonzero vector α_1, we obtain the set $\{\alpha_1\}$, which is linearly independent and spans the line $[\alpha_1]$ through the origin along α_1. To extend this to a larger linearly independent set, we must choose $\alpha_2 \notin [\alpha_1]$. Then $\{\alpha_1, \alpha_2\}$ is linearly independent and spans the plane $[\alpha_1, \alpha_2]$ through the origin that contains α_1 and α_2. To extend this to a still larger linearly independent set, we adjoin any α_3 that does not lie in the plane $[\alpha_1, \alpha_2]$. Geometrically, each stage in this extension process is accomplished in $\mathbb{R}^n$ by adjoining a vector that points in a direction in $\mathbb{R}^n$ that does not exist in the subspace spanned by the vectors of the previous set. In this sense the notion of linear independence seems to be related to our intuitive ideas about the dimension of a space; we shall see in the next section that this impression is accurate.

Theorem 2.13 Let $S = \{\alpha_1, \ldots, \alpha_k\}$ be a finite set of nonzero vectors. Then S is linearly dependent if and only if

$$\alpha_m \in [\alpha_1, \ldots, \alpha_{m-1}]$$

for some $m < k$.

Proof If for some m, $\alpha_m \in [\alpha_1, \ldots, \alpha_{m-1}]$, then $\{\alpha_1, \ldots, \alpha_m\}$ is dependent and so is S (Theorem 2.11). Conversely, suppose S is dependent and let m be the least integer such that $\{\alpha_1, \ldots, \alpha_m\}$ is dependent. Then for suitable scalars $c_1, \ldots, c_m$, not all zero,

$$\sum_{i=1}^{m} c_i \alpha_i = \theta.$$

If $c_m = 0$, then $\{\alpha_1, \ldots, \alpha_{m-1}\}$ is dependent, contradicting the definition of m. Hence

$$\alpha_m = -c_m^{-1}(c_1 \alpha_1 + \cdots + c_{m-1} \alpha_{m-1}) \in [\alpha_1, \ldots, \alpha_{m-1}].$$

The next theorem illustrates two methods of selecting from a finite linearly dependent set S a linearly independent subset that spans $[S]$. We can either begin with S and cast aside vectors that are redundant in generating the subspace $[S]$, or else we can begin with a small linearly independent subset of S and enlarge that subset, retaining linear independence, until the enlarged set spans $[S]$.

Theorem 2.14   If a subspace $[S]$ is spanned by a finite set given by $S = \{\alpha_1, \ldots, \alpha_k\}$, then there exists a linearly independent subset of S that also spans $[S]$.

First Proof   If S is independent, there is nothing more to prove. Otherwise by Theorem 2.13, there is a least integer i such that $\alpha_i \in [\alpha_1, \ldots, \alpha_{i-1}]$. Remove α_i from S and let $S_1 = S - \alpha_i$. Then $[S_1] = [S]$, and the argument can be repeated on S_1: either S_1 is independent, in which case the proof is complete, or for some $\alpha_j \in S_1$, the set $S_1 - \alpha_j$ spans $[S]$. Because S is finite, the theorem follows after a finite number of repetitions of the argument. Observe that this argument is valid even when S is void, because the void set is independent.

Second Proof   The void set Φ is independent and spans $[\theta]$. If $[S] = [\theta]$, the proof is complete. Otherwise choose any nonzero vector α_1 from S. The set $S_1 = \{\alpha_1\}$ is linearly independent, and $[\alpha_1] \subseteq [S]$. If equality holds, the proof is complete. Otherwise, there exists $\alpha_2 \in S$ such that $\alpha_2 \notin [\alpha_1]$. By Theorem 2.13 the set $\{\alpha_1, \alpha_2\}$ is linearly independent and spans a subspace of $[S]$. This argument can be repeated until we obtain a linearly independent subset of S that spans $[S]$.

Actually Theorem 2.14 remains valid even if the word "finite" is omitted, but we have no need for the more general result and shall not pursue that idea here.

Theorem 2.14 is significant because it relates the idea of linear independence to the fundamental algebraic process of vector spaces—the formation of linear combinations of vectors. The theorem establishes the existence of a set of vectors that is large enough to generate a given space, but small enough to do so without any redundancy. The first proof shows that such a set can be regarded as a *minimal spanning subset* of $[S]$, whereas the second proof shows that it is also a *maximal linearly independent subset* of $[S]$. In the next section we formally adopt the term *basis*.

The concept of linear independence is so fundamental to the theory and applications linear algebra that you are urged to acquire a good understanding of Definition 2.7 and the ability to apply it in practice. To illustrate the related computational techniques, we consider the following question in the concrete setting of $\mathbb{R}^m$.

> Given a finite set $S = \{\alpha_1, \ldots, \alpha_n\}$ in $\mathbb{R}^m$, how do we determine whether S is linearly independent?

We want to determine whether or not there exist n scalars, $x_1, \ldots, x_n$ (not all of which are zero) such that

$$x_1 \alpha_1 + x_2 \alpha_2 + \cdots + x_n \alpha_n = \theta.$$

Because each α_i is an m-tuple of real numbers, we can write it as a column vector to obtain

$$x_1 \begin{pmatrix} a_{11} \\ a_{21} \\ \cdot \\ \cdot \\ \cdot \\ a_{m1} \end{pmatrix} + x_2 \begin{pmatrix} a_{12} \\ a_{22} \\ \cdot \\ \cdot \\ \cdot \\ a_{m2} \end{pmatrix} + \cdots + x_n \begin{pmatrix} a_{1n} \\ a_{2n} \\ \cdot \\ \cdot \\ \cdot \\ a_{mn} \end{pmatrix} = \begin{pmatrix} 0 \\ 0 \\ \cdot \\ \cdot \\ \cdot \\ 0 \end{pmatrix}.$$

This vector equation can be written as a homogeneous linear system of m scalar equations in n variables, which has precisely the same form as (1.3). Hence the method of Gaussian elimination can be applied to reduce the m-by-n matrix

$$A = \begin{pmatrix} a_{11} & a_{12} & \cdots & a_{1n} \\ a_{21} & a_{22} & \cdots & a_{2n} \\ \cdot & \cdot & & \cdot \\ \cdot & \cdot & & \cdot \\ \cdot & \cdot & & \cdot \\ a_{m1} & a_{m2} & \cdots & a_{mn} \end{pmatrix}$$

to row echelon form, from which we can tell whether the system has a nontrivial solution (and hence S is dependent), or the system has only the trivial solution (S is independent).

Alternately, we can write the given vectors as row vectors (instead of column vectors) to obtain the n-by-m matrix B, shown below, which is called the *transpose* of A:

$$B = \begin{pmatrix} a_{11} & a_{21} & \cdots & a_{m1} \\ a_{12} & a_{22} & \cdots & a_{m2} \\ \cdot & \cdot & & \cdot \\ \cdot & \cdot & & \cdot \\ \cdot & \cdot & & \cdot \\ a_{1n} & a_{2n} & \cdots & a_{mn} \end{pmatrix}.$$

If the set S of row vectors is linearly dependent, some row is a linear combination of the preceding rows by Theorem 2.13:

$$c_1 \alpha_1 + c_2 \alpha_2 + \cdots + c_p \alpha_p = \theta$$

for suitable c_i, where $c_p \neq 0$. Elementary row operations then permit us to replace row p by this linear combination, producing zeros throughout row p. When we reduce B to row echelon form, the number of nonzero rows turns out to be the number of vectors in a largest linearly independent subset of S.

Of course the rows of B are the columns of A. Later on we shall prove that the largest number of linearly independent row vectors of a matrix is equal to the largest number of linearly independent column vectors of the same matrix. Hence one can use Gaussian elimination equally well with A or with B to investigate questions concerning linear independence. Of course, in some cases a quick eye can observe a simple relation between given vectors that establishes linear dependence without doing the computation that Gaussian elimination involves.

EXERCISES 2.5

1. Refer to the vectors of Exercise 2.4-3.

    (i) Write a nontrivial linear relation that shows that the set $\{\alpha, \beta, \gamma, \delta\}$ is linearly dependent.

    (ii) Write a three-vector subset that is linearly independent.

2. Select a maximal linearly independent subset of each of the following sets of vectors.

    (i) $(1, 0, 1, 0)$, $(0, 1, 0, 1)$, $(1, 1, 1, 1)$, $(-1, 0, 2, 0)$.

    (ii) $(0, 1, 2, 3)$, $(3, 0, 1, 2)$, $(2, 3, 0, 1)$, $(1, 2, 3, 0)$.

    (iii) $(1, -1, 1, -1)$, $(-1, 1, -1, -1)$, $(1, -1, 1, -2)$, $(0, 0, 0, 1)$.

3. For each example of the preceding exercise in which the largest independent subset contains fewer than four vectors, adjoin the vectors (a_1, a_2, a_3, a_4) to obtain a linearly independent set of four vectors.

4. Prove Theorem 2.11.

5. In the space of real n-tuples prove that the vectors

$$\begin{aligned}
\alpha_1 &= (1, 1, 1, \quad \ldots, \quad 1, 1), \\
\alpha_2 &= (0, 1, 1, \quad \ldots, \quad 1, 1), \\
\alpha_3 &= (0, 0, 1, \quad \ldots, \quad 1, 1), \\
&\quad \cdot \\
&\quad \cdot \\
&\quad \cdot \\
\alpha_n &= (0, 0, 0, \quad \ldots, \quad 0, 1),
\end{aligned}$$

are linearly independent.

6. The following exercise illustrates that the scalar field $\mathscr{F}$ of a vector space $\mathscr{V}(\mathscr{F})$ plays an important role in the concept of linear independence, even though that role is not usually emphasized.

(i) In the vector space $\mathbb{R}(\mathbb{Q})$ of real numbers over the field of rational numbers, show that the set $\{1, \sqrt{2}\}$ is linearly independent.

(ii) In the vector space $\mathbb{R}$ of real numbers over the field of real numbers, show that the set $\{1, \sqrt{2}\}$ is linearly dependent.

7. Is the following set of vectors linearly independent in $\mathbb{R}^3$? Justify your answer.

$$\alpha = (\quad 1, 4, \ -1),$$
$$\beta = (\quad 2, 3, \quad 6),$$
$$\gamma = (-1, 2, \ -1).$$

8. Prove that if $m > n$, every set of m vectors in $\mathbb{R}^n$ is linearly dependent.

9. Let $\{\alpha, \beta, \gamma\}$ be linearly independent in $\mathscr{V}(\mathscr{F})$. Determine whether or not each of the following sets is linearly independent.

(i) $\{\alpha + \beta, \beta + \gamma, \gamma + \alpha\}$.

(ii) $\{\alpha - \beta, \beta - \gamma, \gamma - \alpha\}$.

(iii) $\{\alpha, \alpha + \beta, \alpha + \beta + \gamma\}$.

(iv) $\{\alpha, \beta, c_1 \alpha + c_2 \beta + c_3 \gamma\}$.

10. Let $\mathscr{V} = \mathscr{S} \oplus \mathscr{T}$, let S be any linearly independent subset of $\mathscr{S}$, and let T be any linearly independent subset of $\mathscr{T}$. Prove that $S \cup T$ is linearly independent.

11. In the space $\mathscr{P}_k(\mathbb{R})$ of real polynomials of degree not exceeding k show that the set $S = \{1, x, x^2, \ldots, x^k\}$ is linearly independent and spans $\mathscr{P}_k(\mathbb{R})$.

12. In the real space of functions that are differentiable on the interval $[0, 1]$, determine all functions f such that the set $\{f, f'\}$ is linearly dependent, where f' is the derivative of f.

2.6 BASES AND DIMENSION

To explore further the significance of Theorem 2.14, we first adopt some terminology.

Definition 2.8   A *basis B* for a vector space $\mathcal{V}(\mathcal{F})$ is a subset of $\mathcal{V}$ that is linearly independent and spans $\mathcal{V}$. If $\mathcal{V}$ has a finite basis, $\mathcal{V}$ is said to be *finite-dimensional*; otherwise $\mathcal{V}$ is said to be *infinite-dimensional*.

In these terms Theorem 2.14 shows that every finite-dimensional vector space (even the zero space) has a basis. We have already remarked that the same is true for infinite-dimensional spaces, although a proof of that fact requires a type of argument that is logically more sophisticated. In this book our attention will be focused on finite-dimensional spaces; we shall comment from time to time on some important differences between finite and infinite-dimensional spaces, but a given space $\mathcal{V}(\mathcal{F})$ can be interpreted as being finite-dimensional unless we specifically state otherwise.

Examples of Bases

(A) In $\mathbb{R}^n$ let ε_i denote the ordered n-tuple that has zero for each component except component i, which is one. That is,

$$\varepsilon_1 = (1, 0, 0, \quad \ldots, \quad 0),$$
$$\varepsilon_2 = (0, 1, 0, \quad \ldots, \quad 0),$$

.

.

.

$$\varepsilon_n = (0, 0, 0, \quad \ldots, \quad 1).$$

Then $\{\varepsilon_1, \varepsilon_2, \ldots, \varepsilon_n\}$ is a basis; it is linearly independent and spans $\mathbb{R}^n$:

$$(a_1, a_2, \ldots, a_n) = a_1 \varepsilon_1 + a_2 \varepsilon_2 + \cdots + a_n \varepsilon_n.$$

This particular basis for $\mathbb{R}^n$ will be called the *standard basis*.

(B) Another basis for $\mathbb{R}^n$ is exhibited in Exercise 2.5-5. There are, of course, many bases for $\mathbb{R}^n$ and for any vector space.

(C) Exercise 2.5-11 shows that the set $\{1, x, x^2, \ldots, x^k\}$ is a basis for the space $\mathcal{P}_k(\mathbb{R})$ of all real polynomials of degree not exceeding k.

(D) The space $\mathcal{P}(\mathbb{R})$ of *all* real polynomials has the infinite set $\{1, x, x^2, \ldots, x^n, \ldots\}$ as a basis. It is an infinite-dimensional space and is a subspace of the space $\mathcal{D}(a, b)$ of all real functions that are differentiable on the interval $a < x < b$. Hence $\mathcal{D}(a, b)$ is also infinite-dimensional. The same conclusion applies to the space $\mathcal{C}(a, b)$ of continuous functions.

(E) Any field $\mathscr{F}$, considered as a vector space over itself, can be generated by forming all scalar multiples of any nonzero $a \in \mathscr{F}$. Hence $\{a\}$ is a basis for $\mathscr{F}$.

(F) The field $\mathbb{C}$ of complex numbers, considered as a vector space $\mathbb{C}(\mathbb{R})$ over the real numbers, has $\{1, i\}$ as a basis, where $i^2 = -1$ in $\mathbb{C}$.

Theorem 2.15   Let $B = \{\alpha_1, \ldots, \alpha_k\}$ be any basis for $\mathscr{V}(\mathscr{F})$. Each vector $\xi \in \mathscr{V}$ has a *unique* representation as a linear combination of vectors of B.

Proof   Because B spans $\mathscr{V}$, each $\xi \in \mathscr{V}$ is some linear combination of the vectors of B, say

$$\xi = c_1\alpha_1 + c_2\alpha_2 + \cdots + c_k\alpha_k, \; c_i \in \mathscr{F}.$$

To show that the scalars c_i are uniquely determined, suppose that

$$\xi = b_1\alpha_1 + b_2\alpha_2 + \cdots + b_k\alpha_k.$$

Then by subtraction we obtain

$$\theta = \xi - \xi = (c_1 - b_1)\alpha_1 + \cdots + (c_k - b_k)\alpha_k.$$

Because the set $\{\alpha_1, \ldots, \alpha_k\}$ is linearly independent, *each* coefficient in this last equation must be zero. Hence

$$c_i = b_i \quad \text{for } i = 1, 2, \ldots, k,$$

and the uniqueness of the scalars is established.

This theorem reveals a significant interpretation of bases. Given any basis for $\mathscr{V}(\mathscr{F})$, each vector in $\mathscr{V}$ can be expressed in one and only one way as a linear combination of the basis vectors. If we agree always to write the basis vectors in a fixed order, thus obtaining an *ordered basis* of k vectors, then the vector ξ determines uniquely an ordered k-tuple $(c_1, \ldots, c_k)$ of scalars in $\mathscr{F}$, where

$$\xi = c_1\alpha_1 + \cdots + c_k\alpha_k.$$

The ordered k-tuple $(c_1, \ldots, c_k)$ exhibits the *components* of ξ relative to the ordered basis $\{\alpha_1, \ldots, \alpha_k\}$, or the *coordinates* of ξ relative to the coordinate system $\{\alpha_1, \ldots, \alpha_k\}$. In this sense an ordered basis for $\mathscr{V}$ provides a coordinate system for $\mathscr{V}$. In the future we shall interpret the word basis to mean an *ordered basis*.

Now we come to the important result that permits us to define the dimension of a finite-dimensional space to be the number of vectors in a basis for that space.

Theorem 2.16   Every basis for a finite-dimensional vector space $\mathscr{V}$ has the same number of elements.

Proof   Let $A = \{\alpha_1, \ldots, \alpha_k\}$ and $B = \{\beta_1, \ldots, \beta_m\}$ be bases for $\mathscr{V}$. Each set is a maximal independent set, so

$$B_1 = \{\alpha_1, \beta_1, \beta_2, \ldots, \beta_m\}$$

is dependent. By Theorem 2.13 some β_i is a linear combination of the vectors that precede it, and there exists a subset B'_1 of B_1 that contains α_1 as the first vector and that is a basis for $\mathscr{V}$. Then $B_2 = \{\alpha_2, B'_1\}$ is dependent, and some vector is a linear combination of the ones that precede it. This vector cannot be α_1 or α_2 because A is linearly independent. Hence there exists a subset B'_2 of B_2 that contains α_2 and α_1 as the first and second vectors and that is a basis for $\mathscr{V}$. Let $B_3 = \{\alpha_3, B'_2\}$ and repeat the argument. If all the β_i are removed in this way before k steps, we obtain the basis $B_j = \{\alpha_j, \alpha_{j-1}, \ldots, \alpha_1\}$ for $j < k$, which contradicts the independence of A, because $\alpha_k \in [B_j]$. Hence k steps are required to remove all of the β_i, one or more at a time, so $k \leq m$. Reversing the roles of A and B in the replacement process, we obtain $m \leq k$, so the proof is complete.

Definition 2.9   The *dimension* dim $\mathscr{V}$ of a finite-dimensional vector space $\mathscr{V}$ is the number of vectors in any basis for $\mathscr{V}$.

It now follows that $\mathbb{R}^n$ is an n-dimensional space because the standard basis for $\mathbb{R}^n$ has n vectors. Thus the definition of dimension coincides with the familiar practice of regarding a line as one-dimensional, a plane as two-dimensional, and so on. The concept of basis provides us with a precise definition of dimension in higher dimensional spaces, where strong geometric intuition is more difficult to acquire. When we need to specify that a given space is n-dimensional we shall usually denote that space by $\mathscr{V}_n$.

Theorem 2.17   Any linearly independent subset of $\mathscr{V}_n$ can be extended to a basis.

Proof   Exercise. Adapt the second proof of Theorem 2.14.

Recall that by definition a basis B for $\mathscr{V}$ must satisfy two properties: B must be linearly independent, and B must span $\mathscr{V}$. However, if $\mathscr{V}$ is n-dimensional and if B contains n vectors, either of these properties implies the other. This result simplifies the task of verifying that a given set of vectors is a basis.

Theorem 2.18   Let $A = \{\alpha_1, \ldots, \alpha_n\}$ be an arbitrary set of n vectors of an n-dimensional space $\mathscr{V}_n$.

(a) A is a basis for $\mathscr{V}_n$ if and only if A is linearly independent.

(b) A is a basis for $\mathscr{V}_n$ if and only if $[A] = \mathscr{V}_n$.

Proof   If A is linearly independent, A can be extended to a basis by Theorem 2.17. But a basis contains only n vectors, so A is a basis. To prove the second statement, suppose $\mathscr{V}_n = [A]$. By Theorem 2.14 a linearly independent subset of A also spans $\mathscr{V}_n$ and hence is a basis. But any basis contains n vectors, so A itself must be that subset. The "only if" statements of (a) and (b) are valid by definition of a basis.

We conclude this section with a theorem that relates the dimensions of two finite-dimensional subspaces $\mathscr{S}$ and $\mathscr{T}$ to the dimensions of $\mathscr{S} + \mathscr{T}$ and $\mathscr{S} \cap \mathscr{T}$.

Theorem 2.19   If $\mathscr{S}$ and $\mathscr{T}$ are finite-dimensional subspaces of $\mathscr{V}$, then $\dim(\mathscr{S} + \mathscr{T}) + \dim(\mathscr{S} \cap \mathscr{T}) = \dim \mathscr{S} + \dim \mathscr{T}$.

Proof   Of the four subspaces involved in this theorem, $\mathscr{S} \cap \mathscr{T}$ is a subspace of each of the others, and both $\mathscr{S}$ and $\mathscr{T}$ are subspaces of $\mathscr{S} + \mathscr{T}$. Our proof begins with the choice of a basis $\{\alpha_1, \ldots, \alpha_k\}$ for $\mathscr{S} \cap \mathscr{T}$, where $k = \dim(\mathscr{S} \cap \mathscr{T})$. Then $\dim \mathscr{S} = k + i$ and $\dim \mathscr{T} = k + j$ for some nonnegative i and j. The basis for $\mathscr{S} \cap \mathscr{T}$ can be extended to a basis $\{\alpha_1, \ldots, \alpha_k, \beta_1, \ldots, \beta_i\}$ for $\mathscr{S}$. A different extension similarly produces a basis $\{\alpha_1, \ldots, \alpha_k, \gamma_1, \ldots, \gamma_j\}$ for $\mathscr{T}$. Combining these two bases gives a set

$$\{\alpha_1, \ldots, \alpha_k, \beta_1, \ldots, \beta_i, \gamma_1, \ldots, \gamma_j\}$$

of $k + i + j$ vectors. The theorem follows immediately when it is proved that this set is linearly independent and spans $\mathscr{S} + \mathscr{T}$. The latter statement is easy because any vector in $\mathscr{S} + \mathscr{T}$ has the form $\sigma + \tau$ for $\sigma \in \mathscr{S}$ and $\tau \in \mathscr{T}$. But σ is a linear combination of the α's and β's, while τ is a linear combination of the α's and γ's. Hence $\sigma + \tau$ is a linear combination of the α's, β's, and γ's. To prove linear independence suppose that

$$a_1\alpha_1 + \cdots + a_k\alpha_k + b_1\beta_1 + \cdots + b_i\beta_i + c_1\gamma_1 + \cdots + c_j\gamma_j = 0.$$

Let

$$\alpha = \sum_1^k a_m\alpha_m, \qquad \beta = \sum_1^i b_m\beta_m, \qquad \gamma = \sum_1^j c_m\gamma_m.$$

Then

$$\alpha + \beta + \gamma = 0, \text{ where } \alpha \in \mathscr{S} \cap \mathscr{T}, \beta \in \mathscr{S}, \gamma \in \mathscr{T}.$$

Then $\gamma \in \mathscr{S}$ because $\gamma = -(\alpha + \beta) \in \mathscr{S}$. But $\gamma \in \mathscr{T}$, so $\gamma \in \mathscr{S} \cap \mathscr{T}$. Then $\gamma = c_1\gamma_1 + \cdots + c_j\gamma_j$ is a linear combination of the α's, so each $c_m = 0$ because the set $\{\alpha_1, \ldots, \alpha_k, \gamma_1, \ldots, \gamma_j\}$ is a basis for $\mathscr{T}$ and thus is linearly independent. But then $\gamma = 0$ and $\alpha + \beta = 0$. It follows that each a_m and each b_m must be zero because $\{\alpha_1, \ldots, \alpha_k, \beta_1, \ldots, \beta_i\}$ is a basis for $\mathscr{S}$.

This theorem reveals why two distinct planes through the origin in $\mathbb{R}^3$ must intersect in a line through the origin, whereas in $\mathbb{R}^4$ a pair of two-dimensional subspaces might intersect only at 0. Also two distinct three-dimensional subspaces $\mathscr{S}$ and $\mathscr{T}$ in $\mathbb{R}^4$ must intersect in a two dimensional subspace, because

$$\dim(\mathscr{S} \cap \mathscr{T}) = \dim \mathscr{S} + \dim \mathscr{T} - \dim(\mathscr{S} + \mathscr{T})$$
$$= 3 + 3 - 4 = 2.$$

EXERCISES 2.6

1. For each of the following sets of three vectors in $\mathbb{R}^3$ determine the dimension of the subspace $[\alpha, \beta, \gamma]$.

(i) $\alpha = (1, -1, 2)$, $\beta = (0, -1, 1)$, $\gamma = (3, -2, 5)$.

(ii) $\alpha = (0, 1, 1)$, $\beta = (1, 0, 1)$, $\gamma = (1, 1, 0)$.

2. In $\mathbb{R}^3$ show that any three vectors that do not lie on the same plane through the origin determine a basis for $\mathbb{R}^3$.

3. Let $\{\alpha_1, \ldots, \alpha_n\}$ be a basis for $\mathscr{V}_n$. Prove that each of the following sets is also a basis for $\mathscr{V}_n$:

(i) $\{c_1\alpha_1, \ldots, c_n\alpha_n\}$ where $c_1, c_2, \ldots, c_n$ are nonzero scalars,

(ii) $\{\beta_1, \ldots, \beta_n\}$ where $\beta_i = \alpha_i + \alpha_1$ for $i = 1, \ldots, n$.

4. Show that if subspaces $\mathcal{S}$ and $\mathcal{T}$ of $\mathcal{V}$ satisfy $\mathcal{S} \subseteq \mathcal{T}$ and $\dim(\mathcal{S}) = \dim(\mathcal{T})$, then $\mathcal{S} = \mathcal{T}$.

5. Prove Theorem 2.17.

6. Given the vectors $\alpha_1 = (1, 1, 1)$, $\alpha_2 = (0, 1, 1)$,

$$\alpha_3 = (0, 0, 1) \text{ in } \mathbb{R}^3:$$

(i) Prove that $\{\alpha_1, \alpha_2, \alpha_3\}$ is a basis for $\mathbb{R}^3$.

(ii) Express each vector ε_i of the standard basis as a linear combination of the vectors $\alpha_1, \alpha_2,$ and α_3.

7. Let $\mathcal{V}$ be a finite-dimensional vector space.

(i) Prove that if $\{\alpha_1, \ldots, \alpha_n\}$ is a basis for $\mathcal{V}$, if $\mathcal{S} = [\alpha_1, \ldots, \alpha_k]$, and if $\mathcal{T} = [\alpha_{k+1}, \ldots, \alpha_n]$, then $\mathcal{V} = \mathcal{S} \oplus \mathcal{T}$.

(ii) Prove, conversely, that if $\mathcal{S}$ and $\mathcal{T}$ are any subspaces of $\mathcal{V}$ such that $\mathcal{V} = \mathcal{S} \oplus \mathcal{T}$, if $\{\alpha_1, \ldots, \alpha_k\}$ is a basis for $\mathcal{S}$, and if $\{\beta_1, \ldots, \beta_m\}$ is a basis for $\mathcal{T}$, then $\{\alpha_1, \ldots, \alpha_k, \beta_1, \ldots, \beta_m\}$ is a basis for $\mathcal{V}$.

8. Let $\mathcal{S}$ and $\mathcal{T}$ be subspaces of an n-dimensional space $\mathcal{V}$ such that $\mathcal{S} \cap \mathcal{T} = [\theta]$ and $\dim(\mathcal{S}) + \dim(\mathcal{T}) = n$. Prove that $\mathcal{V} = \mathcal{S} \oplus \mathcal{T}$.

9. If $\mathcal{F}_1$ and $\mathcal{F}_2$ are fields such that $\mathcal{F}_1 \subseteq \mathcal{F}_2$, explain how $\mathcal{F}_2$ can be regarded as a vector space over $\mathcal{F}_1$.

10. If $\mathbb{Q}$, $\mathbb{R}$, and $\mathbb{C}$ denote the fields of rational, real, and complex numbers, respectively, what is the dimension of each of the following vector spaces?

(i) $\mathbb{Q}$ regarded as a vector space over $\mathbb{Q}$.

(ii) $\mathbb{C}$ regarded as a vector space over $\mathbb{R}$.

(iii) $\mathbb{C}$ regarded as a vector space over $\mathbb{C}$.

(iv) $\mathbb{R}$ regarded as a vector space over $\mathbb{Q}$.

2.7   ISOMORPHISMS OF VECTOR SPACES

In this section we propose to look more carefully at the question raised in Exercise 2.3-7, which suggested that the vector space $\mathcal{P}_2(\mathbb{R})$ of real polynomials of degree not exceeding two is somehow related to the space $\mathbb{R}^3$. We can just as easily discuss the similarity between the vector spaces $\mathscr{P}_k(\mathbb{R})$ and $\mathbb{R}^{k+1}$.

To begin with, the scalar field is $\mathbb{R}$ in both cases. But a vector in $\mathscr{P}_k(\mathbb{R})$ is a polynomial with real coefficients,

$$p(x) = a_0 + a_1 x + \cdots + a_k x^k,$$

whereas a vector in $\mathbb{R}^{k+1}$ is an ordered $(k+1)$-tuple of reals,

$$\alpha = (a_0, a_1, \ldots, a_k).$$

From Example C of Section 2.6, we know that $\{1, x, x^2, \ldots, x^k\}$ is a basis for $\mathscr{P}_k(\mathbb{R})$, so we know that both of these vector spaces have dimension $k + 1$. The form in which we have written typical elements of each space gives a clue about how we can formalize a relationship between the vectors of these two spaces. With each polynomial $p(x) = a_0 + a_1 x + \cdots + a_k x^k$ we associate the real ordered $(k+1)$-tuple $(a_0, a_1, \ldots, a_k)$ in $\mathbb{R}^{k+1}$. This association defines a function or mapping f from the set of all vectors in $\mathscr{P}_k(\mathbb{R})$ into $\mathbb{R}^{k+1}$:

$$f(p(x)) = (a_0, a_1, \ldots, a_k).$$

The domain of f is the entire space $\mathscr{P}_k(\mathbb{R})$; the range of f is the entire space $\mathbb{R}^{k+1}$, because each vector $\beta = (b_0, b_1, \ldots, b_k)$ in $\mathbb{R}^{k+1}$ is the f-image of some polynomial,

$$q(x) = b_0 + b_1 x + \cdots + b_k x^k.$$

Hence f maps $\mathscr{P}_k(\mathbb{R})$ *onto* $\mathbb{R}^{k+1}$. Furthermore, the mapping f is *one-to-one*, which means that distinct polynomials are mapped by f into distinct $(k+1)$-tuples. (If you are not familiar with this terminology for functions, see Appendix A.4.)

Next we examine the behavior of f relative to the two vector space operations. Let p and q denote the polynomials exhibited in the previous paragraph. Then for each $c \in \mathbb{R}$

$$cp(x) = ca_0 + ca_1 x + \cdots + ca_k x^k,$$

$$p(x) + q(x) = (a_0 + b_0) + (a_1 + b_1)x + \cdots + (a_k + b_k)x^k.$$

Hence

$$f(cp(x)) = (ca_0, ca_1, \ldots, ca_k) = c\alpha = cf(p(x)),$$

$$f(p(x) + q(x)) = (a_0 + b_0, a_1 + b_1, \ldots, a_k + b_k)$$

$$= \alpha + \beta = f(p(x)) + f(q(x)).$$

Observe carefully that these two equations reveal that the function f *preserves* the two operations of a linear space, because the f-image of a scalar multiple of p is the same scalar multiple of the f-image of p, and the f-image of the sum of p and q is the sum of the f-image of p and the f-image of q. Any function having these properties is called a *linear* mapping. The linearity of f confirms formally that the vector space operations with polynomials are a thinly disguised (but exact) replication of the vector space operations in $\mathbb{R}^{k+1}$. If we add two vectors in $\mathscr{P}_k(\mathbb{R})$ and then apply f to the sum, we obtain precisely the same $(k+1)$-tuple as though we had first applied f to each of the two vectors and then added the results in $\mathbb{R}^{k+1}$. And the corresponding property holds for scalar multiples.

In summary, we have shown that for each $k \geq 0$ the vector space $\mathscr{P}_k(\mathbb{R})$ is an identical twin of the vector space $\mathbb{R}^{k+1}$, in the sense that the vectors of $\mathscr{P}_k(\mathbb{R})$ can be renamed in such a way that the resulting system is algebraically indistinguishable from $\mathbb{R}^{k+1}$. The renaming process is performed by a function f with these properties:

The domain of f is $\mathscr{P}_k(\mathbb{R})$,

The range of f is $\mathbb{R}^{k+1}$,

f is one-to-one,

$f(p + q) = f(p) + f(q)$ for all $p, q \in \mathscr{P}_k(\mathbb{R})$.

$f(cp) = cf(p)$ for all $p \in \mathscr{P}_k(\mathbb{R})$ and all $c \in \mathbb{R}$.

Such a mapping is called a vector space *isomorphism*, meaning that the two spaces related in this way have the *same form*; more precisely, they are algebraically indistinguishable except for the notation and terminology used for the vectors of the two spaces. We now restate these ideas formally.

Definition 2.10   Let $\mathscr{V}(\mathscr{F})$ and $\mathscr{W}(\mathscr{F})$ be two vector spaces over the same field $\mathscr{F}$, and let **H** be a function whose domain is $\mathscr{V}$ and whose range is a subset of $\mathscr{W}$. Then **H** is called a vector space *homomorphism* of $\mathscr{V}(\mathscr{F})$ into $\mathscr{W}(\mathscr{F})$ if and only if for all α, β in $\mathscr{V}$ and all c in $\mathscr{F}$

$$\mathbf{H}(\alpha + \beta) = \mathbf{H}(\alpha) + \mathbf{H}(\beta),$$

$$\mathbf{H}(c\alpha) = c\mathbf{H}(\alpha).$$

If every vector of $\mathscr{W}$ is in the range of **H**, then **H** is called a homomorphism of $\mathscr{V}(\mathscr{F})$ *onto* $\mathscr{W}(\mathscr{F})$. A vector space homomorphism that is one-to-one and onto is called an *isomorphism*. $\mathscr{V}(\mathscr{F})$ and $\mathscr{W}(\mathscr{F})$ are said to be *isomorphic* if and only if there exists an isomorphism from $\mathscr{V}(\mathscr{F})$ onto $\mathscr{W}(\mathscr{F})$.

Restated, a vector space homomorphism is simply a linear mapping from a vector space $\mathscr{V}(\mathscr{F})$ to a vector space $\mathscr{W}(\mathscr{F})$, and an isomorphism is a homomorphism that is both onto and one-to-one. Linearity means that the image of each linear combination of vectors is the *same* linear combination of the images of those vectors. We are now ready to prove a general theorem that includes the example of $\mathscr{P}_k(\mathbb{R})$ as a special case.

Theorem 2.20 Any n-dimension space $\mathscr{V}(\mathscr{F})$ is isomorphic to the space $\mathscr{F}^n$ of all ordered n-tuples of scalars in $\mathscr{F}$.

Proof Let $\{\alpha_1, \ldots, \alpha_n\}$ be a basis for $\mathscr{V}$. By Theorem 2.15 each $\gamma \in \mathscr{V}$ has a unique representation as a linear combination of basis vectors:

$$\gamma = c_1\alpha_1 + c_2\alpha_2 + \cdots + c_n\alpha_n.$$

Let $\mathbf{H}$ be the mapping from $\mathscr{V}$ to $\mathscr{F}^n$ defined by

$$\mathbf{H}(\gamma) = (c_1, c_2, \ldots, c_n).$$

Distinct vectors in $\mathscr{V}$ have distinct images in $\mathscr{F}^n$, so $\mathbf{H}$ is *one-to-one*. Each vector in $\mathscr{F}^n$ is the image of some vector in $\mathscr{V}$, so $\mathbf{H}$ maps $\mathscr{V}$ *onto* $\mathscr{F}^n$. To show that $\mathbf{H}$ is linear we let $a \in \mathscr{F}$ and $\beta, \gamma \in \mathscr{V}$, where γ is defined above and where

$$\beta = b_1\alpha_1 + b_2\alpha_2 + \cdots + b_n\alpha_n.$$

Therefore,

$$\mathbf{H}(\beta) = (b_1, b_2, \ldots, b_n),$$

$$\beta + \gamma = (b_1 + c_1)\alpha_1 + \cdots + (b_n + c_n)\alpha_n,$$

$$\mathbf{H}(\beta + \gamma) = (b_1 + c_1, b_2 + c_2, \ldots, b_n + c_n)$$

$$= (b_1, b_2, \ldots, b_n) + (c_1, c_2, \ldots, c_n)$$

$$= \mathbf{H}(\beta) + \mathbf{H}(\gamma).$$

Also,

$$a\beta = ab_1\alpha_1 + ab_2\alpha_2 + \cdots + ab_n\alpha_n,$$

$$\mathbf{H}(a\beta) = (ab_1, ab_2, \ldots, ab_n)$$

$$= a(b_1, b_2, \ldots, b_n) = a\mathbf{H}(\beta).$$

We conclude that $\mathbf{H}$ is an isomorphism from $\mathscr{V}(\mathscr{F})$ onto $\mathscr{F}^n$.

In particular, this theorem demonstrates that any real vector space of dimension n is a copy of $\mathbb{R}^n$. It permits us to think of any such space in terms of the specific and familiar example of real n-tuples. However, for much of our work the n-tuple notation is needlessly cumbersome, so we shall continue to use the less specific notation that has been introduced for vectors in an arbitrary vector space. A more basic reason for retaining the semblance of generality is inherent in the proof of Theorem 2.20; to represent a vector of $\mathscr{V}_n(\mathbb{R})$ as a real n-tuple we must first choose a basis for $\mathscr{V}_n(\mathbb{R})$, and different choices of bases lead to different representations. In many situations the choice of a basis tends to obscure, rather than to clarify, the essence of a result.

In the following chapters most of our study will be directed toward vector space homomorphisms (linear mappings). Some of the basic properties of homomorphisms are given in the first four following exercises, and you should give these results your careful attention.

EXERCISES 2.7

1. Let **H** be a homomorphism from $\mathscr{V}(\mathscr{F})$ to $\mathscr{W}(\mathscr{F})$.

   (i) Prove that $\mathbf{H}(\theta) = \theta$ in $\mathscr{W}$.

   (ii) Prove that **H** is one-to-one if and only if $\mathbf{H}(\alpha) \neq \theta$ in $\mathscr{W}$ whenever $\alpha \neq \theta$ in $\mathscr{V}$.

2. Let **H** be a homomorphism from $\mathscr{V}(\mathscr{F})$ to $\mathscr{W}(\mathscr{F})$. Let **J** be a homomorphism from $\mathscr{W}(\mathscr{F})$ to $\mathscr{X}(\mathscr{F})$.

   (i) Show how the mappings **H** and **J** can be combined to produce a homomorphism **K** from $\mathscr{V}(\mathscr{F})$ to $\mathscr{X}(\mathscr{F})$.

   (ii) Show that if **H** and **J** are isomorphisms onto $\mathscr{W}(\mathscr{F})$ and $\mathscr{X}(\mathscr{F})$, respectively, then $\mathscr{V}(\mathscr{F})$ and $\mathscr{X}(\mathscr{F})$ are isomorphic.

3. Let **H** be a homomorphism from $\mathscr{V}(\mathscr{F})$ to $\mathscr{W}(\mathscr{F})$, and let S be a subset of $\mathscr{V}$.

   (i) Show that if S is linearly dependent in $\mathscr{V}(\mathscr{F})$ then $\mathbf{H}(S)$ is linearly dependent in $\mathscr{W}(\mathscr{F})$, where $\mathbf{H}(S)$ denotes the set of all images $\mathbf{H}(\sigma)$ as σ varies over S.

   (ii) Show that if S is linearly independent in $\mathscr{V}(\mathscr{F})$, then $\mathbf{H}(S)$ is not necessarily linearly independent in $\mathscr{W}(\mathscr{F})$.

   (iii) Show that if S is linearly independent in $\mathscr{V}(\mathscr{F})$, and if **H** is one-to-one, then $\mathbf{H}(S)$ is linearly independent in $\mathscr{W}(\mathscr{F})$.

(iv) Deduce that if **H** is an isomorphism, if $\mathscr{V}(\mathscr{F})$ and $\mathscr{W}(\mathscr{F})$ have the same dimension n, and if B is a basis for $\mathscr{V}(\mathscr{F})$, then $\mathbf{H}(B)$ is a basis for $\mathscr{W}(\mathscr{F})$.

4. Let **H** be a homomorphism from $\mathscr{V}(\mathscr{F})$ to $\mathscr{W}(\mathscr{F})$ and let $\mathscr{S}$ be a subspace of $\mathscr{V}(\mathscr{F})$. Prove that $\mathbf{H}(\mathscr{S})$ is a subspace of $\mathscr{W}(\mathscr{F})$, where $\mathbf{H}(\mathscr{S}) = \{\gamma \in \mathscr{W} \mid \gamma = \mathbf{H}(\sigma) \text{ for some } \sigma \in \mathscr{S}\}$.

5. Show that the space $\mathscr{P}(\mathbb{Q})$ of all polynomials with rational coefficients is isomorphic to the space $\mathscr{S}(\mathbb{Q})$ of all infinite sequences of rational numbers in which all but a finite number of terms are zero (Exercise 2.4-8).

6. Let $\mathscr{V}$ and $\mathscr{W}$ be vector spaces over the same field $\mathscr{F}$, and let $\mathscr{H}$ denote the collection of all homomorphisms from $\mathscr{V}$ into $\mathscr{W}$. The sum of two homomorphisms is defined by

$$(\mathbf{H_1} + \mathbf{H_2})(\xi) = \mathbf{H_1}(\xi) + \mathbf{H_2}(\xi) \text{ for all } \xi \in \mathscr{V}.$$

Also, the product of a scalar and a homomorphism is defined by

$$(a\mathbf{H})(\xi) = a\mathbf{H}(\xi) \text{ for all } a \in \mathscr{F}, \xi \in \mathscr{V}.$$

Show that $\mathbf{H_1} + \mathbf{H_2}$ and $a\mathbf{H}$ are also homomorphisms from $\mathscr{V}$ into $\mathscr{W}$ and that $\mathscr{H}$ forms a vector space over $\mathscr{F}$ relative to these operations.

7. Referring to Exercise 2.7-6, if $\mathscr{V} = \mathscr{W}$, then it is possible to define the product of two homomorphisms in $\mathscr{H}$ in terms of the rule for successive mappings:

$$(\mathbf{H_1}\mathbf{H_2})(\xi) = \mathbf{H_1}(\mathbf{H_2}(\xi)) \text{ for all } \xi \in \mathscr{V}.$$

Show that $\mathbf{H_1}\mathbf{H_2}$ is a homomorphism from $\mathscr{V}$ into $\mathscr{V}$.

8. If we specialize Exercise 2.7-6 by choosing $\mathscr{W}$ to be the field $\mathscr{F}$, considered as a vector space over itself, then the members of $\mathscr{H}$ are scalar-valued functions defined on $\mathscr{V}$ and having the properties of a homomorphism. Any such function is called a *linear functional*, and the vector space $\mathscr{H}$ of all linear functionals from $\mathscr{V}$ to $\mathscr{F}$ is called the *dual space* of $\mathscr{V}$. Determine whether each of the following mappings is a linear functional:

(i) In $\mathscr{F}^n$, the mappings

$$\mathbf{H}(a_1, \ldots, a_n) = 2a_1,$$

$$\mathbf{H}(a_1, \ldots, a_n) = a_1 + 2.$$

(ii) In the space of all real-valued functions that are differentiable on the interval $-1 \le x \le 1$, the mappings

$$\mathbf{H}(f) = f'(0),$$

$$\mathbf{H}(f) = f(0)f'(0).$$

(iii) In the space of all real-valued functions that are continuous on the interval $0 \le x \le 1$, the mappings

$$\mathbf{H}(f) = \int_0^1 f(t)e^{-t}\, dt,$$

$$\mathbf{H}(f) = \int_0^1 f(te^{-t})\, dt,$$

$$\mathbf{H}(f) = \int_0^1 e^{-f(t)}\, dt.$$

CHAPTER 3
LINEAR MAPPINGS

3.1 ALGEBRA OF HOMOMORPHISMS

In the preceding section we defined a vector space homomorphism to be a linear function (or mapping) $\mathbf{H}$ from one vector space $\mathscr{V}$ to another space $\mathscr{W}$, both spaces having the same field $\mathscr{F}$ of scalars. Conceivably $\mathscr{W}$ could be the same space as $\mathscr{V}$, but at this point there is no need to impose that restriction. $\mathbf{H}$ maps each vector α in $\mathscr{V}$ into some vector $\mathbf{H}(\alpha)$ in $\mathscr{W}$ in such a way that for all $\alpha, \beta \in \mathscr{V}$ and all $a, b \in \mathscr{F}$,

$$\mathbf{H}(a\alpha + b\beta) = a\mathbf{H}(\alpha) + b\mathbf{H}(\beta) \text{ in } \mathscr{W}.$$

In particular,

$$\mathbf{H}(\alpha + \beta) = \mathbf{H}(\alpha) + \mathbf{H}(\beta),$$

and

$$\mathbf{H}(a\alpha) = a\mathbf{H}(\alpha).$$

The use of boldface capital letters for linear mappings allows us to simplify the usual notation for functions by writing $\mathbf{H}\alpha$ instead of $\mathbf{H}(\alpha)$ whenever there is no possibility of ambiguity. For example, parentheses are needed in $\mathbf{H}(\alpha + \beta)$ to indicate that the function $\mathbf{H}$ is to be applied to the sum of α and β.

In this section we shall study some examples of linear mappings and observe some of their algebraic properties.

Examples of Linear Mappings

(A) For each $\alpha \in \mathscr{V}$, let $\mathbf{Z}\alpha = \theta$ in $\mathscr{W}$. This mapping is easily seen to be linear, and $\mathbf{Z}$ is called the *zero* linear mapping.

(B) Consider the system (1.1) of m linear equations in n variables. Each vector $\xi = (x_1, \ldots, x_n)$ in $\mathbb{R}^n$ determines a vector $\eta = (y_1, \ldots, y_m)$ in $\mathbb{R}^m$, where

$$y_i = a_{i1}x_1 + a_{i2}x_2 + \cdots + a_{in}x_n$$

$$= \sum_{j=1}^{n} a_{ij}x_j \text{ for } i = 1, \ldots, m.$$

Hence the coefficients a_{ij} of (1.1) define a mapping $\mathbf{T}$ such that $\mathbf{T}\xi = \eta \in \mathbb{R}^m$ for each $\xi \in \mathbb{R}^n$. If we denote by δ the vector $(d_1, \ldots, d_m) \in \mathbb{R}^m$ of constants in the right-hand members of (1.1), then the system can be written in the form $\mathbf{T}\xi = \delta$. To solve (1.1) means to determine all vectors ξ in $\mathbb{R}^n$ that are mapped by $\mathbf{T}$ into the given vector δ. We need to verify that $\mathbf{T}$ is linear; let $b, c \in \mathbb{R}$, and let $\beta = (b_1, \ldots, b_n)$ and $\gamma = (c_1, \ldots, c_n)$ in $\mathbb{R}^n$. Then

$$\mathbf{T}(b\beta + c\gamma) = \mathbf{T}(bb_1 + cc_1, bb_2 + cc_2, \ldots, bb_n + cc_n)$$

$$= \left(\sum_{j=1}^{n} a_{1j}(bb_j + cc_j), \ldots, \sum_{j=1}^{n} a_{mj}(bb_j + cc_j) \right)$$

$$= \left(b\sum_{j=1}^{n} a_{1j}b_j + c\sum_{j=1}^{n} a_{1j}c_j, \ldots, b\sum_{j=1}^{n} a_{mj}b_j + c\sum_{j=1}^{n} a_{mj}c_j \right)$$

$$= b\left(\sum_{j=1}^{n} a_{1j}b_j, \ldots, \sum_{j=1}^{n} a_{mj}b_j \right) + c\left(\sum_{j=1}^{n} a_{1j}c_j, \ldots, \sum_{j=1}^{n} a_{mj}c_j \right)$$

$$= b\mathbf{T}\beta + c\mathbf{T}\gamma.$$

(C) Let $\mathscr{V}_n$ and $\mathscr{W}_m$ be vector spaces over $\mathscr{F}$, and let $B = \{\alpha_1, \ldots, \alpha_n\}$ be a fixed basis for $\mathscr{V}_n$. With each α_j in B we first associate some vector γ_j in $\mathscr{W}_m$, $j = 1, 2, \ldots, n$. Then we define a mapping $\mathbf{T}$ from $\mathscr{V}_n$ to $\mathscr{W}_m$ by the two rules:

$$\mathbf{T} \text{ is linear, and } \mathbf{T}\alpha_j = \gamma_j \text{ for } j = 1, \ldots, n.$$

Each vector ξ in $\mathscr{V}_n$ is uniquely expressible as a linear combination of vectors in B,

$$\xi = \sum_{j=1}^{n} x_j \alpha_j.$$

Because **T** is linear,

$$\mathbf{T}\xi = \mathbf{T}\left(\sum_{j=1}^{n} x_j \alpha_j \right) = \sum_{j=1}^{n} x_j \mathbf{T}\alpha_j = \sum_{j=1}^{n} x_j \gamma_j,$$

so **T** is defined on all of $\mathscr{V}_n$, and its value at each vector ξ in $\mathscr{V}_n$ is a uniquely determined linear combination of its values on the vectors of any basis. We express this fact by saying that a *linear mapping is completely determined by its effect on a basis of its domain space.*

We now turn to the special case in which $\mathscr{V} = \mathscr{W}$; that is, **T** is a linear mapping whose domain is $\mathscr{V}$ and whose range is a subset of $\mathscr{V}$. In this case **T** can be regarded as transforming each vector ξ in into another vector **T**ξ in the same space $\mathscr{V}$. Thus it is convenient to reserve the special term *linear transformation* to refer to a linear mapping from a vector space $\mathscr{V}$ into itself. Thus we shall consider a linear transformation to be a special type of linear mapping, but in many books the two terms are used interchangeably.

Examples of Linear Transformations

(D) In $\mathscr{E}^2$ let $\mathbf{T}(x, y) = (ax + by, cx + dy)$. As an exercise you may verify that **T** is linear. We now give geometric interpretations of **T** for some special choices of a, b, c, and d, using the standard basis to represent $\mathscr{E}^2$.

If $d = a$ and $b = c = 0$, $\mathbf{T}(x, y) = (ax, ay) = a(x, y)$. **T** maps each vector of $\mathscr{E}^2$ into a scalar multiple of itself. That is, **T** maps each point P into a point Q that is collinear with P and the origin but is $|a|$ as far from the origin as P is. **T** is called a *dilation* (or dilatation) of $\mathscr{E}^2$ with center at the origin. If $a = 1$, **T** is the *identity* transformation.

If $a = 1 = -d$ and $b = c = 0$, $\mathbf{T}(x, y) = (x, -y)$; **T** maps each point of $\mathscr{E}^2$ into its mirror image with respect to the x axis. **T** is a *reflection* across the x-axis.

If $d = 1 = -a$ and $b = c = 0$, $\mathbf{T}(x, y) = (-x, y)$; **T** is a reflection across the y-axis.

If $a = -1 = d$ and $b = c = 0$, $\mathbf{T}(x, y) = (-x, -y)$; $\mathbf{T}$ is a reflection through the origin.

If $a = 0 = d$ and $b = c = 1$, $\mathbf{T}(x, y) = (y, x)$; $\mathbf{T}$ is a reflection across the line $y = x$.

If $a = 1$ and $b = c = d = 0$, $\mathbf{T}(x, y) = (x, 0)$; $\mathbf{T}$ projects each point P onto the x-axis, along a direction parallel to the y-axis. $\mathbf{T}$ is called a *projection* onto the x-axis, along the y-axis.

If $d = 1$ and $a = b = c = 0$, $\mathbf{T}(x, y) = (0, y)$; $\mathbf{T}$ is a projection onto the y-axis, along the x-axis.

If $a = \cos \Psi = d$ and $c = \sin \Psi = -b$ for some fixed angle Ψ, then

$$\mathbf{T}(x, y) = (x \cos \Psi - y \sin \Psi, \, x \sin \Psi + y \cos \Psi).$$

The geometric effect of $\mathbf{T}$ is not so readily recognized in this case but one can verify (or perhaps recall some formulas of analytic geometry for rotating axes) that $\mathbf{T}$ rotates each point of $\mathscr{E}^2$ counterclockwise around the origin through the angle Ψ, as in Figure 3.1.

(E) For any real polynomial p, let $\mathbf{D}p$ denote the derivative of p; that is, if

$$p(x) = a_0 + a_1 x + \cdots + a_k x^k,$$

$$(\mathbf{D}p)(x) = a_1 + 2a_2 x + \cdots + ka_k x^{k-1}.$$

For any fixed n we can interpret $\mathbf{D}$ as a linear transformation from the space $\mathscr{P}_n(\mathbb{R})$ onto the proper subspace $\mathscr{P}_{n-1}(\mathbb{R})$. But if we consider $\mathbf{D}$ as a transformation on the space $\mathscr{P}(\mathbb{R})$ of real polynomials of all degrees, then $\mathbf{D}$ maps $\mathscr{P}(\mathbb{R})$ onto the entire space $\mathscr{P}(\mathbb{R})$, because any polynomial is the derivative of a polynomial.

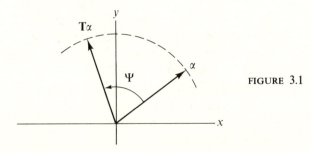

FIGURE 3.1

We now turn from examples of linear mappings and linear transformations to algebraic properties of such functions.

Definition 3.1 Let $\mathscr{V}$ and $\mathscr{W}$ be vector spaces over the same field $\mathscr{F}$, and let $\mathscr{L}(\mathscr{V}, \mathscr{W})$ denote the set of all linear mappings from $\mathscr{V}$ to $\mathscr{W}$. Let **S** and **T** be in $\mathscr{L}(\mathscr{V}, \mathscr{W})$ and $c \in \mathscr{F}$.

(a) $\mathbf{S} = \mathbf{T}$ if and only if for each $\alpha \in \mathscr{V}$, $\mathbf{S}\alpha = \mathbf{T}\alpha$ in $\mathscr{W}$.

(b) $\mathbf{S} + \mathbf{T}$ is the mapping from $\mathscr{V}$ to $\mathscr{W}$ defined by

$$(\mathbf{S} + \mathbf{T})\alpha = \mathbf{S}\alpha + \mathbf{T}\alpha \text{ for each } \alpha \in \mathscr{V}.$$

(c) $c\mathbf{T}$ is the mapping from $\mathscr{V}$ to $\mathscr{W}$ defined by

$$(c\mathbf{T})\alpha = c\mathbf{T}\alpha \text{ for each } \alpha \text{ in } \mathscr{V}.$$

It is a simple matter to verify that the mappings $\mathbf{S} + \mathbf{T}$ and $c\mathbf{T}$ are linear whenever **S** and **T** are linear and c is in $\mathscr{F}$. Hence the set $\mathscr{L}(\mathscr{V}, \mathscr{W})$ of all linear mappings from $\mathscr{V}$ to $\mathscr{W}$ is endowed with the two basic operations of a vector space. The zero linear transformation **Z** is the zero element of that space, because for any $\mathbf{T} \in \mathscr{L}(\mathscr{V}, \mathscr{W})$

$$(\mathbf{T} + \mathbf{Z})\alpha = \mathbf{T}\alpha + \mathbf{Z}\alpha = \mathbf{T}\alpha + \theta = \mathbf{T}\alpha$$

for all $\alpha \in \mathscr{V}$. Hence $\mathbf{T} + \mathbf{Z} = \mathbf{T}$ for all **T**. Addition in $\mathscr{L}(\mathscr{V}, \mathscr{W})$ is commutative and associative and $(-1)\mathbf{T}$ is the negative of **T**. Furthermore, scalar multiples satisfy Postulates 5–8 of the definition of a vector space. Thus we have proved the following result.

Theorem 3.1 The set $\mathscr{L}(\mathscr{V}, \mathscr{W})$ forms a linear space over $\mathscr{F}$ relative to the operations given in Definition 3.1.

Under certain conditions it is also possible to define a "product" of linear mappings. Let $\mathbf{T} \in \mathscr{L}(\mathscr{V}, \mathscr{W})$ and $\mathbf{S} \in \mathscr{L}(\mathscr{W}, \mathscr{X})$. Then **T** maps each $\alpha \in \mathscr{V}$ into a vector $\mathbf{T}\alpha$ in $\mathscr{W}$. And **S** maps each vector β in $\mathscr{W}$ into a vector $\mathbf{S}\beta$ in $\mathscr{X}$. Hence the composition (or successive application) of mappings $\mathbf{S} \circ \mathbf{T}$ can be defined from $\mathscr{V}$ into $\mathscr{X}$ by the rule

$$(\mathbf{S} \circ \mathbf{T})\alpha = \mathbf{S}(\mathbf{T}\alpha) \text{ for each } \alpha \in \mathscr{V}.$$

See Figure 3.2. We shall write **ST** to denote $\mathbf{S} \circ \mathbf{T}$. It is easily verified that **ST** is linear when **S** and **T** are linear.

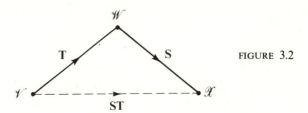

FIGURE 3.2

Now we restrict our attention to the linear space $\mathscr{L}(\mathscr{V}, \mathscr{V})$ of all linear transformations of a space $\mathscr{V}$ into itself. In addition to the two usual operations of sum and scalar multiple in $\mathscr{L}(\mathscr{V}, \mathscr{V})$, a " product " (the successive mapping operation) also is defined on $\mathscr{L}(\mathscr{V}, \mathscr{V})$. It is not hard to verify that this operation is closed, associative, and bilinear; that is,

Closed: $\mathbf{ST} \in \mathscr{L}$ whenever $\mathbf{S}, \mathbf{T} \in \mathscr{L}$,

Associative: $\mathbf{R(ST)} = \mathbf{(RS)T}$ for all $\mathbf{R}, \mathbf{S}, \mathbf{T} \in \mathscr{L}$,

Bilinear: $\mathbf{R}(a\mathbf{S} + b\mathbf{T}) = a\mathbf{RS} + b\mathbf{RT}$, and

$\qquad (a\mathbf{R} + b\mathbf{S})\mathbf{T} = a\mathbf{RT} + b\mathbf{ST}$.

A vector space, such as $\mathscr{L}$, on which there also is defined a closed, associative, and bilinear product of vectors, is called a *linear algebra*. However, the product in $\mathscr{L}(\mathscr{V}, \mathscr{V})$ is *not commutative*; that is, the equation $\mathbf{ST} = \mathbf{TS}$ does not hold in general for the composition of linear transformations. For example, in $\mathscr{E}^2$ let $\mathbf{T}$ be a reflection in the y-axis and let $\mathbf{S}$ be a counterclockwise rotation through 90 degrees. Then

$$\mathbf{ST}(1, 0) = \mathbf{S}(-1, 0) = (0, -1),$$
$$\mathbf{TS}(1, 0) = \mathbf{T}(0, 1) = (0, 1).$$

We also note the existence of a special linear transformation $\mathbf{I}$ from $\mathscr{V}$ to $\mathscr{V}$, defined by

$$\mathbf{I}\alpha = \alpha \text{ for every } \alpha \in \mathscr{V}.$$

$\mathbf{I}$ is clearly linear, and $\mathbf{TI} = \mathbf{T} = \mathbf{IT}$ for every $\mathbf{T}$ in $\mathscr{L}(\mathscr{V}, \mathscr{V})$. Thus, $\mathbf{I}$ acts as an identity element relative to the product operation, and it is called the *identity* linear transformation.

We have already observed that the algebra of linear transformations is noncommutative. Some other curious properties of this algebra are worth noting. In $\mathscr{E}^2$, neither the projection $\mathbf{T}_1$ onto the x-axis nor the projection $\mathbf{T}_2$

onto the y-axis is the zero transformation but their product is the zero transformation; that is,

$$\mathbf{T}_1 \neq \mathbf{Z} \text{ and } \mathbf{T}_2 \neq \mathbf{Z} \text{ but } \mathbf{T}_1\mathbf{T}_2 = \mathbf{Z} = \mathbf{T}_2\mathbf{T}_1.$$

Hence a *product of nonzero linear transformations can be the zero transformation.* We also note that

$$\mathbf{T}_1^2 = \mathbf{T}_1$$

where $\mathbf{T}_1^2$ denotes $\mathbf{T}_1\mathbf{T}_1$. Such a linear transformation is said to be *idempotent* because all powers of $\mathbf{T}_1$ are the same. In the algebra of numbers, zero and one are the only idempotent elements, but in the algebra of linear transformations there are many idempotent elements other than $\mathbf{Z}$ and $\mathbf{I}$; indeed, any projection is idempotent. Finally we note that the derivative transformation $\mathbf{D}$, acting on the space $\mathscr{P}_5(\mathbb{R})$, has the peculiar property that $\mathbf{D}, \mathbf{D}^2, \ldots,$ $\mathbf{D}^5$ are nonzero transformations, but $\mathbf{D}^6 = \mathbf{Z}$ because the sixth derivative of any polynomial of degree not exceeding five is the zero polynomial. A linear transformation $\mathbf{T}$ that has the property that

$$\mathbf{T}^{p-1} \neq \mathbf{Z} \text{ but } \mathbf{T}^p = \mathbf{Z}$$

is said to be *nilpotent of index p*, signifying that some finite power $p > 1$ of $\mathbf{T}$ is zero but that all lower powers of $\mathbf{T}$ are nonzero. We note, however, that on the space $\mathscr{P}(\mathbb{R})$ of all real polynomials, $\mathbf{D}$ is not nilpotent because no fixed number of differentiations maps *every* polynomial into the zero polynomial.

EXERCISES 3.1

1. Let $\mathbf{S}, \mathbf{T} \in \mathscr{L}(\mathscr{V}, \mathscr{W})$ and let $c \in \mathscr{F}$. Verify that $\mathbf{S} + \mathbf{T}$ and $c\mathbf{T}$ are linear mappings.

2. Verify that the linear transformation defined in Example D is linear for every choice of a, b, c, and d.

3. Let $\mathbb{C}(\mathbb{R})$ be the vector space of complex numbers over the field of real numbers. The *conjugate* transformation in $\mathbb{C}$ maps each complex number $a + ib$ into its complex conjugate $a - ib$. Is this mapping linear? How can it be regarded as a special case of Example D?

4. Which of the following transformations on $\mathscr{E}^3$ are linear? Describe the geometric effect of each.

(i) $\mathbf{T}(a_1, a_2, a_3) = (a_1, a_2, 0)$.

(ii) $\mathbf{T}(a_1, a_2, a_3) = (a_1, a_2, 1)$.

(iii) $\mathbf{T}(a_1, a_2, a_3) = (a_2, a_1, a_3)$.

(iv) $\mathbf{T}(a_1, a_2, a_3) = (-a_1, -a_2, a_3)$.

(v) $\mathbf{T}(a_1, a_2, a_3) = (a_1 + 1, a_2 + 2, a_3 + 3)$.

5. The nature of the solution set of a homogeneous system of m linear equations in n variables was discussed in Section 1.3. Writing the system (1.3) in vector notation as $\mathbf{T}\xi = 0$, use the language and notation of vectors and linear mappings to rephrase and to prove Statements A, B, and C of Section 1.3. (Observe how this notation facilitates the writing of proofs.)

6. Prove that $\mathbf{TZ} = \mathbf{Z} = \mathbf{ZT}$ in $\mathscr{L}(\mathscr{V}, \mathscr{V})$.

7. Let $\mathscr{V}$, $\mathscr{W}$, $\mathscr{X}$, and $\mathscr{Y}$ be vector spaces over $\mathscr{F}$. Let $\mathbf{R}$, $\mathbf{S}$, and $\mathbf{T}$ be linear mappings, $\mathbf{T}$ going from $\mathscr{V}$ to $\mathscr{W}$, $\mathbf{S}$ going from $\mathscr{W}$ to $\mathscr{X}$, and $\mathbf{R}$ going from $\mathscr{X}$ to $\mathscr{Y}$. Show that the mappings $\mathbf{R}(\mathbf{ST})$ and $(\mathbf{RS})\mathbf{T}$ go from $\mathscr{V}$ to $\mathscr{Y}$, are linear, and are equal.

8. In the space $\mathscr{P}$ of all real polynomials define mappings $\mathbf{D}$ and $\mathbf{M}$ by

$$\mathbf{D}p(x) = \frac{d}{dx}p(x),$$

$$\mathbf{M}p(x) = xp(x).$$

(i) Prove that both $\mathbf{D}$ and $\mathbf{M}$ are linear transformations.

(ii) Is $\mathbf{D}$ nilpotent on this space?

(iii) Prove that $\mathbf{DM} - \mathbf{MD} = \mathbf{I}$.

(iv) Deduce that $(\mathbf{MD})^2 = \mathbf{M}^2\mathbf{D}^2 + \mathbf{MD}$.

9. Associated with any subspace $\mathscr{S}$ of $\mathscr{V}$ there is a family of linear transformations from $\mathscr{V}$ to $\mathscr{S}$ called *projections*. Each such projection $\mathbf{P}_{\mathscr{S}}$ is determined by the choice of a subspace $\mathscr{T}$ such that $\mathscr{V} = \mathscr{S} \oplus \mathscr{T}$. Because each $\xi \in \mathscr{V}$ has a unique representation $\xi = \sigma + \tau$, where $\sigma \in \mathscr{S}$ and $\tau \in \mathscr{T}$, the mapping $\mathbf{P}_{\mathscr{S}}$, defined by

$$\mathbf{P}_{\mathscr{S}}\xi = \sigma,$$

is called the projection of $\mathscr{V}$ onto $\mathscr{S}$ along $\mathscr{T}$. Similarly, the mapping $\mathbf{P}_{\mathscr{T}}$,

$$\mathbf{P}_{\mathscr{T}}\xi = \tau,$$

is called the projection of $\mathscr{V}$ onto $\mathscr{T}$ along $\mathscr{S}$. Prove the following statements.

(i) $\mathbf{P}_{\mathscr{S}}$ and $\mathbf{P}_{\mathscr{T}}$ are linear and map $\mathscr{V}$ onto $\mathscr{S}$ and onto $\mathscr{T}$, respectively.

(ii) $\mathbf{P}_{\mathscr{S}}$ and $\mathbf{P}_{\mathscr{T}}$ are idempotent.

(iii) $\mathbf{P}_{\mathscr{S}}\mathbf{P}_{\mathscr{T}} = \mathbf{Z} = \mathbf{P}_{\mathscr{T}}\mathbf{P}_{\mathscr{S}}$.

(iv) $\mathbf{P}_{\mathscr{S}} + \mathbf{P}_{\mathscr{T}} = \mathbf{I}$.

10. Prove that the product (composition) of linear transformations on $\mathscr{V}$ is bilinear.

11. Apply the results of Exercises 7 and 10 to conclude that $\mathscr{L}(\mathscr{V}, \mathscr{V})$ is a linear algebra.

12. An important example of a linear algebra of dimension four, given a century ago by Hamilton, was a forerunner of the study of matrices. The elements of the algebra are called *quaternions*, and the scalars are the real numbers. In a notation similar to that of the complex numbers, a quaternion is an expression of the form

$$a_1 1 + a_2 i + a_3 j + a_4 k, \qquad a_i \in \mathbb{R}.$$

Equality, sum, and scalar multiple are defined component by component; quaternion product is defined by bilinearity and the following multiplication table for the basis elements, wherein the product xy appears in the row labeled x at the left and in the column labeled y at the top.

	1	i	j	k
1	1	i	j	k
i	i	-1	k	$-j$
j	j	$-k$	-1	i
k	k	j	$-i$	-1

(i) Verify that this product is closed. All other postulates of a linear algebra are also satisfied.

(ii) Show that every quaternion except $0 + 0i + 0j + 0k$ has an inverse relative to this product; that is,

$$(a_1 + a_2 i + a_3 j + a_4 k)(b_1 + b_2 i + b_3 j + b_4 k) = 1 + 0i + 0j + 0k$$

for suitable b_1, b_2, b_3, b_4.

(iii) Is the product commutative?

From this we conclude that the quaternions form a noncommutative "division algebra." An important theorem of Frobenius proves that the quaternions form the *only* noncommutative division algebra over the real numbers.

3.2 RANK AND NULLITY
OF A LINEAR MAPPING

Quite a lot of useful information about a given linear mapping **T** from $\mathscr{V}$ to $\mathscr{W}$ can be obtained by studying the effect of **T** on the subspaces of $\mathscr{V}$ and the relationship of **T** to various subspaces of $\mathscr{W}$.

Theorem 3.2   A linear mapping **T** from $\mathscr{V}$ to $\mathscr{W}$ carries each subspace $\mathscr{S}$ of $\mathscr{V}$ into a subspace **T**$\mathscr{S}$ of $\mathscr{W}$.

Proof   This result was previously stated as Exercise 2.7-4. We must show that $\mathbf{T}(\mathscr{S})$ is a subspace of $\mathscr{W}$, where

$$\mathbf{T}(\mathscr{S}) = \{\eta \in \mathscr{W} \,|\, \eta = \mathbf{T}\sigma \text{ for some } \sigma \in \mathscr{S}\}.$$

By Theorem 2.7, we need only show that $\mathbf{T}(\mathscr{S})$ is closed with respect to the formation of linear combinations.

Let $\eta_1, \eta_2 \in \mathbf{T}(\mathscr{S})$ and let $a, b \in \mathscr{F}$. There are vectors σ_1 and σ_2 in $\mathscr{S}$ such that $\mathbf{T}(\sigma_1) = \eta_1$ and $\mathbf{T}(\sigma_2) = \eta_2$. Hence

$$a\eta_1 + b\eta_2 = a\mathbf{T}\sigma_1 + b\mathbf{T}\sigma_2 = \mathbf{T}(a\sigma_1 + b\sigma_2).$$

But $a\sigma_1 + b\sigma_2 \in \mathscr{S}$ because $\mathscr{S}$ is a subspace of $\mathscr{V}$. Hence $a\eta_1 + b\eta_2$ is in $\mathbf{T}(\mathscr{S})$, so $\mathbf{T}(\mathscr{S})$ is a subspace of $\mathscr{W}$.

In particular, we note that $\mathbf{T}(\mathscr{V})$ is a subspace of $\mathscr{W}$ and that **T** maps $\mathscr{V}$ onto the vector space $\mathbf{T}(\mathscr{V})$. Now we turn things around to ask what we can say about the set of all vectors in $\mathscr{V}$ that are mapped by **T** into a given subspace $\mathscr{M}$ of $\mathbf{T}(\mathscr{V})$.

Theorem 3.3   Let $\mathbf{T} \in \mathscr{L}(\mathscr{V}, \mathscr{W})$, let $\mathscr{M}$ be a subspace of $\mathbf{T}(\mathscr{V})$, and let $\mathscr{S} = \{\alpha \in \mathscr{V} \,|\, \mathbf{T}\alpha \in \mathscr{M}\}$. Then $\mathscr{S}$ is a subspace of $\mathscr{V}$.

Proof   Let $\sigma_1, \sigma_2 \in \mathscr{S}$ and $a, b \in \mathscr{F}$. All that we need to show is that $a\sigma_1 + b\sigma_2 \in \mathscr{S}$, which is the same as showing that $\mathbf{T}(a\sigma_1 + b\sigma_2) \in \mathscr{M}$. But $\mathbf{T}(a\sigma_1 + b\sigma_2) = a\mathbf{T}\sigma_1 + b\mathbf{T}\sigma_2$, which is in $\mathscr{M}$, because $\mathscr{M}$ is a subspace that contains $\mathbf{T}\sigma_1$ and $\mathbf{T}\sigma_2$.

In particular $[\theta]$ is a subspace of $\mathbf{T}(\mathscr{V})$, so the set of all vectors in $\mathscr{V}$ that are mapped by $\mathbf{T}$ into $\theta \in \mathscr{W}$ is a subspace of $\mathscr{V}$.

Definition 3.2   Let $\mathbf{T}$ be a linear mapping from $\mathscr{V}$ into $\mathscr{W}$.

(a) The *range space* (or *image space*) of $\mathbf{T}$ is the subspace

$$\mathscr{R}(\mathbf{T}) = \mathbf{T}(\mathscr{V}) \subseteq \mathscr{W}.$$

If $\mathscr{R}(\mathbf{T})$ is finite-dimensional, the dimension of $\mathscr{R}(\mathbf{T})$ is called the *rank* of $\mathbf{T}$, denoted $r(\mathbf{T})$.

(b) The *null space* (or *kernel*) of $\mathbf{T}$ is the subspace

$$\mathscr{N}(\mathbf{T}) = \{\alpha \in \mathscr{V} \,|\, \mathbf{T}\alpha = \theta\} \subseteq \mathscr{V}.$$

If $\mathscr{N}(\mathbf{T})$ is finite-dimensional, the dimension of $\mathscr{N}(\mathbf{T})$ is called the *nullity* of $\mathbf{T}$, denoted $n(\mathbf{T})$.

Thus each linear mapping $\mathbf{T}$ from $\mathscr{V}$ to $\mathscr{W}$ automatically identifies a subspace $\mathscr{N}(\mathbf{T})$ of its domain $\mathscr{V}$ and a subspace $\mathscr{R}(\mathbf{T})$ of $\mathscr{W}$. These subspaces and their dimensions are related in a very interesting manner as described in the next two theorems. First we consider an example: the second derivative mapping $\mathbf{D}^2$ carries the space $\mathscr{P}_k$ of real polynomials onto the space $\mathscr{P}_{k-2}$, where $k \geq 2$. Hence $r(\mathbf{D}^2) = k - 1$ (the dimension of $\mathscr{P}_{k-2} = \mathscr{R}(\mathbf{D}^2)$). The null space of $\mathbf{D}^2$ is the space $\mathscr{P}_1$, because the polynomials that $\mathbf{D}^2$ carries into the zero polynomial are precisely those with degree not exceeding one. Then $n(\mathbf{D}^2) = 2$ (the dimension of $\mathscr{P}_1$). Furthermore

$$r(\mathbf{D}^2) + n(\mathbf{D}^2) = k + 1$$

(the dimension of $\mathscr{P}_k$).

Theorem 3.4   Let $\mathbf{T}$ be a linear mapping from an n-dimensional vector space $\mathscr{V}$ to a vector space $\mathscr{W}$. If a basis $\{\alpha_1, \ldots, \alpha_k\}$ for the null space $\mathscr{N}(\mathbf{T})$ is extended in any manner to a basis $\{\alpha_1, \ldots, \alpha_k, \alpha_{k+1}, \ldots, \alpha_n\}$ for $\mathscr{V}$, then $\{\mathbf{T}\alpha_{k+1}, \ldots, \mathbf{T}\alpha_n\}$ is a basis for the range space $\mathscr{R}(\mathbf{T})$.

Proof   Because we don't know the dimension of $\mathscr{R}(\mathbf{T})$, we must show that $\{\mathbf{T}\alpha_{k+1}, \ldots, \mathbf{T}\alpha_n\}$ is linearly independent and spans $\mathscr{R}(\mathbf{T})$. Suppose that

$$c_{k+1}\mathbf{T}\alpha_{k+1} + \cdots + c_n\mathbf{T}\alpha_n = \theta.$$

Then $\mathbf{T}(c_{k+1}\alpha_{k+1} + \cdots + c_n\alpha_n) = \theta$, so $c_{k+1}\alpha_{k+1} + \cdots + c_n\alpha_n \in \mathscr{N}(\mathbf{T})$.

But $\{\alpha_1, \ldots, \alpha_k\}$ is a basis for $\mathcal{N}(\mathbf{T})$ so there exist unique scalars $a_1, \ldots, a_k$ such that

$$c_{k+1}\alpha_{k+1} + \cdots + c_n\alpha_n = a_1\alpha_1 + \cdots + a_k\alpha_k.$$

But because $\{\alpha_1, \ldots, \alpha_n\}$ is linearly independent, each coefficient in this equation must be zero; in particular, each $c_i = 0$, so $\{\mathbf{T}\alpha_{k+1}, \ldots, \mathbf{T}\alpha_n\}$ is linearly independent.

Any vector η of $\mathcal{R}(\mathbf{T})$ is of the form $\eta = \mathbf{T}\xi$ for some $\xi \in \mathcal{V}$. Let $\xi = x_1\alpha_1 + \cdots + x_n\alpha_n$. Then

$$\eta = \mathbf{T}\xi = \mathbf{T}(x_1\alpha_1 + \cdots + x_k\alpha_k + x_{k+1}\alpha_{k+1} + \cdots + x_n\alpha_n)$$

$$= x_1\mathbf{T}\alpha_1 + \cdots + x_k\mathbf{T}\alpha_k + x_{k+1}\mathbf{T}\alpha_{k+1} + \cdots + x_n\mathbf{T}\alpha_n$$

$$= \theta + \cdots + \theta + x_{k+1}\mathbf{T}\alpha_{k+1} + \cdots + x_n\mathbf{T}\alpha_n.$$

Hence $\mathcal{R}(\mathbf{T}) = [\mathbf{T}\alpha_{k+1}, \ldots, \mathbf{T}\alpha_k]$, as claimed.

Theorem 3.5 If $\mathbf{T} \in \mathcal{L}(\mathcal{V}, \mathcal{W})$ and if $\mathcal{V}$ is finite-dimensional, then

$$r(\mathbf{T}) + n(\mathbf{T}) = \dim \mathcal{V}.$$

Proof This is an immediate consequence of Theorem 3.4.

Thus if $\mathbf{T}$ is a linear mapping whose domain space $\mathcal{V}$ is n-dimensional, the dimension of its range space never exceeds n, and the sum of the dimension of its range space and its null space must equal n. The dimension of the space $\mathcal{W}$ (which contains $\mathcal{R}(\mathbf{T})$ as a subspace) does not enter into these relations.

As an extension of these considerations, we now investigate relationships between the range and null spaces of a composition $\mathbf{ST}$ of two linear mappings and the range and null spaces of the individual mappings $\mathbf{S}$ and $\mathbf{T}$, where $\mathbf{T}$ maps $\mathcal{V}$ into $\mathcal{W}$, $\mathbf{S}$ maps $\mathcal{W}$ into $\mathcal{X}$, and $\mathbf{ST}$ maps $\mathcal{V}$ into $\mathcal{X}$, as shown in Figure 3.3. It is not difficult to verify formally the facts that this mapping diagram suggests:

(1) $[\theta] \subseteq \mathcal{N}(\mathbf{T}) \subseteq \mathcal{N}(\mathbf{ST}) \subseteq \mathcal{V}$,

(2) $\mathcal{X} \supseteq \mathcal{R}(\mathbf{S}) \supseteq \mathcal{R}(\mathbf{ST}) \supseteq [\theta]$.

We must be cautious, however, about drawing conclusions from mapping diagrams without proving them formally. For example, in Figure 3.3 it is tempting to diagram the subspaces of $\mathcal{W}$ to look like the sketches for $\mathcal{V}$ and $\mathcal{X}$; that is, to make $\mathcal{N}(\mathbf{S})$ appear to be a subset of $\mathcal{R}(\mathbf{T})$. But clearly by

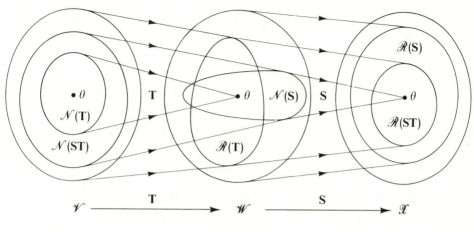

FIGURE 3.3

varying the mappings **S** and **T** we can obtain any subspace of $\mathcal{W}$ for $\mathcal{N}(\mathbf{S})$ and any subspace of $\mathcal{W}$ for $\mathcal{R}(\mathbf{T})$.

We conclude this section by extending these considerations to obtain a theorem that will be of great importance in Chapter 7 but will not be needed before then. Therefore no harm will be done if the remainder of this section is omitted at this time.

Relations 1 and 2 are particularly useful for studying the effect of iterating a linear transformation on $\mathcal{V}$. For that special case we have $\mathcal{V} = \mathcal{W} = \mathcal{X}$ and $\mathbf{S} = \mathbf{T}$. As the transformation **T** is repeatedly applied to $\mathcal{V}$ we obtain a sequence of transformations,

$$\mathbf{I, T, T^2, \ldots, T^k, \ldots.}$$

The corresponding null spaces form an increasing chain of subspaces of $\mathcal{V}$,

$$[\theta] \subseteq \mathcal{N}(\mathbf{T}) \subseteq \mathcal{N}(\mathbf{T}^2) \subset \cdots \subseteq \mathcal{N}(\mathbf{T}^k) \subseteq \cdots,$$

and the range spaces form a decreasing chain of subspaces of $\mathcal{V}$,

$$\mathcal{V} \supseteq \mathcal{R}(\mathbf{T}) \supseteq \mathcal{R}(\mathbf{T}^2) \supseteq \cdots \supseteq \mathcal{R}(\mathbf{T}^k) \supseteq \cdots.$$

If $\mathcal{V}$ is n-dimensional, equality must hold somewhere in each chain. Furthermore, the dimension relation of Theorem 3.5 guarantees that

$$r(\mathbf{T}^k) + n(\mathbf{T}^k) = n,$$

for each k. Hence, if equality holds at some position in either chain, it must hold at the same position in the other chain. Furthermore, we claim that in either chain if equality holds at some position, it holds at all subsequent positions. That is, strict inequality holds in each chain until the first equality appears, and it must appear sometime. Thereafter, there is no change in the terms in either chain:

$$[\theta] \subset \mathcal{N}(\mathbf{T}) \subset \cdots \subset \mathcal{N}(\mathbf{T}^k) = \mathcal{N}(\mathbf{T}^{k+1}) = \cdots,$$

$$\mathcal{V} \supset \mathcal{R}(\mathbf{T}) \supset \cdots \supset \mathcal{R}(\mathbf{T}^k) = \mathcal{R}(\mathbf{T}^{k+1}) = \cdots.$$

To prove this, suppose that $\mathcal{N}(\mathbf{T}^k) = \mathcal{N}(\mathbf{T}^{k+1})$, and let $\xi \in \mathcal{N}(\mathbf{T}^{k+p})$ for any $p \geq 1$. Then

$$\theta = \mathbf{T}^{k+p}\xi = \mathbf{T}^{k+1}(\mathbf{T}^{p-1}\xi).$$

Hence $\mathbf{T}^{p-1}\xi \in \mathcal{N}(\mathbf{T}^{k+1}) = \mathcal{N}(\mathbf{T}^k)$, so $\mathbf{T}^k(\mathbf{T}^{p-1}\xi) = \theta$; that is, $\xi \in \mathcal{N}(\mathbf{T}^{k+p-1})$. Thus $\mathcal{N}(\mathbf{T}^{k+p-1}) = \mathcal{N}(\mathbf{T}^{k+p})$. Because p was arbitrary, we have $\mathcal{N}(\mathbf{T}^k) = \mathcal{N}(\mathbf{T}^{k+1}) = \mathcal{N}(\mathbf{T}^{k+2})$, and so on, as claimed.

Theorem 3.6   If **T** is a linear transformation on $\mathcal{V}_n$, then

$$[\theta] \subseteq \mathcal{N}(\mathbf{T}) \subseteq \mathcal{N}(\mathbf{T}^2) \subseteq \cdots \subseteq \mathcal{N}(\mathbf{T}^m) \subseteq \cdots,$$

$$\mathcal{V}_n \supseteq \mathcal{R}(\mathbf{T}) \supseteq \mathcal{R}(\mathbf{T}^2) \supseteq \cdots \supseteq \mathcal{R}(\mathbf{T}^m) \supseteq \cdots.$$

Furthermore, if $\mathcal{N}(\mathbf{T}^k) = \mathcal{N}(\mathbf{T}^{k+1})$, then

$$\mathcal{N}(\mathbf{T}^{k+p}) = \mathcal{N}(\mathbf{T}^k) \text{ for all } p \geq 1,$$

$$\mathcal{R}(\mathbf{T}^{k+p}) = \mathcal{R}(\mathbf{T}^k) \text{ for all } p \geq 1,$$

$$\mathcal{V}_n = \mathcal{R}(\mathbf{T}^k) \oplus \mathcal{N}(\mathbf{T}^k).$$

Proof   The foregoing discussion has established all of these assertions except the final one, which states that $\mathcal{V}_n$ is the direct sum of $\mathcal{R}(\mathbf{T}^k)$ and $\mathcal{N}(\mathbf{T}^k)$. We have already observed that the sum of the dimensions of these two spaces is n, so by Exercise 2.6-8 we need only to show that $\mathcal{R}(\mathbf{T}^k) \cap \mathcal{N}(\mathbf{T}^k) = [\theta]$. Let $\eta \in \mathcal{R}(\mathbf{T}^k) \cap \mathcal{N}(\mathbf{T}^k)$. Then $\eta = \mathbf{T}^k\xi$ for some $\xi \in \mathcal{V}_n$, and $\theta = \mathbf{T}^k\eta = \mathbf{T}^{2k}\xi$. Hence $\xi \in \mathcal{N}(\mathbf{T}^{2k}) = \mathcal{N}(\mathbf{T}^k)$. Thus $\mathbf{T}^k\xi = \theta$, so $\eta = \mathbf{T}^k\xi = \theta$ and $\mathcal{R}(\mathbf{T}^k) \cap \mathcal{N}(\mathbf{T}^k) = [\theta]$.

EXERCISES 3.2

1. Determine the range space and the null space (and their dimensions) for each of the linear transformations on $\mathcal{E}^2$, as described in Example D, Section 3.1.

(i) A dilation in which $a = -2$.

(ii) A reflection through the origin.

(iii) A projection onto the y-axis.

(iv) A rotation through 30 degrees.

2. Determine the range space and the null space (and their dimensions) for each of the linear transformations on $\mathscr{E}^3$ described in Exercise 3.1-4(i), (iv).

3. Determine the range space and the null space of each of the linear transformations **D**, **M**, **DM**, and **MD** defined in Exercise 3.1-8.

4. Prove that if **T** is a linear transformation for which $r(\mathbf{T}) = 1$, then $\mathbf{T}^2 = c\mathbf{T}$ for some scalar c. Is the converse true? Explain.

5. Suppose that the linear transformation **T** is nilpotent of index p on $\mathscr{V}$.

(i) Show that there exists a vector $\xi \in \mathscr{V}$ such that $\mathbf{T}^{p-1}\xi \neq \theta$ but $\mathbf{T}^p\xi = \theta$.

(ii) Show that if ξ satisfies $\mathbf{T}^{p-1}\xi \neq \theta$, then the set B is linearly independent, where $B = \{\xi, \mathbf{T}\xi, \mathbf{T}^2\xi, \ldots, \mathbf{T}^{p-1}\xi\}$.

(iii) Let α be any vector in the subspace $[B]$, where B is defined in (ii). Prove that $\mathbf{T}\alpha \in [B]$. (A subspace $\mathscr{S}$ of $\mathscr{V}$ that is mapped into itself by **T** is said to be **T**-*invariant*.)

6. Prove the two relations $\mathscr{N}(\mathbf{T}) \subseteq \mathscr{N}(\mathbf{ST})$ and $\mathscr{R}(\mathbf{S}) \supseteq \mathscr{R}(\mathbf{ST})$ that are illustrated in Figure 3.3.

7. Apply the results of Exercise 6 and Theorem 3.5 to deduce that

(i) $n(\mathbf{T}) \leq n(\mathbf{ST})$ and $r(\mathbf{ST}) \leq r(\mathbf{S})$,

(ii) $n(\mathbf{S}) \leq n(\mathbf{ST})$ and $r(\mathbf{ST}) \leq r(\mathbf{T})$.

8. Let **S** and **T** be linear transformations on $\mathscr{V}_n$. Prove that

(i) $\mathscr{R}(\mathbf{S} + \mathbf{T}) \subseteq \mathscr{R}(\mathbf{S}) + \mathscr{R}(\mathbf{T})$,

(ii) $r(\mathbf{S} + \mathbf{T}) \leq r(\mathbf{S}) + r(\mathbf{T})$,

(iii) $n(\mathbf{S} + \mathbf{T}) \geq n(\mathbf{S}) + n(\mathbf{T}) - n$.

9. Let **S** and **T** be linear transformations on $\mathscr{V}_n$. Prove that

(i) $n(\mathbf{S}) + n(\mathbf{T}) \geq n(\mathbf{ST}) \geq \max(n(\mathbf{S}), n(\mathbf{T}))$,

(ii) $r(\mathbf{S}) + r(\mathbf{T}) - n \leq r(\mathbf{ST}) \leq \min(r(\mathbf{S}), r(\mathbf{T}))$.

10. Illustrate each statement in Exercises 7 and 8 for the particular transformations **T** and **S**, where **T** is described by Exercise 3.1-4(i) and **S** is described by Exercise 3.1-4(iv).

3.3 NONSINGULAR MAPPINGS

The range space and null space of a linear mapping **T** have an intrinsic connection with the concept of *nonsingularity* of **T**, which can be characterized in numerous ways. The essential idea is one-to-one-ness. Thus the distinction between a nonsingular linear mapping and an arbitrary linear mapping is precisely the distinction between an isomorphism and a homomorphism. This means, of course, that the null space of a nonsingular mapping must be $[\theta]$, which implies that the mapping is reversible.

> ***Definition 3.3*** A linear mapping **T** from $\mathscr{V}$ into $\mathscr{W}$ is said to be *nonsingular* if and only if there exists a mapping **S** from $\mathscr{R}(\mathbf{T})$ onto $\mathscr{V}$ such that
>
> $$\mathbf{ST} = \mathbf{I}, \text{ where } \mathbf{I} \text{ is the identity mapping on } \mathscr{V}.$$
>
> Otherwise, **T** is said to be *singular*.

Although this definition does not specifically require **S** to be linear, it is necessarily so. To see this, let **S** be any mapping from $\mathscr{R}(\mathbf{T})$ onto $\mathscr{V}$ such that $\mathbf{ST} = \mathbf{I}$ on $\mathscr{V}$. For any $\xi, \eta \in \mathscr{R}(\mathbf{T})$ there exists vectors $\alpha, \beta \in \mathscr{V}$ such that $\mathbf{T}\alpha = \xi$ and $\mathbf{T}\beta = \eta$. Then $\mathbf{S}\xi = \mathbf{ST}\alpha = \alpha$ and $\mathbf{S}\eta = \mathbf{ST}\beta = \beta$. For any a, b in $\mathscr{F}$

$$\mathbf{S}(a\xi + b\eta) = \mathbf{S}(a\mathbf{T}\alpha + b\mathbf{T}\beta) = \mathbf{ST}(a\alpha + b\beta)$$

$$= \mathbf{I}(a\alpha + b\beta) = a\mathbf{S}\xi + b\mathbf{S}\eta.$$

Thus the linearity of **T** and the property $\mathbf{ST} = \mathbf{I}$ imply that **S** is linear. Also, **S** is uniquely determined by **T**, because if $\mathbf{S}_1\mathbf{T} = \mathbf{I} = \mathbf{S}_2\mathbf{T}$, then

$$\mathbf{S}_1\mathbf{T}\alpha = \mathbf{S}_2\mathbf{T}\alpha = \alpha \text{ for all } \alpha \in \mathscr{V}.$$

But by the definition of $\mathscr{R}(\mathbf{T})$, we then have $\mathbf{S}_1\xi = \mathbf{S}_2\xi$ for all $\xi \in \mathscr{R}(\mathbf{T})$, so $\mathbf{S}_1 = \mathbf{S}_2$.

It is also important to observe that a nonsingular linear mapping **T** must be one-to-one because otherwise there exist vectors α and β in $\mathscr{V}$ such that $\alpha \neq \beta$ but $\mathbf{T}\alpha = \mathbf{T}\beta$. In that case any mapping **S** from $\mathscr{R}(\mathbf{T})$ to $\mathscr{V}$ has the property that $\mathbf{ST}\alpha = \mathbf{ST}\beta$, so $\mathbf{ST} \neq \mathbf{I}$ on $\mathscr{V}$. Conversely, any one-to-one mapping **T** is reversible, so a linear, one-to-one mapping is nonsingular.

Theorem 3.7 The following statements are equivalent for any linear mapping **T** from $\mathscr{V}_n$ into $\mathscr{W}$.

(a) **T** is nonsingular.

(b) If $\mathbf{T}(\alpha) = \mathbf{T}(\beta)$, then $\alpha = \beta$.

(c) $\mathscr{N}(\mathbf{T}) = [\theta]$.

(d) $n(\mathbf{T}) = 0$.

(e) $r(\mathbf{T}) = n$.

(f) If $\{\alpha_1, \ldots, \alpha_n\}$ is any basis for $\mathscr{V}_n$, then $\{\mathbf{T}\alpha_1, \ldots, \mathbf{T}\alpha_n\}$ is a basis for $\mathscr{R}(\mathbf{T})$.

Proof To say that the given statements are equivalent means that each statement implies and is implied by each of the others. Often we can establish such a claim by proving a cycle of implications. In this case previous definitions and theorems show that each statement is an immediate consequence of the preceding statement. That is,

$$(a) \Rightarrow (b) \Rightarrow (c) \Rightarrow (d) \Rightarrow (e) \Rightarrow (f).$$

To complete the cycle we show that (f) implies (a).

Let $\{\alpha_1, \ldots, \alpha_n\}$ be a basis for $\mathscr{V}_n$. By hypothesis (f), $\{\mathbf{T}\alpha_1, \ldots, \mathbf{T}\alpha_n\}$ is a basis for $\mathscr{R}(\mathbf{T})$, so each $\eta \in \mathscr{R}(\mathbf{T})$ has a unique representation

$$\eta = \sum_{i=1}^{n} b_i \mathbf{T}\alpha_i.$$

Let **S** be the mapping from $\mathscr{R}(\mathbf{T})$ into $\mathscr{V}_n$ defined by

$$\mathbf{S}\eta = \sum_{i=1}^{n} b_i \alpha_i.$$

To show that $\mathbf{ST} = \mathbf{I}$ on $\mathscr{V}_n$, let $\xi \in \mathscr{V}_n$ have the representation

$$\xi = \sum_{i=1}^{n} a_i \alpha_i.$$

Then

$$\mathbf{T}\xi = \mathbf{T}\left(\sum_{i=1}^{n} a_i \alpha_i \right) = \sum_{i=1}^{n} a_i \mathbf{T}\alpha_i.$$

By the definition of **S**,

$$\mathbf{ST}\xi = \mathbf{S}\left(\sum_{i=1}^{n} a_i \mathbf{T}\alpha_i \right) = \sum_{i=1}^{n} a_i \alpha_i = \xi.$$

That is, $\mathbf{ST} = \mathbf{I}$ on $\mathscr{V}_n$, so **T** is nonsingular.

As an exercise you may prove another useful characterization of nonsingularity: **T** is nonsingular if and only if **T** preserves the property of linear independence. In particular, Statement (f) of Theorem 3.7 assures us that any change of basis in $\mathscr{V}_n$ can be performed by a nonsingular linear transformation on $\mathscr{V}_n$, and any nonsingular linear transformation is a change of basis.

Because a linear mapping **T** is a vector space homomorphism of $\mathscr{V}$ onto $\mathscr{R}(\mathbf{T})$, a nonsingular linear mapping is simply an isomorphism between $\mathscr{V}$ and $\mathscr{R}(\mathbf{T})$. This observation lends intuitive feeling to the statements of Theorem 3.7, because an isomorphism is one-to-one, its kernel is trivial, and it ought to preserve dimension. If we allow $\mathscr{V}$ to be infinite-dimensional, we obviously must delete Statement (e) in Theorem 3.7, but otherwise the theorem and most of its proof remain valid.

The definition of nonsingularity of **T** is sometimes stated a little differently to mean that there exists a mapping **R** from $\mathscr{W}$ to $\mathscr{V}$ (instead of from $\mathscr{R}(\mathbf{T})$ onto $\mathscr{V}$) such that **RT** is the identity mapping on $\mathscr{V}$ and **TR** is the identity mapping on $\mathscr{W}$ (instead of on $\mathscr{R}(\mathbf{T})$). The two definitions are equivalent when both $\mathscr{V}$ and $\mathscr{W}$ are n-dimensional spaces, but Exercise 3.3-7 shows that they are not equivalent for general vector spaces.

Theorem 3.8 If **T** is a linear mapping from $\mathscr{V}$ to $\mathscr{W}$ and if **S** is a mapping from $\mathscr{R}(\mathbf{T})$ to $\mathscr{V}$ such that **ST** = **I** on $\mathscr{V}$, then **TS** = **I** on $\mathscr{R}(\mathbf{T})$.
Proof Because **T** is nonsingular, each $\beta \in \mathscr{R}(\mathbf{T})$ is the **T**-image of one and only one vector $\alpha \in \mathscr{V}$. Then $\mathbf{TS}\beta = \mathbf{TS}(\mathbf{T}\alpha) = \mathbf{T}(\mathbf{ST})\alpha = \mathbf{TI}\alpha = \mathbf{T}\alpha = \beta$, as desired.

In view of this result and our earlier observation that **S** is uniquely determined by **T**, we are justified in calling **S** the *inverse* of the nonsingular linear mapping **T** and in denoting the inverse of **T** by $\mathbf{T}^{-1}$.

Now consider the composition of a nonsingular linear mapping with an arbitrary linear mapping. We let $\mathscr{V}$, $\mathscr{W}_n$, $\mathscr{X}_n$, and $\mathscr{Y}$ be vector spaces with **S** a linear mapping from $\mathscr{V}$ to $\mathscr{W}_n$, **T** a linear mapping from $\mathscr{W}_n$ to $\mathscr{X}_n$ and **U** a linear mapping from $\mathscr{X}_n$ to $\mathscr{Y}$:

$$\mathscr{V} \underset{\mathbf{S}}{\to} \mathscr{W}_n \underset{\mathbf{T}}{\to} \mathscr{X}_n \underset{\mathbf{U}}{\to} \mathscr{Y}.$$

If **T** is nonsingular, then dim $\mathscr{R}(\mathbf{TS})$ = dim $\mathscr{R}(\mathbf{S})$ because a nonsingular mapping preserves dimension. Similarly, dim $\mathscr{R}(\mathbf{UT})$ = dim $\mathscr{R}(\mathbf{U})$ because **T** carries a basis for $\mathscr{W}_n$ into a basis for $\mathscr{X}_n$. Hence where a given linear mapping is composed (on either side) with a nonsingular linear mapping, the rank of the product mapping is the same as the rank of the given mapping. Later we shall be concerned with a special application of this result, which we now state formally.

Theorem 3.9   Let $\mathbf{T}$ be a linear mapping from $\mathscr{V}_n$ to $\mathscr{W}_m$. Let $\mathbf{R}$ be a nonsingular linear transformation on $\mathscr{V}_n$ and let $\mathbf{S}$ be a nonsingular linear transformation on $\mathscr{W}_m$. Then

$$r(\mathbf{STR}) = r(\mathbf{TR}) = r(\mathbf{ST}) = r(\mathbf{T}).$$

Proof   By our previous remarks the rank of $\mathbf{T}$ is unchanged when $\mathbf{T}$ is either preceded or followed by a nonsingular linear transformation.

Theorem 3.10   Let $\mathbf{T}$ be a nonsingular linear mapping from $\mathscr{V}$ to $\mathscr{W}$. Then

(a) $\mathbf{T}^{-1}$ is nonsingular,

(b) $(\mathbf{T}^{-1})^{-1} = \mathbf{T}$,

(c) $(c\mathbf{T})^{-1} = c^{-1}\mathbf{T}^{-1}$ for each nonzero scalar c.

Proof   Exercise.

Theorem 3.11   Let $\mathbf{T}$ be a linear mapping from $\mathscr{V}$ onto $\mathscr{W}$, and let $\mathbf{S}$ be a linear mapping from $\mathscr{W}$ to $\mathscr{X}$. Then

(a) $\mathbf{ST}$ is nonsingular if and only if both $\mathbf{S}$ and $\mathbf{T}$ are nonsingular;

(b) if $\mathbf{ST}$ is nonsingular, $(\mathbf{ST})^{-1} = \mathbf{T}^{-1}\mathbf{S}^{-1}$.

Proof   Exercise. Use the facts that a nonsingular mapping is one-to-one and has a unique inverse.

EXERCISES 3.3

1. Which of the linear transformations listed in Exercise 3.2-1 are nonsingular? Justify your answers.

2. Let $\mathbf{T}$ be any linear transformation on $\mathbb{R}^2$, and suppose that $\mathbf{T}(1, 0) = (a, b)$ and $\mathbf{T}(0, 1) = (c, d)$. Show algebraically that $\mathbf{T}$ is nonsingular if and only if $ad - bc \neq 0$. Interpret this result geometrically.

3. Cite a specific theorem or definition that establishes each implication of Theorem 3.7 from (a) to (f) in sequence.

4. Prove Theorem 3.10.

5. Prove Theorem 3.11.

6. Prove that a linear mapping $\mathbf{T}$ from $\mathscr{V}_n$ to $\mathscr{W}$ is nonsingular if and only if for every linearly independent set $\{\alpha_1, \ldots, \alpha_k\}$ in $\mathscr{V}_n$ the image set $\{\mathbf{T}\alpha_1, \ldots, \mathbf{T}\alpha_k\}$ is linearly independent in $\mathscr{W}$.

7. Let $\mathscr{P}$ be the space of all real polynomials, let $\mathscr{P}_0$ be the subspace of all $p \in \mathscr{P}$ such that $p(0) = 0$, and let $\mathbf{J}$, $\mathbf{D}$, and $\mathbf{D}_0$ be the mappings defined as follows:

$$\mathbf{J}p(x) = \int_0^x p(t)\, dt, \text{ defined on } \mathscr{P},$$

$$\mathbf{D}p(x) = \frac{d}{dx} p(x), \text{ defined on } \mathscr{P},$$

$$\mathbf{D}_0 p(x) = \frac{d}{dx} p(x), \text{ defined on } \mathscr{P}_0.$$

(i) Determine the domain and range of each of the seven mappings $\mathbf{J}$, $\mathbf{D}$, $\mathbf{D}_0$, $\mathbf{DJ}$, $\mathbf{JD}$, $\mathbf{D}_0\mathbf{J}$, $\mathbf{JD}_0$.

(ii) Which of the seven mappings is the identity mapping on its domain?

(iii) Which of the seven mappings is nonsingular?

8. Let $\mathscr{S}$ be any subspace of $\mathscr{V}_n$. Show that $\mathscr{S}$ is the null space of a suitably defined linear transformation on $\mathscr{V}_n$.

3.4   MATRIX REPRESENTATION OF A LINEAR MAPPING

Up to this point in the study of linear mappings our major tools have been the fundamental concepts of function, linearity, and subspace applied to vector spaces with very little reference to the underlying scalar fields or to the choice of particular bases for the spaces involved. For this reason our terminology has acquired a geometric flavor, with almost no hint of any computational techniques that can be used in the study of linear mappings. Because the choice of a basis for $\mathscr{V}_n(\mathscr{F})$ provides a specific representation of each vector as an ordered n-tuple of elements of $\mathscr{F}$, and because a linear mapping is determined by its effect on a basis, we might expect that numerical representations of linear mappings are related to choices of bases. In this section we shall work out some of the details of that connection.

Given a linear mapping **T** from $\mathscr{V}_n(\mathscr{F})$ to $\mathscr{W}_m(\mathscr{F})$, we choose any basis $\{\alpha_1, \ldots, \alpha_n\}$ for $\mathscr{V}_n$ and any basis $\{\beta_1, \ldots, \beta_m\}$ for $\mathscr{W}_m$. Each vector ξ in $\mathscr{V}_n$ has a unique representation as a linear combination of the vectors in the α-basis, and each vector η in $\mathscr{W}_m$ has a unique representation as a linear combination of the vectors in the β-basis:

$$\xi = \sum_{j=1}^{n} x_j \alpha_j \quad \text{and} \quad \eta = \sum_{i=1}^{m} y_i \beta_i.$$

We agree to write these ordered sets of coefficients as column vectors,

$$\xi \to X = \begin{pmatrix} x_1 \\ x_2 \\ \cdot \\ \cdot \\ \cdot \\ x_n \end{pmatrix} \quad \text{and} \quad \eta \to Y = \begin{pmatrix} y_1 \\ y_2 \\ \cdot \\ \cdot \\ \cdot \\ y_m \end{pmatrix},$$

as the numerical representations of ξ and η, respectively, relative to the pair of chosen bases. We ask how these two ordered sets of scalars are related.

Letting η denote $\mathbf{T}\xi$ and using the linearity of $\mathbf{T}$, we have

$$\mathbf{T}\xi = \mathbf{T}\left(\sum_{j=1}^{n} x_j \alpha_j \right) = \sum_{j=1}^{n} x_j(\mathbf{T}\alpha_j) = \sum_{i=1}^{m} y_i \beta_i.$$

But each $\mathbf{T}\alpha_j$ is a linear combination of the vectors of the β-basis, which we can write as $\mathbf{T}(\alpha_j) = \sum_{i=1}^{m} a_{ij} \beta_i$:

$$\mathbf{T}\alpha_1 = a_{11} \beta_1 + a_{21} \beta_2 + \cdots + a_{m1} \beta_m,$$
$$\mathbf{T}\alpha_2 = a_{12} \beta_1 + a_{22} \beta_2 + \cdots + a_{m2} \beta_m,$$
$$\vdots$$
$$\mathbf{T}\alpha_n = a_{1n} \beta_1 + a_{2n} \beta_2 + \cdots + a_{mn} \beta_m.$$

Then from the preceding expression for $\mathbf{T}\xi$ we have

$$\mathbf{T}\xi = \sum_{j=1}^{n} x_j \mathbf{T}\alpha_j = \sum_{j=1}^{n} x_j \left(\sum_{i=1}^{m} a_{ij} \beta_i \right) = \sum_{i=1}^{m} \left(\sum_{j=1}^{n} a_{ij} x_j \right) \beta_i,$$

and the uniqueness of this representation implies that

$$y_i = \sum_{j=1}^{n} a_{ij}x_j \text{ for } i = 1, \ldots, m.$$

The mn scalars a_{ij} can be arranged in an m-by-n matrix A by writing the coefficients that represent $T\alpha_1, T\alpha_2, \ldots, T\alpha_n$ as *column* vectors:

$$T\alpha_1 \rightarrow \begin{pmatrix} a_{11} \\ a_{21} \\ \cdot \\ \cdot \\ \cdot \\ a_{m1} \end{pmatrix}, T\alpha_2 \rightarrow \begin{pmatrix} a_{12} \\ a_{22} \\ \cdot \\ \cdot \\ \cdot \\ a_{m2} \end{pmatrix}, \cdots, T\alpha_n \rightarrow \begin{pmatrix} a_{1n} \\ a_{2n} \\ \cdot \\ \cdot \\ \cdot \\ a_{mn} \end{pmatrix}.$$

Thus the linear mapping T is represented relative to the α-basis for $\mathscr{V}_n$ and the β-basis for $\mathscr{W}_m$ by the m-by-n matrix A in which column j is the column vector that represents $T\alpha_j$ relative to the β-basis for $\mathscr{W}_m$:

$$T \rightarrow A = \begin{pmatrix} a_{11} & a_{12} & \cdots & a_{1n} \\ a_{21} & a_{22} & \cdots & a_{2n} \\ \cdot & \cdot & & \cdot \\ \cdot & \cdot & & \cdot \\ \cdot & \cdot & & \cdot \\ a_{m1} & a_{m2} & \cdots & a_{mn} \end{pmatrix}.$$

As in Chapter 1, the entry in row i and column j of a matrix A is denoted by a_{ij}.

Theorem 3.12 Let $\{\alpha_1, \ldots, \alpha_n\}$ be a basis for $\mathscr{V}_n$, and let $\{\beta_1, \ldots, \beta_m\}$ be a basis for $\mathscr{W}_m$.

(a) Each linear mapping T from $\mathscr{V}_n$ to $\mathscr{W}_m$ determines uniquely an m-by-n matrix $A = (a_{ij})$ of scalars such that, for each α_j in the α-basis, column j of A is the representation of $T\alpha_j$ as a column vector, relative to the β-basis.

(b) Conversely, each m-by-n matrix $A = (a_{ij})$ of scalars determines uniquely a linear mapping T such that for each $j = 1, \ldots, n$, column j of A is the vector representation of $T\alpha_j$ relative to the β-basis.

Proof Statement (a) was established by the preceding remarks, and Statement (b) follows from Example C in Section 3.1.

Now consider the linear space $\mathscr{L}(\mathscr{V}_n, \mathscr{W}_m)$ of all linear mappings from $\mathscr{V}_n$ to $\mathscr{W}_m$. Relative to a fixed choice of bases, each $\mathbf{T} \in \mathscr{L}(\mathscr{V}_n, \mathscr{W}_m)$ determines a unique m-by-n matrix $A = (a_{ij})$ of scalars. Let $\mathbf{S} \in \mathscr{L}(\mathscr{V}_n, \mathscr{W}_m)$ determine the m-by-n matrix B. Because $\mathbf{T} + \mathbf{S} \in \mathscr{L}(\mathscr{V}_n, \mathscr{W}_m)$ determines an m-by-n matrix C, we would like to define a method of adding m-by-n matrices so that $C = A + B$. We also would like to define the scalar multiple cA of a matrix in such a way that cA represents $c\mathbf{T}$ whenever A represents $\mathbf{T}$. In short, we want to define operations on the set $\mathscr{M}_{m \times n}$ of all m-by-n matrices so that it becomes a linear space that is *isomorphic* to $\mathscr{L}(\mathscr{V}_n, \mathscr{W}_m)$. Then we shall be able to investigate geometric properties of linear mappings by means of algebraic computations with m-by-n matrices. The principal aim of Section 4.1 is to see how this can be accomplished.

A linear transformation from $\mathscr{V}_n$ to itself, of course, is simply a linear mapping from $\mathscr{V}_n$ to $\mathscr{V}_n$. In that case we need only the α-basis to represent each vector involved. The discussion preceding Theorem 3.12 remains valid with $m = n$ and $\beta_i = \alpha_i$ for $i = 1, \ldots, n$. Then $\mathbf{T}$ and $\mathbf{S}$ are represented by square n-by-n matrices A and B. Because the composition transformations $\mathbf{TS}$ and $\mathbf{ST}$ are linear, we will want to define a product of square matrices so that $\mathbf{TS}$ will be represented by AB whenever $\mathbf{T}$ is represented by A and $\mathbf{S}$ is represented by B. To see what is involved, we have

$$\mathbf{S}\alpha_j = \sum_{k=1}^{n} b_{kj}\alpha_k,$$

$$\mathbf{T}\alpha_k = \sum_{i=1}^{n} a_{ik}\alpha_i.$$

Then

$$(\mathbf{TS})\alpha_j = \mathbf{T}\left(\sum_{k=1}^{n} b_{kj}\alpha_k\right) = \sum_{k=1}^{n} b_{kj}\mathbf{T}\alpha_k$$

$$= \sum_{k=1}^{n}\left(b_{kj}\sum_{i=1}^{n} a_{ik}\alpha_i\right) = \sum_{i=1}^{n}\left(\sum_{k=1}^{n} a_{ik}b_{kj}\right)\alpha_i.$$

Hence if $\mathbf{TS}$ is represented by $C = (c_{ij})$, we should define the product of matrices so that $C = AB$; that is,

$$c_{ij} = \sum_{k=1}^{n} a_{ik}b_{kj}.$$

Examples of Matrix Representations

(A) In $\mathscr{E}^2$ with standard basis $\{\varepsilon_1, \varepsilon_2\}$ let $\mathbf{T}$ denote the rotation through 90 degrees, and let $\mathbf{S}$ denote the reflection across the x-axis. Then

$$\mathbf{T}\varepsilon_1 = \mathbf{T}(1, 0) = (0, 1) = 0\varepsilon_1 + 1\varepsilon_2 = \varepsilon_2,$$
$$\mathbf{T}\varepsilon_2 = \mathbf{T}(0, 1) = (-1, 0) = -1\varepsilon_1 + 0\varepsilon_2 = -\varepsilon_1.$$

Thus $\mathbf{T}$ is represented by the 2-by-2 matrix

$$A = \begin{pmatrix} 0 & -1 \\ 1 & 0 \end{pmatrix}.$$

Similarly,

$$\mathbf{S}\varepsilon_1 = \mathbf{S}(1, 0) = (1, 0) = 1\varepsilon_1 = 0\varepsilon_2 = \varepsilon_1,$$
$$\mathbf{S}\varepsilon_2 = \mathbf{S}(0, 1) = (0, -1) = 0\varepsilon_1 + (-1)\varepsilon_2 = -\varepsilon_2,$$

so $\mathbf{S}$ is represented by

$$B = \begin{pmatrix} 1 & 0 \\ 0 & -1 \end{pmatrix}.$$

The product transformation $\mathbf{ST}$ affects the basis vectors as follows:

$$\mathbf{ST}\varepsilon_1 = \mathbf{S}(\varepsilon_2) = -\varepsilon_2,$$
$$\mathbf{ST}\varepsilon_2 = \mathbf{S}(-\varepsilon_1) = -\varepsilon_1,$$

so $\mathbf{ST}$ is represented by

$$C = \begin{pmatrix} 0 & -1 \\ -1 & 0 \end{pmatrix}.$$

For the product transformation $\mathbf{TS}$ we have

$$\mathbf{TS}\varepsilon_1 = \mathbf{T}\varepsilon_1 = \varepsilon_2,$$
$$\mathbf{TS}\varepsilon_2 = \mathbf{T}(-\varepsilon_2) = \varepsilon_1,$$

so $\mathbf{TS}$ is represented by

$$D = \begin{pmatrix} 0 & 1 \\ 1 & 0 \end{pmatrix}.$$

Geometrically, $\mathbf{TS}$ is a reflection across the line $y = x$, whereas $\mathbf{ST}$ is a reflection across the line $y = -x$. In particular, $\mathbf{TS} \neq \mathbf{ST}$.

(B) Let $\mathbf{T}$ be a linear transformation that is nilpotent of index n on $\mathscr{V}_n$. We use the nature of $\mathbf{T}$ itself to choose a basis for $\mathscr{V}_n$, relative to which $\mathbf{T}$ can be represented very simply in matrix form. From Exercise 3.2-5 we know that for any ξ such that $\mathbf{T}^{n-1}\xi \neq \theta$, the set $B = \{\xi, \mathbf{T}\xi, \ldots, \mathbf{T}^{n-1}\xi\}$ is linearly independent and hence is a basis for $\mathscr{V}_n$. If we let $\alpha_1 = \xi, \alpha_2 = \mathbf{T}\xi, \ldots,$ $\alpha_n = \mathbf{T}^{n-1}\xi$, then we have

$$\mathbf{T}\alpha_1 = \alpha_2 = 0\alpha_1 + 1\alpha_2 + 0\alpha_3 + \cdots + 0\alpha_n,$$
$$\mathbf{T}\alpha_2 = \alpha_3 = 0\alpha_1 + 0\alpha_2 + 1\alpha_3 + \cdots + 0\alpha_n,$$
$$\cdot$$
$$\cdot$$
$$\cdot$$
$$\mathbf{T}\alpha_n = \theta = 0\alpha_1 + 0\alpha_2 + 0\alpha_3 + \cdots + 0\alpha_n.$$

Hence relative to the basis B, $\mathbf{T}$ is represented by the n-by-n matrix

$$N = \begin{pmatrix} 0 & 0 & \cdots & 0 & 0 \\ 1 & 0 & \cdots & 0 & 0 \\ 0 & 1 & \cdots & 0 & 0 \\ \cdot & \cdot & & \cdot & \cdot \\ \cdot & \cdot & & \cdot & \cdot \\ \cdot & \cdot & & \cdot & \cdot \\ 0 & 0 & \cdots & 1 & 0 \end{pmatrix}$$

in which $n_{ij} = 1$ if $i = j + 1$, and $n_{ij} = 0$ otherwise.

(C) Let $\mathbf{T}$ be a linear transformation that is idempotent on $\mathscr{V}_n$; $\mathbf{T}^2 = \mathbf{T}$. For each vector η in the range space $\mathscr{R}(\mathbf{T})$ we know that $\eta = \mathbf{T}\xi$ for some $\xi \in \mathscr{V}_n$. Then $\mathbf{T}\eta = \mathbf{T}^2\xi = \mathbf{T}\xi = \eta$. In other words, $\mathbf{T}$ maps each vector η in $\mathscr{R}(\mathbf{T})$ onto itself. Considered as a linear transformation on the space $\mathscr{R}(\mathbf{T})$, $\mathbf{T}$ is the identity mapping. Considered as a linear transformation on the null space $\mathscr{N}(\mathbf{T})$, $\mathbf{T}$ is the zero mapping. Now let $\{\alpha_1, \ldots, \alpha_k\}$ be a basis for $\mathscr{R}(\mathbf{T})$ where $k = r(\mathbf{T})$, and let $\{\alpha_{k+1}, \ldots, \alpha_n\}$ be a basis for $\mathscr{N}(\mathbf{T})$. The basic dimension relation of Theorem 3.5 assures us that each basis has the proper number of vectors in it. Suppose $\beta \in \mathscr{R}(\mathbf{T}) \cap \mathscr{N}(\mathbf{T})$. Then $\mathbf{T}\beta = \theta$ because $\beta \in \mathscr{N}(\mathbf{T})$. And $\mathbf{T}\beta = \beta$ because $\beta \in \mathscr{R}(\mathbf{T})$. Hence $\mathscr{R}(\mathbf{T}) \cap \mathscr{N}(\mathbf{T}) = [\theta]$. It follows that $\{\alpha_1, \ldots, \alpha_n\}$ is a basis for $\mathscr{V}_n$, and $\mathscr{V}_n$ is the direct sum of $\mathscr{R}(\mathbf{T})$ and $\mathscr{N}(\mathbf{T})$. Let us find the matrix A that represents $\mathbf{T}$ relative to the α-basis. Because

$$\mathbf{T}\alpha_i = \alpha_i \text{ for } i = 1, \ldots, k,$$

$$\mathbf{T}\alpha_i = \theta \text{ for } i = k + 1, \ldots, n,$$

we have determined that $a_{ii} = 1$ if $i \leq k$, and $a_{ij} = 0$ otherwise:

$$A = \begin{pmatrix} 1 & 0 & 0 & \cdots & 0 \\ 0 & 1 & 0 & \cdots & 0 \\ 0 & 0 & 1 & \cdots & 0 \\ \cdot & \cdot & \cdot & & \cdot \\ \cdot & \cdot & \cdot & & \cdot \\ \cdot & \cdot & \cdot & & \cdot \\ 0 & 0 & 0 & \cdots & 0 \end{pmatrix}.$$

Again we have seen how a strategic choice of basis can lead to a simple matrix representation of a linear transformation.

EXERCISES 3.4

1. Let $\mathbf{T}$ be the linear mapping from $\mathbb{R}^2$ to $\mathbb{R}^3$ defined by

$$\mathbf{T}\varepsilon_1 = \varepsilon_1 + 2\varepsilon_2 + 3\varepsilon_3,$$

$$\mathbf{T}\varepsilon_2 = 4\varepsilon_1 + 5\varepsilon_2 - \varepsilon_3.$$

(i) Write the matrix that represents $\mathbf{T}$ relative to standard bases.

(ii) Determine the $\mathbf{T}$-image in $\mathbb{R}^3$ of an arbitrary vector $\xi = x\varepsilon_1 + y\varepsilon_2$ in $\mathbb{R}^2$.

2. A linear transformation $\mathbf{T}$ of $\mathbb{R}^2$ is defined by

$$\mathbf{T}\alpha_1 = \alpha_1 - \alpha_2 \quad \text{and} \quad \mathbf{T}\alpha_2 = \alpha_1 + \alpha_2$$

where

$$\alpha_1 = \varepsilon_1 + 2\varepsilon_2 \quad \text{and} \quad \alpha_2 = \varepsilon_1 - \varepsilon_2.$$

(i) Write the matrix A that represents $\mathbf{T}$ relative to the α-basis.

(ii) Write the matrix B that represents $\mathbf{T}$ relative to the ε-basis.

3. A linear transformation $\mathbf{S}$ of $\mathbb{R}^2$ carries the point $P(1, 1)$ into the point $P'(-2, 0)$ and carries the point $Q(0, 1)$ into $Q'(-1, 1)$ where these coordinates are computed relative to the standard basis $\{\varepsilon_1, \varepsilon_2\}$.

(i) Determine $\mathbf{S}\varepsilon_1$ and $\mathbf{S}\varepsilon_2$.

(ii) Represent $\mathbf{S}$ by a matrix A relative to the standard basis.

(iii) Compute the S-image R' of an arbitrary point $R(a, b)$.

(iv) Let $\beta_1 = \varepsilon_1 + \varepsilon_2$ and $\beta_2 = \varepsilon_2$. Show that $\{\beta_1, \beta_2\}$ is a basis.

(v) Represent $\mathbf{S}$ by a matrix B relative to the β-basis.

(vi) Show that $\mathbf{S}$ is nonsingular, and write a matrix C that represents $\mathbf{S}^{-1}$ relative to the standard basis.

4. In $\mathbb{R}^2$ let $\alpha_1 = \varepsilon_1 + 2\varepsilon_2$ and $\alpha_2 = -\varepsilon_1 + \varepsilon_2$, and let $\mathbf{T}$ be the linear transformation defined by

$$\mathbf{T}\varepsilon_1 = \alpha_1,$$

$$\mathbf{T}\varepsilon_2 = \alpha_2.$$

(i) Show that $\{\alpha_1, \alpha_2\}$ is a basis for $\mathbb{R}^2$ (which shows that $\mathbf{T}$ is nonsingular).

(ii) Represent $\mathbf{T}$ by a matrix D relative to the ε-basis.

(iii) Represent $\mathbf{T}$ by a matrix E relative to the α-basis.

(iv) Represent $\mathbf{T}^{-1}$ by a matrix F relative to the α-basis.

(v) Compute the T-image of an arbitrary vector $a\varepsilon_1 + b\varepsilon_2$.

5. Referring to Exercises 3 and 4 for the definitions of the linear transformations $\mathbf{S}$ and $\mathbf{T}$, calculate the matrix that represents the transformation $\mathbf{ST}$ relative to the ε-basis. Similarly calculate the matrix that represents $\mathbf{TS}$.

6. Let $\mathbf{D}$ be the derivative transformation on the space $\mathscr{P}_k$ of all real polynomials of degree not exceeding k. Using $\{1, x, \ldots, x^k\}$ as a basis for $\mathscr{P}_k$, write the matrix that represents $\mathbf{D}$ relative to that basis.

7. Let $\mathscr{V}_n$, $\mathscr{W}_m$, and $\mathscr{X}_p$ be vector spaces over $\mathscr{F}$. Let $\mathbf{T}$ be a linear mapping from $\mathscr{V}_n$ into $\mathscr{W}_m$, and let S be a linear mapping from $\mathscr{W}_m$ into $\mathscr{X}_p$. Then $\mathbf{ST}$ is a linear mapping from $\mathscr{V}_n$ into $\mathscr{X}_p$. Let $\{\alpha_1, \ldots, \alpha_n\}$ be a basis for $\mathscr{V}_n$, $\{\beta_1, \ldots, \beta_m\}$ a basis for $\mathscr{W}_m$, and $\{\gamma_1, \ldots, \gamma_p\}$ a basis for $\mathscr{X}_p$. Suppose that

$$\mathbf{T}\alpha_j = \sum_{k=1}^{m} a_{kj}\beta_k \text{ for } j = 1, \ldots, n,$$

$$\mathbf{S}\beta_k = \sum_{i=1}^{p} b_{ik}\gamma_i \text{ for } k = 1, \ldots, m.$$

Letting

$$\mathbf{ST}\alpha_j = \sum_{i=1}^{p} c_{ij}\gamma_i \text{ for } j = 1, \ldots, n,$$

apply the previous expressions to deduce that

$$c_{ij} = \sum_{k=1}^{m} b_{ik}a_{kj}$$

for $i = 1, \ldots, p$ and $j = 1, \ldots, n$. This calculation motivates the definition of the product of a p-by-m matrix B by an m-by-n matrix A to produce a p-by-n matrix BA that represents the linear mapping $\mathbf{ST}$ whenever B represents $\mathbf{S}$ and A represents $\mathbf{T}$.

CHAPTER 4
MATRICES

4.1 ALGEBRA OF MATRICES

We have seen that any linear mapping $\mathbf{T}$ from $\mathscr{V}_n(\mathscr{F})$ to $\mathscr{W}_m(\mathscr{F})$ can be represented uniquely, relative to a fixed basis $\{\alpha_1, \ldots, \alpha_n\}$ for $\mathscr{V}_n$ and a fixed basis $\{\beta_1, \ldots, \beta_m\}$ for $\mathscr{W}_m$, by a rectangular matrix A of scalars having m rows and n columns. The first column of A is the scalar m-tuple that represents $\mathbf{T}\alpha_1$ relative to the β-basis, the second column of A is the m-tuple that represents $\mathbf{T}\alpha_2$, and so on. If we keep these bases fixed, each linear mapping in $\mathscr{L}(\mathscr{V}_n, \mathscr{W}_m)$ determines a unique m-by-n matrix of scalars; different mappings determine different matrices, and each m-by-n matrix is associated in this way with a linear mapping. Expressed concisely, this method of representing linear mappings establishes a one-to-one correspondence between the linear space $\mathscr{L}(\mathscr{V}_n, \mathscr{W}_m)$ and the set $\mathscr{M}_{m \times n}$ of all m-by-n matrices of elements of $\mathscr{F}$.

We now want to define operations for matrices that will make $\mathcal{M}_{m \times n}$ a linear space, isomorphic to $\mathscr{L}(\mathscr{V}_n, \mathscr{W}_m)$. But equality, sum, and scalar multiple in $\mathscr{L}(\mathscr{V}_n, \mathscr{W}_m)$ satisfy

$$\mathbf{T} = \mathbf{S} \quad \text{if and only if } \mathbf{T}\xi = \mathbf{S}\xi,$$

$$(\mathbf{T} + \mathbf{S})\xi = \mathbf{T}\xi + \mathbf{S}\xi,$$

$$(k\mathbf{T})\xi = k(\mathbf{T}\xi),$$

for all $\mathbf{T}, \mathbf{S}$ in $\mathscr{L}(\mathscr{V}_n, \mathscr{W}_m)$, all ξ in $\mathscr{V}_n$, and all k in $\mathscr{F}$. Because equality, sum, and scalar multiple of vectors are defined component by component, the corresponding concepts for matrices are defined in the same way.

Definition 4.1 Let $A = (a_{ij})$ and $B = (b_{ij})$ be two m-by-n matrices of scalars, and let k be a scalar. *Equality*, *sum*, and *scalar multiple* are defined for m-by-n matrices as follows:

(a) $A = B$ if and only if $\alpha_{ij} = b_{ij}$ for each $i = 1, \ldots, m$ and each $j = 1, \ldots, n$.

(b) $A + B = (c_{ij})$, where $c_{ij} = a_{ij} + b_{ij}$ for each $i = 1, \ldots, m$ and each $j = 1, \ldots, n$.

(c) $kA = (d_{ij})$, where $d_{ij} = ka_{ij}$ for each $i = 1, \ldots, m$ and each $j = 1, \ldots, n$.

For example, let

$$A = \begin{pmatrix} 1 & -2 & 0 \\ 0 & 1 & 2 \end{pmatrix}, \quad B = \begin{pmatrix} 0 & 2 & 1 \\ 0 & -1 & -1 \end{pmatrix}.$$

Then

$$A + B = \begin{pmatrix} 1+0 & -2+2 & 0+1 \\ 0+0 & 1-1 & 2-1 \end{pmatrix} = \begin{pmatrix} 1 & 0 & 1 \\ 0 & 0 & 1 \end{pmatrix},$$

and

$$3A = \begin{pmatrix} 3 & -6 & 0 \\ 0 & 3 & 6 \end{pmatrix}.$$

Theorem 4.1 The set $\mathcal{M}_{m \times n}$ of all m-by-n matrices of elements of a field $\mathscr{F}$ forms a linear space of dimension mn, relative to the operations of matrix sum and scalar multiple.

Proof Of the eight properties of a vector space listed in Definition 2.2, all but Postulates 3 and 4 are inherited by $\mathcal{M}_{m \times n}$ from the properties of $\mathcal{F}$ and the component-by-component definition of matrix sum and scalar multiple. Clearly the *m*-by-*n* matrix Z in which each entry is zero satisfies Postulate 3, and the matrix $(-1)A$ satisfies Postulate 4. Hence $\mathcal{M}_{m \times n}$ is a vector space over $\mathcal{F}$. To show that its dimension is *mn*, let $U_{ij} = (u_{ij})$ denote the *m*-by-*n* matrix for which $u_{ij} = 1$ and $u_{rs} = 0$ if $r \neq i$ or if $s \neq j$. The set $B = \{U_{ij} \,|\, i = 1, \ldots, m; j = 1, \ldots, n\}$ spans $\mathcal{M}_{m \times n}$, because if $A = (a_{ij})$ then

$$A = \sum_{q=1}^{n} \left(\sum_{p=1}^{m} a_{pq} U_{pq} \right).$$

Also B is linearly independent, because if a linear combination $\sum c_{pq} U_{pq}$, $p = 1, \ldots, m$ and $q = 1, \ldots, n$, equals the zero matrix Z, then each entry of that linear combination must equal zero, and therefore each c_{pq} must equal zero.

Theorem 4.2 The linear space $\mathcal{M}_{m \times n}$ of all *m*-by-*n* matrices of elements of $\mathcal{F}$ is isomorphic to the linear space $\mathcal{L}(\mathcal{V}_n, \mathcal{W}_m)$ of all linear mappings from $\mathcal{V}_n(\mathcal{F})$ to $\mathcal{W}_m(\mathcal{F})$.
Proof Choose any basis for $\mathcal{V}_n$ and any basis for $\mathcal{W}_m$. As previously described, each $\mathbf{T}$ in $\mathcal{L}(\mathcal{V}_n, \mathcal{W}_m)$ determines uniquely a matrix $A_{\mathbf{T}}$, and the correspondence $\mathbf{T} \to A_{\mathbf{T}}$ is one-to-one from $\mathcal{L}(\mathcal{V}_n, \mathcal{W}_m)$ onto $\mathcal{M}_{m \times n}$. Definition 4.1 shows that this correspondence preserves vector sum and scalar multiple, so the correspondence is an isomorphism.

From Theorems 4.1 and 4.2 we conclude that the dimension of the space $\mathcal{L}(\mathcal{V}_n, \mathcal{W}_m)$ is *mn*. This is an example of a result that seems to be easier to establish in terms of matrices than in terms of linear mappings. Other results appear to be more natural when expressed in terms of linear mappings instead of in terms of matrices. The real significance of Theorem 4.2 is that it guarantees that every theorem about $\mathcal{L}(\mathcal{V}_n, \mathcal{W}_m)$ can be translated into a theorem about $\mathcal{M}_{m \times n}$ and conversely. Thus it provides us with a choice of two ways to formulate and to analyze a given question about matrices or linear mappings.

Next we consider the matrix analogue of the composition $\mathbf{ST}$ of two linear mappings, which is possible only when the range space of $\mathbf{T}$ is a subspace of the domain of $\mathbf{S}$. If $\mathbf{T}$ maps $\mathcal{V}_n$ to $\mathcal{W}_m$ and $\mathbf{S}$ maps $\mathcal{W}_m$ to $\mathcal{X}_p$, we fix a basis for each of the three spaces. Relative to the chosen bases, $\mathbf{S}$ is represented by a *p*-by-*m* matrix $B = (b_{ik})$; $\mathbf{T}$ is represented by an *m*-by-*n* matrix $A = (a_{kj})$; and **ST** is represented by a p-by-n matrix $C = (c_{ij})$. The scalar c_{ij} can be computed from the entries of B and A by the formula

$$c_{ij} = \sum_{k=1}^{m} b_{ik} a_{kj}, \text{ for } i = 1, \ldots, p \text{ and } j = 1, \ldots, n,$$

as described in Exercise 3.4-7.

Definition 4.2 The *product BA* of the p-by-m matrix $B = (b_{ik})$ and the m-by-n matrix $A = (a_{kj})$ is the p-by-n matrix $C = (c_{ij})$, where

$$c_{ij} = \sum_{k=1}^{m} b_{ik} a_{kj}, \text{ for } i = 1, \ldots, p \text{ and } j = 1, \ldots, n.$$

Observe that the product BA of two matrices is defined only when the number of columns of the left-hand matrix B in the product equals the number of rows of the right-hand matrix A in the product. The product matrix BA has as many rows as B and as many columns as A. Furthermore, the entry c_{ij} in row i and column j of the product BA is simply the dot product of the m-tuple in row i of B with the m-tuple in column j of A:

$$c_{ij} = b_{i1} a_{1j} + b_{i2} a_{2j} + \cdots + b_{im} a_{mj},$$

$$
i \rightarrow
\begin{pmatrix}
\cdot & \cdot & & \cdot \\
\cdot & \cdot & & \cdot \\
b_{i1} & b_{i2} & \cdots & b_{im} \\
\cdot & \cdot & & \cdot \\
\cdot & \cdot & & \cdot
\end{pmatrix}
\begin{pmatrix}
\cdots & a_{1j} & \cdots \\
\cdots & a_{2j} & \cdots \\
 & \cdot & \\
 & \cdot & \\
\cdots & a_{mj} & \cdots
\end{pmatrix}
= \; i \rightarrow
\begin{pmatrix}
\cdot & & \\
\cdot & & \\
\cdots & c_{ij} & \cdots \\
\cdot & & \\
\cdot & &
\end{pmatrix}.
$$

In practice we can perform this computation easily by using the left index finger to run across row i of the left-hand matrix and simultaneously using the right index finger to run down column j of the right-hand matrix, multiplying elements in corresponding positions and adding successively the products obtained. An example will help to clarify the procedure.

$$
B = \begin{pmatrix} 1 & 0 & -1 \\ 2 & 4 & 7 \\ 5 & 3 & 0 \end{pmatrix}, \quad A = \begin{pmatrix} 6 & 1 \\ 0 & 4 \\ -2 & 3 \end{pmatrix};
$$

$$BA = \begin{pmatrix} (1)(6) + (0)(0) + (-1)(-2) & (1)(1) + (0)(4) + (-1)(3) \\ (2)(6) + (4)(0) + (7)(-2) & (2)(1) + (4)(4) + (7)(3) \\ (5)(6) + (3)(0) + (0)(-2) & (5)(1) + (3)(4) + (0)(3) \end{pmatrix}$$

$$= \begin{pmatrix} 8 & -2 \\ -2 & 39 \\ 30 & 17 \end{pmatrix}.$$

We note also that the matrix product AB is not defined.

Another technique for computing the entries of row i of BA is to write the entries of row i of B as a column to the left of A:

$$\begin{matrix} b_{i1} \\ b_{i2} \\ \cdot \\ \cdot \\ \cdot \\ b_{im} \end{matrix} \begin{pmatrix} a_{11} & \cdots & a_{1j} & \cdots & a_{1n} \\ a_{21} & \cdots & a_{2j} & \cdots & a_{2n} \\ \cdot & & \cdot & & \cdot \\ \cdot & & \cdot & & \cdot \\ \cdot & & \cdot & & \cdot \\ a_{m1} & \cdots & a_{mj} & \cdots & a_{mn} \end{pmatrix}.$$

Then the (i, j) entry of BA is easily computed as

$$b_{i1}a_{1j} + b_{i2}a_{2j} + \cdots + b_{im}a_{mj}.$$

After a little practice you will be able to multiply matrices without such special techniques.

In discussing the algebraic properties of the product of rectangular matrices we shall assume that the matrices involved have the proper number of rows and columns so that the indicated operations are defined. We claim that

$$A(bB + cC) = bAB + cAC,$$

$$(aA + bB)C = aAC + bBC,$$

$$(AB)C = A(BC),$$

for any rectangular matrices for which the corresponding sums and products are defined. One way to prove these assertions is by computing with the entries of the matrices according to the rules established in Definitions 4.1 and 4.2. But that method is tedious and unnecessary. Instead, we can interpret each matrix as a linear mapping of a pair of vector spaces. We know that if A represents **T** and B represents **S**, then $A + B$ represents **T** + **S**. And if C represents **R**, then BC represents **SR**. Because linear mappings have the algebraic properties claimed above, and because matrix algebra precisely mimics the algebra of linear mappings, we can conclude that these properties are valid for computations with rectangular matrices.

Square n-by-n matrices, of course, represent linear transformations on $\mathscr{V}_n$. Matrix sum, scalar multiple, and product are all defined universally on the set $\mathscr{M}_{n \times n}$ of all n-by-n matrices over $\mathscr{F}$. Furthermore, we know that $\mathscr{L}(\mathscr{V}_n, \mathscr{V}_n)$ is a linear space of dimension n^2. Thus we can extend Theorem 4.2 for the case in which $m = n$ to obtain the following result.

Theorem 4.3 The set $\mathscr{M}_{n \times n}$ of all n-by-n matrices of elements in $\mathscr{F}$ forms a linear algebra of dimension n^2, relative to the operations of the sum, scalar multiple, and product of n-by-n matrices. Furthermore, $\mathscr{M}_{n \times n}$ is isomorphic to $\mathscr{L}(\mathscr{V}_n, \mathscr{V}_n)$.

We summarize explicitly some properties of matrix algebra that are implied by the isomorphisms of Theorems 4.2 and 4.3.

Theorem 4.4 Let A, B, and C denote m-by-n matrices, let Z denote the m-by-n matrix of zeros, and let a and b denote scalars. Then

(a) $A + B = B + A$,

(b) $(A + B) + C = A + (B + C)$,

(c) $A + Z = A$,

(d) $A + (-1)A = Z$,

(e) $(a + b)A = aA + bA$,

(f) $a(A + B) = aA + aB$,

(g) $(ab)A = a(bA)$,

(h) $0A = Z = aZ$.

Furthermore, if $m = n$, then

(i) $(AB)C = A(BC)$,

(j) $(aA + bB)C = aAC + bBC$,

(k) $C(aA + bB) = aCA + bCB$.

Properties (i), (j), and (k) also are valid for rectangular matrices, provided that the dimensions of the matrices are such that each of the indicated sums and products is defined.

Finally we observe that matrix multiplication can be used to compute the image η in $\mathscr{W}_m$ of an arbitrary vector ξ in $\mathscr{V}_n$ under a linear mapping **T** from $\mathscr{V}_n$ to $\mathscr{W}_m$. The essential calculations were made at the beginning of Section 3.4. Relative to an α-basis for $\mathscr{V}_n$ and a β-basis for $\mathscr{W}_m$, the vectors ξ in $\mathscr{V}_n$ and η in $\mathscr{W}_m$ were represented by column vectors X and Y with n and

m components respectively, and the linear mapping **T** was represented by an m-by-n matrix A:

$$A = \begin{pmatrix} a_{11} & a_{12} & \cdots & a_{1n} \\ a_{21} & a_{22} & \cdots & a_{2n} \\ \cdot & \cdot & & \cdot \\ \cdot & \cdot & & \cdot \\ \cdot & \cdot & & \cdot \\ a_{m1} & a_{m2} & \cdots & a_{mn} \end{pmatrix}, \quad X = \begin{pmatrix} x_1 \\ x_2 \\ \cdot \\ \cdot \\ x_n \end{pmatrix}, \quad Y = \begin{pmatrix} y_1 \\ y_2 \\ \cdot \\ \cdot \\ y_m \end{pmatrix}.$$

Then $\mathbf{T}\xi = \eta$ if and only if $AX = Y$. In particular, the linear system (1.1) can be expressed in matrix form as

$$AX = D, \text{ where } D = \begin{pmatrix} d_1 \\ d_2 \\ \cdot \\ \cdot \\ \cdot \\ d_m \end{pmatrix},$$

and therefore can be regarded as a single vector equation,

$$\mathbf{T}\xi = \delta.$$

We shall return to this observation in Section 4.5 when we reexamine linear systems from a geometric point of view.

EXERCISES 4.1

1. Compute $A + B$, $A - B$, $2A$, and BA for the square matrices

$$A = \begin{pmatrix} -2 & 0 & 1 \\ 0 & 2 & 3 \\ 1 & 4 & 0 \end{pmatrix} \text{ and } B = \begin{pmatrix} 3 & 0 & -1 \\ 0 & -1 & -3 \\ -1 & -4 & 1 \end{pmatrix}.$$

2. Compute AB, AC, B^2, BC, CA, given that

$$A = \begin{pmatrix} 1 & 0 & -1 \\ 0 & 2 & 3 \end{pmatrix}, B = \begin{pmatrix} 2 & -1 & 4 \\ 1 & 0 & -2 \\ 0 & 3 & 1 \end{pmatrix}, C = \begin{pmatrix} 0 & 2 \\ -1 & 0 \\ 3 & 1 \end{pmatrix}.$$

Are any other binary products possible for these three matrices?

3. Given the matrices

$$A = \begin{pmatrix} 1 & 0 \\ 0 & -1 \end{pmatrix}, B = \begin{pmatrix} 0 & 1 \\ 1 & 0 \end{pmatrix}, C = \begin{pmatrix} 1 & 0 \\ 2 & 0 \end{pmatrix}:$$

(i) Describe geometrically the linear transformation on $\mathbb{R}^2$ that each matrix represents relative to the standard basis.

(ii) Calculate the matrix products AB, BA, A^2, B^2, C^2 and interpret each as a geometric transformation.

4. The derivative transformation $\mathbf{D}$ on $\mathscr{P}_k$ can be represented as described in Exercise 3.4-6 by the $(k + 1)$-by-$(k + 1)$ matrix

$$A = \begin{pmatrix} 0 & 1 & 0 & \cdots & 0 \\ 0 & 0 & 2 & \cdots & 0 \\ & \cdot & \cdot & \cdot & \cdot \\ \cdot & \cdot & \cdot & & \cdot \\ & \cdot & \cdot & \cdot & \cdot \\ 0 & 0 & 0 & \cdots & k \\ 0 & 0 & 0 & \cdots & 0 \end{pmatrix}.$$

The $(i, i + 1)$ entry is i and every other entry is zero. Because $\mathbf{D}$ is nilpotent of index $(k + 1)$ on $\mathscr{P}_k$, we expect that $A^{k+1} = Z$ but $A^k \neq Z$. For the special case $k = 3$ compute A^2, A^3, and A^4 to confirm this expectation.

5. Show that the system of all one-by-one matrices over a field $\mathscr{F}$, together with matrix addition and multiplication, is a field which is isomorphic to $\mathscr{F}$.

6. Prove that the set of all real two-by-two matrices of the form

$$\begin{pmatrix} a & b \\ -b & a \end{pmatrix}$$

forms a system that is isomorphic to the field of complex numbers.

7. Prove that the set of all complex two-by-two matrices of the form

$$\begin{pmatrix} a + ib & c + id \\ -c + id & a - ib \end{pmatrix}, \text{ where } i^2 = -1,$$

forms a system that is isomorphic to the algebra of quaternions as described in Exercise 3.1-12.

8. Let I_m denote the m-by-m matrix in which the (i, i) entry is one for each i and the (i, j) entry is zero whenever $i \neq j$. Show that for every m-by-n matrix A,

(i) $I_m A = A$.

(ii) $AI_n = A$.

Interpret I_m as a linear transformation on $\mathbb{R}^m$.

9. Let $\{\alpha_1, \alpha_2, \alpha_3\}$ be any basis for $\mathbb{R}^3$, and let

$$\beta_1 = \alpha_1 - 2\alpha_2 \quad ,$$
$$\beta_2 = \alpha_1 + \alpha_2 + \alpha_3,$$
$$\beta_3 = \alpha_2 - \alpha_3.$$

(i) Prove that $\{\beta_1, \beta_2, \beta_3\}$ is a basis, and express each α_i as a linear combination of the β_j.

(ii) Let **T** be defined on $\mathbb{R}^3$ by $\mathbf{T}(\alpha_i) = \beta_i$ for $i = 1, 2, 3$. Determine the matrix A that represents **T** relative to the α-basis.

(iii) Let **S** be defined on $\mathbb{R}^3$ by $\mathbf{S}(\beta_i) = \alpha_i$ for $i = 1, 2, 3$. Determine the matrix B that represents **S** relative to the β-basis.

(iv) By matrix computations show that $BA = I_3 = AB$, where I_3 is defined in Exercise 8.

10. Let X and A be n-by-n matrices such that $XA = I_n$, where I_n is defined in Exercise 8.

(i) Use Theorems 4.3 and 3.8 to deduce that $AX = I_n$.

(ii) For the special case $n = 2$ write a system of four scalar equations that is equivalent to the single matrix equation $XA = I_2$, where $X = (x_{ij})$ and $A = (a_{ij})$. Write an analogous system that is equivalent to $AX = I_2$. Observe that it is not obvious that a solution of the first system is also a solution of the second system.

11. An n-by-n Markov matrix is defined to be any n-by-n real matrix $A = (a_{ij})$ that satisfies the two properties

$$0 \leq a_{ij} \leq 1,$$

$$\sum_{j=1}^{n} a_{ij} = 1 \quad \text{for } i = 1, 2, \ldots, n.$$

Prove that the product of two Markov matrices is a Markov matrix.

12. In quantum mechanics the Pauli theory of electron spin makes use of linear transformations $\mathbf{T}_x$, $\mathbf{T}_y$, $\mathbf{T}_z$, whose complex matrices in the preferred coordinate system are, respectively,

$$X = \begin{pmatrix} 0 & 1 \\ 1 & 0 \end{pmatrix},$$

$$Y = \begin{pmatrix} 0 & -i \\ i & 0 \end{pmatrix}, \text{ where } i^2 = -1,$$

$$Z = \begin{pmatrix} 1 & 0 \\ 0 & -1 \end{pmatrix}.$$

(i) Show that $X^2 = Y^2 = Z^2 = I$, and therefore that each is nonsingular.

(ii) Form a multiplication table of the four matrices I, X, Y, Z, and observe that any product of these matrices is a scalar times one of these matrices.

4.2 SPECIAL TYPES OF SQUARE MATRICES

In this section we shall examine a few types of n-by-n matrices that play important roles in the study of matrices and linear transformations. From the calculations already made with matrices and linear mappings, we are warned that the algebra of matrices is different from the algebra of real numbers. For example, matrix multiplication is noncommutative, and the product of nonzero matrices can be the zero matrix.

Unit matrices   For each $i = 1, \ldots, n$ and each $j = 1, \ldots, n$ the unit matrix U_{ij} is defined to have 1 in the (i, j) position and 0 in every other position. As we observed in Theorem 4.1 the set of all such unit matrices is a basis for the linear space $\mathcal{M}_{n \times n}$. As an exercise you may verify that for any n-by-n matrix A, the product $U_{ij} A$ is the matrix having zero everywhere except in row i, while row i of $U_{ij} A$ is the same as row j of A. The product $A U_{ij}$ has zero everywhere except in column j, while column j of $A U_{ij}$ is the same as column i of A.

Identity matrix   The n-by-n identity matrix I can be defined as $I = U_{11} + U_{22} + \cdots + U_{nn}$. That is, I has 0 in position (i, j) when $j \neq i$ and 1 in position (i, j) when $i = j$. The positions (i, i) in a square matrix are said to comprise the *main diagonal* of A. Hence I has 1 in each position along the main diagonal, and 0 elsewhere. We can also write $I = (\delta_{ij})$ where the number δ_{ij} is called the *Kronecker delta* and is defined by

$$\delta_{ij} = \begin{cases} 1 & \text{if } i = j, \\ 0 & \text{if } i \neq j. \end{cases}$$

It is easy to verify that $IA = A = AI$ for every n-by-n matrix A. For each n there is an n-by-n identity matrix, and we write I_n when there is need to identify the size of I.

Observe that the (r, s) entry in U_{ij} can be expressed in terms of the Kronecker delta by the equation $u_{rs} = \delta_{ri}\delta_{sj}$.

Scalar matrices   Although matrix multiplication is not commutative, the n-by-n identity matrix commutes with every n-by-n matrix. Are there other n-by-n matrices $A = (a_{ij})$ that have the property that $AX = XA$ for every n-by-n matrix $X = (x_{ij})$? A straightforward approach would be to determine scalars a_{ij} which satisfy the n^2 equations,

$$\sum_{k=1}^{n} a_{ik}x_{kj} = \sum_{k=1}^{n} x_{ik}a_{kj}, \quad i, j = 1, 2, \ldots, n.$$

This is a somewhat fearful task. Instead, we argue as follows: If A commutes with all n-by-n matrices, then in particular A commutes with each unit matrix U_{rs},

$$AU_{rs} = \begin{pmatrix} 0 & \cdots & a_{1r} & \cdots & 0 \\ 0 & \cdots & a_{2r} & \cdots & 0 \\ \cdot & & \cdot & & \cdot \\ \cdot & & \cdot & & \cdot \\ \cdot & & \cdot & & \cdot \\ 0 & \cdots & a_{nr} & \cdots & 0 \end{pmatrix} = \begin{pmatrix} 0 & 0 & \cdots & 0 \\ \cdot & \cdot & & \cdot \\ \cdot & \cdot & & \cdot \\ \cdot & \cdot & & \cdot \\ a_{s1} & a_{s2} & \cdots & a_{sn} \\ \cdot & \cdot & & \cdot \\ \cdot & \cdot & & \cdot \\ 0 & 0 & \cdots & 0 \end{pmatrix} = U_{rs}A,$$

where every element not in column s of the first matrix is zero, and every element not in row r of the second matrix is zero. Hence we have $a_{rr} = a_{ss}$, $a_{ir} = 0$ if $i \neq r$, and $a_{sj} = 0$ if $j \neq s$. Thus if A commutes with all unit matrices,

A must have the same element k in every position of the *main diagonal* and zeros elsewhere:

$$a_{ij} = k\delta_{ij}, \quad i, j = 1, 2, \ldots, n,$$

$$A = \begin{pmatrix} k & 0 & \cdots & & 0 \\ 0 & k & \cdots & & 0 \\ \cdot & \cdot & & & \cdot \\ \cdot & \cdot & & & \cdot \\ \cdot & \cdot & & & \cdot \\ 0 & 0 & \cdots & & k \end{pmatrix} = kI.$$

Such a matrix is called a *scalar matrix*, being merely a scalar multiple of I. Clearly, a scalar matrix commutes with every matrix, so the question is answered completely.

Diagonal matrices   Scalar matrices form a subclass of the class of n-by-n *diagonal matrices*, which are defined by the property that $a_{ij} = 0$ if $i \neq j$. Thus zeros appear everywhere except possibly on the main diagonal. Clearly, the sum of diagonal matrices is diagonal; so is the product, for if A and B are diagonal and $AB = C$, then

$$c_{ij} = \sum_{k=1}^{n} a_{ik}b_{kj} = a_{ii}b_{ij} = \begin{cases} a_{ii}b_{ii} & \text{if } j = i, \\ 0 & \text{if } j \neq i. \end{cases}$$

Diagonal matrices are especially nice to work with, and a major problem in the study of linear transformations is to determine whether a given transformation can be represented by a diagonal matrix relative to a suitably chosen basis.

Triangular matrices   A still more inclusive class of square matrices is that for which $a_{ij} = 0$ whenever $i < j$. Such a matrix is called *lower triangular* because all the nonzero elements lie on or below the main diagonal:

$$\begin{pmatrix} a_{11} & 0 & 0 & \cdots & 0 \\ a_{21} & a_{22} & 0 & \cdots & 0 \\ \cdot & \cdot & \cdot & & \cdot \\ \cdot & \cdot & \cdot & & \cdot \\ \cdot & \cdot & \cdot & & \cdot \\ a_{n1} & a_{n2} & a_{n3} & \cdots & a_{nn} \end{pmatrix}.$$

A square matrix $A = (a_{ij})$ is called *upper triangular* if $a_{ij} = 0$ whenever $i > j$. A square matrix in echelon form (Gaussian elimination) is upper triangular. An upper triangular or lower triangular matrix for which $a_{ii} = 0$ for $i = 1, \ldots, n$ is called *strictly triangular*. Clearly, any triangular matrix is the sum of a strictly triangular matrix and a diagonal matrix.

Idempotent matrices   A square matrix A is said to be *idempotent* if and only if $A^2 = A$. An example other than Z and I is

$$\begin{pmatrix} 1 & -1 \\ 0 & 0 \end{pmatrix}.$$

Idempotent matrices represent geometric projections.

Nilpotent matrices   A square matrix A is said to be *nilpotent of index p* if $A^p = Z$ but $A^{p-1} \neq Z$. Any strictly triangular matrix is nilpotent.

Nonsingular matrices   This type of matrix is of great importance, because any such matrix represents a nonsingular linear transformation on $\mathscr{V}_n$. An n-by-n matrix A is *nonsingular* if and only if there exists an n-by-n matrix B such that

$$BA = I.$$

Otherwise A is said to be *singular*. From Theorem 4.3 and the corresponding property of linear transformations we deduce:

$$\text{If } BA = I, \text{ then } AB = I.$$

(See Exercise 4.1-10.) Hence B is called the *inverse* of A and is denoted A^{-1}. Later we shall develop several computational schemes for computing the inverse of a nonsingular matrix.

> **Theorem 4.5**   If A and B are nonsingular n-by-n matrices, then
>   (a) A^{-1} is nonsingular and $(A^{-1})^{-1} = A$,
>   (b) (cA) is nonsingular when $c \neq 0$, and $(cA)^{-1} = c^{-1}A^{-1}$,
>   (c) AB is nonsingular and $(AB)^{-1} = B^{-1}A^{-1}$.
> **Proof**   Exercise.

To describe the next two types of matrices considered here we need the notion of the transpose of a matrix, which is simply the matrix A^t obtained by writing the rows of A as the columns of A^t.

Definition 4.3 Let $A = (a_{ij})$ be any m-by-n matrix. The *transpose* of A, denoted A^t, is the n-by-m matrix (b_{ij}), where $b_{ij} = a_{ji}$.

Theorem 4.6 $(A^t)^t = A$, $(A + B)^t = A^t + B^t$, and $(cA)^t = cA^t$. If AB is defined, then $(AB)^t = B^t A^t$.

Proof Exercise. Note in particular that the transpose of a product is the product of the transposes *in the reverse order*.

Symmetric matrices An n-by-n matrix A is said to be *symmetric* if and only if $A = A^t$. Obviously any diagonal matrix is symmetric.

Skew-symmetric matrices An n-by-n matrix A is said to be *skew-symmetric* (or simply *skew*) if and only if $A^t = -A$. This implies that if $1 + 1 \neq 0$ in the base field, then every diagonal element of a skew matrix is zero. We exclude from our consideration any field in which $1 + 1 = 0$. Now let A be any square matrix. We have

$$A = \tfrac{1}{2}(A + A^t) + \tfrac{1}{2}(A - A^t).$$

Because $A + A^t$ is symmetric and $A - A^t$ is skew (see Exercise 4.2-14), this expresses A as a sum of two matrices, the first of which is symmetric and the second skew. Furthermore, this decomposition is unique (see Exercise 4.2-13).

Row vectors and column vectors Relative to a fixed basis, every vector of $\mathscr{V}_n$ has a unique representation as an n-tuple of scalars, $(a_1, \ldots, a_n)$. Except for the presence of commas, this is formally the same as a matrix of one row and n columns. Accordingly, a one-by-n matrix is called a *row vector*. The transpose of a row vector is an n-by-one matrix of n rows and one column and is called a *column vector*. If A is a row vector, and if B is a column vector, then both AB and BA are defined, but AB is a one-by-one matrix (a scalar), and BA is an n-by-n matrix.

EXERCISES 4.2

1. Verify that the matrices $U_{ij} A$ and $A U_{ij}$ have the forms described in the text discussion of unit matrices.

2. Prove that $IA = AI = A$ for every square matrix A.

3. Show that for each n the set of all n-by-n scalar matrices over $\mathscr{F}$ forms a field that is isomorphic to $\mathscr{F}$.

4. (i) Prove that all n-by-n diagonal matrices commute.

(ii) Prove that if A commutes with all n-by-n diagonal matrices, then A is diagonal.

5. Prove that the set of all n-by-n lower triangular matrices is closed under matrix sum and product and scalar multiple.

6. Prove that if A is idempotent and $A \neq I$, then A is singular.

7. Determine all real two-by-two idempotent matrices.

8. Let A be a four-by-four strictly lower triangular matrix. By calculating successive powers of A, show that A is nilpotent. Is the same conclusion valid for any n-by-n strictly lower triangular matrix?

9. Determine all real two-by-two matrices that are nilpotent of index two.

10. Determine all real two-by-two nonsingular matrices.

11. Prove Theorem 4.5.

12. Prove Theorem 4.6.

13. Show that if $A = S + K$, where S is symmetric and K is skew, then $S = \frac{1}{2}(A + A^t)$ and $K = \frac{1}{2}(A - A^t)$.

14. Prove for every square matrix A that
   (i) AA^t is symmetric.
   (ii) $A + A^t$ is symmetric.
   (iii) $A - A^t$ is skew.

15. Prove that A^2 is symmetric if either A is symmetric or A is skew.

16. If A and B are both symmetric, prove that
   (i) $A + B$ is symmetric.
   (ii) AB is symmetric if and only if A and B commute.

17. If A and B are both skew, prove that $A + B$ is skew.

18. Prove that if A is nonsingular, so is A^t, and $(A^t)^{-1} = (A^{-1})^t$.

19. Let $A = (a_1 \ldots a_n)$ be a row vector and

$$B = \begin{pmatrix} b_1 \\ \cdot \\ \cdot \\ \cdot \\ b_n \end{pmatrix}$$

a column vector. Compute AB and BA.

20. In the special theory of relativity, use is made of the Lorentz transformation,

$$x' = b(x - vt),$$

$$t' = b\left(\frac{-vx}{c^2} + t\right),$$

where $|v|$ represents the speed of a moving object, c the speed of light, and $b = c(c^2 - v^2)^{-1/2}$. The corresponding matrix is

$$L(v) = b \begin{pmatrix} 1 & -v \\ \dfrac{-v}{c^2} & 1 \end{pmatrix}.$$

(i) Show that $L(v)$ is nonsingular for $|v| < c$.

(ii) Show that the set of all $L(v)$ for $|v| < c$ forms a multiplicative group. (This group is called the *Lorentz* group.)

4.3 ELEMENTARY MATRICES

We now are prepared to reexamine Gaussian elimination and the reduction of a matrix to echelon form, introduced in Chapter 1 as a computational scheme for solving a system of linear equations. This time we shall describe those processes in terms of matrix algebra; that is, we start with an m-by-n matrix A, which represents a system $\mathscr{S}_1$ of m linear equations in $n - 1$ variables, and seek to describe a sequence of matrix operations that replaces A by a matrix B in echelon form, which represents a system $\mathscr{S}_2$ of linear equations that has precisely the same solutions as $\mathscr{S}_1$.

The three elementary row operations described in Section 1.2 were denoted

$M_i(c)$: Multiply row i by a nonzero scalar c, $(R_i \to cR_i)$,

$R_{i,\,i+cj}$: Replace row i by the sum of row i and a scalar multiple c of row j, $(R_i \rightarrow R_i + cR_j)$,

P_{ij}: Permute (interchange) row i and row j, $(R_i \leftrightarrow R_j)$.

We use these same symbols to denote the three matrices obtained by applying each elementary row operation to the identity matrix.

Definition 4.4   An *m-by-m elementary matrix* is any matrix obtained by applying a single elementary row operation to the *m-by-m* identity matrix I.

Each of the three types of elementary matrices can be expressed in terms of the unit matrices U_{ij} described in Section 4.2:

$$M_i(c) = I + (c - 1)U_{ii} = \quad \text{row } i \begin{pmatrix} 1 & & & & & & & & \\ & \cdot & & & & & & & \\ & & \cdot & & & & & & \\ & & & \cdot & & & & & \\ \cdot & \cdot & \cdot & \cdot & c & \cdot & \cdot & \cdot & \cdot \\ & & & & 1 & & & & \\ & & & & & 1 & & & \\ & & & & & & \cdot & & \\ & & & & & & & \cdot & \\ & & & & & & & & 1 \end{pmatrix}, \; c \neq 0.$$

$$R_{i,\,i+cj} = I + cU_{ij} = \quad \begin{matrix} \\ \text{row } i \\ \text{row } j \end{matrix} \begin{pmatrix} 1 & & & & & \\ & \cdot & & & & \\ \cdot & \cdot & 1 & c & \cdot & \cdot \\ & & \cdot & & & \\ & & 1 & & \\ & & & \cdot & \\ & & & & 1 \end{pmatrix}, \; i \neq j.$$

$$P_{ij} = I - U_{ii} + U_{ij} - U_{jj} + U_{ji} = \quad \begin{matrix} \\ \\ \text{row } i \\ \\ \text{row } j \end{matrix} \begin{pmatrix} 1 & & & & & & & \\ & \cdot & & & & & & \\ & & 1 & & & & & \\ \cdot & \cdot & \cdot & 0 & \cdot & \cdot & 1 & \cdot & \cdot \\ & & & & 1 & & & \\ & & & & & 1 & & \\ \cdot & \cdot & \cdot & 1 & \cdot & \cdot & 0 & \cdot & \cdot \\ & & & & & & & 1 & \\ & & & & & & & & \cdot \\ & & & & & & & & 1 \end{pmatrix}.$$

The algebraic significance of elementary matrices is that each elementary row operation can be performed on an arbitrary m-by-n matrix A by multiplying A *on the left* by the corresponding m-by-m elementary matrix. An analogous result holds for the corresponding operations on the columns of A except that to perform column operations on A we must multiply A *on the right* by an appropriate elementary matrix.

Theorem 4.7 Let A be any m-by-n matrix. If $M_i(c)$, $R_{i,\,i+cj}$ and P_{ij} are the elementary m-by-m matrices, then

(a) $M_i(c)A$ is A with row i multiplied by $c \neq 0$.

(b) $R_{i,\,i+cj} A$ is A with row i replaced by the sum of row i and c times row j.

(c) $P_{ij} A$ is A with row i and row j interchanged.

(d) $AM_i(c)$ is A with column i multiplied by $c \neq 0$.

(e) $AR_{i,\,i+cj}$ is A with column j replaced by the sum of column j and c times column i.

(f) AP_{ij} is A with column i and column j interchanged.

Proof Exercise. Note that as a row (left-hand) operator $R_{i,\,i+cj}$ replaces row i by the sum of row i and c times row j, but as a column (right-hand) operator $R_{i,\,i+cj}$ replaces column j by the sum of column j and c times column i. Also see Exercise 4.3-7.

The information contained in Theorem 4.7 is restated in the table below.

Elementary matrix	Effect on rows of A	Transposed matrix	Effect on columns of A
$M_i(c)$	$R_i \to cR_i$	$M_i(c)$	$C_i \to cC_i$
$R_{i,\,i+cj}$	$R_i \to R_i + cR_j$	$R_{j,\,j+ci}$	$C_i \to C_i + cC_j$
P_{ij}	$R_i \leftrightarrow R_j$	P_{ij}	$C_i \leftrightarrow C_j$

Pay particular attention to the second line in this table. As a row operator $R_{i,\,i+cj}$ denotes "replace row i by the sum of row i and c times row j." The transpose of $R_{i,\,i+cj}$ is $R_{j,\,j+ci}$. But the effect of $R_{j,\,j+ci}$ as a column operator is to replace column i by the sum of column i and c times column j. Hence for each elementary matrix E, the transpose E^t performs on the columns of A the same elementary column operation as E performs on the corresponding rows of A.

Theorem 4.8   Each elementary matrix is nonsingular:

$$M_i(c)^{-1} = M_i(c^{-1}), \ c \neq 0,$$

$$\left(R_{i,\,i+cj}\right)^{-1} = R_{i,\,i-cj},$$

$$P_{ij}^{-1} = P_{ij}.$$

Proof   This theorem is simply the observation that each elementary row operation can be reversed by an elementary row operation of the same type. The stated results are easily verified by considering the effect that each of these expressions performs on the corresponding elementary matrix. For example if we apply the row operation $R_{i,\,i-cj}$ to the matrix $R_{i,\,i+cj}$, the result is I. Similarly $M_i(c^{-1})M_i(c) = I$, and $P_{ij}P_{ij} = I$.

The process of Gaussian elimination uses a sequence of elementary row operations to replace a given m-by-n matrix A by an m-by-n matrix B in echelon form, which means that

(1) The first k rows of B are nonzero, and the last $m-k$ rows of B are zero, for some $k \leq m$, and

(2) The first nonzero entry in each nonzero row is 1, and it occurs in a column to the right of the first nonzero entry in the preceding row.

Thus the four-by-six matrix B, shown below, is in echelon form:

$$B = \begin{pmatrix} 1^* & 2 & 3 & 1 & 0 & 1 \\ 0 & 0 & 0 & 1^* & 5 & 2 \\ 0 & 0 & 0 & 0 & 1^* & -1 \\ 0 & 0 & 0 & 0 & 0 & 0 \end{pmatrix}.$$

Now consider the first nonzero entry in each nonzero row, marked with an asterisk. Each such element is the last nonzero element in its column: by using a sequence of additional elementary row operations, we can replace each preceding entry in each such column by zero, obtaining

$$C = \begin{pmatrix} 1^* & 2 & 3 & 0 & 0 & -6 \\ 0 & 0 & 0 & 1^* & 0 & 7 \\ 0 & 0 & 0 & 0 & 1^* & -1 \\ 0 & 0 & 0 & 0 & 0 & 0 \end{pmatrix}.$$

The matrix C is in echelon form and has this additional property:

(3) the first nonzero entry in each nonzero row is the *only* nonzero entry in its column.

Any matrix that possesses Properties 1, 2, and 3 is said to be in *reduced echelon form*. This form results when a system of linear equations is solved by the Gauss-Jordan method of elimination.

> **Theorem 4.9**   For any m-by-n matrix A there exists a finite sequence of elementary matrices $E_1, \ldots, E_r$ such that $E_r E_{r-1}, \ldots, E_2 E_1 A$ is in reduced echelon form.
>
> **Proof**   We have seen that a sequence of elementary row operations transforms A into a matrix in echelon form and that additional elementary row operations then transform that matrix into a matrix in reduced echelon form. Each elementary row operation is performed by multiplying a given matrix on the left by an elementary matrix.

The reduced echelon form of a matrix A is of particular interest because it is *uniquely* determined by A; that means that no matter what sequence of elementary row operations might be used to transform A into reduced echelon form, the resulting reduced echelon matrix will always be the same. By contrast, different matrices in echelon (but not reduced echelon) form can be derived from A by different sequences of elementary row operations. A proof of the uniqueness of the reduced echelon form is a bit technical and will be deferred until Section 6.1. But the idea of the proof is simple: two different matrices in reduced echelon form can be shown to represent two systems of equations whose solutions differ. But each elementary row operation replaces a given system of linear equations by another system having exactly the same solutions. We state this result as a theorem.

> **Theorem 4.10**   For each m-by-n matrix A there is one and only one m-by-n matrix in reduced echelon form that can be obtained by performing a finite number of elementary row operations on A.

This theorem has a practical application to the problem of solving systems of linear equations. Suppose we are given two such systems, represented by two m-by-n matrices A and B. We ask whether the two systems are equivalent; that is, do they have exactly the same solutions? We can transform both A and B separately to matrices E_1 and E_2 in reduced echelon form. If E_1 and E_2 are identical then the systems represented by A and B are equivalent. Otherwise they are not equivalent.

Theorem 4.11   Let B be an m-by-n matrix in reduced echelon form. The maximum number of vectors in a linearly independent set of row vectors of B equals the maximum number of vectors in a linearly independent set of column vectors of B.

Proof   For some $k \leq m$ the first k row vectors of B are nonzero and all other row vectors are zero. Each nonzero row vector has 1 as its first nonzero entry and that entry occurs in a position to the right of the first nonzero entry in any preceding row vector. Hence the first k row vectors form a linearly independent set; any larger set of row vectors must contain a zero row vector and therefore be linearly dependent. Now consider the column vectors; those k column vectors that contain the first nonzero entry of the k nonzero rows are the vectors $\{\varepsilon_1, \varepsilon_2, \ldots, \varepsilon_k\}$, the first k vectors of the standard basis for $\mathscr{F}^m$. This set is linearly independent, and every column vector of B is a linear combination of these vectors. Hence it is a basis for the subspace of $\mathscr{F}^m$ that is spanned by the column vectors, and therefore is a maximal linearly independent subset of the column vectors of B.

We observe further that if the reduced echelon form of A has exactly k nonzero rows, then A itself has a linearly independent set of k row vectors, and any set of $k + 1$ row vectors of A is linearly dependent. To see this we examine the effect of each elementary row operation on the set $\{\mathbf{R}_1, \ldots, \mathbf{R}_m\}$ of row vectors of A. P_{ij} simply permutes these vectors and hence does not affect linear independence. $M_i(c)$ replaces row i by a nonzero multiple of itself, which again has no effect on linear independence. $R_{i,\, i+cj}$ replaces row i by the sum of row i and c times row j. $\{\mathbf{R}_1, \ldots, \mathbf{R}_{i-1}, \mathbf{R}_i + c\mathbf{R}_j, \mathbf{R}_{i+1}, \ldots, \mathbf{R}_m\}$ spans the same subspace of $\mathscr{F}^n$ as does $\{\mathbf{R}_1, \ldots, \mathbf{R}_m\}$, and the dimension of that space is the number of vectors in a maximal linearly independent subset of each set. To summarize, the row vectors of A and the row vectors of the reduced echelon form of A span precisely the same subspace of $\mathscr{F}^n$. We shall use these observations in the next section.

Example   We illustrate a technique for finding the reduced echelon form of a given matrix, using notation that records each elementary row operation as it occurs. Given

$$A = \begin{pmatrix} 3 & -6 & -2 & 0 & 5 \\ 2 & -4 & 0 & 4 & 2 \\ 1 & -2 & -1 & -1 & 2 \end{pmatrix},$$

there are many ways to proceed. To minimize the possibility of arithmetic errors we try to keep the numbers simple; to make it possible for
us to retrace a sequence of row operations, we carefully record each
operation when it is performed. For example, suppose we first interchange the first row and the third row to bring 1 into the (1, 1) position.
Then it is easy to obtain a unit vector in column one. This and similar
steps are recorded below. Because row equivalent matrices are not
necessarily equal, we avoid using the equality symbol between successive matrices, and instead use the symbol $\sim$, with the row operation
being performed written below it.

$$A \underset{(R_1 \leftrightarrow R_3)}{\sim} \begin{pmatrix} 1 & -2 & -1 & -1 & 2 \\ 2 & -4 & 0 & 4 & 2 \\ 3 & -6 & -2 & 0 & 5 \end{pmatrix}$$

$$\underset{(R_2 - 2R_1)}{\sim} \begin{pmatrix} 1 & -2 & -1 & -1 & 2 \\ 0 & 0 & 2 & 6 & -2 \\ 3 & -6 & -2 & 0 & 5 \end{pmatrix}$$

$$\underset{(R_3 - 3R_1)}{\sim} \begin{pmatrix} 1 & -2 & -1 & -1 & 2 \\ 0 & 0 & 2 & 6 & -2 \\ 0 & 0 & 1 & 3 & -1 \end{pmatrix}$$

$$\underset{(R_2 \leftrightarrow R_3)}{\sim} \begin{pmatrix} 1 & -2 & -1 & -1 & 2 \\ 0 & 0 & 1 & 3 & -1 \\ 0 & 0 & 2 & 6 & -2 \end{pmatrix}$$

$$\underset{(R_3 - 2R_2)}{\sim} \begin{pmatrix} 1 & -2 & -1 & -1 & 2 \\ 0 & 0 & 1 & 3 & -1 \\ 0 & 0 & 0 & 0 & 0 \end{pmatrix}$$

$$\underset{(R_1 + R_2)}{\sim} \begin{pmatrix} 1 & -2 & 0 & 2 & 1 \\ 0 & 0 & 1 & 3 & -1 \\ 0 & 0 & 0 & 0 & 0 \end{pmatrix}.$$

This final matrix is the reduced echelon form of A. As you gain experience with row operations you will be able to reduce the amount of
writing by performing two or possibly three operations in one step. But
such shortcuts are also the most frequent source of errors, in which case
they turn out not to be shortcuts at all.

EXERCISES 4.3

1. Use elementary row operations to determine the reduced echelon form of each of the following matrices. Then from that form write the solution of the nonhomogeneous system of three linear equations in three unknowns that each of the given matrices represents.

(i) $\begin{pmatrix} 1 & 1 & 1 & 0 \\ -1 & 1 & 2 & 1 \\ 1 & 1 & 4 & 4 \end{pmatrix}$.   (ii) $\begin{pmatrix} 2 & 1 & 5 & 4 \\ 3 & -2 & 2 & 2 \\ 5 & -8 & -4 & 1 \end{pmatrix}$.

2. Solve each of the following systems by first finding the reduced echelon form of the corresponding matrix. Are any of the systems equivalent?

(i) $5x_1 + 3x_2 = 8,$
       $3x_1 + x_2 = 4,$
       $-x_1 + 3x_2 = 2.$

(ii) $-x_1 - 3x_2 = -2,$
       $5x_1 + 7x_2 = 2,$
       $-3x_1 + x_2 = 4.$

(iii) $-x_1 + 5x_2 = 6,$
       $x_1 + 2x_2 = 1,$
       $-x_1 - 3x_2 = -2.$

3. Show that $U_{ih} U_{kj} = \delta_{hk} U_{ij}$ where δ_{hk} is the Kronecker delta (defined in Section 4.2). Deduce from this that U_{ij} is idempotent when $j = i$, but U_{ij} is nilpotent of index 2 when $i \neq j$.

4. Prove Statements (a), (c), and (e) of Theorem 4.7.

5. Prove Statements (b), (d), and (f) of Theorem 4.7.

6. Calculate the transpose of each type of elementary matrix and observe that it is again an elementary matrix of the same type.

7. For each type of elementary matrix E verify that E^t performs the same operations on the columns of A as E performs on corresponding rows.

8. Show that P_{ij} can be expressed as a product of elementary matrices of the form $R_{r, r+cs}$ and $M_t(c)$.

4.4 RANK OF A MATRIX

Let A be an m-by-n matrix, interpreted as a linear mapping $\mathbf{T}$ from $\mathscr{V}_n$ to $\mathscr{W}_m$, relative to a basis $\{\alpha_1, \ldots, \alpha_n\}$ for $\mathscr{V}_n$ and a basis $\{\beta_1, \ldots, \beta_m\}$ for $\mathscr{W}_m$. For $j = 1, \ldots, n$, column j of A is the m-tuple of scalars that represents $\mathbf{T}(\alpha_j)$ relative to the β-basis. The rank of $\mathbf{T}$ was defined to be the dimension of the range space $\mathscr{R}(\mathbf{T})$, which is the space spanned by the vectors $\mathbf{T}\alpha_1, \ldots, \mathbf{T}\alpha_n$ (the column vectors of A). Hence the number $r(\mathbf{T})$ specifies the number of vectors in any maximal linearly independent subset of column vectors of A. If we choose a different pair of bases, A will determine a different linear mapping $\mathbf{T}_1$ from $\mathscr{V}_n$ to $\mathscr{W}_m$, but $r(\mathbf{T}_1)$ will be the same as $r(\mathbf{T})$ because each of these numbers is the dimension of the space spanned by the columns of A. This discussion motivates the following definition and establishes Theorem 4.12.

Definition 4.5   The *rank* $r(A)$ of an m-by-n matrix A is the maximum number of linearly independent column vectors of A.

Theorem 4.12   If $\mathbf{T}$ is any linear mapping represented by a matrix A, then $r(\mathbf{T}) = r(A)$.

Having established that the rank of a matrix A is the rank of any linear mapping that is represented by A, we can use the isomorphism between linear mappings and matrices to deduce for the rank of matrices some information that we already know for the rank of linear mappings. One distinction that we should point out concerns the two meanings of nonsingularity. A linear mapping $\mathbf{T}$ from $\mathscr{V}_n$ into $\mathscr{W}_m$ is nonsingular if and only if $r(\mathbf{T}) = n$, whereas nonsingularity of matrices was defined only for n-by-n matrices. This discrepancy can be removed by regarding a nonsingular mapping $\mathbf{T}$ as being from $\mathscr{V}_n$ onto $\mathscr{R}(\mathbf{T})$, where $\mathscr{R}(\mathbf{T})$ is an n-dimensional subspace of $\mathscr{W}_m$. Then $\mathbf{T}$ can be represented by a square matrix that is nonsingular.

Theorem 4.13   Let A be an m-by-n matrix, let P be a nonsingular m-by-m matrix, and let Q be a nonsingular n-by-n matrix. Then

$$r(PAQ) = r(AQ) = r(PA) = r(A).$$

Proof   This is simply a restatement of Theorem 3.9 for matrices. It shows that the rank of a matrix is unchanged by multiplying that matrix on either side by a nonsingular matrix.

Theorem 4.14   The maximum number of linearly independent row vectors of an *m*-by-*n* matrix *A* equals the maximum number of linearly independent column vectors of *A*.

Proof   Let $s(A)$ be the maximum number of linearly independent row vectors of *A*, and let *E* be the reduced echelon form of *A*. By Theorem 4.11 $s(E) = r(E)$, and by the argument following the proof Theorem 4.11, $s(E) = s(A)$. Then by Theorem 4.13, $r(E) = r(A)$ because $E = PA$, where *P* is a product of the elementary matrices that transforms *A* into *E*. Because each elementary matrix is nonsingular, and a product of nonsingular matrices is nonsingular, *P* is nonsingular. Hence $s(A) = r(A)$, as claimed.

Theorem 4.15   For any *m*-by-*n* matrix *A*, $r(A) = r(A^t)$.

Proof   The rank of A^t is the maximum number of linearly independent columns of A^t. But the columns of A^t are the rows of *A*. Theorem 4.14 then provides the desired conclusion.

Theorem 4.16   The following statements are equivalent for any *n*-by-*n* matrix *A*:

(a)   *A* is nonsingular.

(b) $r(A) = n$.

(c)   The column vectors of *A* are linearly independent.

(d)   The row vectors of *A* are linearly independent.

(e)   The reduced echelon form of *A* is I_n.

(f)   *A* is the product of elementary matrices.

Proof   Statements (a) and (b) are equivalent because of Theorem 3.7 (with $\mathscr{W} = \mathscr{V}_n$) and the isomorphism of Theorem 4.3. Statements (b) and (c) are equivalent by Definition 4.5. Statements (c) and (d) are equivalent by Theorem 4.14 (with $m = n$). Now let *E* be the reduced echelon form of *A*, $E = PA$ for some nonsingular *n*-by-*n* matrix *P*. If the *n* row vectors of *A* are linearly independent, so are the *n* row vectors of *E*, so there are no zero rows in *E*. Each nonzero row of *E* has 1 as its first nonzero entry, and it is the only nonzero entry in its column. Hence $E = I_n$. Conversely if $E = I_n$, $n = r(A) = r(E)$, so the *n* row vectors of *A* are linearly independent. Hence (d) and (e) are equivalent. To show that Statements (e) and (f) are equivalent, we first assume that a sequence of elementary row operations transforms *A* into *I*. Then

$$E_k E_{k-1} \cdots E_2 E_1 A = I,$$

where each E_i is elementary and hence nonsingular. Then by multiplying successively on the left by E_k^{-1}, E_{k-1}^{-1}, and so on, we have

$$A = E_1^{-1}E_2^{-1} \cdots E_{k-1}^{-1}E_k^{-1}.$$

Because the inverse of each elementary matrix is an elementary matrix (Theorem 4.8), A is a product of elementary matrices. A similar argument shows that (f) implies (e).

The preceding proof provides a simple computational scheme for calculating the inverse of a nonsingular matrix. Beginning with the equation

$$A = E_1^{-1}E_2^{-1} \cdots E_k^{-1},$$

we use Theorem 4.5 to compute

$$A^{-1} = E_k \cdots E_2 E_1 I.$$

This shows that any sequence of elementary row operations that transforms A to I also transforms I to A^{-1}. Hence we can write I beside A,

$$(I \mid A),$$

to obtain an n-by-$2n$ matrix. We perform row operations on this new matrix to transform A to I, obtaining

$$(B \mid I).$$

Then $B = A^{-1}$. An example will illustrate the method.

Example   To calculate the inverse of

$$A = \begin{pmatrix} 1 & 2 & 3 \\ 2 & 3 & 0 \\ 0 & 1 & 2 \end{pmatrix}$$

we write the block form $(I \mid A)$ and perform on this three-by-six matrix a sequence of row operations that reduces A to I, yielding $(B|I)$. Then $B = A^{-1}$.

$$\begin{pmatrix} 1 & 0 & 0 & \vline & 1 & 2 & 3 \\ 0 & 1 & 0 & \vline & 2 & 3 & 0 \\ 0 & 0 & 1 & \vline & 0 & 1 & 2 \end{pmatrix}$$

$(R_2 - 2R_1)$
$$\begin{pmatrix} 1 & 0 & 0 & \vline & 1 & 2 & 3 \\ -2 & 1 & 0 & \vline & 0 & -1 & -6 \\ 0 & 0 & 1 & \vline & 0 & 1 & 2 \end{pmatrix}$$

$(R_3 + R_2)$
$$\begin{pmatrix} 1 & 0 & 0 & \vline & 1 & 2 & 3 \\ -2 & 1 & 0 & \vline & 0 & -1 & -6 \\ -2 & 1 & 1 & \vline & 0 & 0 & -4 \end{pmatrix}$$

$(-1R_2; -\frac{1}{4}R_3)$
$$\begin{pmatrix} 1 & 0 & 0 & \vline & 1 & 2 & 3 \\ 2 & -1 & 0 & \vline & 0 & 1 & 6 \\ \frac{1}{2} & -\frac{1}{4} & -\frac{1}{4} & \vline & 0 & 0 & 1 \end{pmatrix}$$

$(R_2 - 6R_3)$
$$\begin{pmatrix} & 0 & 0 & \vline & 1 & 2 & 3 \\ -1 & \frac{1}{2} & \frac{3}{2} & \vline & 0 & 1 & 0 \\ \frac{1}{2} & -\frac{1}{4} & -\frac{1}{4} & \vline & 0 & 0 & 1 \end{pmatrix}$$

$(R_1 - 2R_2 - 3R_3)$
$$\begin{pmatrix} \frac{3}{2} & -\frac{1}{4} & -\frac{9}{4} & \vline & 1 & 0 & 0 \\ -1 & \frac{1}{2} & \frac{3}{2} & \vline & 0 & 1 & 0 \\ \frac{1}{2} & -\frac{1}{4} & -\frac{1}{4} & \vline & 0 & 0 & 1 \end{pmatrix}.$$

Thus

$$A^{-1} = \frac{1}{4} \begin{pmatrix} 6 & -1 & -9 \\ -4 & 2 & 6 \\ 2 & -1 & -1 \end{pmatrix}.$$

A word of warning is in order. Although this method for calculating A^{-1} is conceptually simple, quite a few arithmetic computations are involved. Gremlins and human imperfections frequently conspire to introduce errors. These can be reduced by working carefully, one operation at each step. Also you will save time in the long run if you *make a practice of checking your final answer* by multiplying $A^{-1}A$ to obtain I.

EXERCISES 4.4

1. Use the method described in this section to find the inverse of each of the following nonsingular matrices.

(i) $A = \begin{pmatrix} -2 & 1 & 3 \\ 0 & -1 & 1 \\ 1 & 2 & 0 \end{pmatrix}.$

(ii) $B = \begin{pmatrix} 1 & -1 & 1 & -1 \\ 0 & 1 & 0 & 1 \\ 1 & 0 & -1 & 0 \\ 0 & 1 & 0 & -1 \end{pmatrix}.$

(iii) $C = \begin{pmatrix} 2 & -1 & 3 \\ 0 & 1 & 0 \\ 2 & 1 & 1 \end{pmatrix}.$

2. Use elementary row operations to determine the rank of each of the following matrices.

(i) $A = \begin{pmatrix} 3 & 1 & -2 & 4 \\ 2 & 0 & -5 & 1 \\ 1 & -1 & 2 & 6 \end{pmatrix}.$

(ii) $B = \begin{pmatrix} -2 & 1 & 4 & -2 & 3 \\ 1 & -5 & 2 & -3 & -2 \\ -4 & -7 & 16 & -12 & 5 \end{pmatrix}.$

(iii) $C = \begin{pmatrix} 1 & -1 & 2 & 1 \\ 4 & 3 & -1 & 0 \\ -2 & 2 & 1 & 7 \\ 2 & -9 & 3 & -10 \\ 9 & -2 & 4 & -4 \end{pmatrix}.$

3. Use elementary row operations to determine the rank of each of the following matrices and to calculate the inverse of any that is nonsingular.

(i) $A = \begin{pmatrix} 1 & 1 & 2 \\ 1 & 2 & 5 \\ 2 & 1 & 1 \end{pmatrix}.$

(ii) $B = \begin{pmatrix} 4 & 2 & -1 \\ -5 & -3 & 1 \\ 3 & 2 & 0 \end{pmatrix}.$

$$(iii) \quad C = \begin{pmatrix} 1 & -2 & -3 \\ -2 & 0 & 4 \\ 1 & 1 & -1 \end{pmatrix}.$$

4. Prove that an upper triangular matrix is singular if and only if some diagonal entry is zero.

5. If A and B are n-by-n matrices, what statements can you make about $r(A + B)$ and $r(AB)$?

6. Write the details of that part of the proof of Theorem 4.16 that asserts that Statement (f) implies Statement (e).

4.5 GEOMETRY OF LINEAR SYSTEMS

Now that we have developed some basic terminology and notation for vector spaces, linear mappings, and matrices, we have a much more efficient way to look at a linear system and to describe its solution geometrically. As noted at the end of Section 4.1, the linear system

$$
\begin{aligned}
a_{11}x_1 + a_{12}x_2 + \cdots + a_{1n}x_n &= y_1, \\
a_{21}x_1 + a_{22}x_2 + \cdots + a_{2n}x_n &= y_2,
\end{aligned}
$$

(4.1)

$$
a_{m1}x_1 + a_{m2}x_2 + \cdots + a_{mn}x_n = y_m,
$$

can be written as a single equation $AX = Y$, where

$$
A = \begin{pmatrix} a_{11} & a_{12} & \cdots & a_{1n} \\ a_{21} & a_{22} & \cdots & a_{2n} \\ \vdots & \vdots & & \vdots \\ a_{m1} & a_{m2} & \cdots & a_{mn} \end{pmatrix}, \quad X = \begin{pmatrix} x_1 \\ x_2 \\ \vdots \\ x_n \end{pmatrix}, \quad Y = \begin{pmatrix} y_1 \\ y_2 \\ \vdots \\ y_m \end{pmatrix}.
$$

The matrix equation $AX = Y$ can be written in vector notation as $\mathbf{T}\xi = \eta$, where relative to a fixed basis for $\mathscr{V}_n$ and a fixed basis for $\mathscr{W}_m$, ξ in $\mathscr{V}_n$ is represented by X, η in $\mathscr{W}_m$ is represented by Y, and the linear mapping $\mathbf{T}$ from $\mathscr{V}_n$ to $\mathscr{W}_m$ is represented by A.

From Chapter 1 we recall that it is convenient to distinguish two cases; the system (4.1) is

(1) homogeneous if $\eta = 0$,
(2) nonhomogeneous if $\eta \neq 0$.

We now apply vector space methods to describe conditions for the existence of solutions and to characterize the set of solutions. The results will depend upon the three numbers n, $r(A)$, and $r(A \,|\, Y)$, where A is the matrix of coefficients of the system (4.1) and $(A \,|\, Y)$ is the *augmented matrix* with the column vector Y appended to the matrix A as a final column.

Homogeneous case   The solution of the vector equation $\mathbf{T}\xi = 0$ is the null space of $\mathbf{T}$. It is a subspace of $\mathscr{V}_n$ having dimension $n(\mathbf{T}) = n - r(\mathbf{T}) = n - r(A)$. The zero vector is always a solution, called the *trivial* solution. Nontrivial solutions exist if and only if $r(A) < n$. If $n(\mathbf{T}) = 0$, the solution space is the origin $[\theta]$; if $n(\mathbf{T}) = 1$, the solution space is a line through the origin; if $n(\mathbf{T}) = 2$, it is a plane through the origin, and so on.

Nonhomogeneous case   A solution ξ of $\mathbf{T}\xi = \eta$ exists if and only if $\eta \in \mathscr{R}(\mathbf{T})$. In terms of the matrix A, this means that Y is in the subspace of $\mathscr{W}_m$ that is spanned by the column vectors of A; equivalently $r(A \,|\, Y) = r(A)$. Hence a solution exists if and only if the rank of the augmented matrix $(A \,|\, Y)$ equals the rank of the coefficient matrix A. That equality is called the *consistency condition*, and a system that satisfies it is said to be *consistent*, meaning that at least one solution exists. For a consistent system let ξ_0 be a known solution. For any solution ξ, $\mathbf{T}(\xi - \xi_0) = \mathbf{T}\xi - \mathbf{T}\xi_0 = \eta - \eta = 0$. Hence $\xi - \xi_0 \in \mathscr{N}(\mathbf{T})$. Conversely let ξ_0 be a known solution and let ξ be such that $\xi - \xi_0 \in \mathscr{N}(\mathbf{T})$. Then $\mathbf{T}\xi = \mathbf{T}(\xi - \xi_0 + \xi_0) = \mathbf{T}(\xi - \xi_0) + \mathbf{T}\xi_0 = 0 + \eta = \eta$. Hence the solution of $\mathbf{T}\xi = \eta$ is the set of vectors

$$\xi_0 + \mathscr{N}(\mathbf{T}) = \{\xi_0 + v \,|\, v \in \mathscr{N}(\mathbf{T})\},$$

where ξ_0 is any known solution. Because $\mathscr{N}(\mathbf{T})$ is a subspace of $\mathscr{V}_n$, the solution can be described as a translation of the subspace $\mathscr{N}(\mathbf{T})$ by the vector ξ_0. In summary, *if* $r(A \,|\, Y) > r(A)$, *the solution of* $\mathbf{T}\xi = \eta \neq 0$ *is the void set (the system is inconsistent). If* $r(A \,|\, Y) = r(A)$, *the system is consistent; the solution is a translated subspace of* $\mathscr{V}_n$, *of dimension* $n(\mathbf{T}) = n - r(A)$. The solution is a point if $n(\mathbf{T}) = 0$, a line if $n(\mathbf{T}) = 1$, a plane if $n(\mathbf{T}) = 2$, and so on.

Finally, if $n = r(A)$ then $\mathcal{N}(\mathbf{T}) = [\theta]$, so the equation $\mathbf{T}\xi = \eta$ has exactly one solution if $\eta \in \mathcal{R}(\mathbf{T})$, but no solution if $\eta \notin \mathcal{R}(\mathbf{T})$. If, further, $m = n = r(A)$, then for each $\eta \in \mathcal{W}_m$ there exists a unique solution, $\xi = \mathbf{T}^{-1}\eta$, or $X = A^{-1}Y$. Conversely, if there exists a solution for each $\eta \in \mathcal{W}_m$, then $\mathcal{R}(\mathbf{T}) = \mathcal{W}_m$, so

$$m = r(\mathbf{T}) = r(A) = r(A \mid Y).$$

If there exists a unique solution for each $\eta \in \mathcal{W}_m$, then $n(\mathbf{T}) = 0$, so $n = r(\mathbf{T}) = r(A) = m$.

We also note that the three numbers n, $r(A)$, and $r(A \mid Y)$ can be determined by transforming $(A \mid Y)$ to echelon form, because n is the number of columns of A, $r(A)$ is the number of nonzero rows in any echelon form for A, and $r(A \mid Y)$ is the number of nonzero rows in any echelon form for $(A \mid Y)$.

Further insight into the geometry of a linear system can be obtained by reconsidering a single linear equation in n variables

$$b_1 x_1 + b_2 x_2 + \cdots + b_n x_n = c.$$

The left-hand side has the form of the dot product in $\mathscr{E}^n$ of the vectors $\beta = (b_1, \ldots, b_n)$ and $\xi = (x_1, \ldots, x_n)$. If $c = 0$ (homogeneous case), the solution consists of all vectors ξ that are orthogonal to β, a subspace of dimension $n - 1$, called a *hyperplane* in $\mathscr{E}^n$. If $c \neq 0$, the solution vectors ξ are of the form $\xi_0 + \gamma$, where $\beta \cdot \xi_0 = c$, and $\beta \cdot \gamma = 0$. Hence the solution set is a translation by some vector ξ_0 of the hyperplane of all vectors orthogonal to β. For a system of m such equations, the solutions set consists of all points of intersection of m hyperplanes. If no point is common to all the m hyperplanes, the system is inconsistent. If at least one point ξ_0 is common to all the m hyperplanes, then the solution set is a translation by ξ_0 of a subspace of dimension $n - r(A)$, where A is the coefficient matrix of the system.

EXERCISES 4.5

1. For each of the following linear systems determine the values of m, n, $r(A)$, and $r(A \mid Y)$. Then solve that system and reconcile the nature of the solution with the values of those numbers.

(i)
$$\begin{aligned}
x_1 - x_2 + x_3 - x_4 + x_5 &= 1, \\
2x_1 - x_2 + 3x_3 \quad\quad + 4x_5 &= 2, \\
3x_1 - 2x_2 + 2x_3 + x_4 + x_5 &= 1, \\
x_1 \quad\quad + x_3 + 2x_4 + x_5 &= 0.
\end{aligned}$$

(ii) $\begin{aligned} x_1 + 2x_2 + \ x_3 &= -1, \\ 6x_1 + \ x_2 + \ x_3 &= -4, \\ 2x_1 - 3x_2 - \ x_3 &= \ 0, \\ -x_1 - 7x_2 - 2x_3 &= \ 7, \\ x_1 - \ x_2 \ &= \ 1. \end{aligned}$

(iii) $\begin{aligned} 2x_1 + \ x_2 + 5x_3 &= 4, \\ 3x_1 - 2x_2 + 2x_3 &= 2, \\ 5x_1 - 8x_2 - 4x_3 &= 1. \end{aligned}$

2. Find a necessary and sufficient condition on $r(A)$ that the system (4.1) will have a solution for all possible choices of Y. Prove your result.

3. Prove that if $m > n = r(A)$, then $AX = Y$ has either no solution or exactly one solution. Describe a computational process to distinguish these cases.

4. What can you deduce about the number of solutions of a system of m linear equations in n variables if $m < n$? Take the consistency condition into account.

5. (i) Describe geometrically the solutions of the single equation

$$a_{i1}x_1 + a_{i2}x_2 + a_{i3}x_3 = y_i.$$

Distinguish the cases $y_i = 0$ and $y_i \neq 0$.

(ii) Describe geometrically the solutions of the system of two equations ($i = 1, 2$) of the type in (i). Must solutions exist? Discuss fully.

(iii) Given three such equations ($i = 1, 2, 3$), discuss the geometric meaning of $r(A) = 1, 2, 3$. Include in your discussion both the consistency and nonconsistency of the system for each value of $r(A)$.

4.6 BLOCK MULTIPLICATION OF MATRICES

Although the primary emphasis in this book is on the theory of matrices rather than on the practical problems that arise in applications, it would be misleading to pretend that such problems do not exist. For example, the form of the product of two matrices may be unfamiliar to a beginner, but it is conceptually simple. In the product of an m-by-n matrix and an n-by-p matrix there are mp terms to be calculated, and each term requires n binary products and $n - 1$ sums. Hence, there are altogether mpn products and $mp(n - 1)$ sums to be performed. For square matrices, this reduces to n^3 products and $n^3 - n^2$ sums.

In matrices that arise from experimental work the individual entries are decimal numbers, so that multiplication is considerably more tedious than addition. For this reason the amount of work required for a matrix calculation usually is expressed in terms of the number of multiplications involved. Because the product of two *n*-by-*n* matrices requires n^3 multiplications, it is clear that a tremendous amount of computation is required when *n* is large. Even for $n = 5$ the work is sufficiently long to discourage mental computation. The development of high-speed computers has reduced this problem considerably and thereby has opened to solution by matrix methods many applied problems for which theoretical solutions previously were known but were computationally unfeasible. But even a large electronic computer has a limited storage space, and the practical question of computational technique remains.

We now indicate a device, known as *block multiplication* of matrices, that can be used to decompose the product of two large matrices into numerous products of smaller matrices. Let *A* be *m*-by-*n* and *B* be *n*-by-*p*. Write $n = n_1 + n_2 + \cdots + n_k$, where each n_i is a positive integer; partition the *columns* of *A* by putting the first n_1 columns in the first block, the next n_2 columns in the second block, and so on. Partition the *rows* of *B* in *exactly the same way*. Then

$$A = (A_1 \,|\, A_2 \,|\, \cdots \,|\, A_k), \qquad B = \begin{pmatrix} B_1 \\ \hline B_2 \\ \hline \cdot \\ \cdot \\ \cdot \\ \hline B_k \end{pmatrix},$$

where A_i is the *m*-by-n_i matrix consisting of columns of *A* beginning with column $n_1 + \cdots + n_{i-1} + 1$ and ending with column $n_1 + \cdots + n_i$, and where B_j is the n_j-by-*p* matrix consisting of rows of *B* beginning with row $n_1 + \cdots + n_{j-1} + 1$ and ending with row $n_1 + \cdots + n_j$. Then the method of block multiplication asserts that

$$AB = A_1 B_1 + A_2 B_2 + \cdots + A_k B_k.$$

More generally, suppose that having partitioned the columns of *A* and the rows of *B* as described above, we partition the *rows* of *A* in any manner and the *columns* of B in any manner. We obtain

$$
A = \begin{pmatrix} A_{11} & A_{12} & \cdots & A_{1k} \\ A_{21} & A_{22} & \cdots & A_{2k} \\ \cdot & \cdot & & \cdot \\ \cdot & \cdot & & \cdot \\ \cdot & \cdot & & \cdot \\ A_{r1} & A_{r2} & \cdots & A_{rk} \end{pmatrix}, \qquad B = \begin{pmatrix} B_{11} & B_{12} & \cdots & B_{1s} \\ B_{21} & B_{22} & \cdots & B_{2s} \\ \cdot & \cdot & & \cdot \\ \cdot & \cdot & & \cdot \\ \cdot & \cdot & & \cdot \\ B_{k1} & B_{k2} & \cdots & B_{ks} \end{pmatrix},
$$

where A_{it} is a matrix (rectangular array) having r_i rows and n_t columns and B_{tj} is a matrix having n_t rows and s_j columns. Then for fixed i, j the product $A_{it} B_{tj}$ is defined and yields an r_i-by-s_j matrix; therefore $\sum_{t=1}^{k} A_{it} B_{tj}$ is an r_i-by-s_j matrix. The method of block multiplication asserts that

$$
AB = \begin{pmatrix} C_{11} & C_{12} & \cdots & C_{1s} \\ C_{21} & C_{22} & \cdots & C_{2s} \\ \cdot & \cdot & & \cdot \\ \cdot & \cdot & & \cdot \\ \cdot & \cdot & & \cdot \\ C_{r1} & C_{r2} & \cdots & C_{rs} \end{pmatrix},
$$

where $C_{ij} = \sum_{t=1}^{k} A_{it} B_{tj}$.

This is in the same form as the element-by-element definition of the product of matrices in which each element is considered as a one-by-one block. The important thing to remember in block multiplication of AB is that the column partition of A must coincide with the row partition of B so that all the matrix products $A_{it} B_{tj}$ are defined. Because matrix multiplication is noncommutative, it is essential that the proper order be maintained in forming products of blocks.

The proof of this result is not difficult, but it does require care in choosing notation and manipulating indices. Because we do not require the result for the development of theory, a general proof is omitted. To understand the application of block multiplication to the problem of large-scale computations, consider two 50-by-50 matrices. There are 2,500 elements in each matrix; the multiplication of two such matrices requires 125,000 multiplications and almost as many additions. One method of performing such calculations on a computer whose storage capacity is exceeded by the magnitude of the problem would be to partition each of the matrices into smaller matrices, perhaps into four 25-by-25 blocks,

$$
A = \left(\begin{array}{c|c} A_{11} & A_{12} \\ \hline A_{21} & A_{22} \end{array} \right), \qquad B = \left(\begin{array}{c|c} B_{11} & B_{12} \\ \hline B_{21} & B_{22} \end{array} \right).
$$

If the blocks are suitably small, the machine can successively compute the products $A_{11}B_{11}$, $A_{11}B_{12}$, $A_{12}B_{21}$, $A_{12}B_{22}$, and so on, record the results on tape or punched cards to clear the machine storage for the next block of calculations, and finally compute $C_{11} = A_{11}B_{11} + A_{12}B_{21}$, and so on.

Even when the matrices are small enough not to strain the storage capacity of the computer (machine or human), block multiplication with two-by-two matrices is very efficient computationally, requiring fewer arithmetic operations than entry-by-entry multiplication. Block multiplication also is useful in case special patterns appear in the matrix. For example, let

$$A = \left(\begin{array}{cc|ccc} 2 & 0 & 0 & 0 & 0 \\ 0 & 2 & 0 & 0 & 0 \\ \hline 1 & 0 & a & b & c \\ 0 & 1 & d & e & f \end{array}\right) = \left(\begin{array}{c|c} 2I & Z \\ \hline I & A_0 \end{array}\right),$$

and let

$$B = \left(\frac{B_0}{C_0}\right),$$

where B_0 has two rows and C_0 has three rows. Then

$$AB = \left(\frac{2B_0}{B_0 + A_0 C_0}\right).$$

Hence the only nontrivial computation required for AB is the product of the two-by-three matrix A_0 with the three-by-p matrix C_0.

To observe that it is sometimes possible to introduce convenient patterns in a matrix by judicious selection of bases, consider a linear mapping **T** from $\mathscr{V}_n$ to $\mathscr{W}_m$. Choose a basis for $\mathscr{N}(\mathbf{T})$ and extend it to a basis $\{\alpha_1, \dots, \alpha_n\}$ for $\mathscr{V}_n$, numbered so that the basis $\{\alpha_{r+1}, \dots, \alpha_n\}$ for $\mathscr{N}(\mathbf{T})$ appears last. Now in $\mathscr{W}_m$ choose any basis $\{\beta_1, \dots, \beta_r\}$ for $\mathscr{R}(\mathbf{T})$ and extend it to a basis $\{\beta_1, \dots, \beta_m\}$ for $\mathscr{W}_m$. Then relative to these bases **T** is represented by an m-by-n block matrix

$$A = \left(\begin{array}{c|c} B & Z_1 \\ \hline Z_2 & Z_3 \end{array}\right),$$

where B is r-by-r, and Z_1, Z_2, Z_3 are zero matrices of suitable dimensions.

In particular, **T** is nonsingular if and only if $r = n$. In that case we have

$$A = \left(\frac{B}{Z_2}\right).$$

The n-by-n matrix B is nonsingular because its n columns represent the linearly independent vectors $T\alpha_i$, $i = 1, \ldots, n$. Let C be the n-by-m matrix defined by

$$C = (B^{-1} | Z),$$

where Z is the n-by-$(m - n)$ zero matrix. Then we compute

$$CA = I_n,$$

$$AC = \left(\begin{array}{c|c} I_n & Z \\ \hline Z & Z \end{array} \right).$$

These calculations should help to make clear the distinction between our definition of nonsingular linear mappings and that of nonsingular matrices. If T is a nonsingular linear mapping from $\mathscr{V}_n$ to $\mathscr{W}_m$, then $m \geq n$, and T can be represented by a block column consisting of an n-by-n block which is a nonsingular matrix and an $(m - n)$-by-n block of zeros.

EXERCISES 4.6

1. Prove the first of the two assertions in the text concerning block multiplication: If $A = (A_1 | \cdots | A_k)$ and

$$B = \left(\begin{array}{c} B_1 \\ \hline \cdot \\ \cdot \\ \cdot \\ \hline B_k \end{array} \right),$$

then $AB = A_1 B_1 + \cdots + A_k B_k$.

2. Calculate AB in three ways: directly without partition; with the partition indicated; with a different partition of your own choosing.

$$A = \left(\begin{array}{cc|c|cc} 2 & 3 & 4 & 0 & 0 \\ 3 & 1 & 0 & 0 & 0 \\ \hline 1 & 0 & 1 & 0 & 4 \\ \hline -1 & 0 & 0 & 1 & 0 \\ 0 & -1 & 4 & 0 & 1 \end{array} \right).$$

$$B = \left(\begin{array}{c|cc} 1 & 0 & 0 \\ 3 & 0 & 0 \\ \hline 0 & 2 & 1 \\ \hline -1 & 0 & 0 \\ -1 & 0 & 0 \end{array}\right).$$

3. Suppose an n-by-n matrix A is of the form

$$A = \left(\begin{array}{c|c} A_1 & A_2 \\ \hline Z & A_4 \end{array}\right),$$

where Z is an $(n-k)$-by-k block of zeros.

(i) Consider the linear transformation **T** determined by A relative to a chosen basis $\{\alpha_1, \ldots, \alpha_n\}$. What is the geometric meaning of the block of zeros?

(ii) Suppose an n-by-n matrix B is of the same form described above for the matrix A. Prove by block multiplication that AB has this same property.

(iii) Prove the result of (ii) by a geometric argument.

4. An n-by-n matrix A of the form

$$A = \left(\begin{array}{c|c} A_1 & Z \\ \hline Z & A_4 \end{array}\right),$$

where A_1 and A_4 are square matrices along the main diagonal of A and Z denotes a matrix of zeros, is said to be in *block diagonal* form. If A and B are n-by-n matrices in block diagonal form and if A_1 and B_1 are of the same size, show that AB also has the same form.

CHAPTER 5
DETERMINANTS

5.1 BASIC PROPERTIES
OF DETERMINANTS

It is quite likely that you encountered determinants during secondary school in connection with the problem of solving a system of n linear equations in n variables, at least for the case $n = 2$ and perhaps for $n = 3$. As we shall see, the determinant of an n-by-n matrix exists for any n. But as a computational method, determinants become increasingly inefficient as n increases, and their principal importance stems from their use as a theoretical tool and a notational convenience in subjects such as linear algebra and multivariable calculus. Determinants also have a very simple and important geometric meaning, as we shall now illustrate.

Although the case $n = 1$ is trivial, it is the obvious starting point. The determinant of a one-by-one matrix is defined by

$$\det(a) = a.$$

We observe two key facts, one algebraic and one geometric. First, the linear equation $ax = b$ has a unique solution if and only if $\det(a) \neq 0$. Second, the length (one-dimensional geometric content) of the vector $\alpha = a\varepsilon_1$ is $|\det(a)|$.

For $n = 2$ the determinant of a two-by-two matrix is defined by

$$\det\begin{pmatrix} a_{11} & a_{12} \\ a_{21} & a_{22} \end{pmatrix} = a_{11}a_{22} - a_{21}a_{12}.$$

It is easy to verify that the linear system

$$a_{11}x + a_{12}y = e$$
$$a_{21}x + a_{22}y = f$$

has a unique solution if and only if $\det A \neq 0$, where A is the coefficient matrix of this system. It is also easy to verify that $|\det A|$ is the *area* (two-dimensional geometric content) of the parallelogram having the column vectors of A as adjacent edges (see Figure 5.1).

For $n = 3$ the determinant of a three-by-three matrix is defined by

$$\det A = \det\begin{pmatrix} a_{11} & a_{12} & a_{13} \\ a_{21} & a_{22} & a_{23} \\ a_{31} & a_{32} & a_{33} \end{pmatrix} = \begin{array}{l} a_{11}a_{22}a_{33} + a_{21}a_{32}a_{13} + a_{31}a_{12}a_{23} \\ - a_{11}a_{32}a_{23} - a_{21}a_{12}a_{33} - a_{31}a_{22}a_{13}. \end{array}$$

By transforming the coefficient matrix of the linear system

$$a_{11}x + a_{12}y + a_{13}z = d_1,$$
$$a_{21}x + a_{22}y + a_{22}z = d_2,$$
$$a_{31}x + a_{32}y + a_{33}z = d_3,$$

to echelon form, it can be verified that the system has a unique solution if and only if $\det A \neq 0$. It can also be verified that $|\det A|$ is the *volume* (three-dimensional geometric content) of the parallelepiped having the column vectors of A as mutually adjacent edges. Obviously the computation required to verify these statements for $n = 3$ is much longer than that required to verify the corresponding statements for $n = 2$. It is equally obvious that a computational definition of $\det A$ for an n-by-n matrix will be quite awkward, so we shall take a different approach.

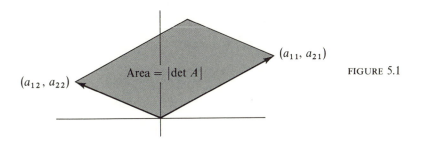

FIGURE 5.1

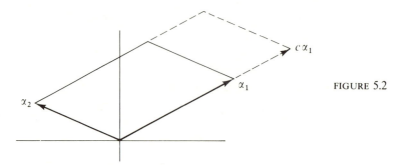

FIGURE 5.2

We wish to extract from these examples a few properties to guide us in formulating a definition for det A, where A is any square matrix. To begin with, a determinant of order n is a scalar-valued function defined on the set of all n-by-n matrices: for each A in $\mathcal{M}_{n \times n}$, det A is a scalar. If $\{\alpha_1, \ldots, \alpha_n\}$ are the column vectors of A, then the absolute value of det A is the n-dimensional geometric content of the parallelotope having the vectors $\alpha_1, \ldots, \alpha_n$ as mutually adjacent edges. Hence the following properties are reasonable expectations:

(1) Multiplying any single edge of a parallelotope by a scalar c should multiply the n-dimensional geometric content by $|c|$. See Figure 5.2. Hence for each $i = 1, \ldots, n$ and each c,

$$\det(\alpha_1, \ldots, c\alpha_i, \ldots, \alpha_n) = \pm c \, \det(\alpha_1, \ldots, \alpha_i, \ldots, \alpha_n).$$

(2) The area of the parallelogram having $\{\alpha_i, \alpha_j\}$ as adjacent sides is the same as the area of the parallelogram having $\{\alpha_i + c\alpha_j, \alpha_j\}$ as adjacent sides. See Figure 5.3. Hence the n-dimensional geometric content of the parallelotope having

$$\{\alpha_1, \ldots, \alpha_i, \ldots, \alpha_j, \ldots, \alpha_n\}$$

as adjacent sides should equal the n-dimensional geometric content of the parallelotope having $\{\alpha_1, \ldots, \alpha_i + c\alpha_j, \ldots, \alpha_j, \ldots, \alpha_n\}$. Hence for each c and for all $i, j = 1, \ldots, n$ such that $j \neq i$,

$$\det(\alpha_1, \ldots, \alpha_i, \ldots, \alpha_j, \ldots, \alpha_n)$$
$$= \pm \det(\alpha_1, \ldots, \alpha_i + c\alpha_j, \ldots, \alpha_j, \ldots, \alpha_n).$$

(3) Because the n-dimensional content of the unit n-cube is one,

$$\det(\varepsilon_1, \ldots, \varepsilon_n) = \pm 1.$$

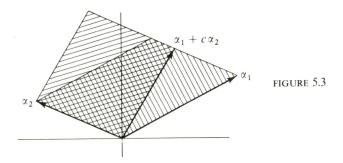

$\alpha_1 + c\alpha_2$

α_1

α_2

FIGURE 5.3

Properties 1 and 2 are reminiscent of elementary column operations on matrices. Property 3 is needed only as a scaling factor to make the value of $|\det A|$ coincide with the standard unit of geometric content in $\mathscr{E}^n$.

Definition 5.1 A *determinant* is a function, denoted det, that assigns to each *n-by-n* matrix A having column vectors $A_1, \ldots, A_n$ a scalar value det A that has the following three properties: For each scalar c and each $i = 1, \ldots, n$

(1) $\det(A_1, \ldots, cA_i, \ldots, A_n) = c \det(A_1, \ldots, A_i, \ldots, A_n)$,

(2) $\det(A_1, \ldots, A_i, \ldots, A_j, \ldots, A_n)$
$= \det(A_1, \ldots, A_i + cA_j, \ldots, A_j, \ldots, A_n)$ for each $j \neq i$,

(3) $\det(I) = 1$.

Observe that there is no a priori guarantee that at least one such function exists for each n. However, the previous examples of det for $n = 1, 2,$ and 3 encourage us to proceed to investigate the properties of *n-by-n* determinants on the tentative assumption that at least one such function exists for each n.

Theorem 5.1 If det is a function having Properties 1 and 2 of Definition 5.1, then the following statements hold for each $i, j = 1, \ldots, n$ such that $j \neq i$.

(a) $\det(A_1, \ldots, A_i, \ldots, A_j, \ldots, A_n) = -\det(A_1, \ldots, A_j, \ldots, A_i, \ldots, A_n)$. In words, *the interchange of any two columns of A reverses the sign of the determinant*; or, det is an *alternating* function of the columns of A.

(b) *If the columns of A are linearly dependent, then* det $A = 0$.

(c) *If the columns of B are a permutation of the columns of A, then* det $B = \pm\det A$, where the plus sign applies if that permutation can be performed by an *even* number of transpositions (interchanges of pairs of columns), and the minus sign applies if the permutation can be performed by an *odd* number of transpositions.

(d) $\det(A_1, \ldots, B_i + C_i, \ldots, A_n) = \det(A_1, \ldots, B_i, \ldots, A_n)$
$$+ \det(A_1, \ldots, C_i, \ldots, A_n).$$

In words, if column i is expressed as the sum of two column vectors, $A_i = B_i + C_i$, then det A is the sum of the two indicated determinants; or det A is *an additive function of each column.*

Proof of (a)   We use Property 2 with $c = 1$, then with $c = -1$, then with $c = 1$, to obtain

$$\det A = \det(A_1, \ldots, A_i, \ldots, A_j, \ldots, A_n)$$
$$= \det(A_1, \ldots, A_i + A_j, \ldots, A_j, \ldots, A_n)$$
$$= \det(A_1, \ldots, A_i + A_j, \ldots, A_j - A_i - A_j, \ldots, A_n)$$
$$= \det(A_1, \ldots, A_i + A_j - A_i, \ldots, -A_i, \ldots, A_n)$$
$$= \det(A_1, \ldots, A_j, \ldots, -A_i, \ldots, A_n)$$
$$= -\det(A_1, \ldots, A_j, \ldots, A_i, \ldots, A_n),$$

where the last equality results from Property 1.

Proof of (b)   If the columns of A are linearly dependent, some column can be expressed as a linear combination of the other columns. By using Properties 1 and 2, that column can be replaced by a column of zeros, and then Property 1 shows that det $A = 0$.

Proof of (c)   Statement (a) tells us that each interchange of a pair of columns merely changes the sign of the determinant. An even number of interchanges of pairs of columns therefore leaves the determinant unchanged, whereas an odd number changes the sign. (Later we shall prove that a determinant function exists; then (c) will show that any permutation can be classified either *even* or *odd*, according to whether that permutation can be expressed as an even number of interchanges of pairs or an odd number of interchanges of pairs.)

Proof of (d)   In view of Statement (a) we need only prove the case where $i = 1$. (Be sure you understand why.) Hence consider $\det(B_1 + C_1, A_2, \ldots, A_n)$. If the columns $A_2, \ldots, A_n$ are linearly dependent, then $\det(X, A_2, \ldots, A_n) = 0$ for any X, so (d) is valid. When

$\{A_2, \ldots, A_n\}$ is linearly independent, we can choose a vector A_1 such that $\{A_1, A_2, \ldots, A_n\}$ is a basis for the space of n-tuples. Then

$$B_1 = \sum_{i=1}^{n} b_i A_i \text{ and } C_1 = \sum_{i=1}^{n} c_i A_i.$$

Then from Properties 1 and 2

$$\det(B_1 + C_1, A_2, \ldots, A_n) = \det\left(\sum_{i=1}^{n} (b_i + c_i)A_i, A_2, \ldots, A_n\right)$$

$$= \det((b_1 + c_1)A_1, A_2, \ldots, A_n)$$

$$= (b_1 + c_1)\det(A_1, A_2, \ldots, A_n)$$

$$= b_1 \det(A_1, A_2, \ldots, A_n) + c_1 \det(A_1, A_2, \ldots, A_n)$$

$$= \det(b_1 A_1, A_2, \ldots, A_n) + \det(c_1 A_1, A_2, \ldots, A_n)$$

$$= \det\left(\sum_{i=1}^{n} b_i A_i, A_2, \ldots, A_n\right) + \det\left(\sum_{i=1}^{n} c_i A_i, A_2, \ldots, A_n\right)$$

$$= \det(B_1, A_2, \ldots, A_n) + \det(C_1, A_2, \ldots, A_n).$$

Property 1 of Definition 5.1 and Statement (d) of Theorem 5.1 show that det is a linear function of each of the n column vectors and hence is called a *multilinear* or *n-linear* function. Together with Statement (a) and Property 3 this means that det is an alternating, multilinear function of the column vectors having the property that det $I = 1$. Actually those three properties characterize det; that is, we could *define* det to be an alternating, multilinear function of the column vectors such that det $I = 1$, and then we could prove the three statements of Definition 5.1. We preferred to start with Definition 5.1 because of the geometric significance of these statements. There are numerous other ways to define det.

Properties 1 and 2 of Definition 5.1 and Statement (a) of Theorem 5.1 describe the effect of elementary column operations on a determinant:

$M_i(c)$ multiplies the determinant by c,

$R_{i, i+cj}$ leaves the determinant unchanged,

P_{ij} changes the sign of the determinant.

EXERCISES 5.1

1. Show that in $\mathscr{E}^2$ the absolute value of det A is the area of the parallelogram having the column vectors of A as adjacent edges.

2. Consider the linear system

$$a_{11}x + a_{12}y = e,$$

$$a_{21}x + a_{22}y = f.$$

(i) Show that the consistency condition can be expressed as follows:

$$\text{if } \det \begin{pmatrix} a_{11} & a_{12} \\ a_{21} & a_{22} \end{pmatrix} = 0, \text{ then } \det \begin{pmatrix} a_{11} & e \\ a_{21} & f \end{pmatrix} = 0.$$

(ii) Assuming that a unique solution exists, express that solution in determinant form.

3. In $\mathscr{E}^3$ the *cross product* (or *vector product*) $\alpha \times \beta$ of two vectors $\alpha = a_1 \varepsilon_1 + a_2 \varepsilon_2 + a_3 \varepsilon_3$ and $\beta = b_1 \varepsilon_1 + b_2 \varepsilon_2 + b_3 \varepsilon_3$ is defined to be the vector

$$\alpha \times \beta = (a_2 b_3 - a_3 b_2)\varepsilon_1 + (a_3 b_1 - a_1 b_3)\varepsilon_2 + (a_1 b_2 - a_2 b_1)\varepsilon_3.$$

(i) Show that $\alpha \times \beta$ is orthogonal to both α and β.

(ii) Show that $\|\alpha \times \beta\| = \|\alpha\|\|\beta\| \sin \Psi(\alpha, \beta)$, where $\Psi(\alpha, \beta)$ is the angle between α and β.

4. Referring to Exercise 3, show that if $\gamma = c_1 \varepsilon_1 + c_2 \varepsilon_2 + c_3 \varepsilon_3$, then the triple scalar product $(\alpha \times \beta) \cdot \gamma$ has the value

$$\det \begin{pmatrix} a_1 & b_1 & c_1 \\ a_2 & b_2 & c_2 \\ a_3 & b_3 & c_3 \end{pmatrix}.$$

(The results of Exercises 3 and 4 provide a method for showing that the absolute value of a three-by-three determinant is the volume of the parallelepiped having the column vectors as adjacent edges.)

5. (i) Show that a Cartesian equation for the line through two points (a, b) and (c, d) in $\mathbb{R}^2$ is

$$\det \begin{pmatrix} x & y & 1 \\ a & b & 1 \\ c & d & 1 \end{pmatrix} = 0.$$

(ii) Generalize this result to obtain a corresponding form for a Cartesian equation in $\mathbb{R}^3$ for the plane through three given points in space.

6. Let $A(x)$ be the three-by-three matrix

$$A(x) = \begin{pmatrix} 3-x & 2 & 2 \\ 1 & 4-x & 1 \\ -2 & -4 & -1-x \end{pmatrix}.$$

(i) Evaluate det $A(x)$.

(ii) For what values of x is det $A(x) = 0$?

(iii) Show that $A(x)$ is singular for each value of x for which det $A(x) = 0$.

7. Give a geometric argument to prove Statement (b) of Theorem 5.1.

8. Let A and B be any two-by-two matrices.

(i) Compute BA.

(ii) Using *only* the properties of determinants listed in Definition 5.1 and Theorem 5.1 calculate $\det(BA)$, writing your answer in the form k det B for some scalar k that is a combination of the entries of A.

(iii) Specialize the result in (ii) by letting $B = I$, thus showing that the specific form stated in the text for two-by-two determinants is actually a consequence of Definition 5.1.

5.2   AN EXPLICIT FORMULA FOR det A

We shall now apply the properties of det A, derived in the previous section, to express det A in terms of the entries of A, from which a number of additional properties will follow. The method we shall use was described for the special case $n = 2$ in Exercise 5.1-8.

Let A and B be any n-by-n matrices, and let $C = BA$. We compute the elements in C_k, column k of C:

$$c_{1k} = b_{11}a_{1k} + b_{12}a_{2k} + \cdots + b_{1n}a_{nk},$$
$$c_{2k} = b_{21}a_{1k} + b_{22}a_{2k} + \cdots + b_{2n}a_{nk},$$
$$\vdots$$
$$c_{nk} = b_{n1}a_{1k} + b_{n2}a_{2k} + \cdots + b_{nn}a_{nk}.$$

Letting B_j denote column j of B, we can write

$$C_k = B_1 a_{1k} + B_2 a_{2k} + \cdots + B_n a_{nk} = \sum_{j=1}^{n} B_j a_{jk}.$$

Hence

$$\det C = \det(C_1, C_2, \ldots, C_n) = \det\left(\sum_{j=1}^{n} B_j a_{j1}, \sum_{j=1}^{n} B_j a_{j2}, \ldots, \sum_{j=1}^{n} B_j a_{jn}\right),$$

where each index of summation runs independently. Because each column is the sum of n column vectors, we use Statement (d) in Theorem 5.1 to expand $\det C$ to a sum of n^n determinants:

$$\det C = \sum \det(B_{j(1)} a_{j(1)1}, B_{j(2)} a_{j(2)2}, \ldots, B_{j(n)} a_{j(n)n}),$$

where the summation is extended over all values from one to n for *each* of the indices $j(i)$ independently. There are n^n such sets of indices, but by Statement (b), Theorem 5.1, whenever $j(i) = j(k)$ for $i \neq k$, the corresponding determinant is zero. Hence the only nonzero determinants in this sum are those for which $j(1), \ldots, j(n)$ are all different; in other words, those determinants in which the n subscripts of the B's form a permutation p of the ordered set $\{1, \ldots, n\}$.

Hence

$$\det C = \sum \det(B_{p(1)} a_{p(1)1}, B_{p(2)} a_{p(2)2}, \ldots, B_{p(n)} a_{p(n)n})$$

$$= \sum a_{p(1)1} a_{p(2)2} \cdots a_{p(n)n} \det(B_{p(1)}, B_{p(2)}, \ldots, B_{p(n)}),$$

because for each j, $a_{p(j)j}$ is a common factor of column j. Here the summation is extended over all of the $n!$ permutations of $\{1, \ldots, n\}$. But for each such permutation p

$$\det(B_{p(1)}, B_{p(2)}, \ldots, B_{p(n)}) = \pm \det B,$$

where the sign, $+$ or $-$, is determined entirely by p. (See Exercise 5.2-10). Then

$$\det C = \det(BA) = \det B \sum_{\text{all } p} \pm a_{p(1)1} a_{p(2)2} \cdots a_{p(n)n}$$

where the sum is extended over all permutations p of $1, \ldots, n$ and where each $\pm$ sign depends only on p. Now *for the first time* we use Property 3 of Definition 5.1. Let $B = I$ so that $C = A$. Then

(5.1) $$\det A = \sum_{\text{all } p} \pm a_{p(1)1} a_{p(2)2} \cdots a_{p(n)n}.$$

We note that this equation expresses $\det A$ as an algebraic sum of $n!$ terms, each term being a signed product of n entries of A. We note further that the column indices of the entries in each product term are $\{1, \ldots, n\}$ and so are the row indices because p is a permutation of $\{1, \ldots, n\}$. And each such permutation p determines one of the two numbers $+1$ or -1, providing the sign indicated in (5.1).

If you are already familiar with permutations you will know that any permutation is a product of transpositions, a transposition being an interchange of two symbols. Furthermore, a permutation p is classified as *even* if an even number of transpositions transforms the ordered set $\{p(1), \ldots, p(n)\}$ into $\{1, \ldots, n\}$; p is *odd* if an odd number of transpositions transforms $\{p(1), \ldots, p(n)\}$ into $\{1, \ldots, n\}$. What must (and can) be proved to make this a useful definition is that no permutation can be both even and odd. Actually that fact can be seen to be a consequence of the existence of a determinant function, which we shall establish by a method that is suggested by, but logically independent of, Formula 5.1. We first state formally the result derived above and several immediate consequences.

Theorem 5.2 Any function det that has the three properties of Definition 5.1 must have the form

$$\det A = \sum_{p} (-1)^q a_{p(1)1} a_{p(2)2} \cdots a_{p(n)n},$$

where the sum is extended over all permutations p of $\{1, \ldots, n\}$ and where q is an integer that is determined by p.

Theorem 5.3 If A and B are n-by-n matrices, then

$$\det(BA) = (\det B)(\det A).$$

Proof Apply Theorem 5.2 to the formula preceding Formula 5.1 in the preceding derivation.

Theorem 5.4 An n-by-n matrix is nonsingular if and only if det $A \neq 0$. If A is nonsingular, $\det(A^{-1}) = (\det A)^{-1}$.
Proof Exercise. Apply Theorem 5.3.

Theorem 5.5 If A^t is the transpose of A, then $\det(A^t) = \det A$.
Proof $\det(A^t)$ is the sum of all signed products of n entries of A^t, one from each row and each column. Because the rows and columns of A^t are the columns and rows of A, the terms in Formula 5.1 for det A are the same as the terms in the corresponding formula for $\det(A^t)$, except

possibly for the attached sign. For example, if $A^t = (b_{ij})$, then a typical product in the expansion of $\det(A^t)$ is

$$\pm b_{p(1)1} b_{p(2)2} \cdots b_{p(n)n} = \pm a_{1p(1)} a_{2p(2)} \cdots a_{np(n)}$$

$$= \pm a_{r(1)1} a_{r(2)2} \cdots a_{r(n)n},$$

where the permutation r is the inverse of p. (All we have done in the last step is to arrange the factors $a_{kp(k)}$ of the previous step to put the column indices, rather than the row indices, in natural order.) Because a permutation and its inverse are both even or both odd, the signs attached to corresponding terms in the two sums for $\det A$ and $\det(A^t)$ also agree.

An important consequence of Theorem 5.5 is that any statement about rows in a determinant (as in Theorem 5.1, for example) is also valid for columns.

Theorem 5.2 should not be used to evaluate $\det A$ when $n > 3$ because it is too inefficient computationally. That theorem is important, however, because it shows that if there exists a determinant function for n-by-n matrices, then there is only one such function, the one specified by (5.1). Hence the question of uniqueness is settled if we can prove existence. We can verify that the determinant formulas stated in Section 5.1 for $n = 1, 2,$ and 3 do satisfy Definition 5.1, so we proceed by induction. Assuming that a determinant function $\det$ exists for all $(n-1)$-by-$(n-1)$ matrices, we use $\det$ to define a function D on all n-by-n matrices, and then verify that D satisfies Definition 5.1. Some preliminary definitions and notation are needed.

Let A be any n-by-n matrix. We denote by A_{rs} the $(n-1)$-by-$(n-1)$ matrix obtained by deleting row r and column s of A. The *cofactor* of a_{rs} in A is defined by

$$\operatorname{cof} a_{rs} = (-1)^{r+s} \det A_{rs}.$$

That is,

$$\operatorname{cof} a_{rs} = (-1)^{r+s} \det \begin{pmatrix} a_{11} & \cdots & a_{1s} & \cdots & a_{1n} \\ & & & & \\ & \cdot & & & \\ & & & & \\ a_{r1} & \cdots & a_{rs} & \cdots & a_{rn} \\ & & & & \\ & \cdot & & & \\ & & & & \\ a_{n1} & \cdots & a_{ns} & \cdots & a_{nn} \end{pmatrix}.$$

For a *fixed* row index r we define a function D on all n-by-n matrices by

(5.2) $$D(A) = \sum_{s=1}^{n} a_{rs} \text{ cof } a_{rs} = \sum_{s=1}^{n} (-1)^{r+s} a_{rs} \det A_{rs}.$$

We want to show that D satisfies the three properties of Definition 5.1. The proof is technical, and you might prefer just to accept the fact that a determinant function exists. If so, you may skip to Theorem 5.6.

(1) Let A_i denote column i of A, so that $A = (A_1, \ldots, A_i, \ldots, A_n)$ and let $B = (A_1, \ldots, cA_i, \ldots, A_n)$. To verify Property 1 we want to show that $D(B) = cD(A)$. By definition,

$$D(B) = \sum_{s=1}^{n} (-1)^{r+s} b_{rs} \det B_{rs}.$$

(In referring to the columns of B_{rs}, we shall use the index k to refer to the column of B_{rs} that is a reduced form of column k of B; thus "column k of B_{rs}" identifies the kth column of B_{rs} when $k < r$ but the $(k - 1)$st column of B_{rs} when $k > r$, because B_{rs} has one fewer columns than B.) If $s \neq i$, $b_{rs} = a_{rs}$ and column i of B_{rs} is c times column i of A_{rs}. Hence $\det B_{rs} = c \det A_{rs}$, because det satisfies Property 1. If $s = i$, $b_{rs} = ca_{rs}$ and $B_{rs} = A_{rs}$. Hence

$$D(B) = cD(A).$$

(2) Let $B = (A_1, \ldots, A_i + cA_j, \ldots, A_j, \ldots, A_n)$. Then $b_{rs} = a_{rs}$ if $s \neq i$, and $b_{ri} = a_{ri} + ca_{rj}$. To express $D(B)$, we write individually the terms for $s = i$ and $s = j$:

$$D(B) = \sum_{s \neq i, j} (-1)^{r+s} b_{rs} \det B_{rs} + (-1)^{r+i} b_{ri} \det B_{ri} + (-1)^{r+j} b_{rj} \det B_{rj}.$$

For $s \neq i, j$, column i of B_{rs} is column i of A_{rs} plus a scalar multiple of column j of A_{rs}, and column j of A_{rs} is column j of B_{rs}. Because det satisfies Property 2,

$$\det B_{rs} = \det A_{rs}, \text{ when } s \neq i, j.$$

If $s = i$, $B_{ri} = A_{ri}$, so $\det B_{ri} = \det A_{ri}$. If $s = j$, we have

$$B_{rj} = \begin{pmatrix} a_{11} & \cdots & a_{1i} + ca_{1j} & \cdots & a_{1j} & \cdots & a_{1n} \\ \vdots & & \vdots & & \vdots & & \vdots \\ a_{r1} & \cdots & a_{ri} + ca_{rj} & \cdots & a_{rj} & \cdots & a_{rn} \\ \vdots & & \vdots & & \vdots & & \vdots \\ a_{n1} & \cdots & a_{n1} + ca_{nj} & \cdots & a_{nj} & \cdots & a_{nn} \end{pmatrix}.$$

Because det satisfies Statement (d) of Theorem 5.2,

$$\det B_{rj} = \det A_{rj} + \det A_{rj}^*$$

where A_{rj}^* agrees with A_{rj} except that column i of A_{rj}^* is c times column j of A_{ri}. With this information $D(B)$ can be written

$$D(B) = \sum_{s \neq i,\, j} (-1)^{r+s} a_{rs} \det A_{rs} + (-1)^{r+i}(a_{ri} + ca_{rj}) \det A_{ri}$$

$$+ (-1)^{r+j} a_{rj}(\det A_{rj} + \det A_{rj}^*)$$

$$= D(A) + (-1)^{r+i} ca_{rj} \det A_{ri} + (-1)^{r+j} a_{rj} \det A_{rj}^*,$$

where

$$A_{rj}^* = \begin{pmatrix} a_{11} & \cdots & ca_{1j} & \cdots & a_{1j} & \cdots & a_{1n} \\ \cdot & & \cdot & & \cdot & & \cdot \\ \cdot & & \cdot & & \cdot & & \cdot \\ a_{r1} & \cdots & ca_{rj} & \cdots & a_{rj} & \cdots & a_{rn} \\ \cdot & & \cdot & & \cdot & & \cdot \\ \cdot & & \cdot & & \cdot & & \cdot \\ a_{n1} & \cdots & ca_{nj} & \cdots & a_{nj} & \cdots & a_{nn} \end{pmatrix} .$$

This last matrix can be made to coincide with A_{ri} by moving column i to the position of column j, which can be done by transposing column i successively $|j - i - 1|$ times with its next neighboring column. Because det satisfies Statement (c) of Theorem 5.2, we have

$$\det A_{rj}^* = (-1)^{i+1-j} c \det A_{ri} .$$

Then

$$D(B) = D(A) + (-1)^{r+i} ca_{rj} \det A_{ri} + (-1)^{r+i+1} ca_{rj} \det A_{ri}$$

$$= D(A).$$

(3) In row r of I only the diagonal entry is nonzero; hence

$$D(I) = \sum_{s=1}^{n} (-1)^{r+s} \delta_{rs} \det I_{rs} = (-1)^{2r} \det I_{rr} = \det I_{rr} = 1.$$

Hence D satisfies the three defining properties of a determinant function, and the proof by induction is complete.

Theorem 5.6   For each value of $n = 1, 2, \ldots$, there exists one and only one determinant function for n-by-n matrices. Its value is expressed by Formula 5.1) and also by 5.2.

Formula 5.2 is called the *Laplace expansion* of a determinant by the entries of row r. Because r was arbitrary in the previous derivation, a determinant can be evaluated using the Laplace expansion by the entries of any row. Also, because det $(A^t) = $ det A, the Laplace expansion is valid also for the entries of any column.

Theorem 5.7

$$\sum_{s=1}^{n} a_{rs} \text{ cof } a_{ts} = \delta_{rt} \text{ det } A,$$

$$\sum_{r=1}^{n} a_{rs} \text{ cof } a_{rt} = \delta_{st} \text{ det } A.$$

Proof   The first statement is the Laplace expansion of det A by the entries of row r when $r = t$. For $r \neq t$ the given sum can be seen to be a Laplace expansion by the elements of row t of a determinant in which rows r and t are identical. The second statement expresses for columns the relation that the first statement asserts for rows.

When it is necessary to compute the value of det A, Theorem 5.7 should be used, but only after taking full advantage of Definition 5.1 and Theorem 5.1 by applying elementary row and column operations to A to simplify the work of evaluation. Note that by Definition 5.1(1), if column i is multiplied by c, the value of the determinant is also multiplied by c. By Definition 5.1(2), however, if column i is replaced by the sum of column i and c times column j, the value of the determinant is unchanged. Finally, by Theorem 5.1a, if two columns are interchanged, the value of the determinant is multiplied by -1. Hence we can use elementary row and column operations, as in Gauss-Jordan elimination, to produce a column or row in which there is *only one nonzero entry*. Then the Laplace expansion formula can be applied to that column or row to reduce by one the order of the determinant to be evaluated. This same process should then be repeated on the smaller determinant, keeping track of the effect of the operations on the value of the determinant, until a two-by-two determinant remains to be evaluated.

Example

$$\det \begin{pmatrix} 3 & 1 & -2 & 4 \\ 2 & 0 & -5 & 1 \\ 1 & -1 & 2 & 6 \\ -2 & 3 & -2 & 3 \end{pmatrix} \underset{(R_3 + R_1)}{=} \det \begin{pmatrix} 3 & 1 & -2 & 4 \\ 2 & 0 & -5 & 1 \\ 4 & 0 & 0 & 10 \\ -2 & 3 & -2 & 3 \end{pmatrix}$$

$$\underset{(R_4 - 3R_1)}{=} \det \begin{pmatrix} 3 & 1 & -2 & 4 \\ 2 & 0 & -5 & 1 \\ 4 & 0 & 0 & 10 \\ -11 & 0 & 4 & -9 \end{pmatrix}.$$

We have produced zero in all but one entry in column two; then the Laplace expansion by entries of column two yields

$$(-1) \det \begin{pmatrix} 2 & -5 & 1 \\ 4 & 0 & 10 \\ -11 & 4 & -9 \end{pmatrix} \underset{(C_3 - \frac{5}{2}C_1)}{=} (-1) \det \begin{pmatrix} 2 & -5 & -4 \\ 4 & 0 & 0 \\ -11 & 4 & 37\frac{1}{2} \end{pmatrix}.$$

Now expand by the entries of the second row and use the two-by-two formula:

$$(-1)(-1)4 \det \begin{pmatrix} -5 & -4 \\ 4 & 37\frac{1}{2} \end{pmatrix} = 4[(-5)(37\frac{1}{2}) - (-4)(4)]$$

$$= [(-10)(37) + 64] = -306.$$

EXERCISES 5.2

1. Prove that the determinant of an n-by-n triangular matrix is the product of the diagonal elements.

2. Compute the determinant of each of the three types of elementary matrices.

3. Prove Theorem 5.4.

4. Evaluate the following determinant by first using properties of determinants to produce a row or column having zero in all but one position and then applying the Laplace expansion:

$$\det \begin{pmatrix} 0 & 3 & -2 & 1 & 0 \\ -3 & 0 & 1 & -4 & 1 \\ 2 & -1 & 0 & 0 & -2 \\ -1 & 4 & 0 & 0 & 1 \\ 0 & -1 & 2 & 1 & 0 \end{pmatrix}.$$

5. Illustrate Theorem 5.7 by calculating $\sum_{s=1}^{3} a_{rs} \operatorname{cof} a_{ts}$ for $r = 1$ and $t = 2$, and then for $r = 2$ and $t = 2$, given that

$$A = \begin{pmatrix} 1 & -1 & 2 \\ -2 & 3 & 1 \\ 2 & -2 & x \end{pmatrix}.$$

6. Prove that if $r \neq t$, then $\sum_{s=1}^{n} a_{rs} \operatorname{cof} a_{ts} = 0$, thus completing the proof of Theorem 5.7.

7. (i) Show that if A is n-by-n then $\det(cA) = c^n \det A$.

(ii) Show that every skew-symmetric matrix of odd dimension is singular.

8. Let an n-by-n matrix A be partitioned into blocks as shown, where B is k-by-k and Z is the k-by-$(n - k)$ zero matrix:

$$A = \begin{pmatrix} B & Z \\ C & D \end{pmatrix}.$$

Deduce that $\det A = (\det B)(\det D)$.

9. The Vandermonde matrix of order n is, by definition,

$$V(x_1, \ldots, x_n) = \begin{vmatrix} 1 & 1 & \cdots & 1 \\ x_1 & x_2 & \cdots & x_n \\ \cdot & \cdot & & \cdot \\ \cdot & \cdot & & \cdot \\ \cdot & \cdot & & \cdot \\ x_1^{n-1} & x_2^{n-1} & \cdots & x_n^{n-1} \end{vmatrix}.$$

(i) For $n = 2, 3$ verify that $\det V = \prod_{1 \le i < j \le n} (x_j - x_i)$, where $\prod$ denotes "product."

(ii) Prove this statement for all $n > 1$.

10. Suppose that a permutation p of $\{1, 2, \ldots, n\}$ can be performed by a sequence of k transpositions and also by a sequence of m transpositions. Assume that an n-by-n determinant function exists, and therefore $\det I = 1$. Let J be the matrix obtained by applying p to the columns of I. Compute $\det J$ to deduce that either k and m are both even, or k and m are both odd.

5.3 SOME APPLICATIONS
   OF DETERMINANTS

Because determinants are so inefficient as a computational tool for $n > 3$, their use in mathematics typically arises either from their ability to represent the n-dimensional geometric content of a parallellotope in $\mathscr{E}^n$, or else from their ability to neatly characterize nonsingularity of a matrix (or equivalently, the linear independence of a set of n vectors in $\mathscr{V}_n$). We now examine a few applications of determinants in linear algebra and calculus.

A second method for computing A^{-1}   The first statement of Theorem 5.7 can be written

$$\sum_{k=1}^{n} a_{ik} \operatorname{cof} a_{jk} = \delta_{ij} \det A.$$

If we let $b_{kj} = \operatorname{cof} a_{jk}$ (observe the transposition of subscripts), we have

$$\sum_{k=1}^{n} a_{ik} b_{kj} = \delta_{ij} \det A,$$

or

$$AB = (\det A)I.$$

If A is nonsingular, $(\det A)^{-1}B = A^{-1}$. That is, the (k, j) entry of A^{-1} is $(\det A)^{-1} \operatorname{cof} a_{jk}$. If $C = (c_{jk})$ is the matrix of cofactors of A, $c_{jk} = \operatorname{cof} a_{jk}$, then the *transpose* C^t of C is called the *comatrix* of A, denoted com A. Thus

$$\operatorname{com} A = (b_{kj}), \text{ where } b_{kj} = c_{jk} = \operatorname{cof} a_{jk},$$

and

$$A \operatorname{com} A = (\det A)I.$$

Hence

$$A^{-1} = (\det A)^{-1} \operatorname{com} A.$$

The term *adjoint of A* is frequently used for the matrix com A, but adjoint will be used later in another context, so the term comatrix is used to avoid ambiguity. We illustrate with an example.
Given

$$A = \begin{pmatrix} 1 & 2 & 3 \\ 2 & 3 & 0 \\ 0 & 1 & 2 \end{pmatrix},$$

we compute det $A = 4$, so that A is nonsingular. The matrix C of cofactors of elements of A is

$$C = \begin{pmatrix} 6 & -4 & 2 \\ -1 & 2 & -1 \\ -9 & 6 & -1 \end{pmatrix},$$

so

$$C^{\mathsf{t}} = \operatorname{com} A = \begin{pmatrix} 6 & -1 & -9 \\ -4 & 2 & 6 \\ 2 & -1 & -1 \end{pmatrix},$$

and

$$A^{-1} = \tfrac{1}{4} \begin{pmatrix} 6 & -1 & -9 \\ -4 & 2 & 6 \\ 2 & -1 & -1 \end{pmatrix},$$

as obtained by elementary row operations at the end of Section 4.4. The comatrix method for computing A^{-1} is conveneient for $n = 2$ and $n = 3$, but not for larger values of n.

Cramer's Rule for solving a linear system If A is a nonsingular n-by-n matrix, the linear system $AX = Y$ has a unique solution, and because the consistency condition is satisfied, Y is a linear combination of the column vectors of A:

$$Y = x_1 A_1 + \cdots + x_n A_n,$$

where the coefficients x_i are the components of the unique solution vector X. For fixed j, consider the determinant of the matrix obtained by replacing A_j by Y in A. Then

$$\det(A_1, \ldots, A_{j-1}, Y, A_{j+1}, \ldots, A_n)$$

$$= \det\left(A_1, \ldots, A_{j-1}, \sum_{i=1}^{n} x_i A_i, A_{j+1}, \ldots, A_n\right)$$

$$= \sum_{i=1}^{n} x_i \det(A_1, \ldots, A_{j-1}, A_i, A_{j+1}, \ldots, A_n)$$

$$= x_j \det A,$$

because whenever $i \neq j$ the corresponding determinant in the sum is zero.

Hence

$$x_j = \frac{\det(A_1, \ldots, Y, \ldots, A_n)}{\det(A_1, \ldots, A_j, \ldots, A_n)}.$$

This result is known as *Cramer's Rule* for solving a nonsingular *n*-by-*n* linear system by means of determinants. Computationally Cramer's Rule offers no advantages over Gaussian elimination and should not be used in practice.

Change of variables in integration   A common technique for evaluating an integral

$$\int_I f(x)dx$$

is to make a substitution, or change of variable,

$$x = g(y)$$

where *g* is a one-to-one function that maps an interval *J* of $\mathbb{R}$ onto the given interval *I* of $\mathbb{R}$ and where *g* has a continuous derivative at each point of *J*. Then the formula

(5.3) $$\int_I f(x)dx = \int_J f(g(y))|g'(y)| \, dy$$

describes the method of integration by substitution and is an integral form of the chain rule for differentiation. Our attention is on the intervals *I* and *J* and the geometric significance of $|g'(y)|$. Let $I = [a, b]$ and let *J* be the interval $[c, d]$ that is mapped by the function *g* onto $[a, b]$. Because *g* is continuously differentiable and one-to-one on *J*, $g'(y)$ does not change sign on *J*. If $g'(y)$ is nonnegative, then $g(c) = a$ and $g(d) = b$, and (5.3) yields the familiar formula

$$\int_a^b f(x) \, dx = \int_c^d f(g(y))g'(y) \, dy.$$

If $g'(y)$ is nonpositive, then $g(c) = b$ and $g(d) = a$, so (5.3) becomes

$$\int_a^b f(x) \, dx = \int_d^c f(g(y))(-g'(y)) \, dy = \int_c^d f(g(y))g'(y) \, dy,$$

as before. The two intervals of integration are generally of different length, because the mapping is not necessarily length-preserving. That is, *g distorts lengths*; indeed, the amount by which *g* distorts length is not necessarily uniform along the interval between *c* and *d*. If *y* is any point of that interval, then $|g'(y)|$ is a linear approximation to the ratio by which g magnifies lengths near y. This means that a small interval of length dy near y is mapped by g into an interval near $g(y)$ having length approximately $|g'(y)|\ dy$.

A similar formula holds for real functions of n variables, which can be written in vector form. Let $X = (x_1, \ldots, x_n)$ be in $\mathscr{E}^n$, let f be a real-valued function on $\mathscr{E}^n$, and let R be a region of $\mathscr{E}^n$ over which f is integrable. To evaluate the multiple integral

$$\int_R f(X)\ dX$$

by the method of substitution, we let G denote a continuously differentiable vector-valued function that defines a one-to-one mapping of a region S of $\mathscr{E}^n$ onto R. For each $Y = (y_1, \ldots, y_n)$ in S we have $X = G(Y)$, which can be expressed in terms of the component functions as

$$x_1 = g_1(y_1, \ldots, y_n),$$
$$x_2 = g_2(y_1, \ldots, y_n),$$
$$\vdots$$
$$x_n = g_n(y_1, \ldots, y_n).$$

As in the case of a single variable, G distorts n-dimensional geometric content. The content of R in general is different from the content of S. Moreover, the distortion ratio varies over the region S. If Y is any point of S, then a linear approximation to the distortion ratio of G near Y can be expressed in terms of the partial derivatives of the component function evaluated at Y, specifically by the absolute value of the *Jacobian determinant*,

$$J = \det \begin{pmatrix} \dfrac{\partial g_1}{\partial y_1} & \cdots & \dfrac{\partial g_1}{\partial y_n} \\[1em] \vdots & & \vdots \\[1em] \dfrac{\partial g_n}{\partial y_1} & \cdots & \dfrac{\partial g_n}{\partial y_n} \end{pmatrix}.$$

The corresponding multivariable formula for integration by substitution then becomes

(5.4)
$$\int_R f(X)\ dX = \int_S f(G(Y))|J|\ dY.$$

Clearly (5.3) is the special form of (5.4) for the case $n = 1$.

Linear independence of functions Consider the set $\mathscr{S}$ of all solutions of a linear, homogeneous differential equation of order n,

(5.5) $y^{(n)} + a_1(x)y^{(n-1)} + \cdots + a_{n-1}(x)y' + a_n(x)y = 0,$

where $y^{(k)}$ denotes the kth derivative of y at x and where $a_1(x), \ldots, a_n(x)$ are functions that are continuous over some interval J of the real line. As we observed in Chapter 2, the special case $n = 2$, $\mathscr{S}$ forms a vector space over $\mathbb{R}$. Furthermore, it can be shown that dim $\mathscr{S} = n$. Therefore, a basis for $\mathscr{S}$ is a set of n linearly independent real valued functions y_i, each of which is a solution of (5.5), so we are interested in determining when a set $\{y_1, \ldots, y_k\}$ of functions is linearly independent. We assume that there exist scalars c_i, $i = 1, \ldots, k$, such that the function

$$\sum_{i=1}^{k} c_i y_i$$

is the *zero function* over J, and ask whether or not each of scalars c_i must be zero. If each y_i is sufficiently differentiable for all x in J, we have

$$c_1 y_1(x) \quad + c_2 y_2(x) \quad + \cdots + c_k y_k(x) \quad = 0,$$
$$c_1 y_1'(x) \quad + c_2 y_2'(x) \quad + \cdots + c_k y_k'(x) \quad = 0,$$

(5.6)

$$c_1 y_1^{(k-1)}(x) + c_2 y_2^{(k-1)}(x) + \cdots + c_k y_k^{(k-1)}(x) = 0.$$

This is a homogeneous system of k equations in the k variables $c_1, \ldots, c_k$, the coefficients being functions of x. The determinant of the matrix of coefficients is called the *Wronskian determinant* of the functions $y_1, \ldots, y_k$:

$$W(y_1, \ldots, y_k) = \det \begin{pmatrix} y_1 & y_2 & \cdots & y_k \\ y_1' & y_2' & \cdots & y_k' \\ \cdot & \cdot & & \cdot \\ \cdot & \cdot & & \cdot \\ \cdot & \cdot & & \cdot \\ y_1^{(k-1)} & y_2^{(k-1)} & \cdots & y_k^{(k-1)} \end{pmatrix}$$

Suppose that for some $x_0 \in J$, $W(x_0) \neq 0$. Then the system (5.6) evaluated at x_0 has *only* the trivial solution. Therefore, if $\{y_1, \ldots, y_k\}$ is linearly *dependent* over J, the Wronskian determinant must be zero at each point of J. Furthermore, if $k \leq n$ and if each y_i is a solution of (5.5), the converse can be proved: if the Wronskian vanishes identically over J, then the set $\{y_1, \ldots, y_k\}$ is linearly dependent. Exercise 5.3-10 provides an example of a set of functions that are not solutions of an equation of the form (5.5), the Wronskian of these functions vanishes identically over an interval J, yet the functions are linearly independent.

EXERCISES 5.3

1. Verify the computations of the text illustration for calculating C, com A, and A^{-1} for the given example.

2. Solve the following system of equations,

$$2x_1 - x_2 + 3x_3 = 3,$$
$$x_2 = -2,$$
$$2x_1 + x_2 + x_3 = 1,$$

(i) by the comatrix method of calculating A^{-1},

(ii) by Cramer's rule,

(iii) by direct algebraic elimination.

3. Do the same as in Exercise 2 for the system

$$x + 2y + 8z = 1,$$
$$-x + y + 5z = -3,$$
$$2x + 2y + 4z = 5.$$

4. Determine the comatrix of

$$A = \begin{pmatrix} a & b \\ c & d \end{pmatrix},$$

and verify that $A \text{ com } A = (\det A)I$.

5. Prove that $\det (\text{com } A) = (\det A)^{n-1}$, where A is n-by-n. Deduce that A is singular if and only if com A is singular.

6. Use Exercise 5.2-9 to show that A is nonsingular, where

$$A = \begin{pmatrix} 1^0 & 1^1 & \cdots & 1^{n-1} \\ 2^0 & 2^1 & \cdots & 2^{n-1} \\ \cdot & \cdot & & \cdot \\ \cdot & \cdot & & \cdot \\ \cdot & \cdot & & \cdot \\ n^0 & n^1 & \cdots & n^{n-1} \end{pmatrix}.$$

7. An integration formula for changing from rectangular to polar coordinates in $\mathscr{E}^2$ is

$$\iint_R f(x, y)\, dxdy = \iint_S f(r \cos \theta, r \sin \theta)\, r\, dr\, d\theta.$$

Demonstrate in detail how this formula is a special case of (5.4).

8. Apply the substitution $x = u^2 - v^2$, $y = 2uv$ to express

$$\iint_R \sqrt{x^2 + y^2}\, dxdy$$

as a double integral over some region S of the u-v plane.

9. Calculate the Wronskian of each of the following sets of functions. Which sets are linearly independent over $\mathbb{R}$?

(i) $\{e^{ax}, e^{bx}\}$, $a \neq b$.

(ii) $\{\sin bx, \cos bx\}$.

(iii) $\{2, \sin^2 x, \cos 2x\}$.

10. Let $y_1 = x^2$ and $y_2 = x|x|$. Show that y_1 and y_2 are differentiable on $[-1, 1]$, the Wronskian $W(y_1, y_2)$ vanishes identically on $[-1, 1]$, and yet $\{y_1, y_2\}$ is linearly independent.

CHAPTER 6
EQUIVALENCE
RELATIONS
ON RECTANGULAR
MATRICES

6.1 ROW EQUIVALENCE

We have already seen that an m-by-n matrix can be used to represent linear mappings from $\mathscr{V}_n$ to $\mathscr{W}_m$ and also to represent systems of linear equations. Later we shall use matrices to represent other mathematical structures, each of which has its distinctive problems and methods, which in turn generate corresponding questions about matrices.

In the case of a linear mapping $\mathbf{T}$, a matrix A that represents $\mathbf{T}$ is determined by a choice of basis for $\mathscr{V}_n$ and a basis for $\mathscr{W}_m$. A different choice of bases presumably yields a different matrix B that also represents $\mathbf{T}$. We know that A and B have the same rank because the rank of each is $r(\mathbf{T})$. We wonder what other properties A and B share. In particular, given two matrices A and B, we wonder whether they both might represent the same linear mapping and how we can use matrix methods to settle the question. Also among all the matrices that represent **T** we would like to discover one which is simplest for computational purposes or for descriptive purposes.

In the case of a linear system we have already skirmished tentatively with such questions. Algorithms for solving linear systems typically replace a given system by another system that has the same set of solutions but is easier in some sense to solve. For example, the Gauss-Jordan algorithm can be translated into computational operations on matrices that lead from a given matrix A through a sequence of matrices to the reduced echelon form E of A. All the matrices obtained in proceeding from A to E represent equivalent linear systems, but the system represented by E is particularly easy to solve.

To deal effectively with such questions we must apply the general concept of *equivalence relation*. (See Appendix A.3.) If $\mathcal{M}$ is any nonvoid set, a binary relation $\sim$ between the elements of $\mathcal{M}$ is called an equivalence relation if and only if three properties are satisfied for all A, B, C in $\mathcal{M}$:

(1) $\sim$ is *reflexive*; $A \sim A$,
(2) $\sim$ is *symmetric*; if $A \sim B$, then $B \sim A$,
(3) $\sim$ is *transitive*; if $A \sim B$ and $B \sim C$, then $A \sim C$.

Any equivalence relation on $\mathcal{M}$ separates the elements of $\mathcal{M}$ into mutually disjoint subsets, called *equivalence classes*. Each such class $\mathcal{E}$ has the property that all "equivalent" elements of $\mathcal{M}$ are in the same class $\mathcal{E}$, and any two elements in the same class are equivalent:

$\qquad$ if $A \in \mathcal{E}$, then $B \in \mathcal{E}$ if and only if $B \sim A$.

Many different equivalence relations arise naturally in the study of matrices, as we shall see. For each such equivalence relation we shall want to describe the equivalence classes and, if possible, to describe a simple *canonical form*. A form is said to be *canonical* for a given notion of equivalence when there is *one and only one* matrix of that form in each equivalence class, whereas the form is said to be *standard* when there is *at least one* matrix of that form in each equivalence class.

We illustrate this discussion by considering *row equivalence*. Many of the details concerning row equivalence were developed in Sections 4.3 and 4.4 without using that terminology.

Definition 6.1 An m-by-n matrix A is said to be *row equivalent* to an m-by-n matrix B if and only if A can be obtained by performing a finite number of elementary row operations on B.

The relation of row equivalence of m-by-n matrices is easily seen to be an equivalence relation: it is reflexive and transitive by nature, and it is symmetric because if a sequence of elementary row operations produces A from B then the reversed sequence of row operations applied to A will produce B. (Recall Theorem 4.8.) Hence the relation of row equivalence separates $\mathcal{M}_{m \times n}$ into disjoint equivalence classes. In each equivalence class, each matrix in that class can be obtained from any other matrix in that class by applying elementary row operations, and any matrix that can be so obtained must be in that class.

Theorem 6.1   Let A and B be m-by-n matrices.

(a) A and B are row equivalent if and only if $A = PB$ for some nonsingular matrix P.
(b) If A and B are row equivalent, then $r(A) = r(B)$.
(c) When $m = n$, A is nonsingular if and only if A is row equivalent to I_n.

Proof   These statements follow directly from Theorems 4.16 and 4.13.

We also have shown that each matrix A is row equivalent to a matrix B in echelon form. However, B is not the only matrix in echelon form that is row equivalent to A. This illustrates that echelon form is a standard form but not a canonical form for row equivalence. However in Theorem 4.10 we stated without proof that there is one and only one matrix in *reduced* echelon form that is row equivalent to A. We now restate that result and supply a proof.

Theorem 6.2   Two matrices B and C in reduced echelon form are row equivalent if and only if $B = C$; that is, the reduced echelon form is canonical with respect to row equivalence.

Proof   Suppose reduced echelon matrices B and C are row equivalent, and let $r = r(B) = r(C)$. Then the first r rows of B and of C are nonzero and the last $(m - r)$ rows are zero. The first nonzero entry in each nonzero row is 1, and it is the only nonzero entry in its column. Let $b(i)$ denote the column of B in which the first nonzero entry of row i appears, and similarly let $c(i)$ denote the column of C in which the first nonzero entry of row i appears. Recalling that row equivalent matrices represent homogeneous linear systems that have exactly the same solution set, we shall show that $b(i) = c(i)$ for $i = 1, \ldots, r$. If $b(i) < c(i)$ for some i, we can assume that i is the smallest such index. Then $b(k) = c(k)$ for each $k < i$.

Column $b(i)$ of C has $c_{kb(i)}$ in row k for $k < i$ and 0 in row k for $k \geq i$. We use the numbers in column $b(i)$ of C to form an n-tuple X that has -1 in position $b(i)$, has $c_{kb(i)}$ in position $b(k)$ for $k < i$, and has 0 elsewhere. Thus with $k < i$ we have the following vectors.

$$
\begin{array}{ccccc}
 b(1) & b(2) & b(k) & b(i) & c(i)
\end{array}
$$

$$
\begin{array}{l}
X\colon (0 \cdots c_{1b(i)} \cdots c_{2b(i)} \cdots c_{kb(i)} \cdots -1 \cdots 0 \cdots) \\
\text{Row } k \text{ of } C\colon (0 \cdots 0 \cdots 0 \cdots 1 \cdots c_{kb(i)} \cdots 0 \cdots) \\
\text{Row } k \text{ of } B\colon (0 \cdots 0 \cdots 0 \cdots 1 \cdots 0 \cdots b_{kc(i)} \cdots) \\
\text{Row } i \text{ of } B\colon (0 \cdots 0 \cdots 0 \cdots 0 \cdots 1 \cdots b_{ic(i)} \cdots)
\end{array}
$$

The dot product of X and row t of C is 0 for every t, so $CX = Z$. The dot product of X and row i of B is -1, so $BX \neq Z$. This contradicts the assumption that B and C represent equivalent homogeneous linear systems. Hence $b(i) \geq c(i)$. By reversing the roles of B and C we deduce that $b(i) = c(i)$ for $i = 1, \ldots, r$.

Now let $c(1), c(2), \ldots, c(r)$ denote the columns in which the first nonzero element appears in rows $1, 2, \ldots, r$ in both B and C. Consider the form of row s in B and in C, where $s \leq r$. Row s of B is a string of $(c(s) - 1)$ zeros, followed by 1, followed by $(n - c(s))$ other entries b_{sj} about which we know nothing except that $b_{sc(t)} = 0$ when $t \neq s$. We know that columns $c(i)$ of B and C coincide for $i = 1, \ldots, r$. Let $k \neq c(i)$ for $i = 1, 2, \ldots, r$, and let X be the n-tuple that has -1 in position k, has b_{ik} in position $c(i)$ for $i = 1, \ldots, r$, and has 0 elsewhere. We can display these vectors as follows:

$$
\begin{array}{ccccc}
 c(1) & c(s) & k & c(r)
\end{array}
$$

$$
\begin{array}{l}
\text{Row } s \text{ of } B\colon (0 \cdots 0 \cdots 1 \cdots b_{sk} \cdots 0 \cdots b_{sn}), \\
\text{Row } s \text{ of } C\colon (0 \cdots 0 \cdots 1 \cdots c_{sk} \cdots 0 \cdots c_{sn}), \\
X\colon (0 \cdots b_{1k} \cdots b_{sk} \cdots -1 \cdots b_{rk} \cdots 0).
\end{array}
$$

The dot product of X with row s of B is zero for each $s = 1, \ldots, m$; that is, X is a solution of the homogeneous linear system represented by B. Hence it must be a solution of the corresponding system represented by C. But this means that $b_{sk} - c_{sk} = 0$ for $s = 1, \ldots, m$. Hence column k of B is the same, entry for entry, as column k of C for $k = 1, \ldots, n$. Thus $B = C$.

Now we have answered some important questions. Two homogeneous linear systems of m equations in n variables are equivalent if and only if their two coefficient matrices are row equivalent. Two m-by-n matrices are row equivalent if and only if they have the same reduced echelon form. Because we know how to find the reduced echelon form of any matrix, this provides a means of determining whether two given matrices are in the same row-equivalence class.

In Sections 6.3, 6.4, and 9.4 other equivalence relations will be introduced. In each case we shall determine a canonical form that is distinctive for that equivalence relation.

EXERCISES 6.1

1. Determine which of the three properties of an equivalence relation are satisfied by the following relations.

(i) Similarity in the set of all triangles in $\mathscr{E}^2$.

(ii) Parallelism in the set of all lines in $\mathscr{E}^2$.

(iii) Strong inequality in the set of all real numbers.

(iv) Divisibility (with remainder zero) in the set of all integers.

(v) Perpendicularity in the set of all planes in $\mathscr{E}^3$.

2. Describe a simple canonical form for the equivalence relation in Exercise 1(i).

3. On the set $\mathscr{M}_{n \times n}$ of all n-by-n matrices, we define $A \sim B$ to mean that $r(A) = r(B)$.

(i) Verify that $\sim$ is an equivalence relation.

(ii) How many distinct equivalence classes are there? Describe each.

(iii) Describe a simple canonical form for this equivalence relation.

4. Fill in the details of the proof of Theorem 6.1.

5. Are any two of the following matrices row equivalent? Explain.

(i)
$$A = \begin{pmatrix} 1 & 2 & 2 & 7 \\ 2 & 4 & 3 & 11 \\ 0 & 0 & 1 & 3 \end{pmatrix},$$

(ii)
$$B = \begin{pmatrix} 1 & 2 & 2 & 5 \\ 2 & 4 & 3 & 7 \\ 0 & 0 & 1 & 3 \end{pmatrix},$$

(iii)
$$C = \begin{pmatrix} 1 & 2 & 2 & 5 \\ -2 & -4 & 3 & 11 \\ 0 & 0 & -1 & -3 \end{pmatrix}.$$

6. Let A be an m-by-$(m + n)$ matrix of rank m, and let E denote the reduced echelon form of A.

(i) Show that a permutation of the columns of E transforms E into block form $(I | B)$, where I is the m-by-m identity matrix and B is an m-by-n matrix.

(ii) Recalling Theorem 4.7, deduce that there exist nonsingular matrices P and Q such that $PAQ = (I | B)$.

6.2   CHANGE OF BASIS

Until now our study of linear mappings has been carried out without much concern for the choice of bases for the spaces involved. From Section 2.6 we recall that a basis for $\mathscr{V}_n$ determines a coordinate system for $\mathscr{V}_n$ in which the axes are the lines along the basis vectors, which need not be mutually orthogonal even in those spaces where orthogonality is defined. Thus by choosing a basis $\{\alpha_1, \ldots, \alpha_n\}$ for $\mathscr{V}_n$, we obtain a specific representation of each vector ξ of $\mathscr{V}_n$ as an ordered n-tuple X of scalars:

$$\text{If } \xi = \sum_{j=1}^{n} x_j \alpha_j, \text{ then } X = \begin{pmatrix} x_1 \\ \cdot \\ \cdot \\ \cdot \\ x_n \end{pmatrix}.$$

Relative to a different basis $\{\gamma_1, \ldots, \gamma_n\}$ for $\mathscr{V}_n$, ξ is represented by a column vector Y:

$$\text{If } \xi = \sum_{k=1}^{n} y_k \gamma_k, \text{ then } Y = \begin{pmatrix} y_1 \\ \cdot \\ \cdot \\ \cdot \\ y_n \end{pmatrix}.$$

It is natural to ask how the n-tuples X and Y are related when they represent the same vector relative to different bases. Stated geometrically, suppose that a point ξ in $\mathscr{V}_n$ has coordinates $(x_1, \ldots, x_n)$ in one coordinate system and that we change coordinates by means of a nonsingular linear transformation on $\mathscr{V}_n$, giving ξ a new set of coordinates $(y_1, \ldots, y_n)$. How can y_k be computed if we know the value of each x_j? Not surprisingly, an answer depends on the linear transformation that performs the change of basis. Specifically, if we express each vector α_j of the *original* basis as a linear combination of the vectors γ_k of the *new* basis, we have

(6.1)
$$\alpha_j = \sum_{k=1}^{n} m_{kj} \gamma_k, \qquad j = 1, \ldots, n.$$

Then

$$\xi = \sum_{j=1}^{n} x_j \alpha_j = \sum_{j=1}^{n} x_j \left(\sum_{k=1}^{n} m_{kj} \gamma_k \right) = \sum_{k=1}^{n} \left(\sum_{j=1}^{n} m_{kj} x_j \right) \gamma_k .$$

But relative to the γ-basis ξ is *uniquely* represented as

$$\xi = \sum_{k=1}^{n} y_k \gamma_k .$$

Hence

$$y_k = \sum_{j=1}^{n} m_{kj} x_j,$$

or in matrix notation

$$Y = MX,$$

where M is the nonsingular n-by-n matrix that expresses the *original* basis vectors in terms of the *new* basis vectors.

Although our question about the relation between X and Y has been answered, it is useful to look more closely at the linear transformation that performs the change from the α-basis to the γ-basis. We take the point of view that ξ is a point (or arrow) in $\mathscr{V}_n$ that is described by the column vector X relative to the original coordinate system. When we change coordinate systems, the geometric point ξ remains fixed but acquires a new set Y of coordinates. In the sense that X and Y are two names for the same point ξ, this interpretation of a change of coordinates is called the *alias* (or *static*) interpretation to suggest that the process merely gives ξ a new name. But then the linear transformation that carries out the change of coordinates must be the identity transformation, because each vector ξ is mapped into itself. Hence we can regard (6.1) as defining the linear transformation I, where

(6.2)
$$I\alpha_j = \alpha_j = \sum_{k=1}^{n} m_{kj} \gamma_k, \qquad j = 1, \ldots, n.$$

Previously we have interpreted linear transformations in terms of a single basis for $\mathscr{V}_n$, taking the view that each vector ξ is carried by a linear transformation $\mathbf{T}$ into the vector $\mathbf{T}\xi$. This interpretation is called the *alibi* (or *dynamic*) interpretation to suggest that $\mathbf{T}$ moves the vector ξ into the vector $\mathbf{T}\xi$. Whenever $\mathbf{I}$ is represented relative to a single basis, the corresponding matrix is the identity matrix I, but whenever $\mathbf{I}$ is represented relative to a pair of bases as in (6.2), the corresponding matrix M can be any nonsingular matrix.

It is sometimes helpful to use a notation that displays the bases involved in a particular representation of a linear mapping $\mathbf{T}$. For example, when $\mathbf{T}$ maps $\mathscr{V}_n$ into $\mathscr{W}_m$, a choice of basis for each space allows us to represent $\mathbf{T}$ uniquely by an m-by-n matrix A. We shall write

$$\mathbf{T} \to {}^\beta A_\alpha$$

to indicate that the columns of A express the $\mathbf{T}\alpha_j$ in terms of the β_i. That is, $\mathbf{T}$ is represented by the matrix A relative to the α-basis for its domain $\mathscr{V}_n$ and the β-basis for $\mathscr{W}_m$. In this notation, the identity transformation $\mathbf{I}$ is represented by ${}^\alpha I_\alpha$ (the identity matrix I) when a single basis is used, but by ${}^\gamma M_\alpha$ (some nonsingular matrix M) when two bases are used and each α_j is expressed in terms of the γ_k. Of course to describe a change of basis, the use of two bases is inevitable.

Now let us proceed with questions concerning linear mappings. If $\mathbf{T}$ maps $\mathscr{V}_n$ into $\mathscr{W}_m$ and $\mathbf{S}$ maps $\mathscr{W}_m$ into $\mathscr{X}_p$, then $\mathbf{ST}$ maps $\mathscr{V}_n$ into $\mathscr{X}_p$. Suppose that a basis is chosen for each space, so that

$$\mathbf{T} \to {}^\beta A_\alpha \quad \text{and} \quad \mathbf{S} \to {}^\gamma B_\beta .$$

Then

$$\mathbf{ST} \to {}^\gamma C_\alpha = {}^\gamma B_\beta \, {}^\beta A_\alpha .$$

This is simply a restatement of what we already know, but in a notation that records the bases used. The composite mapping $\mathbf{ST}$ (which is read from right to left as the mapping $\mathbf{T}$ followed by the mapping $\mathbf{S}$) carries each α_j into a linear combination $\sum_{i=1}^p c_{ij}\gamma_i$ of the γ's, because $\mathbf{T}$ carries each α_j into a linear combination $\sum_{k=1}^m a_{kj}\beta_k$ of the β's and $\mathbf{S}$ carries each β_k into a linear combination $\sum_{i=1}^p b_{ik}\gamma_i$ of the γ's and $c_{ij} = \sum_{k=1}^m b_{ik}a_{kj}$.

Let us focus on the linear mapping $\mathbf{T}$ from $\mathscr{V}_n$ to $\mathscr{W}_m$, which is represented by the m-by-n matrix ${}^\beta A_\alpha$ relative to an α-basis for $\mathscr{V}_n$ and a β-basis for $\mathscr{W}_m$. Suppose we choose a new γ-basis in $\mathscr{V}_n$ and a new δ-basis in $\mathscr{W}_m$. Then $\mathbf{T}$ is represented by the m-by-n matrix ${}^\delta B_\gamma$:

$$\mathbf{T} \to {}^\beta A_\alpha \quad \text{and} \quad \mathbf{T} \to {}^\delta B_\gamma .$$

We wonder how two matrices A and B are related when both represent the same linear mapping $\mathbf{T}$ relative to two pairs of bases.

When we considered the effect of a change of basis in $\mathscr{V}_n$ on the coordinates of ξ, we expressed each vector α_j of the original basis as a linear combination of the vectors γ_k of the new basis, as in (6.1). For the present problem it turns out to be more convenient to turn things around and to express each vector γ_k of the *new* basis as a linear combination of the vectors of the *original* basis:

$$\text{In } \mathscr{V}_n, \; \gamma_k = \sum_{j=1}^{n} q_{jk}\alpha_j, \qquad k = 1, \ldots, n.$$

$$\text{In } \mathscr{W}_m, \; \delta_j = \sum_{i=1}^{m} p_{ij}\beta_i, \qquad j = 1, \ldots, m.$$

We also have

$$\mathbf{T} \to {}^{\beta}\!A_{\alpha}, \text{ so } \mathbf{T}\alpha_j = \sum_{i=1}^{m} a_{ij}\beta_i, \qquad j = 1, \ldots, n,$$

$$\mathbf{T} \to {}^{\delta}\!B_{\gamma}, \text{ so } \mathbf{T}\gamma_k = \sum_{j=1}^{m} b_{jk}\delta_j, \qquad k = 1, \ldots, n.$$

Then

$$\mathbf{T}\gamma_k = \sum_{j=1}^{m} b_{jk}\left(\sum_{i=1}^{m} p_{ij}\beta_i\right) = \sum_{i=1}^{m}\left(\sum_{j=1}^{m} p_{ij}b_{jk}\right)\beta_i.$$

But also

$$\mathbf{T}\gamma_k = \mathbf{T}\left(\sum_{j=1}^{n} q_{jk}\alpha_j\right) = \sum_{j=1}^{n} q_{jk}\mathbf{T}\alpha_j$$

$$= \sum_{j=1}^{n} q_{jk}\left(\sum_{i=1}^{m} a_{ij}\beta_i\right) = \sum_{i=1}^{m}\left(\sum_{j=1}^{n} a_{ij}q_{jk}\right)\beta_i.$$

Because the vector $\mathbf{T}\gamma_k$ is a unique linear combination of the β's, the coefficients of β_i in these two calculations of $\mathbf{T}\gamma_k$ must agree; that is,

$$\sum_{j=1}^{m} p_{ij}b_{jk} = \sum_{j=1}^{n} a_{ij}q_{jk} \quad \text{for} \quad i = 1, \ldots, m \quad \text{and} \quad k = 1, \ldots, n.$$

In matrix notation, this becomes

$$PB = AQ,$$

or more specifically,

$$ {}^{\beta}\!P_{\delta}\,{}^{\delta}\!B_{\gamma} = {}^{\beta}\!A_{\alpha}\,{}^{\alpha}\!Q_{\gamma}.$$

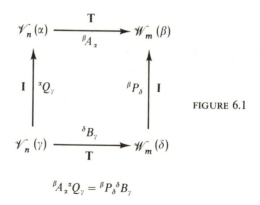

$$^{\beta}A_{\alpha}{}^{\alpha}Q_{\gamma} = {}^{\beta}P_{\delta}{}^{\delta}B_{\gamma}$$

FIGURE 6.1

The matrix $^{\alpha}Q_{\gamma}$ expresses each vector γ_k of the γ-basis for $\mathscr{V}_n$ in terms of the α-basis for $\mathscr{V}_n$. It therefore represents the identity linear transformation **I** on $\mathscr{V}_n$, where

$$\mathbf{I}\gamma_k = \gamma_k = \sum_{j=1}^{n} q_{jk}\alpha_j.$$

We recall that the matrix M of our earlier calculation also represented **I** in the form

$$\mathbf{I}\alpha_j = \alpha_j = \sum_{k=1}^{n} m_{kj}\gamma_k.$$

Hence we write M as $^{\gamma}M_{\alpha}$. It is easy to verify that

$$^{\gamma}M_{\alpha}{}^{\alpha}Q_{\gamma} = {}^{\gamma}I_{\gamma},$$

the n-by-n identity matrix. Hence $Q = M^{-1}$.

These relationships can be summarized conveniently by the mapping diagram in Figure 6.1. Starting in the lower left corner, we interpret Figure 6.1 as indicating that a change of basis **I** in $\mathscr{V}_n$ from the γ-basis to the α-basis, followed by the mapping **T** from $\mathscr{V}_n$ to $\mathscr{W}_m$, produces the same effect on each vector of the γ-basis as the mapping **T** from $\mathscr{V}_n$ to $\mathscr{W}_m$, followed by a change of basis **I** in $\mathscr{W}_m$ from the δ-basis to the β-basis. In terms of matrices,

$$^{\beta}A_{\alpha}{}^{\alpha}Q_{\gamma} = {}^{\beta}P_{\delta}{}^{\delta}B_{\gamma}.$$

Hence the m-by-n matrices B and A are related by the equation

$$B = P^{-1}AQ$$

for some nonsingular m-by-m matrix P and some nonsingular n-by-n matrix Q.

The results of this section are summarized in the following theorem, which we shall use in the next section.

Theorem 6.3   Let $\{\alpha_1, \ldots, \alpha_n\}$ be a basis for $\mathscr{V}_n$, $\{\beta_1, \ldots, \beta_m\}$ a basis for $\mathscr{W}_m$, and **T** a linear mapping from $\mathscr{V}_n$ to $\mathscr{W}_m$. Let $\{\gamma_1, \ldots, \gamma_n\}$ and $\{\delta_1, \ldots, \delta_m\}$ be bases for $\mathscr{V}_n$ and $\mathscr{W}_m$, respectively. Let ξ be a vector in $\mathscr{V}_n$.

(a) If ξ is represented by column vectors X and Y relative to the α-basis and γ-basis, respectively, then

$$Y = MX,$$

where $M = (m_{kj})$ is the nonsingular n-by-n matrix defined by

$$\alpha_j = \sum_{k=1}^{n} m_{kj} \gamma_k, \qquad j = 1, \ldots, n.$$

(b) If **T** is represented by the m-by-n matrix A relative to the α-basis for $\mathscr{V}_n$ and the β-basis for $\mathscr{W}_m$, and **T** is also represented by the m-by-n matrix B relative to the γ-basis for $\mathscr{V}_n$ and the δ-basis for $\mathscr{W}_m$, then

$$PB = AQ,$$

where $P = (p_{ij})$ and $Q = (q_{jk})$ are the nonsingular matrices defined by

$$\delta_j = \sum_{i=1}^{m} p_{ij} \beta_i, \qquad j = 1, \ldots, m,$$

$$\gamma_k = \sum_{j=1}^{n} q_{jk} \alpha_j, \qquad k = 1, \ldots, n.$$

(c) The matrix Q defined in (b) is the inverse of the matrix M defined in (a).

EXERCISES 6.2

1. In $\mathbb{R}^3$ let $\xi = a\varepsilon_1 + b\varepsilon_2 + c\varepsilon_3$. Let $\gamma_1 = (1, 1, 0)$, $\gamma_2 = (1, 0, 1)$, and $\gamma_3 = (1, -1, 1)$.

(i) Verify that $\{\gamma_1, \gamma_2, \gamma_3\}$ is a basis.

(ii) Express ξ as a linear combination of the γ's.

(iii) Write a matrix $^\gamma M_\varepsilon$ that expresses the change of basis from $\{\varepsilon_1, \varepsilon_2, \varepsilon_3\}$ to $\{\gamma_1, \gamma_2, \gamma_3\}$; that is,

$$\varepsilon_j = \sum_{k=1}^{3} m_{kj} \gamma_k .$$

(iv) Verify your answer to (ii) by computing MX where X is the column vector that represents ξ relative to the ε-basis.

2. Let $\mathbf{T}$ be the linear mapping from $\mathbb{R}^3$ to $\mathbb{R}^2$, defined relative to the standard bases by the matrix

$$A = \begin{pmatrix} 2 & 1 & 0 \\ 0 & 2 & 1 \end{pmatrix}.$$

Let $\{\gamma_1, \gamma_2, \gamma_3\}$ be the basis for $\mathbb{R}^3$ defined in Exercise 1.

(i) By computing $\mathbf{T}\gamma_k$ for $k = 1, 2, 3$, write the matrix B that represents $\mathbf{T}$ relative to the γ-basis for $\mathbb{R}^3$ and the standard ε-basis for $\mathbb{R}^2$.

(ii) Check your answer to (i) by matrix multiplication, using Theorem 6.3(b) with $P = I$, and $Q = M^{-1}$, where M is the matrix determined in Exercise 1(iii).

3. Let $\mathbf{T}$ be the linear mapping of $\mathscr{V}_3$ into $\mathscr{W}_2$ whose matrix relative to bases $\{\alpha_1, \alpha_2, \alpha_3\}$ for $\mathscr{V}_3$ and $\{\beta_1, \beta_2\}$ for $\mathscr{W}_2$ is

$$^\beta A_\alpha = \begin{pmatrix} 1 & 0 & -3 \\ 2 & 1 & 1 \end{pmatrix}.$$

Let new bases be defined for $\mathscr{V}_3$ and $\mathscr{W}_2$ by

$$\begin{aligned} \gamma_1 &= \alpha_1 + \alpha_2 + \alpha_3, \\ \gamma_2 &= \alpha_2 + \alpha_3, \quad \text{and} \\ \gamma_3 &= \alpha_1 + \alpha_3, \end{aligned} \qquad \begin{aligned} \delta_1 &= \beta_1 - 2\beta_2, \\ \delta_2 &= \beta_1 + \beta_2. \end{aligned}$$

(i) By expressing $\mathbf{T}\gamma_k = \sum_{j=1}^{2} b_{jk} \delta_j$, determine the matrix $^\delta B_\gamma$ that represents $\mathbf{T}$ relative to the pair (γ, δ) of bases.

(ii) Compute the matrices $^\beta P_\delta$ and $^\alpha Q_\gamma$ of Theorem 6.3.

(iii) Verify your answer to (i) by computing $^\beta A_\alpha \, ^\alpha Q_\gamma$ and $^\beta P_\delta \, ^\delta B_\gamma$.

4. Prove the general assertion made in the text prior to Figure 6.1 that $Q = M^{-1}$.

5. Reversing the roles of matrices and linear mappings that was discussed in the text, suppose that we are given an m-by-n matrix A, two bases

(denoted α and γ) for $\mathscr{V}_n$ and two bases (denoted β and δ) for $\mathscr{W}_m$. Relative to the pair (α, β) of bases, A represents a linear mapping $\mathbf{T}_1$ from $\mathscr{V}_n$ to $\mathscr{W}_m$. Relative to the pair (γ, δ) the same matrix A represents another linear mapping $\mathbf{T}_2$. How are $\mathbf{T}_1$ and $\mathbf{T}_2$ related? (Hint: Let $\mathbf{R}$ denote the linear transformation $\mathbf{R}\delta_k = \beta_k$ on $\mathscr{W}_m$, and let $\mathbf{S}$ be the linear transformation $\mathbf{S}\gamma_j = \alpha_j$.)

6.3 EQUIVALENCE

Now we are ready to reap some major benefits from the detailed computations of the previous section. The basic conclusion of that work was that two m-by-n matrices A and B represent the same linear mapping only if there exist a nonsingular m-by-m matrix N and a nonsingular n-by-n matrix Q such that

$$B = NAQ.$$

There is no need here to be concerned with the specific definition and geometric significance of the matrices N and Q, although we can easily recover such information from Section 6.2. We should realize that matrices which represent the same linear mapping must have many properties in common, and we plan to explore that idea in this section.

Instead of interpreting matrices as linear mappings, we can characterize the relation $B = NAQ$ purely in terms of matrix operations. We recall that any nonsingular matrix is a product of elementary matrices. Furthermore, if we multiply A on the left by any elementary matrix, we perform an elementary row operation on A, whereas if we multiply A on the right by any elementary matrix, we perform an elementary column operation on A. Hence we make the following definition and formally record some direct consequences of that definition.

Definition 6.2 Let A and B be m-by-n matrices. B is said to be *equivalent* to A if and only if B can be obtained by performing a finite sequence of elementary row and column operations on A.

Theorem 6.4 B is equivalent to A if and only if there exist nonsingular matrices N and Q such that

$$B = NAQ.$$

Proof Exercise.

Theorem 6.5 B is equivalent to A if and only if B and A represent the same linear mapping relative to suitable pairs of bases.
Proof   Exercise.

Theorem 6.6   Equivalence of matrices is an equivalence relation on the set of all m-by-n matrices.
Proof   Exercise.

Thus we know that the relation of matrix equivalence separates the set $\mathcal{M}_{m \times n}$ of all m-by-n matrices into mutually disjoint equivalence classes. Each m-by-n matrix appears in precisely one of those classes, all matrices in a single class are equivalent to each other, and any matrix not in that class is not equivalent to any matrix in that class. Naturally we would like to describe a canonical form for matrix equivalence—a simple form such that one and only one matrix in that form appears in each equivalence class. We recall that the reduced echelon form is canonical for row equivalence. A and B are row equivalent if and only if $B = NA$ for some nonsingular matrix N, so row equivalent matrices are equivalent. But not conversely; equivalent matrices are not necessarily row equivalent.

Theorem 6.7   An m-by-n matrix of rank k is equivalent to the m-by-n matrix B in which $b_{11} = b_{22} = \cdots = b_{kk} = 1$, and $b_{ij} = 0$ otherwise.
Proof   Let A be of rank k. If $k = 0$, then $A = Z$, and there is nothing to prove. Otherwise, by row operations we obtain the reduced echelon form of A, with k nonzero rows, the first nonzero element of each of which is one, and it is the only nonzero element in its column. By permuting the columns we place these ones in the first k diagonal positions, obtaining the block form

$$\left(\begin{array}{c|c} I_k & M \\ \hline Z & Z \end{array} \right).$$

Column operations are then used to produce zeros in the last $n - k$ columns of the first k rows.

Theorem 6.8   Two m-by-n matrices are equivalent if and only if they have the same rank.
Proof   If A and B are equivalent, $r(A) = r(B)$ because $B = NAQ$. Conversely, if A and B have rank k, each is equivalent to the matrix described in Theorem 6.7.

From Theorem 6.8 we derive two immediate corollaries. The first is that the form described in Theorem 6.7 is canonical with respect to equivalence. The second we state formally.

Theorem 6.9 A square matrix is nonsingular if and only if it is equivalent to the identity matrix.
Proof Apply Theorem 6.8.

Therefore, if A is nonsingular, there exist nonsingular matrices N and Q such that

$$I = NAQ,$$

$$N^{-1}Q^{-1} = A,$$

$$A^{-1} = QN.$$

Recall that N is obtained by performing on I the same row operations that were performed on A, and that Q is obtained by performing on I the same column operations that were performed on A, the combination of row and column operations transforming A into I. This gives us another method of computing A^{-1}, and the following scheme simplifies the calculation of N and Q. Write I to the left of A and below A:

$$\begin{array}{c|c} I & A \\ \hline & I \end{array}.$$

Perform row and column operations on A as needed to transform A into I. As each row operation is performed on A, perform the same row operation on the matrix at the upper left of the array. Similarly, as each column operation is performed on A, perform the same column operation on the lower right matrix. Then the final array is

$$\begin{array}{c|c} N & I \\ \hline & Q \end{array},$$

and $A^{-1} = QN$.

Example Let

$$A = \begin{pmatrix} 2 & -1 & 0 \\ 1 & 2 & 1 \\ -1 & 0 & 3 \end{pmatrix}.$$

$$\left(\frac{I \mid A}{I}\right) = \begin{pmatrix} 1 & 0 & 0 & \vline & 2 & -1 & 0 \\ 0 & 1 & 0 & \vline & 1 & 2 & 1 \\ 0 & 0 & 1 & \vline & -1 & 0 & 3 \\ \hline & & & \vline & 1 & 0 & 0 \\ & & & \vline & 0 & 1 & 0 \\ & & & \vline & 0 & 0 & 1 \end{pmatrix}$$

$$\underbrace{(R_1 \leftrightarrow R_2)} \begin{pmatrix} 0 & 1 & 0 & \vline & 1 & 2 & 1 \\ 1 & 0 & 0 & \vline & 2 & -1 & 0 \\ 0 & 0 & 1 & \vline & -1 & 0 & 3 \\ \hline & & & \vline & 1 & 0 & 0 \\ & & & \vline & 0 & 1 & 0 \\ & & & \vline & 0 & 0 & 1 \end{pmatrix}$$

$$\underbrace{\begin{matrix}(C_2 - 2C_1) \\ (C_3 - C_1)\end{matrix}} \begin{pmatrix} 0 & 1 & 0 & \vline & 1 & 0 & 0 \\ 1 & 0 & 0 & \vline & 2 & -5 & -2 \\ 0 & 0 & 1 & \vline & -1 & 2 & 4 \\ \hline & & & \vline & 1 & -2 & -1 \\ & & & \vline & 0 & 1 & 0 \\ & & & \vline & 0 & 0 & 1 \end{pmatrix}$$

$$\underbrace{\begin{matrix}(R_2 - 2R_1) \\ (R_3 + R_1)\end{matrix}} \begin{pmatrix} 0 & 1 & 0 & \vline & 1 & 0 & 0 \\ 1 & -2 & 0 & \vline & 0 & -5 & -2 \\ 0 & 1 & 1 & \vline & 0 & 2 & 4 \\ \hline & & & \vline & 1 & -2 & -1 \\ & & & \vline & 0 & 1 & 0 \\ & & & \vline & 0 & 0 & 1 \end{pmatrix}$$

$$\underbrace{\begin{matrix}\frac{1}{2}R_3 \\ (R_2 \leftrightarrow R_3)\end{matrix}} \begin{pmatrix} 0 & 1 & 0 & \vline & 1 & 0 & 0 \\ 0 & \frac{1}{2} & \frac{1}{2} & \vline & 0 & 1 & 2 \\ 1 & -2 & 0 & \vline & 0 & -5 & -2 \\ \hline & & & \vline & 1 & -2 & -1 \\ & & & \vline & 0 & 1 & 0 \\ & & & \vline & 0 & 0 & 1 \end{pmatrix}$$

$(C_3 - 2C_2)$
$$\left(\begin{array}{ccc|ccc} 0 & 1 & 0 & 1 & 0 & 0 \\ 0 & \frac{1}{2} & \frac{1}{2} & 0 & 1 & 0 \\ 1 & -2 & 0 & 0 & -5 & 8 \\ \hline & & & 1 & -2 & 3 \\ & & & 0 & 1 & -2 \\ & & & 0 & 0 & 1 \end{array}\right)$$

$(R_3 + 5R_2)$
$$\left(\begin{array}{ccc|ccc} 0 & 1 & 0 & 1 & 0 & 0 \\ 0 & \frac{1}{2} & \frac{1}{2} & 0 & 1 & 0 \\ 1 & \frac{1}{2} & \frac{5}{2} & 0 & 0 & 8 \\ \hline & & & 1 & -2 & 3 \\ & & & 0 & 1 & -2 \\ & & & 0 & 0 & 1 \end{array}\right)$$

$\frac{1}{8}R_3$
$$\left(\begin{array}{ccc|ccc} 0 & 1 & 0 & 1 & 0 & 0 \\ 0 & \frac{1}{2} & \frac{1}{2} & 0 & 1 & 0 \\ \frac{1}{8} & \frac{1}{16} & \frac{5}{16} & 0 & 0 & 1 \\ \hline & & & 1 & -2 & 3 \\ & & & 0 & 1 & -2 \\ & & & 0 & 0 & 1 \end{array}\right)$$

We have

$$N = \frac{1}{16}\begin{pmatrix} 0 & 16 & 0 \\ 0 & 8 & 8 \\ 2 & 1 & 5 \end{pmatrix} \quad \text{and} \quad Q = \begin{pmatrix} 1 & -2 & 3 \\ 0 & 1 & -2 \\ 0 & 0 & 1 \end{pmatrix},$$

and therefore

$$A^{-1} = QN = \frac{1}{16}\begin{pmatrix} 6 & 3 & -1 \\ -4 & 6 & -2 \\ 2 & 1 & 5 \end{pmatrix}.$$

This method of calculating A^{-1} is tedious to write, but all calculations are easy.

EXERCISES 6.3

1. Which of the matrices of Exercise 6.1-5 are equivalent?

2. Use the method illustrated in this section to calculate the inverse of each of the matrices A and C in Exercise 4.4-1. (Be sure to use both row and column operations to illustrate this method.)

3. Prove Theorem 6.4.

4. Prove Theorem 6.6.

5. Explain in detail why the form described by Theorem 6.7 is canonical with respect to equivalence of matrices.

6. Relative to matrix equivalence how many equivalence classes of m-by-n matrices are there? Explain. Answer the same question for row equivalence.

7. If A and B are equivalent, determine whether each of the following pairs of matrices are equivalent.

    (i) A^t and B^t.

    (ii) A^2 and B^2.

    (iii) AB and BA if either A or B is nonsingular.

    (iv) AB and BA if neither A nor B is nonsingular.

8. Prove the "only if" statement of Theorem 6.5. (The "if" statement was proved in Section 6.2.)

6.4   SIMILARITY

To summarize the work of this chapter until now, we have introduced two different equivalence relations on the set of all m-by-n matrices, called row equivalence and equivalence. Row equivalence was motivated by the representation of a linear system by a matrix, and we obtained these key results, stated more precisely in Section 6.1.

Row equivalent matrices correspond to equivalent linear systems.

The reduced echelon form is canonical for row equivalence.

Equivalence was motivated by the representation of a linear mapping by matrices relative to various pairs of bases, and we obtained these key results, stated more precisely in Section 6.3.

> *Equivalent matrices represent the same linear mapping.*

> *The rank of a matrix determines its equivalence class.*

> *The block form* $\left(\begin{array}{c|c} I_r & Z \\ \hline Z & Z \end{array}\right)$ *is canonical for equivalence.*

In this section we shall investigate a special form of matrix equivalence, called *similarity*, which arises by studying matrices that represent linear transformations on $\mathscr{V}_n$ rather than matrices that represent linear mappings from $\mathscr{V}_n$ to $\mathscr{W}_m$. Hence we now consider only *square* matrices. Fortunately, most of the hard work has already been done because we can make use of the calculations in Section 6.2.

Let **T** be a linear transformation on $\mathscr{V}_n$. For each choice of basis $\{\alpha_1, \ldots, \alpha_n\}$ for $\mathscr{V}_n$ we obtain a unique *n*-by-*n* matrix $A = (a_{ij})$, where

$$\mathbf{T}\alpha_j = \sum_{i=1}^{n} a_{ij}\alpha_i.$$

If **T** is represented relative to $\{\gamma_1, \ldots, \gamma_n\}$ by the matrix B, we wonder how A and B are related. This is a special form of the question investigated in Section 6.2, simplified in the sense that only one vector space is involved instead of two. Thus in Theorem 6.3 we can specify that $\mathscr{W}_m = \mathscr{V}_n$ and that $\beta_i = \alpha_i$ and $\gamma_i = \delta_i$ for $i = 1, \ldots, n$. It follows that the matrices P and Q are equal, so we obtain an answer in the form

$$PB = AP \text{ for some nonsingular matrix } P.$$

Definition 6.3 Let A and B be *n*-by-*n* matrices. B is said to be *similar* to A if and only if

$$B = P^{-1}AP$$

for some nonsingular matrix P.

Theorem 6.10 Similarity is an equivalence relation on the set of all *n*-by-*n* matrices.
Proof Exercise.

It is important to realize that although both equivalence and similarity are defined for n-by-n matrices, they are different equivalence relations. If two matrices are similar, they are equivalent; but equivalent n-by-n matrices are not necessarily similar.

Theorem 6.11   If A and B are similar, then

$$r(A) = r(B),$$

$$\det(A) = \det(B).$$

The converse, however, is false.

Proof   If $B = P^{-1}AP$, known properties of rank (Theorem 4.13) and determinant (Theorems 5.3 and 5.4) establish the assertions.

To see that equality of rank and determinant do not imply similarity, consider the matrices

$$A = \begin{pmatrix} 1 & 0 \\ 0 & 0 \end{pmatrix} \quad \text{and} \quad B = \begin{pmatrix} 0 & 0 \\ 1 & 0 \end{pmatrix}.$$

Both have rank one and determinant zero. Let A represent $\mathbf{T}$ relative to the standard basis. Then $\mathbf{T}\varepsilon_1 = \varepsilon_1$, so $\mathbf{T}$ maps some nonzero vector into itself. Let B represent $\mathbf{S}$ relative to any basis, and let ξ be any vector represented by a column vector X. Then $\mathbf{S}\xi$ is represented by

$$BX = \begin{pmatrix} 0 & 0 \\ 1 & 0 \end{pmatrix}\begin{pmatrix} a \\ b \end{pmatrix} = \begin{pmatrix} 0 \\ a \end{pmatrix}.$$

For $\mathbf{S}$ to map ξ into itself, the condition is that

$$\begin{pmatrix} 0 \\ a \end{pmatrix} = \begin{pmatrix} a \\ b \end{pmatrix},$$

so $0 = a = b$. That is, $\mathbf{T}$ maps at least one nonzero vector onto itself, but $\mathbf{S}$ maps no nonzero vector onto itself. Hence $\mathbf{T} \neq \mathbf{S}$. From the next theorem it will follow that A and B are not similar, although they are equivalent.

Theorem 6.12   Two n-by-n matrices A and B are similar if and only if they represent the same linear transformation $\mathbf{T}$ on $\mathscr{V}_n$, each relative to a suitable basis for $\mathscr{V}_n$.

Proof   This is a special case of Theorem 6.5, for which a proof was requested in Exercise 6.3-8. We here provide details that will also show how to proceed in that exercise. The "if" statement of this theorem follows from the calculations of Section 6.2 (with $m = n$, $\beta_i = \alpha_i$, and $\delta_i = \gamma_i$ for $i = 1, \ldots, n$). To prove the "only if" assertion, let $B = P^{-1}AP$. Choose any basis $\{\alpha_1, \ldots, \alpha_n\}$ for $\mathscr{V}_n$ and let **T** be the linear transformation represented by B relative to the γ-basis. Consequently, for $j = 1, \ldots, n$,

$$\gamma_k = \sum_{i=1}^{n} p_{ik}\alpha_i, \qquad k = 1, \ldots, n.$$

Because P is nonsingular, the γ's form a basis for $\mathscr{V}_n$. Let **S** be the linear transformation represented by B relative to the γ-basis. Consequently, for $j = 1, \ldots, n$,

$$\mathbf{S}\gamma_j = \sum_{k=1}^{n} b_{kj}\gamma_k = \sum_{k=1}^{n} b_{kj}\left(\sum_{i=1}^{n} p_{ik}\alpha_i\right)$$

$$= \sum_{i=1}^{n} \left(\sum_{k=1}^{n} p_{ik}b_{kj}\right)\alpha_i.$$

But also

$$\mathbf{T}\alpha_k = \sum_{i=1}^{n} a_{ik}\alpha_i,$$

$$\mathbf{T}\gamma_j = \mathbf{T}\left(\sum_{k=1}^{n} p_{kj}\alpha_k\right) = \sum_{k=1}^{n} p_{kj}(\mathbf{T}\alpha_k)$$

$$= \sum_{k=1}^{n} p_{kj}\left(\sum_{i=1}^{n} a_{ik}\alpha_i\right)$$

$$= \sum_{i=1}^{n} \left(\sum_{k=1}^{n} a_{ik}p_{kj}\right)\alpha_i.$$

Because $\sum_{k=1}^{n} a_{ik}p_{kj}$ is the (i, j) entry of AP, and $\sum_{k=1}^{n} p_{ik}b_{kj}$ is the (i, j) entry of $PB = AP$, we see that $\mathbf{S}\gamma_j = \mathbf{T}\gamma_j$ for $j = 1, \ldots, n$, so $\mathbf{S} = \mathbf{T}$.

From this proof we observe the significance of the nonsingular matrix P in the relation $B = P^{-1}AP$. *If A represents* **T** *relative to a given basis and B represents* **T** *relative to a second basis, then P is the matrix whose columns express the vectors of the second basis in terms of the vectors of the given basis.*

The natural next step in studying similarity is to look for a canonical form for similarity, a simple form such that each similarity class of n-by-n matrices has one and only one matrix in that form. It turns out that a canonical form for similarity is much harder to derive than for row equivalence or equivalence, and we shall devote all of the next chapter to that derivation and related ideas. Then in Chapter 9 we shall study two more equivalence relations on *n*-by-*n* matrices, called *congruence* and *conjunctivity*.

Example To illustrate the ideas of this section and to suggest some of the matters that we will address in the next chapter, suppose that we denote by **T** the linear transformation that is represented relative to the standard basis by a given matrix,

$$A = \begin{pmatrix} 2 & 10 & 5 \\ -2 & -4 & -4 \\ 3 & 5 & 6 \end{pmatrix}.$$

Not much useful information about **T** is evident from A, except that with a little work we could see that A, and hence **T**, is nonsingular. But we would like to find a way of representing **T** by a matrix B that yields information more readily. So we change bases. Without concerning ourselves at this time with how we can make **T** tell us how to choose a basis wisely, suppose we choose

$$\alpha_1 = (0, 2, -3),$$

$$\alpha_2 = (5, 2, -5),$$

$$\alpha_3 = (1, 1, -2).$$

To find the matrix B that represents **T** relative to the α-basis, we compute $P^{-1}AP$, where P is the matrix whose column vectors are α_1, α_2, and α_3. You can verify that

$$P^{-1} = \begin{pmatrix} 1 & 5 & 3 \\ 1 & 3 & 2 \\ -4 & -15 & -10 \end{pmatrix} \quad \text{and} \quad B = P^{-1}AP = \begin{pmatrix} 1 & 0 & 0 \\ 1 & 1 & 0 \\ 0 & 0 & 2 \end{pmatrix}.$$

Now we can immediately extract from B some geometrical information about the behavior of **T**; for example,

$$\mathbf{T}\alpha_3 = 2\alpha_3,$$

$$\mathbf{T}\alpha_2 = \alpha_2,$$

$$\mathbf{T}\alpha_1 = \alpha_1 + \alpha_2.$$

That is, **T** merely doubles the vector α_3 (and any vector in the subspace $[\alpha_3]$). **T** leaves α_2 unchanged (and any vector in the subspace $[\alpha_2]$). And **T** maps α_1 into the sum of α_1 and α_2. Clearly we prefer to work with B

rather than with A in studying **T** because B is simpler computationally and geometrically. Is it possible to find still another matrix C that represents **T** even more simply than B? In the next chapter we shall see that the answer is no. B is the best we can do, the canonical form of A relative to similarity. We shall also learn how to use **T** itself to select the new basis, relative to which **T** is represented by a matrix in canonical form.

EXERCISES 6.4

1. Prove Theorem 6.10.

2. Let **T** denote the linear transformation on $\mathbb{R}^2$ that is represented relative to the standard basis by the matrix

$$A = \begin{pmatrix} 1 & 1 \\ 1 & 1 \end{pmatrix}.$$

Let $\alpha_1 = (1, 1)$ and $\alpha_2 = (1, -1)$ define another basis.

(i) By computing $\mathbf{T}\alpha_1$ and $\mathbf{T}\alpha_2$, expressed in terms of α_1 and α_2, determine then matrix B that represents **T** relative to the α-basis.

(ii) Write a matrix P such that $P^{-1}AP = B$, and carry out that computation to verify your answer to (i).

3. Let **T** denote the linear transformation on $\mathbb{R}^3$ that is represented relative to the standard basis by the matrix

$$A = \begin{pmatrix} 1 & 2 & 1 \\ 2 & 0 & -2 \\ -1 & 2 & 3 \end{pmatrix}.$$

Let $\alpha_1 = (1, 1, 0)$, $\alpha_2 = (1, 0, 1)$, $\alpha_3 = (1, -1, 1)$.

(i) Show that $\{\alpha_1, \alpha_2, \alpha_3\}$ is a basis for $\mathbb{R}^3$.

(ii) By computing each $\mathbf{T}\alpha_j$ and expressing each in terms of the α-basis, determine the matrix B that represents **T** relative to the α-basis.

(iii) Write a matrix P that represents the change from the standard basis to the α-basis and verify by computation that $P^{-1}AP = B$.

4. Are AB and BA similar for all n-by-n matrices A, B? What can be said if A is nonsingular?

5. Let A and B be similar matrices. Show that

    (i) cA and cB are similar for each scalar c,

    (ii) A^k and B^k are similar for each positive integer k,

    (iii) $p(A)$ and $p(B)$ are similar for each polynomial p, where if

$$p(x) = \sum_{i=0}^{n} a_i x^i,$$

then

$$p(A) = \sum_{i=0}^{n} a_i A^i.$$

6. Let A and B be similar matrices.

    (i) Are A^{-1} and B^{-1} similar, given that A is nonsingular?

    (ii) Are A^t and B^t similar?

7. Let $\mathbf{T}$ be any idempotent linear transformation on a vector space $\mathscr{V}$, and let $\mathscr{R}(\mathbf{T})$ and $\mathscr{N}(\mathbf{T})$ denote the range space and null space of $\mathbf{T}$.

    (i) Prove that $\mathscr{V}$ is the direct sum of $\mathscr{R}(\mathbf{T})$ and $\mathscr{N}(\mathbf{T})$.

    (ii) Prove that $\mathbf{T}$ maps each vector in $\mathscr{R}(\mathbf{T})$ into itself, and each vector in $\mathscr{N}(\mathbf{T})$ into 0. (Thus $\mathbf{T}$ acts as the identity mapping on the subspace $\mathscr{R}(\mathbf{T})$.)

    (iii) Describe a simple matrix representation of any idempotent linear transformation $\mathbf{T}$ on $\mathscr{V}_n$. The rank of $\mathbf{T}$ should be part of your description.

8. In the analysis of three-phase power systems an impedance matrix occurs in the form

$$C = \begin{pmatrix} c_1 & c_3 & c_2 \\ c_2 & c_1 & c_3 \\ c_3 & c_2 & c_1 \end{pmatrix},$$

where each c_i is a complex number. Let $e = (1/2)(-1 + \sqrt{3}\,i)$, where $i^2 = -1$. (Observe that $e^3 = 1$, so $e^2 + e + 1 = 0$). Let P be the matrix

$$P = \begin{pmatrix} 1 & 1 & 1 \\ 1 & e & e^2 \\ 1 & e^2 & e \end{pmatrix}.$$

Show by computation that $P^{-1}CP$ is diagonal.

9. Explain how Theorems 6.11 and 6.12 can be used to define the determinant of a linear transformation $\mathbf{T}$ on $\mathscr{V}_n$. Interpret $|\det \mathbf{T}|$ in terms of the n-dimensional content of the parallelotope having as adjacent sides the $\mathbf{T}$-images of the vectors of an arbitrary basis for $\mathscr{V}_n$.

CHAPTER 7
A CANONICAL FORM
FOR SIMILARITY

7.1 CHARACTERISTIC VALUES AND VECTORS

Each linear transformation $\mathbf{T}$ on $\mathcal{V}_n$ determines a similarity class of n-by-n matrices. A given matrix A is in the similarity class determined by $\mathbf{T}$ if and only if A represents $\mathbf{T}$ relative to some basis for $\mathcal{V}_n$. Therefore, all matrices in the same similarity class must share in common those properties of $\mathbf{T}$ that are independent of the choice of basis and therefore are valid in any coordinate system. In looking for a canonical matrix representation of $\mathbf{T}$, therefore, it is natural to look for a basis that is intrinsically related to $\mathbf{T}$. Thus we begin by finding vectors that are mapped by $\mathbf{T}$ in the simplest possible way.

We know that $\mathbf{T}v = \theta$ for every $v \in \mathcal{N}(\mathbf{T})$. But if $\mathbf{T}$ is nonsingular, then $v = \theta$. But every linear transformation carries θ into θ, so this is no help in characterizing a given $\mathbf{T}$. We might look instead for a fixed point of $\mathbf{T}$, a vector ξ such that $\mathbf{T}\xi = \xi$. More generally, we could look for a vector ξ that is mapped by $\mathbf{T}$ into a scalar multiple of itself:

$$\mathbf{T}\xi = c\xi \text{ for some scalar } c.$$

If $c = 0$, $\xi \in \mathcal{N}(\mathbf{T})$; and if $c = 1$ then ξ is a fixed point of $\mathbf{T}$.

Definition 7.1   A nonzero vector ξ is called a *characteristic vector* of **T** if and only if there exists a scalar c such that

$$\mathbf{T}\xi = c\xi.$$

The associated scalar c is called a *characteristic value* of **T**.

There are many synonyms for characteristic vectors (*eigenvectors, eigenstates, proper vectors, proper states*) and for characteristic values (*eigenvalues, proper values, spectral numbers, characteristic roots, latent roots*). The set of all characteristic values of **T** is called the *spectrum* of **T**. Traditionally, characteristic values have been denoted by the Greek letter λ, so we shall use that notation even though it is an exception to our convention of using Latin letters for scalars.

Example   In Exercise 6.4-3 **T** was represented relative to the standard basis by

$$A = \begin{pmatrix} 1 & 2 & 1 \\ 2 & 0 & -2 \\ -1 & 2 & 3 \end{pmatrix}.$$

To find a characteristic vector ξ and associated characteristic value λ, we wish to solve the vector equation

$$\mathbf{T}\xi = \lambda\xi.$$

In vector form this vector equation is the homogeneous linear system

$$AX = \lambda X, \text{ or}$$

$$(A - \lambda I)X = Z.$$

This system has a nonzero solution X if and only if

$$\det(A - \lambda I) = 0;$$

$$\det \begin{pmatrix} 1-\lambda & 2 & 1 \\ 2 & 0-\lambda & -2 \\ -1 & 2 & 3-\lambda \end{pmatrix} = 0.$$

By evaluating the determinant we obtain a cubic polynomial in the unknown scalar λ, which can then be factored to obtain the equation

$$\lambda(\lambda - 2)^2 = 0.$$

There are two solutions, $\lambda_1 = 0$ and $\lambda_2 = 2$, and these are the characteristic values of A. There are only two distinct characteristic values because $(\lambda - 2)$ is a repeated factor of the cubic polynomial.

To find characteristic vectors associated with the characteristic value $\lambda_2 = 2$, we solve for X the homogeneous linear system

$$(A - 2I)X = Z,$$

obtaining the solution vector

$$X_2 = a \begin{pmatrix} 1 \\ 0 \\ 1 \end{pmatrix} \text{ for any } a.$$

Hence for any $a \neq 0$, X_2 is a characteristic vector of A, associated with the characteristic value $\lambda_2 = 2$. Similarly, for $\lambda_1 = 0$, we solve for X the linear system

$$(A - 0I)X = Z,$$

obtaining the solution vector

$$X_1 = b \begin{pmatrix} 1 \\ -1 \\ 1 \end{pmatrix} \text{ for any } b.$$

Hence for any $b \neq 0$, X_1 is a characteristic vector of A, associated with the characteristic value $\lambda_1 = 0$.

Now we choose a basis for $\mathbb{R}^3$ containing as many linearly independent characteristic vectors as possible. No matter how we choose a and b, only two linearly independent characteristic vectors can be obtained, so we take $b = 1$ and $a = 1$, and let

$$\alpha_3 = (1, -1, 1),$$

$$\alpha_2 = (1, 0, 1),$$

as described in Exercise 6.4-3. To complete a basis suppose we choose

$$\alpha_1 = (1, 1, 0).$$

Then Exercise 6.4-3 shows that $\mathbf{T}$ is represented relative to the ordered basis $\{\alpha_1, \alpha_2, \alpha_3\}$ by the matrix

$$B = \begin{pmatrix} 2 & 0 & 0 \\ 1 & 2 & 0 \\ 0 & 0 & 0 \end{pmatrix}.$$

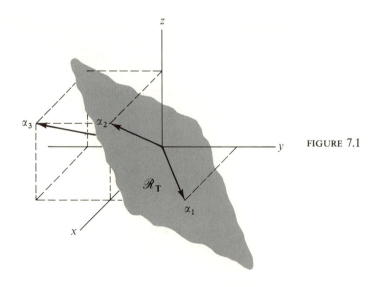

FIGURE 7.1

Observe that the diagonal elements of B are the characteristic values of A, with the repeated value 2 appearing twice. Had we chosen a different vector in place of α_1 the first column would be changed; but, in a sense that will be explained later, the basis $\{\alpha_1, \alpha_2, \alpha_3\}$ produces the simplest possible matrix representation of **T** (see Figure 7.1).

As we proceed to develop a general theory you should refer to this example to illustrate the ideas introduced. Already we observe that much algebraic computation is required to find the characteristic values and vectors of a linear transformation. In this example, we had to factor a cubic polynomial, and you will recall that a real polynomial does not necessarily factor completely into real linear factors. However, any polynomial with real or complex coefficients does factor into a product of *complex* linear factors. Thus we make the simplifying assumption that *during the remainder of this chapter we shall consider the field* $\mathbb{C}$ *of complex numbers to be the scalar field for the vector spaces considered.* For ease of computation, the exercises and examples will be carefully chosen so that complex numbers are rarely if ever needed for numerical work. But because some key theorems depend upon the algebraic completeness of the scalar field, we shall assume that the scalar field is $\mathbb{C}$.

Now let **T** be any linear transformation on an n-dimensional vector space over the complex numbers. Relative to a basis let **T** be represented by the n-by-n matrix A, and let ξ be a nonzero vector represented by the column vector X. Then ξ is a characteristic vector of **T** associated with the characteristic value λ if and only if

$$\mathbf{T}\xi = \lambda\xi,$$

$$AX = \lambda X,$$

$$(A - \lambda I)X = Z.$$

This is the matrix form of a homogeneous linear system of n equations in the n variables $x_1, \ldots, x_n$. The system has a nontrivial solution if and only if the matrix $A - \lambda I$ is singular; that is, if and only if

$$\det(A - \lambda I) = 0.$$

From our knowledge of determinants we see that

$$\det(A - \lambda I) = \det \begin{pmatrix} a_{11} - \lambda & a_{12} & \cdots & a_{1n} \\ a_{21} & a_{22} - \lambda & \cdots & a_{2n} \\ \cdot & \cdot & & \cdot \\ \cdot & \cdot & & \cdot \\ \cdot & \cdot & & \cdot \\ a_{n1} & a_{n2} & \cdots & a_{nn} - \lambda \end{pmatrix}$$

is a polynomial of degree n in the variable λ, of the form

$$p(\lambda) = (-1)^n(\lambda^n + c_1\lambda^{n-1} + \cdots + c_{n-1}\lambda + c_n),$$

where each c_i is a sum of signed products of the entries of A. The fundamental theorem of algebra and its corollaries guarantee that there are n complex numbers $\lambda_1, \ldots, \lambda_n$ (not necessarily distinct from each other) such that p factors into linear factors of the form

$$p(\lambda) = (-1)^n(\lambda - \lambda_1)(\lambda - \lambda_2) \cdots (\lambda - \lambda_n).$$

Then for $i = 1, 2, \ldots, n$, $p(\lambda_i) = 0$; hence $A - \lambda_i I$ is singular, and there exists at least one nonzero vector X_i such that

$$AX_i = \lambda_i X_i.$$

If ξ_i is the vector represented by X_i, then

$$\mathbf{T}\xi_i = \lambda_i\xi_i, \qquad \xi_i \neq \theta.$$

Hence ξ_i is a characteristic vector of **T**, associated with the characteristic value λ_i of **T**.

Definition 7.2 The polynomial $\det(A - xI)$ is called the *characteristic polynomial* of the matrix A. The equation $\det(A - xI) = 0$ is called the *characteristic equation* of A. The roots of the characteristic equation of A are called the *characteristic values* of A, and for each characteristic value λ_i of A any nonzero column vector X_i such that

$$(A - \lambda_i I)X_i = Z$$

is called a *characteristic vector* of A, associated with λ_i.

Our previous discussion shows that if **T** is represented by A, the characteristic values of **T** are the characteristic values of A, and each characteristic vector of **T** is represented by a characteristic vector of A.

Theorem 7.1   If A and B are similar, then A and B have the same characteristic polynomial and hence the same characteristic values.
Proof   Exercise.

Theorem 7.2   The characteristic values of a triangular matrix are the diagonal entries.
Proof   Exercise.

Theorem 7.3   Let **T** be a linear transformation on $\mathscr{V}$, let λ be a characteristic value of **T**, and let $\mathscr{C}(\lambda)$ denote the set consisting of θ and all characteristic vectors ξ of **T** that are associated with λ. Then $\mathscr{C}(\lambda)$ is a subspace of $\mathscr{V}$, and $\mathbf{T}\xi \in \mathscr{C}(\lambda)$ for each ξ in $\mathscr{C}(\lambda)$.
Proof   Let $\xi_1, \xi_2 \in \mathscr{C}(\lambda)$ and let $a, b \in \mathbb{C}$. Then

$$\mathbf{T}(a\xi_1 + b\xi_2) = a\mathbf{T}\xi_1 + b\mathbf{T}\xi_2$$

$$= a\lambda\xi_1 + b\lambda\xi_2$$

$$= \lambda(a\xi_1 + b\xi_2).$$

That is, any linear combination of vectors of $\mathscr{C}(\lambda)$ is in $\mathscr{C}(\lambda)$. By Theorem 3.12, $\mathscr{C}(\lambda)$ is a subspace of the domain of **T**. Also for each $\xi \in \mathscr{C}(\lambda)$, $\mathbf{T}\xi = \lambda\xi \in \mathscr{C}(\lambda)$.

The subspace $\mathscr{C}(\lambda)$, consisting of θ and all characteristic vectors associated with the same characteristic value λ, is called a *characteristic subspace* of **T** or of any matrix that represents **T**. Properties of $\mathscr{C}(\lambda)$ are developed in Exercises 12–14 below. Other exercises explore further properties of the characteristic polynomial, characteristic values, and characteristic vectors. Be sure to read each exercise carefully and to understand its content whether or not you write a solution.

EXERCISES 7.1

1. Determine the characteristic polynomial, characteristic values, characteristic subspaces, and a maximal linearly independent set of characteristic vectors of each of the following matrices:

(i) $A = \begin{pmatrix} 0 & 2 \\ 3 & -1 \end{pmatrix}$.

· (ii) $B = \begin{pmatrix} 3 & 2 & 4 \\ 2 & 0 & 2 \\ 4 & 2 & 3 \end{pmatrix}$.

(iii) $C = \begin{pmatrix} 3 & 2 & 1 & 0 \\ 0 & 1 & 0 & 1 \\ 0 & 2 & 1 & 0 \\ 0 & 0 & 0 & 1 \end{pmatrix}$.

2. Referring to the Example in Section 6.4, determine the characteristic values of A and show that α_2 and α_3 are characteristic vectors. Explain how the matrix B displays this information directly.

3. Prove Theorem 7.1.

4. Show that the converse of Theorem 7.1 is not valid by letting

$$A = \begin{pmatrix} 0 & 0 \\ 1 & 0 \end{pmatrix} \text{ and } B = Z.$$

Show that A and B have the same characteristic polynomial, but that the characteristic subspaces $\mathscr{C}(0)$ for A and B have dimensions one and two, respectively. How does this fact show that A and B are not similar?

5. Prove Theorem 7.1.

6. Determine the possible characteristic values of A if

(i) A is idempotent,

(ii) A is nilpotent,

(iii) A is nonsingular.

7. Prove that if A is nonsingular then the characteristic values of A^{-1} are the reciprocals of the characteristic values of A. What can be said about the corresponding characteristic vectors?

8. Show that if X is a characteristic vector of A associated with the value λ, then for any natural number k, X is a characteristic vector of A^k associated with the characteristic value λ^k.

9. By considering the characteristic polynomial of A,

$$p(x) = \det(A - xI) = (-1)^n(x - \lambda_1)(x - \lambda_2) \cdots (x - \lambda_n),$$

determine the value of the constant term $p(0)$ to deduce that

   (i) $\det A = \lambda_1 \lambda_2 \cdots \lambda_n$,

   (ii) A is singular if and only if some characteristic value of A is zero.

10. By considering the characteristic polynomial of A,

$$p(x) = \det(A - xI) = (-1)^n(x - \lambda_1)(x - \lambda_2) \cdots (x - \lambda_n),$$

show that the coefficient of x^{n-1} is $-(\lambda_1 + \lambda_2 + \cdots + \lambda_n)$. (The sum $(\lambda_1 + \lambda_2 + \cdots + \lambda_n)$ is called the *trace* of A.)

11. A *companion* matrix C is any n-by-n matrix of the form

$$C = \begin{pmatrix} 0 & 0 & \cdots & 0 & c_1 \\ 1 & 0 & \cdots & 0 & c_2 \\ 0 & 1 & \cdots & 0 & c_3 \\ . & . & & . & . \\ . & . & & . & . \\ . & . & & . & . \\ 0 & 0 & \cdots & 1 & c_n \end{pmatrix}.$$

   (i) Show by induction or otherwise that the characteristic polynomial of C is

$$\det(C - xI) = (-1)^n[x^n - c_n x^{n-1} - \cdots - c_2 x - c_1].$$

   (ii) Deduce that any polynomial is a scalar multiple of the characteristic polynomial of some matrix.

12. Referring to Theorem 7.3, suppose that we choose any basis for $\mathscr{C}(\lambda)$ and extend it to a basis for $\mathscr{V}_n$. Describe the matrix that represents $\mathbf{T}$ relative to that basis.

13. Let $\lambda_1, \ldots, \lambda_k$ be distinct characteristic values of $\mathbf{T}$. Let ξ_i be any characteristic vector associated with λ_i. Thus ξ_i is any nonzero vector of $\mathscr{C}(\lambda_i)$. Prove that $\{\xi_1, \xi_2, \ldots, \xi_k\}$ is linearly independent.

14. With λ_i as in the preceding exercise, show that

$$\mathscr{C}(\lambda_k) \cap (\mathscr{C}(\lambda_1) + \cdots + \mathscr{C}(\lambda_{k-1})) = [\theta].$$

Deduce that the sum $\mathscr{C}(\lambda_1) + \cdots + \mathscr{C}(\lambda_k)$ of characteristic subspaces is a *direct* sum. (Recall Definition 2.6.)

15. Let

$$A = \begin{pmatrix} a & b \\ c & d \end{pmatrix}.$$

(i) Show that the characteristic equation of A is

$$p(x) = x^2 - (a + d)x + (ad - bc).$$

(ii) Relate this result to the claims of Exercises 9 and 10.

(iii) Show by matrix computation that $p(A)$ is the zero matrix, where $p(A) = A^2 - (a + d)A + (ad - bc)I$.

16. A Markov matrix was defined in Exercise 4.1-11. Prove that every characteristic value of a Markov matrix satisfies $|\lambda| \leq 1$.

17. Prove that $\lambda = 1$ is a characteristic value of every Markov matrix. This means that any linear transformation that is represented by a Markov matrix must have a fixed point, $\mathbf{T}\xi = \xi$.

18. Let A be an n-by-n matrix in which the sum of the elements in each row is the same number s. Show that s is a characteristic value of A. (See Exercise 17 for a special case of this result.)

19. Let $\mathscr{V}$ be the space of all real functions that are differentiable on the interval $0 \leq x \leq 1$. The derivative linear transformation $\mathbf{D}$ and the integral linear transformation $\mathbf{J}$ are defined on $\mathscr{V}$ by

$$\mathbf{D}f(x) = f'(x),$$

$$\mathbf{J}f(x) = \int_0^x f(t)\, dt.$$

Determine the characteristic values and corresponding characteristic vectors of each of $\mathbf{D}$ and $\mathbf{J}$.

7.2 DIAGONABILITY

It is now apparent that, in seeking a simple matrix representative of $\mathbf{T}$, the best we can hope for is a diagonal matrix with the characteristic values of $\mathbf{T}$ along the main diagonal. Because this is not always possible, it is important to distinguish those linear transformations that can be so represented. Expressed in terms of square matrices, we want to determine whether a given matrix A is similar to a diagonal matrix D. Such a matrix is said to be *diagonable*. We shall derive several different characterizations of diagonability in this and subsequent sections. (The proper words are *diagonalizable* and *diagonalizability*; the contractions adopted here are easier to use.)

To begin, suppose that A is similar to a diagonal matrix D,

$$D = P^{-1}AP = \begin{pmatrix} d_{11} & 0 & \cdots & 0 \\ 0 & d_{22} & \cdots & 0 \\ \cdot & \cdot & & \cdot \\ \cdot & \cdot & & \cdot \\ \cdot & \cdot & & \cdot \\ 0 & 0 & \cdots & d_{nn} \end{pmatrix} = \mathrm{diag}(d_{11}, \ldots, d_{nn}).$$

The characteristic values of D, and hence of A, are the diagonal entries of D. Furthermore, if A represents **T** relative to an α-basis and D represents **T** relative to a δ-basis, then from the form of D we see that

$$\mathbf{T}\delta_i = d_{ii}\delta_i.$$

Hence each of the n vectors in the δ-basis is a characteristic vector of **T** and therefore of D and A. Thus if an n-by-n matrix A is diagonable, then A has a set of n linearly independent characteristic vectors. We now show that the converse is true.

Theorem 7.4   An n-by-n matrix A is similar to a diagonal matrix if and only if A has n linearly independent characteristic vectors, $X_1, \ldots, X_n$. If such vectors exist, $P^{-1}AP$ is diagonal, where P has $X_1, \ldots, X_n$ as column vectors, and where the diagonal entries of $P^{-1}AP$ are the associated characteristic values of A.

Proof   The previous argument shows that if A is similar to a diagonal matrix, then A has n linearly independent characteristic vectors. Conversely, let $\{X_1, \ldots, X_n\}$ be a linearly independent set of characteristic vectors of A, where $AX_i = \lambda_i X_i$, $i = 1, \ldots, n$. Relative to a given basis for $\mathscr{V}_n$, let A represent the linear transformation **T** and let X_i represent the vector ξ_i, $i = 1, \ldots, n$. Then $\{\xi_1, \ldots, \xi_n\}$ is a basis for $\mathscr{V}_n$, and $\mathbf{T}\xi_i = \lambda_i \xi_i$. If we change from the given basis to the ξ-basis, **T** is represented by a matrix D that is similar to A and of the form

$$D = P^{-1}AP = \mathrm{diag}(\lambda_1, \ldots, \lambda_n).$$

Furthermore, from the observation immediately following the proof of Theorem 6.12, we know that P is the matrix whose column vectors are $X_1, X_2, \ldots, X_n$.

Example   We shall find the characteristic values, characteristic vectors, and a diagonalizing matrix P for the matrix

$$A = \begin{pmatrix} 1 & 0 & -2 \\ 0 & 0 & 0 \\ -2 & 0 & 4 \end{pmatrix}.$$

The characteristic equation of A is

$$0 = \det(A - xI) = \det \begin{pmatrix} 1-x & 0 & -2 \\ 0 & -x & 0 \\ -2 & 0 & 4-x \end{pmatrix}$$

$$= (1-x)(-x)(4-x) - (-2)(-x)(-2)$$

$$= -x^3 + 5x^2.$$

Hence the characteristic values of A are $\lambda_1 = 0$, $\lambda_2 = 0$, $\lambda_3 = 5$.

If $X = \begin{pmatrix} x_1 \\ x_2 \\ x_3 \end{pmatrix}$, then $AX = \begin{pmatrix} x_1 - 2x_3 \\ 0 \\ -2x_1 + 4x_3 \end{pmatrix}$ and $\lambda X = \begin{pmatrix} \lambda x_1 \\ \lambda x_2 \\ \lambda x_3 \end{pmatrix}$.

Necessary and sufficient conditions that X be a characteristic vector associated with λ are therefore

$$x_1 - 2x_3 = \lambda x_1,$$

$$0 = \lambda x_2,$$

$$-2x_1 + 4x_3 = \lambda x_3.$$

For $\lambda = 0$ these reduce to

$$x_1 = 2x_3.$$

Hence any vector of the form $(2c, b, c)$ is characteristic. Two such vectors that are linearly independent are

$$X_1 = \begin{pmatrix} 2 \\ 0 \\ 1 \end{pmatrix} \quad \text{and} \quad X_2 = \begin{pmatrix} 0 \\ 1 \\ 0 \end{pmatrix}.$$

Notice in this case that it is possible to select two linearly independent characteristic vectors both of which are associated with the same characteristic value. For $\lambda_3 = 5$ the conditions reduce to

$$-2x_1 = x_3,$$

$$x_2 = 0.$$

Hence any vector of the form $(a, 0, -2a)$ is characteristic. As a simple vector of this form we choose

$$X_3 = \begin{pmatrix} 1 \\ 0 \\ -2 \end{pmatrix}$$

as a characteristic vector associated with $\lambda_3 = 5$. Then

$$P = \begin{pmatrix} 2 & 0 & 1 \\ 0 & 1 & 0 \\ 1 & 0 & -2 \end{pmatrix}$$

and det $P = -5$, which checks the linear independence of X_1, X_2, and X_3. You should check the calculations, which show that

$$P^{-1} = -\tfrac{1}{5} \begin{pmatrix} -2 & 0 & -1 \\ 0 & -5 & 0 \\ -1 & 0 & 2 \end{pmatrix}$$

and

$$P^{-1}AP = \begin{pmatrix} 0 & 0 & 0 \\ 0 & 0 & 0 \\ 0 & 0 & 5 \end{pmatrix}.$$

The geometric interpretation of these calculations is that in $\mathbb{R}^3$ each point η on the line determined by the origin and $\xi_3 = (1, 0, -2)$ is mapped by $\mathbf{T}$ into 5η, whereas each point ζ on the plane determined by the origin, $\xi_1 = (2, 0, 1)$, and $\xi_2 = (0, 1, 0)$ is mapped by $\mathbf{T}$ into $0\zeta = 0$. The three characteristic vectors ξ_1, ξ_2, and ξ_3 are linearly independent and may be chosen as a basis for $\mathbb{R}^3$. Relative to that basis, $\mathbf{T}$ is represented by the diagonal matrix, diag$(0, 0, 5)$. See Figure 7.2.

When we attempt to diagonalize a given matrix A by the preceding method there are three possible outcomes.

(1) If there are n distinct characteristic values, the next theorem shows that there are n linearly independent characteristic vectors, so A is diagonable.

(2) If there are fewer than n distinct characteristic values, a set of n linearly independent characteristic vectors might or might not exist. If such a set exists, A is diagonable.

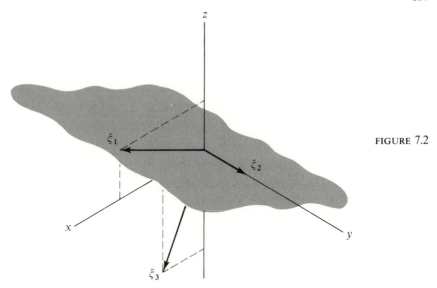

FIGURE 7.2

(3) If no set of n linearly independent characteristic vectors exists, A is not diagonable.

Each of these cases occurs in the examples of Exercise 7.2-1.

> **Theorem 7.5**   If an n-by-n matrix A has n distinct characteristic values, then A is diagonable. The converse, however, is false.
> **Proof**   Exercise 7.1-13 shows that any set $\{\xi_1, \ldots, \xi_k\}$ of characteristic vectors is linearly independent when the corresponding characteristic values $\lambda_1, \ldots, \lambda_k$, are k distinct numbers. When $k = n$, this implies that A is diagonable by Theorem 7.4. The preceding example shows that a matrix can be diagonable even when it has fewer than n distinct characteristic values.

> **Theorem 7.6**   An n-by-n matrix A is diagonable if and only if its characteristic subspaces span $\mathscr{V}_n$.
> **Proof**   This is essentially a restatement of Theorem 7.4 because any nonzero vector in any characteristic subspace is a characteristic vector.

The relation of the characteristic subspaces to diagonability can be clarified by examining the characteristic polynomial of A. A number λ_i is a characteristic value of A if and only if $(x - \lambda_i)$ is a factor of the characteristic polynomial p of A. If there are exactly t distinct characteristic values of A, then

$$p(\lambda) = (-1)^n(x - \lambda_1)^{s_1}(x - \lambda_2)^{s_2} \cdots (x - \lambda_t)^{s_t},$$

where $s_1 + s_2 + \cdots + s_t = n$. The number s_i is called the *algebraic multiplicity* of the characteristic value λ_i. The dimension g_i of the characteristic subspace $\mathscr{C}(\lambda_i)$ is called the *geometric multiplicity* of λ_i. Furthermore,

$$1 \le g_i = \dim \mathscr{C}(\lambda_i) \le s_i;$$

that is, the geometric multiplicity of λ_i never exceeds its algebraic multiplicity. (See Exercise 7.2-6.) If for some i dim $\mathscr{C}(\lambda_i) < s_i$, then the characteristic subspaces fail to span $\mathscr{V}_n$, and A is not diagonable. On the other hand, if equality holds for each i, a basis of characteristic vectors for $\mathscr{V}_n$ can be obtained by choosing a basis for each $\mathscr{C}(\lambda_i)$ and forming the union of these bases for $i = 1, \ldots, t$. Hence A is diagonable if and only if the algebraic and geometric multiplicities of each characteristic value coincide.

To obtain still another characterization of diagonability we start from the characteristic polynomial of A, written as in the previous paragraph. If A is diagonable, it is similar to a block diagonal matrix in which each block also is diagonal:

$$D = \begin{pmatrix} D_1 & & & & \\ & D_2 & & & \\ & & \cdot & & \\ & & & \cdot & \\ & & & & \cdot \\ & & & & & D_t \end{pmatrix}, \quad \text{where } D_i = \begin{pmatrix} \lambda_i & & & & \\ & \lambda_i & & & \\ & & \cdot & & \\ & & & \cdot & \\ & & & & \cdot \\ & & & & & \lambda_i \end{pmatrix}.$$

Thus D_i is a scalar matrix of dimension s_i. For each $i = 1, \ldots, t$, $D_i = \lambda_i I(s_i)$ where $I(s_i)$ is the identity matrix of dimension s_i. Let E_i be the n-by-n matrix that has 1 in each diagonal position in rows $s_1 + \cdots + s_{i-1} + 1$ through $s_1 + \cdots + s_{i-1} + s_i$, and 0 in every other position. In block form E_i looks like D except that D_i is replaced by $I(s_i)$, and D_j is replaced by $Z(s_j)$ when $j \ne i$:

$$E_i = \begin{pmatrix} Z & & & & \\ & \cdot & & & \\ & & \cdot & & \\ & & & I(s_i) & \\ & & & & \cdot \\ & & & & & \cdot \\ & & & & & & Z \end{pmatrix}.$$

Interpreted as a linear transformation, E_i defines the identity transformation on $\mathscr{C}(\lambda_i)$ and the zero transformation on the rest of $\mathscr{V}_n$. For example, in the language of Exercise 3.1-9, E_1 defines a projection on $\mathscr{C}(\lambda_1)$ along $\mathscr{C}(\lambda_2) \oplus \cdots \oplus \mathscr{C}(\lambda_t)$. By means of block multiplication it is easy to verify that

$$E_i^2 = E_i \qquad \text{(each } E_i \text{ defines a } projection),$$

$$E_i E_j = Z \text{ if } j \neq i \quad \text{(the projections are } orthogonal),$$

$$\sum_{i=1}^{t} E_i = I \qquad \text{(the projections are } supplementary),$$

$$\sum_{i=1}^{t} \lambda_i E_i = D \qquad (D \text{ is a linear combination of projections)}.$$

But $D = P^{-1}AP$ for some nonsingular P. If we let

$$F_i = PE_i P^{-1},$$

then the F's represent the same orthogonal and supplementary projections as the E's, and

$$A = \sum_{i=1}^{t} \lambda_i F_i.$$

Such a representation of A is called a *spectral decomposition* of A. Any theorem that asserts the existence of a spectral decomposition is called a *spectral theorem*. The next result shows that the question of diagonability of A is the question of whether or not a spectral decomposition exists for A.

Theorem 7.7 An n-by-n matrix A is diagonable if and only if there exist t distinct scalars $c_1, \ldots, c_t$ and t nonzero idempotent matrices $F_1, \ldots, F_t$ such that

$$F_i F_j = Z \text{ when } i \neq j,$$

$$\sum_{i=1}^{t} F_i = I,$$

$$\sum_{i=1}^{t} c_i F_i = A.$$

Proof Our previous arguments show that if A is diagonable, it has a spectral decomposition. Conversely, suppose that t numbers c_i and t matrices F_i have the properties listed above. We show that for every vector X and for each F_j, either $F_j X$ is the zero vector or else a characteristic vector of A with c_j as the associated characteristic value:

$$A(F_j X) = \left(\sum_{i=1}^{t} c_i F_i \right) F_j X = \left(\sum_{i=1}^{t} c_i F_i F_j \right) X$$

$$= (c_j F_j F_j)X = c_j(F_j X).$$

Thus for all X such that $F_j X \neq Z$, $F_j X$ is a characteristic vector of A, and c_j is a characteristic value of A. Furthermore,

$$X = IX = \sum_{i=1}^{t} F_i X,$$

so every vector is a sum of characteristic vectors, and by Theorem 7.6 A is diagonable.

Another criterion for diagonability, related to the characteristic polynomial of A, is derived in Section 7.4.

EXERCISES 7.2

1. For each of the following matrices determine the characteristic values and corresponding characteristic vectors. If a given matrix is diagonable, find a diagonalizing matrix P and verify that $P^{-1}AP$ is diagonal.

(i) $\begin{pmatrix} -2 & 3 & -1 \\ -6 & 7 & -2 \\ -9 & 9 & -2 \end{pmatrix}$.   (iv) $\begin{pmatrix} 1 & 0 & 0 \\ 1 & 0 & -2 \\ -1 & 1 & -3 \end{pmatrix}$.

(ii) $\begin{pmatrix} 7 & 4 & -4 \\ 4 & 7 & -4 \\ -1 & -1 & 4 \end{pmatrix}$.   (v) $\begin{pmatrix} 1 & 3 & 0 \\ 0 & -2 & 0 \\ 0 & 6 & 1 \end{pmatrix}$.

(iii) $\begin{pmatrix} 9 & 7 & 3 \\ -9 & -7 & -4 \\ 4 & 4 & 4 \end{pmatrix}$.   (vi) $\begin{pmatrix} 2 & 2 & 1 \\ 1 & 3 & 1 \\ 1 & 2 & 2 \end{pmatrix}$.

2. Which matrices in Exercise 7.1-1 are diagonable, and why? Without computing a diagonalizing matrix P, write a diagonal form for each matrix that is diagonable.

3. Which of the following matrices are diagonable, and why? (Exercise 7.1-11 will simplify finding the characteristic polynomial for these matrices.)

(i) $\begin{pmatrix} 0 & 0 & 1 \\ 1 & 0 & -3 \\ 0 & 1 & 3 \end{pmatrix}$.   (ii) $\begin{pmatrix} 0 & 0 & 0 \\ 1 & 0 & -9 \\ 0 & 1 & 6 \end{pmatrix}$.   (iii) $\begin{pmatrix} 0 & 0 & 0 & 0 \\ 1 & 0 & 0 & -4 \\ 0 & 1 & 0 & 4 \\ 0 & 0 & 1 & 1 \end{pmatrix}$.

4. Use Exercise 7.1-14 and Theorem 7.6 to show that an n-by-n matrix A is diagonable if and only if $\mathscr{V}_n$ is the direct sum of the characteristic subspaces of A.

5. Fill in the details of a proof of Theorem 7.6.

6. Let λ_1 be a characteristic value of **T** having algebraic multiplicity s_1 and geometric multiplicity g_1. Choose any basis for $\mathscr{C}(\lambda_1)$ and extend this to a basis for $\mathscr{V}_n$ in any manner.

(i) Show that **T** is represented relative to that basis by a block matrix of the form

$$A = \left(\begin{array}{c|c} \lambda_1 I(g_1) & B \\ \hline Z & C \end{array} \right),$$

where $I(g_1)$ is the g_1-by-g_1 identity matrix.

(ii) Show that the characteristic polynomial of A is

$$p(x) = (\lambda_1 - x)^{g_1} q(x),$$

where $q(x)$ is the characteristic polynomial of C.

(iii) Deduce that $g_1 \le s_1$.

7. Prove that an n-by-n matrix A is diagonable if and only if the sum of the geometric multiplicities of the distinct characteristic values is n.

8. Let the n-by-n matrices $E_1, \ldots, E_t$ satisfy the conditions $E_i E_j = Z$ when $i \ne j$ and $E_1 + \cdots + E_t = I$.

(i) Prove that E_i is idempotent for each i.

(ii) Let $F_i = P E_i P^{-1}$ for $i = 1, \ldots, t$, where P is any nonsingular n-by-n matrix. Show that each F_i is idempotent, and show that the set $\{F_1, \ldots, F_t\}$ is orthogonal and supplementary.

9. Write a spectral decomposition for each of the following matrices.

(i) Exercise 1(iv).

(ii) Exercise 1(v).

(iii) Exercise 1(vi).

10. Given the two matrices A and P such that $P^{-1}AP = \text{diag}(1, 1, -2)$, where

$$A = \begin{pmatrix} 1 & 3 & 0 \\ 0 & -2 & 0 \\ 0 & 6 & 1 \end{pmatrix}, \qquad P = \begin{pmatrix} 1 & 0 & 1 \\ 0 & 0 & -1 \\ 0 & 1 & 2 \end{pmatrix}.$$

(i) Determine a spectral decomposition for A.

(ii) Verify your answer to (i) by matrix computations.

11. Carry out the steps suggested below to prove that any n-by-n complex matrix A is similar to an upper triangular matrix.

(i) Let λ_1 be a characteristic value of A, with X_1 as an associated characteristic vector. Let $\{X_1, \ldots, X_n\}$ be a basis for $\mathbb{C}^n$ where $X_2, \ldots, X_n$ are chosen arbitrarily to form a basis. Describe a matrix P such that

$$P^{-1}AP = \left(\begin{array}{c|c} \begin{matrix} \lambda_1 \\ 0 \\ \cdot \\ \cdot \\ \cdot \\ 0 \end{matrix} & \begin{matrix} B \\ \\ C \end{matrix} \end{array} \right).$$

(ii) Now proceed by induction on n. The induction hypothesis applies to C so $Q^{-1}CQ$ is upper triangular for some $(n-1)$-by-$(n-1)$ matrix Q. Define

$$R = \left(\begin{array}{c|ccc} 1 & 0 & \cdots & 0 \\ \hline 0 & & & \\ \cdot & & & \\ \cdot & & Q & \\ \cdot & & & \\ 0 & & & \end{array} \right)$$

and show that $S^{-1}AS$ is upper triangular, where $S = PR$.

12. Show that A and B are similar, where each entry of A is 1, and where B has n in the $(1, 1)$ position and 0 elsewhere.

13. (i) Show that AB and BA have the same set of characteristic values, where A and B are any n-by-n matrices.

(ii) Show further that AB and BA have the same characteristic polynomial.

14. State a necessary and sufficient condition for the diagonability of the diagonal block matrix

$$A = \left(\begin{array}{c|c} B & Z \\ \hline Z & C \end{array}\right)$$

in terms of the diagonability of the square blocks B and C.

7.3 INVARIANT SUBSPACES

In the previous section the characteristic subspaces $\mathscr{C}(\lambda_i)$ associated with a diagonable matrix A were used to choose a basis of characteristic vectors and thereby to find P such that $P^{-1}AP$ is diagonal with λ_1 in the first s_1 diagonal positions, λ_2 in the next s_2 diagonal positions, and so on, where the characteristic polynomial of A is

$$p(x) = (-1)^n (x - \lambda_1)^{s_1} (x - \lambda_2)^{s_2} \cdots (x - \lambda_t)^{s_t}.$$

Because A is diagonable the algebraic multiplicity s_i and geometric multiplicity g_i are equal for $i = 1, \ldots, t$.

In case A is not diagonable, the sum of the characteristic subspaces is still a direct sum (Exercise 7.1-14), but it does not span $\mathscr{V}_n$. As before, suppose we choose a basis for each characteristic subspace $\mathscr{C}(\lambda_i)$, $i = 1, \ldots, t$, then form the union of those bases, and finally extend that linearly independent set in any manner to form a basis for $\mathscr{V}_n$. Then

$$\mathscr{V}_n = \mathscr{C}(\lambda_1) \oplus \cdots \oplus \mathscr{C}(\lambda_t) \oplus \mathscr{W},$$

where $\mathscr{W}$ is the subspace spanned by the noncharacteristic vectors of that basis. Then A is similar to a block matrix of the form

$$\left(\begin{array}{c|c} D & R \\ \hline Z & S \end{array}\right).$$

The matrix D is in diagonal block form, with t blocks D_i, each of the form $\lambda_i I(g_i)$. The number of columns of R and S is dim $\mathscr{W}$, which is

$$n - \sum_{i=1}^{t} g_i.$$

The diagonal block form of D reflects two special relationships between the characteristic subspaces $\mathscr{C}(\lambda_i)$ of A and any linear transformation **T** that

A represents; first, **T** maps $\mathscr{C}(\lambda_i)$ into itself, and second, the subspace corresponding to *D* is the direct sum of such subspaces,

$$\mathscr{C}(\lambda_1) \oplus \cdots \oplus \mathscr{C}(\lambda_t).$$

Whenever **T** maps a subspace $\mathscr{S}$ into itself we can obtain information about **T** by studying the effect of **T** on $\mathscr{S}$, ignoring its effect on the rest of $\mathscr{V}_n$. Then if $\mathscr{V}_n$ can be constructed as a direct sum of such subspaces, we can describe the effect of **T** on $\mathscr{V}_n$ from a knowledge of **T** on those subspaces. Hence we are led to the following definition.

Definition 7.3   Let **T** be a linear transformation on $\mathscr{V}$, and let $\mathscr{S}$ be a subspace of $\mathscr{V}$. Then $\mathscr{S}$ is said to be **T**-*invariant* if and only if $\mathbf{T}\sigma \in \mathscr{S}$ for every $\sigma \in \mathscr{S}$. When $\mathscr{S}$ is **T**-invariant the linear transformation $\mathbf{T}_{\mathscr{S}}$, called **T** *restricted to* $\mathscr{S}$, is defined to be the mapping from $\mathscr{S}$ into $\mathscr{S}$ with values given by the rule

$$\mathbf{T}_{\mathscr{S}}\sigma = \mathbf{T}\sigma \quad \text{for each } \sigma \in \mathscr{S}.$$

In short, $\mathbf{T}_{\mathscr{S}}$ describes **T** on $\mathscr{S}$, ignoring the fact that **T** is defined on all of $\mathscr{V}_n$. Of course this is possible only when $\mathscr{S}$ is **T**-invariant. Although each characteristic subspace is **T**-invariant, $\mathscr{V}_n$ is not necessarily the direct sum of characteristic subspaces, so we must look elsewhere for a simple description of **T** when **T** is not diagonable.

Recall that in Theorem 3.6 we showed that

$$\mathscr{V}_n = \mathscr{R}(\mathbf{T}^k) \oplus \mathscr{N}(\mathbf{T}^k),$$

where *k* is the smallest number such that $\mathscr{N}(\mathbf{T}^k) = \mathscr{N}(\mathbf{T}^{k+1})$. Furthermore, we can prove that $\mathscr{R}(\mathbf{T}^k)$ and $\mathscr{N}(\mathbf{T}^k)$ are **T**-invariant spaces such that **T** restricted to $\mathscr{R}(\mathbf{T}^k)$ is nonsingular and **T** restricted to $\mathscr{N}(\mathbf{T}^k)$ is nilpotent.

Theorem 7.8   Let **T** be any linear transformation on $\mathscr{V}_n$. There exist subspaces $\mathscr{R}$ and $\mathscr{N}$ of $\mathscr{V}_n$ such that

(a) $\mathscr{V}_n = \mathscr{R} \oplus \mathscr{N}$,

(b) $\mathscr{R}$ and $\mathscr{N}$ are **T**-invariant,

(c) $\mathbf{T}_{\mathscr{R}}$ is nonsingular,

(d) $\mathbf{T}_{\mathscr{N}}$ is nilpotent.

Proof   Let $\mathscr{R} = \mathscr{R}(\mathbf{T}^k)$ and $\mathscr{N} = \mathscr{N}(\mathbf{T}^k)$ as defined in Theorem 3.6. Thus $\mathscr{V}_n = \mathscr{R} \oplus \mathscr{N}$. For each $\rho \in \mathscr{R}$, $\rho = \mathbf{T}^k\xi$ for some $\xi \in \mathscr{V}_n$. Hence

$$\mathbf{T}\rho = \mathbf{T}^{k+1}\xi \in \mathscr{R}(\mathbf{T}^{k+1}) = \mathscr{R}(\mathbf{T}^k) = \mathscr{R}$$

so $\mathscr{R}$ is **T**-invariant. Next let $v \in \mathscr{N}$. Then $\mathbf{T}^k v = \theta$, and $\mathbf{T}^k(\mathbf{T}v) = \theta$, so $\mathbf{T}v \in \mathscr{N}(\mathbf{T}^k) = \mathscr{N}$. To show that $\mathbf{T}_{\mathscr{R}}$ is nonsingular, we note that **T** maps $\mathscr{R}$ onto $\mathscr{R}$ because $\mathscr{R} = \mathscr{R}(\mathbf{T}^{k+1}) = \mathbf{T}\mathscr{R}$. Finally, for any $v \in \mathscr{N}$, $\mathbf{T}^k v = \theta$, so **T** is nilpotent on $\mathscr{N}$. Indeed, because $\mathscr{N}(\mathbf{T}^{k-1})$ is a proper subspace of $\mathscr{N}(\mathbf{T}^k)$, $\mathbf{T}_{\mathscr{N}}$ is nilpotent of index k.

This theorem reveals quite a bit about an arbitrary linear transformation **T** on $\mathscr{V}_n$. In a sense **T** has a split personality—it selects two subspaces of its domain and behaves independently on those subspaces. On one subspace it is nonsingular, and by contrast, it is nilpotent on the other. Of course, one or the other of those subspaces can be $[\theta]$.

We shall make use of this result in Section 7.6 to derive the Jordan canonical form for similarity. For the present, however, we recall from Exercise 3.2-5 a useful fact about any nilpotent linear transformation. If **T** is nilpotent of index k on $\mathscr{V}_n$, then for some vector $\xi \in \mathscr{V}_n$, $\mathbf{T}^{k-1}\xi \neq \theta$ but $\mathbf{T}^k\xi = \theta$. Then the set $\{\xi, \mathbf{T}\xi, \ldots, \mathbf{T}^{k-1}\xi\}$ is linearly independent. In the case where $k = n$ the matrix representation of **T** relative to this basis is given in Example B of Section 3.4.

This idea can be applied generally, whether or not **T** is nilpotent. Let **T** be any linear transformation on $\mathscr{V}_n$, and let ξ be any nonzero vector. In the sequence $\xi, \mathbf{T}\xi, \mathbf{T}^2\xi, \ldots, \mathbf{T}^n\xi$, there must be a largest index k, $1 \leq k \leq n$ such that the set $\{\xi, \mathbf{T}\xi, \ldots, \mathbf{T}^{k-1}\xi\}$ is linearly independent and therefore spans a k-dimensional subspace $\mathscr{S}$ of $\mathscr{V}_n$. As formalized in the following definition, $\mathscr{S}$ is called a **T**-cyclic subspace and the set $\{\xi, \mathbf{T}\xi, \ldots, \mathbf{T}^{k-1}\xi\}$ is called a **T**-cyclic basis for $\mathscr{S}$.

Definition 7.4 Let **T** be any linear transformation on $\mathscr{V}_n$, let ξ be any nonzero vector in $\mathscr{V}_n$, and let k be the largest index such that the set $\{\xi, \mathbf{T}\xi, \ldots, \mathbf{T}^{k-1}\xi\}$ is linearly independent.

  (a) The subspace $[\{\xi, \mathbf{T}\xi, \ldots, \mathbf{T}^{k-1}\xi\}]$ is called the **T**-*cyclic subspace generated by* ξ, and is denoted by $[(\xi)_{\mathbf{T}}]$.

  (b) The basis $\{\xi, \mathbf{T}\xi, \ldots, \mathbf{T}^{k-1}\xi\}$ for $[\{\xi\}_{\mathbf{T}}]$ is called the **T**-*cyclic basis generated by* ξ and is denoted by $\{(\xi)_{\mathbf{T}}\}$.

  (c) More generally, any linearly independent set of the form $\{(\xi_1)_{\mathbf{T}}, (\xi_2)_{\mathbf{T}}, \ldots, (\xi_r)_{\mathbf{T}}\}$ is called a **T**-*cyclic basis* for the space that it spans.

Theorem 7.9 Let $\mathscr{S} = [(\xi)_{\mathbf{T}}]$ be a **T**-cyclic subspace of dimension k in $\mathscr{V}_n$. Then the following statements hold:

(a) $\mathscr{S}$ is **T**-invariant.

(b) Relative to the **T**-cyclic basis $\{(\xi)_\text{T}\}$, $\mathbf{T}_\mathscr{S}$ is represented by the k-by-k companion matrix

$$C = \begin{pmatrix} 0 & 0 & 0 & \cdots & 0 & c_1 \\ 1 & 0 & 0 & \cdots & 0 & c_2 \\ 0 & 1 & 0 & \cdots & 0 & c_3 \\ \cdot & \cdot & \cdot & & \cdot & \cdot \\ \cdot & \cdot & \cdot & & \cdot & \cdot \\ \cdot & \cdot & \cdot & & \cdot & \cdot \\ 0 & 0 & 0 & \cdots & 1 & c_k \end{pmatrix},$$

having $(-1)^k(x^k - c_k x^{k-1} - \cdots - c_2 x - c_1)$ as its characteristic polynomial.

Proof **T** maps each vector of the **T**-cyclic basis into the next vector of that basis with one exception; the last basis vector is mapped by **T** into a linear combination $c_1 \xi + \cdots + c_k \mathbf{T}^{k-1}\xi$ of the basis vectors, because k is the largest index such that $\{\xi, \mathbf{T}\xi, \ldots, \mathbf{T}^{k-1}\xi\}$ is linearly independent. Hence **T** maps each basis vector of $\mathscr{S}$ into some vector in $\mathscr{S}$, which means that $\mathscr{S}$ is **T**-invariant. This also shows that $\mathbf{T}_\mathscr{S}$ is represented by the matrix C as shown. The characteristic polynomial of C is established in Exercise 7.1-11.

Now we combine the results of Theorem 7.9 and Theorem 7.8. We have $\mathscr{V}_n = \mathscr{R} \oplus \mathscr{N}$, where $\mathscr{R}$ and $\mathscr{N}$ are **T**-invariant subspaces; **T** is nonsingular on $\mathscr{R}$ and nilpotent of index k on $\mathscr{N}$. If we choose any basis for $\mathscr{R}$ and any basis for $\mathscr{N}$, and combine them to obtain a basis for $\mathscr{V}_n$, we obtain a matrix A that represents **T** and is of the block diagonal form

$$A = \left(\begin{array}{c|c} R & Z \\ \hline Z & N \end{array} \right),$$

where R is nonsingular and N is nilpotent of index k. If $k = \dim \mathscr{N}$ and if we choose the **T**-cyclic basis for $\mathscr{N}$ as described in the discussion preceding Definition 7.4, then Theorem 7.9 tells us that N will assume the form

$$N = \begin{pmatrix} 0 & 0 & 0 & \cdots & 0 & 0 \\ 1 & 0 & 0 & \cdots & 0 & 0 \\ 0 & 1 & 0 & \cdots & 0 & 0 \\ \cdot & \cdot & \cdot & & \cdot & \cdot \\ \cdot & \cdot & \cdot & & \cdot & \cdot \\ \cdot & \cdot & \cdot & & \cdot & \cdot \\ 0 & 0 & 0 & \cdots & 1 & 0 \end{pmatrix}$$

with 1 in position $(i + 1, i)$ for $i = 1, \ldots, k - 1$ and 0 elsewhere, because the characteristic polynomial of a k-by-k matrix that is nilpotent of index k is $(-\lambda)^k$. (See Exercise 7.1-6(ii).)

We shall use Theorems 7.8 and 7.9 in the next three sections to obtain the famous Hamilton-Cayley Theorem, a canonical form for similarity of nilpotent matrices, and the Jordan canonical form for similarity of arbitrary n-by-n matrices.

EXERCISES 7.3

1. Referring to the example in Section 7.1, write the matrix B that represents **T** relative to the ordered basis $\{\alpha_3, \alpha_2, \alpha_1\}$. In the notation of the second paragraph of this section, identify the space $\mathscr{W}$ and the characteristic subspaces for this example and also the matrices D, R, and S.

2. In $\mathscr{E}^2$ let **T** be a reflection across the line $y = x$.

   (i) Write the matrix A that represents **T** relative to the standard basis.

   (ii) Determine two **T**-invariant subspaces $\mathscr{M}$ and $\mathscr{N}$ such that $\mathscr{E}^2 = \mathscr{M} \oplus \mathscr{N}$, where neither $\mathscr{M}$ nor $\mathscr{N}$ is $[\theta]$.

   (iii) Write a basis $\{\beta_1, \beta_2\}$ for $\mathscr{E}^2$ such that $\mathscr{M} = [\beta_1]$ and $\mathscr{N} = [\beta_2]$.

   (iv) Write the matrix B that represents **T** relative to $\{\beta_1, \beta_2\}$.

   (v) Write a matrix P such that $B = P^{-1}AP$.

3. Let **T** be an idempotent linear transformation of rank r on $\mathscr{V}_n$.

   (i) Show that $\mathscr{R}(\mathbf{T}^2) = \mathscr{R}(\mathbf{T})$, and hence $\mathscr{V}_n = \mathscr{R}(\mathbf{T}) \oplus \mathscr{N}(\mathbf{T})$.

   (ii) Show that $\mathbf{T}_{\mathscr{R}(\mathbf{T})}$ is the identity transformation and $\mathbf{T}_{\mathscr{N}(\mathbf{T})}$ is the zero transformation.

   (iii) Describe a simple matrix representation of **T**.

   (iv) Deduce that **T** is a projection onto $\mathscr{R}(\mathbf{T})$ along $\mathscr{N}(\mathbf{T})$, as described in Exercise 3.1-9.

4. Let $\mathscr{V} = \mathscr{M} \oplus \mathscr{N}$, and let **E** and **T** be linear transformations on $\mathscr{V}$. Prove the following assertions.

   (i) **E** is the projection of $\mathscr{V}$ on $\mathscr{M}$ along $\mathscr{N}$ if and only if $\mathbf{I} - \mathbf{E}$ is the projection of $\mathscr{V}$ on $\mathscr{N}$ along $\mathscr{M}$.

   (ii) $\mathscr{M}$ is **T**-invariant if and only if $\mathbf{ETE} = \mathbf{TE}$, where **E** is the projection of $\mathscr{V}$ onto $\mathscr{M}$ along $\mathscr{N}$.

(iii) $\mathcal{M}$ and $\mathcal{N}$ are **T**-invariant if and only if $\mathbf{ET} = \mathbf{TE}$ where $\mathbf{E}$ is the projection of $\mathcal{V}$ onto $\mathcal{M}$ along $\mathcal{N}$.

5. Let $\mathcal{V} = \mathcal{M}_1 \oplus \mathcal{N}_1 = \mathcal{M}_2 \oplus \mathcal{N}_2$ be two direct sum decompositions of $\mathcal{V}$. Let $\mathbf{E}_1$ be the projection of $\mathcal{V}$ onto $\mathcal{M}_1$ along $\mathcal{N}_1$, and let $\mathbf{E}_2$ be the projection of $\mathcal{V}$ onto $\mathcal{M}_2$ along $\mathcal{N}_2$. Prove that $\mathbf{E}_1 + \mathbf{E}_2$ is a projection if and only if $\mathbf{E}_1\mathbf{E}_2 = \mathbf{Z} = \mathbf{E}_2\mathbf{E}_1$, and in that case $\mathbf{E}_1 + \mathbf{E}_2$ is the projection onto $\mathcal{M}_1 + \mathcal{M}_2$ along $\mathcal{N}_1 \cap \mathcal{N}_2$.

6. Prove that if $\mathcal{M}$ and $\mathcal{N}$ are **T**-invariant subspaces of $\mathcal{V}$, then so are $\mathcal{M} + \mathcal{N}$ and $\mathcal{M} \cap \mathcal{N}$.

7. Let **T** be the linear transformation of $\mathbb{R}^3$ whose matrix A relative to the standard basis is given in the example of Section 7.1:

$$A = \begin{pmatrix} 1 & 2 & 1 \\ 2 & 0 & -2 \\ -1 & 2 & 3 \end{pmatrix}.$$

(i) Determine the dimension of the **T**-cyclic subspace generated by the vector

$$X = \begin{pmatrix} 1 \\ 1 \\ 0 \end{pmatrix}.$$

(ii) Extend $\{(X)_{\mathbf{T}}\}$, if necessary, to obtain a **T**-cyclic basis for $\mathbb{R}^3$.

8. Describe the cyclic subspace $[(\xi)_{\mathbf{T}}]$ generated by a characteristic vector ξ of **T**.

7.4   THE MINIMAL POLYNOMIAL

From Theorem 4.3 the set of all n-by-n matrices forms a linear algebra of dimension n^2. Hence for any n-by-n matrix A the set $\{I, A, A^2, \ldots, A^{n^2}\}$ is linearly dependent in $\mathcal{M}_{n \times n}$. Thus there exists a smallest index s such that $\{I, A, \ldots, A^s\}$ is linearly dependent, and for suitable scalars we have

$$a_0 A^s + a_1 A^{s-1} + \cdots + a_{s-1} A + a_s I = Z,$$

where, by the definition of s, $a_0 \neq 0$. Letting $b_k = a_0^{-1} a_k$ for $k = 0, \ldots, s$, we have

$$A^s + b_1 A^{s-1} + \cdots + b_{s-1} A + b_s I = Z.$$

This is a polynomial equation in the matrix A; the corresponding scalar polynomial is

$$m(x) = x^s + b_1 x^{s-1} + \cdots + b_{s-1} x + b_s.$$

Any polynomial such as m in which the leading coefficient is one is said to be *monic*.

Definition 7.5   The *minimal polynomial m* of the n-by-n matrix A is the monic scalar polynomial of least degree such that $m(A) = Z$.

As an exercise you may prove that m is uniquely determined by A.

The adjective "scalar" in Definition 7.5 signifies that the coefficients of the polynomial are scalars. One can also consider "matrix" polynomials in which the coefficients are matrices. But the algebra of matrix polynomials is more complicated than the algebra of scalar polynomials, principally because matrix multiplication is noncommutative and the product of nonzero matrices can be the zero matrix. Thus from the matrix polynomial equation

$$(X - A)(X - B) = Z,$$

we cannot conclude that $X = A$ or $X = B$. Also the matrix polynomial

$$X^2 - (A + B)X + AB$$

is different from the matrix polynomial

$$(X - A)(X - B)$$

unless $XB = BX$. However, we know that the set of all n-by-n scalar matrices cI is isomorphic to the scalar field. Given a scalar polynomial

$$a_0 x^n + a_1 x^{n-1} + \cdots + a_n,$$

if we replace each a_k by the n-by-n scalar matrix $a_k I$ and replace x by the n-by-n matrix X, we obtain

$$a_0 I X^n + a_1 I X^{n-1} + \cdots + a_n I,$$

which is a special form of matrix polynomial in which all the matrices commute with each other. Hence the factorization of such polynomials coincides with the factorization of scalar polynomials:

$$a_0 x^n + a_1 x^{n-1} + \cdots + a_n = a_0(x - r_1) \cdots (x - r_n)$$

if and only if

$$a_0 X^n + a_1 X^{n-1} + \cdots + a_n I = a_0(X - r_1 I) \cdots (X - r_n I).$$

In particular, the division algorithm for scalar polynomials also holds for these matrix polynomials. This algorithm states that, given any two polynomials s and t, there are unique polynomials q and r such that

$$s(x) = t(x)q(x) + r(x),$$

where r either is the zero polynomial or has degree less than the degree of t.

Theorem 7.10 Let m be the minimal polynomial of A and let s be any scalar polynomial such that $s(A) = Z$. Then

$$s(x) = m(x)q(x)$$

for some scalar polynomial $q(x)$. That is, the minimal polynomial of A is a divisor of every polynomial s for which $s(A) = Z$.
Proof By the division algorithm we can write

$$s(x) = m(x)q(x) + r(x),$$

where either r is the zero polynomial or has degree less than the degree of m. But $s(A) = Z$ by hypothesis and $m(A) = Z$ from the definition of the minimal polynomial. Hence

$$s(A) = m(A)q(A) + r(A),$$

$$Z = r(A).$$

But by definition, m is the polynomial of smallest degree for which $m(A) = Z$, so r must be the zero polynomial.

In Exercise 7.1-15 you were asked to show for each two-by-two matrix A that $p(A) = Z$, where p is the characteristic polynomial of A. The same conclusion holds for any n-by-n matrix.

Theorem 7.11 (Hamilton-Cayley Theorem) If A is any n-by-n matrix with characteristic polynomial p, then $p(A) = Z$.
Proof Relative to any fixed basis let $\mathbf{T}$ be the linear transformation represented by A. Let p be the characteristic polynomial of A,

$$p(x) = (-1)^n(x^n + c_1 x^{n-1} + \cdots + c_{n-1}x + c_n).$$

To show that $p(A) = Z$, it suffices to show that $p(\mathbf{T})$ is the zero linear transformation, because $p(A)$ is the matrix that represents $p(\mathbf{T})$ by Exercise 6.4-5. Thus letting ξ be any nonzero vector, we wish to show that $p(\mathbf{T})\xi = \theta$.

First let $\mathscr{S}$ denote the **T**-cyclic subspace $[(\xi)_T]$ generated by ξ, and let dim $\mathscr{S} = k \geq 1$. As in Theorem 7.9, $\mathbf{T}^k\xi = c_1\xi + \cdots + c_k\mathbf{T}^{k-1}\xi$, so

$$(\mathbf{T}^k - c_k\mathbf{T}^{k-1} - \cdots - c_2\mathbf{T} - c_1\mathbf{I})\xi = \theta.$$

$\mathscr{S}$ is a **T**-invariant subspace and, relative to the basis $\{(\xi)_T\}$, **T** is represented by the k-by-k companion matrix C, whose characteristic polynomial is $c(x) = (-1)^k(x^k - c_k x^{k-1} - \cdots - c_2 x - c_1)$. Now extend this basis for $\mathscr{S}$ to a basis for $\mathscr{V}_n$. Because $\mathscr{S}$ is **T**-invariant, **T** is represented by a matrix B in the block form

$$B = \left(\begin{array}{c|c} C & R \\ \hline Z & S \end{array}\right).$$

Since B is similar to A, the characteristic polynomial of B is

$$p(x) = \det(B - xI) = \det(S - xI)\det(C - xI)$$

$$= s(x)c(x)$$

Hence

$$p(\mathbf{T}) = s(\mathbf{T})c(\mathbf{T}) = (-1)^k s(\mathbf{T})(\mathbf{T}^k - c_k\mathbf{T}^{k-1} - \cdots - c_1\mathbf{I}),$$

and therefore

$$p(\mathbf{T})\xi = (-1)^k s(\mathbf{T})\theta = \theta,$$

from which it follows that $p(\mathbf{T}) = Z$ and $p(\mathbf{A}) = Z$.

The characteristic polynomial of $\mathbf{A}$ has degree n, so the degree of the minimal polynomial does not exceed n. Previously we only knew that its degree does not exceed n^2. But from Theorem 7.10 we can deduce much more about the minimal polynomial.

Theorem 7.12 If $\lambda_1, \ldots, \lambda_t$ are the distinct characteristic values of A and if the characteristic polynomial of A is

$$p(x) = (-1)^n(x - \lambda_1)^{s_1} \cdots (x - \lambda_t)^{s_t},$$

then the minimal polynomial of A has the form

$$m(x) = (x - \lambda_1)^{r_1} \cdots (x - \lambda_t)^{r_t},$$

where $1 \leq r_i \leq s_i$ for $i = 1, \ldots, t$.

Proof By Theorem 7.10, $p(x)$ is evenly divisible by $m(x)$, so $0 \leq r_i \leq s_i$.

Suppose that $r_1 = 0$ and let X_1 be a characteristic vector associated with λ_1. Then we have

$$m(A)X_1 = [(A - \lambda_2 I)^{r_2} \cdots (A - \lambda_t I)^{r_t}]X_1.$$

But

$$(A - \lambda_t I)X_1 = AX_1 - \lambda_t X_1 = (\lambda_1 - \lambda_t)X_1,$$

and this argument can be repeated to obtain

$$m(A)X_1 = [(\lambda_1 - \lambda_2)^{r_2} \cdots (\lambda_1 - \lambda_t)^{r_t}]X_1.$$

Because $X_1 \neq Z$ and λ_1 is distinct from $\lambda_2, \ldots, \lambda_t$, we conclude that $m(A) \neq Z$, a contradiction of the definition of m. Hence $r_1 \neq 0$. Since the factors of $m(A)$ commute, this argument applies to any index, so $r_i \geq 1$ for each $i = 1, \ldots, t$.

Now we are able to relate diagonability of A to the form of the minimal polynomial of A.

Theorem 7.13 A is diagonable if and only if its minimal polynomial is

$$m(x) = (x - \lambda_1)(x - \lambda_2) \cdots (x - \lambda_t),$$

where $\lambda_1, \ldots, \lambda_t$ are the distinct characteristic values of A.
Proof   Let A be diagonable. We can choose P such that $D = P^{-1}AP$ is diagonal with λ_1 in the first s_1 diagonal positions, λ_2 in the next s_2 diagonal positions, and so on. Then $D - \lambda_1 I$ is diagonal with 0 in the first s_1 diagonal positions, $D - \lambda_2 I$ is diagonal with 0 in the next s_2 diagonal positions, and so on. Because the product of diagonal matrices is diagonal, with each diagonal entry being the product of the corresponding diagonal entries of the factors, we obtain

$$(D - \lambda_1 I)(D - \lambda_2 I) \cdots (D - \lambda_t I) = Z.$$

From Theorem 7.12 it follows that the minimum polynomial of D, and hence of A, is

$$m(x) = (x - \lambda_1)(x - \lambda_2) \cdots (x - \lambda_t).$$

To prove the converse we use this form of m to construct a spectral decomposition of A as described in Theorem 7.7. For $i = 1, \ldots, t$, let p_i be the polynomial of degree $t - 1$ defined by

$$m(x) = (x - \lambda_i)p_i(x).$$

Since the λ_i are distinct, $p_i(\lambda_i) \neq 0$ for all i. Let

$$q_i(x) = \frac{p_i(x)}{p_i(\lambda_i)}, \quad i = 1, \ldots, t,$$

and observe that if $i \neq j$ then $q_i q_j$ is a polynomial multiple of m. Each q_i is a polynomial of degree $t - 1$, and $q_i(\lambda_j) = 0$ for $j \neq i$. Now let r be the polynomial defined by

$$r(x) = 1 - q_1(x) - q_2(x) - \cdots - q_t(x),$$

which does not have degree exceeding $t - 1$. For $j = 1, \ldots, t$

$$r(\lambda_j) = 1 - q_j(\lambda_j) = 0.$$

Because the number of zeros of a nonzero polynomial cannot exceed its degree, r must be the zero polynomial, so $r(B) = Z$ for any square matrix B. Hence

$$r(A) = Z = I - \sum_{i=1}^{t} q_i(A).$$

Let $F_1, \ldots, F_t$ be the matrices defined by

$$F_i = q_i(A).$$

Then $F_i \neq Z$ because the degree of q_i is less than the degree of m, the minimal polynomial of A.

As an exercise you may verify that the scalars $\lambda_1, \ldots, \lambda_t$ and the matrices $F_1, \ldots, F_t$ comprise a spectral decomposition for A:

$$\sum_{i=1}^{t} F_i = I,$$

$$F_i F_j = \delta_{ij} F_i,$$

$$\sum_{i=1}^{t} \lambda_i F_i = A.$$

Then by Theorem 7.7 A is diagonable.

The Jordan canonical form for similarity provides an easier proof of Theorem 7.13 and many other results. If we compare the tests for diagonability proved in Section 7.2 with Theorem 7.13, we see that the latter is the most efficient procedure computationally. For any of these tests we need to determine the t distinct characteristic values of A. If $t = n$, Theorem 7.5 tells us that A is diagonable, but if $t < n$ that theorem tells us nothing. We can then compute the product

$$(A - \lambda_1 I) \cdots (A - \lambda_t I).$$

A is diagonable or not according to whether this product is Z or not. Of course, if we want to find a diagonalizing matrix P we should compute all the characteristic vectors of A and use Theorem 7.4.

EXERCISES 7.4

1. Show that the minimal polynomial of A is unique.

2. Verify the Hamilton-Cayley Theorem for each of the following matrices by actually computing $p(A)$:

$$\text{(i)} \quad A = \begin{pmatrix} 4 & 2 & -1 \\ -5 & -3 & 1 \\ 3 & 2 & 0 \end{pmatrix}, \ p(x) = -(x-1)^2(x+1).$$

$$\text{(ii)} \quad A = \begin{pmatrix} 0 & 0 & 1 \\ 1 & 0 & -3 \\ 0 & 1 & 3 \end{pmatrix}, \ p(x) = -(x-1)^3.$$

$$\text{(iii)} \quad A = \begin{pmatrix} -2 & 3 & -1 \\ -6 & 7 & -2 \\ -9 & 9 & -2 \end{pmatrix}, \ p(x) = -(x-1)^3.$$

$$\text{(iv)} \quad A = \begin{pmatrix} 2 & 2 & 1 \\ 1 & 3 & 1 \\ 1 & 2 & 2 \end{pmatrix}, \ p(x) = -(x-1)^2(x-5).$$

3. Determine the minimal polynomial for each of the matrices in Exercise 2 by matrix computation. Use Theorem 7.13 to determine which matrices are diagonable.

4. Prove that A and A^t have the same characteristic polynomial.

5. Prove that if A is similar to a scalar matrix, then A equals that matrix.

6. Prove that if A and B are similar, then A and B have the same minimal polynomial.

7. Use Theorem 7.13 to determine necessary and sufficient conditions on a, b, c, d so that the matrix

$$\begin{pmatrix} a & b \\ c & d \end{pmatrix}$$

is *not* diagonable.

8. (i) Determine all real two-by-two matrices A that satisfy $A^2 = -I$.

(ii) Show that no real three-by-three matrix satisfies $A^2 = -I$.

9. Use the Hamilton-Cayley Theorem to prove the following statements for an n-by-n singular matrix A.

(i) If $n = 1$ or 2, then A^2 is proportional to A.

(ii) If $n > 2$, A^2 is not necessarily proportional to A.

10. Complete the proof of Theorem 7.13 by showing that the matrices $F_1, \ldots, F_t$ have the properties of a spectral decomposition of A.

7.5 NILPOTENT TRANSFORMATIONS

Having obtained various criteria for diagonability and having seen several examples of matrices that do not meet those criteria, we ask this natural question: if a matrix A is not similar to a diagonal matrix, how close can we come to diagonalizing A? Is there a form, somewhat more general than a diagonal form, such that A is similar to one and only one matrix in that form?

The answer that we shall derive in Section 7.6 is that every matrix is similar to a matrix J that has zero in every position except along the diagonal and the *subdiagonal*; that is, the only possibly nonzero entries are in the (i, i) and $(i + 1, i)$ positions. The diagonal entries of J are the characteristic values of A, and each subdiagonal entry is either 0 or 1. Furthermore, the arrangement of the zeros and ones along the subdiagonal can be explicitly described. Thus $J = D + N$, where D is diagonal and N is nilpotent. (Recall Exercise 4.2-8.) The particular form of N, with ones and zeros along the subdiagonal and zero elsewhere might be anticipated in view of the discussion following Theorem 7.9. What we need to show is that for every nilpotent linear transformation **T**, there exists a **T**-cyclic basis of the form described in Definition 7.4(c). This is a fairly intricate task to which this entire section is devoted, using a form of argument given by A. J. Insel.

To assist you in understanding the details of proof we shall first discuss informally the main result to be proved. We shall show that any linear transformation $\mathbf{T}$ that is nilpotent of index p on $\mathscr{V}_n$ can be represented by a matrix N in block diagonal form

$$N = \begin{pmatrix} N_1 & & & & \\ & N_2 & & & \\ & & \cdot & & \\ & & & \cdot & \\ & & & & \cdot \\ & & & & & N_k \end{pmatrix}$$

with k block matrices $N_1, \ldots, N_k$ of dimension $p_1, p_2, \ldots, p_k$ along the diagonal. The number p_1 is the index p of nilpotency of $\mathbf{T}$ and

$$p = p_1 \geq p_2 \geq \cdots \geq p_k,$$
$$p_1 + p_2 + \cdots + p_k = n.$$

Each N_j has one in each of its $(p_j - 1)$ subdiagonal positions, and zero elsewhere.

$$N_j = \begin{pmatrix} 0 & 0 & \cdots & 0 & 0 \\ 1 & 0 & \cdots & 0 & 0 \\ 0 & 1 & \cdots & 0 & 0 \\ \cdot & \cdot & & \cdot & \cdot \\ \cdot & \cdot & & \cdot & \cdot \\ \cdot & \cdot & & \cdot & \cdot \\ 0 & 0 & \cdots & 1 & 0 \end{pmatrix}.$$

Hence the subdiagonal of N consists of a string of $(p_1 - 1)$ ones followed by a zero (between the first and second blocks), then a string of $(p_2 - 1)$ ones followed by a zero (between the second and third blocks), and so on. The form of N in diagonal blocks shows that $\mathscr{V}_n$ is decomposed into a direct sum of T-invariant subspaces $\mathscr{S}_1, \ldots, \mathscr{S}_k$ of dimension $p_1, \ldots, p_k$. The form of each N_j shows that $\mathbf{T}$ is nilpotent of index p_j on N_j and that a T-cyclic basis has been chosen for each $\mathscr{S}_j$ and combined to produce a basis for $\mathscr{V}_n$. Hence all the integers that describe the form of N are determined by the geometric properties of $\mathbf{T}$. We now begin the formal derivation, the first theorem being a tool that will be used in proving the second.

Theorem 7.14 Let $\mathbf{T}$ be a linear transformation that is nilpotent of index p; let $p = p_1 \geq p_2 \geq \cdots \geq p_k$ be positive integers and let $\alpha_1, \ldots, \alpha_k$

be vectors such that $\{T^{p_1-1}\alpha_1, \ldots, T^{p_k-1}\alpha_k\}$ is linearly independent in $\mathcal{N}(T)$. Then the T-cyclic set

$$\{(\alpha_1)_T, (\alpha_2)_T, \ldots, (\alpha_k)_T\}$$

is linearly independent in $\mathcal{V}_n$.

Proof We use the symbol $(\alpha_i)_T$ to denote the set $\{\alpha_i, T\alpha_i, \ldots, T^{p_i-1}\alpha_i\}$. We know that $T^{p_i-1}\alpha_i \neq \theta$ and is in $\mathcal{N}(T)$, so $T^{p_i}\alpha_i = \theta$. Hence $(\alpha_i)_T$ is the T-cyclic set generated by α_i. To show that the T-cyclic set generated by $\alpha_1, \ldots, \alpha_k$ is linearly independent, suppose that

$$\sum_{i=1}^{k}\left(\sum_{j=1}^{p_i} c_{ij}T^{j-1}\alpha_i\right) = \theta.$$

For each i let s_i denote the smallest index j for which $c_{ij} \neq 0$. Then

$$T^{p_i-s_i}\left(\sum_{j=1}^{p_i} c_{ij}T^{i-1}\alpha_i\right) = \sum_{j=s_i}^{p_i} c_{ij}T^{p_i-s_i+j-1}\alpha_i = c_{is_i}T^{p_i-1}\alpha_i.$$

Let s denote the largest of the numbers $p_i - s_i$ for $i = 1, \ldots, k$. Then for each i

$$T^s\left(\sum_{j=1}^{p_i} c_{ij}T^{j-1}\alpha_i\right) = \begin{cases} \theta & \text{if } s > p_i - s_i, \\ c_{is_i}T^{p_i-1}\alpha_i & \text{if } s = p_i - s_i. \end{cases}$$

Hence

$$T^s\left(\sum_{i=1}^{k}\left(\sum_{j=1}^{p_i} c_{ij}T^{j-1}\alpha_i\right)\right) = \theta = \sum c_{is_i}T^{p_i-1}\alpha_i,$$

where the last sum extends only over those values of i such that $p_i - s_i = s$. By the choice of s there is at least one such value of i, and because the set $\{T^{p_1-1}, \ldots, T^{p_k-1}\}$ is linearly independent, $c_{is_i} = 0$ for all such values of i, which contradicts the definition of s_i.

Theorem 7.15 If $\mathcal{V}$ is a finite-dimensional vector space and if T is a linear transformation that is nilpotent of index p on $\mathcal{V}$, then for some integer k there exist k integers $p = p_1 \geq p_2 \geq \cdots \geq p_k \geq 1$ and k vectors $\alpha_1, \ldots, \alpha_k$ such that the set

$$\{(\alpha_1)_T, \ldots, (\alpha_k)_T\}$$

is a T-cyclic basis for $\mathcal{V}$. The numbers $k, p_1, \ldots, p_k$ are uniquely determined by T.

Proof We use induction on p. For $p = 1$ any basis for $\mathcal{V}$ is T-cyclic because $T = Z$. The induction hypothesis guarantees that a T-cyclic basis exists for any finite-dimensional space on which T is nilpotent of index $p - 1$. Clearly $\mathcal{R}(T)$ is such a space, so for some integer t, there exist t integers $p - 1 = q_1 \geq q_2 \geq \cdots \geq q_t$ and t vectors $\eta_1, \ldots, \eta_t$ such that the **T**-cyclic set

$$A = \{(\eta_1)_{\mathbf{T}}, \ldots, (\eta_t)_{\mathbf{T}}\}$$

is a basis for $\mathscr{R}(\mathbf{T})$. By Exercise 3.2-5 and Definition 7.4 we know that q_i is the smallest number such that $\mathbf{T}^{q_i}\eta_i = \theta$. Hence $\{\mathbf{T}^{q_1-1}\eta_1, \ldots, \mathbf{T}^{q_t-1}\eta_t\}$ is a basis for $\mathscr{R}(\mathbf{T}) \cap \mathscr{N}(\mathbf{T})$. We note that $t = \dim(\mathscr{R}(\mathbf{T}) \cap \mathscr{N}(\mathbf{T}))$. By the induction hypothesis $q_1, \ldots, q_t$ are also uniquely determined by **T**. We can extend this basis for $\mathscr{R}(\mathbf{T}) \cap \mathscr{N}(\mathbf{T})$ to a basis B for $\mathscr{N}(\mathbf{T})$:

$$B = \{\mathbf{T}^{q_1-1}\eta_1, \ldots, \mathbf{T}^{q_t-1}\eta_t, \zeta_1, \ldots, \zeta_s\}.$$

Because $\zeta_i \in \mathscr{N}(\mathbf{T})$ we note that $(\zeta_i)_{\mathbf{T}} = \{\zeta_i\}$. Then by adjoining the vectors ζ_i to A we obtain the **T**-cyclic set

$$C = \{(\eta_1)_{\mathbf{T}}, \ldots, (\eta_t)_{\mathbf{T}}, (\zeta_1)_{\mathbf{T}}, \ldots, (\zeta_s)_{\mathbf{T}}\}$$

of vectors in $\mathscr{R}(\mathbf{T}) + \mathscr{N}(\mathbf{T})$. As an exercise you may verify that C is a basis for $\mathscr{R}(\mathbf{T}) + \mathscr{N}(\mathbf{T})$. Because

$$\dim(\mathscr{R}(\mathbf{T}) + \mathscr{N}(\mathbf{T})) = \dim \mathscr{R}(\mathbf{T}) + \dim \mathscr{N}(\mathbf{T}) - \dim(\mathscr{R}(\mathbf{T}) \cap \mathscr{N}(\mathbf{T}))$$

$$= r(\mathbf{T}) + n(\mathbf{T}) - t$$

$$= \dim \mathscr{V} - t,$$

a basis for $\mathscr{V}$ can be obtained by adjoining to C a suitable set of t vectors. For each $i = 1, \ldots, t$ we have $\eta_i \in \mathscr{R}(\mathbf{T})$, so we can choose ξ_i to be any vector such that $\mathbf{T}\xi_i = \eta_i$. Then $(\xi_i)_{\mathbf{T}} = \xi_i \cup (\eta_i)_{\mathbf{T}}$, and therefore the **T**-cyclic set

$$D = \{(\xi_1)_{\mathbf{T}}, \ldots, (\xi_t)_{\mathbf{T}}, (\zeta_1)_{\mathbf{T}}, \ldots, (\zeta_s)_{\mathbf{T}}\}$$

contains n vectors, where $n = \dim \mathscr{V}$. Also in the notation of Theorem 7.14 if we let

$$k = t + s = \dim \mathscr{N}(\mathbf{T}),$$

$$p_i = q_i + 1 \text{ and } \alpha_i = \xi_i \text{ for } i = 1, \ldots, t,$$

$$p_{t+j} = 1 \qquad \text{and } \alpha_{t+j} = \zeta_j \text{ for } j = 1, \ldots, s,$$

we can conclude that D is a **T**-cyclic basis for $\mathscr{V}$. The numbers q_i are uniquely determined by **T** and therefore so are the numbers p_i.

For emphasis we restate this result to indicate the significance of the numbers $k, p_1, \ldots, p_k$ in the matrix that represents **T** relative to D. Let **T** be nilpotent of index p on $\mathscr{V}_n$, and let k be the nullity of $\mathbf{T}$. It is possible to choose k vectors α_i such that

$$\{(\alpha_1)_\mathbf{T}, \ldots, (\alpha_k)_\mathbf{T}\}$$

is a $\mathbf{T}$-cyclic basis for $\mathscr{V}_n$. For each α_i, let p_i be the smallest integer such that $\mathbf{T}^{p_i}\alpha_i = \theta$. We can choose the α_i so that $p = p_1 \geq p_2 \geq \cdots \geq p_k \geq 1$. Relative to this basis $\mathbf{T}$ is represented by a diagonal block matrix N having k diagonal blocks $N_1, \ldots, N_k$ of dimensions $p_1, p_2, \ldots, p_k$. Each block N_i has 1 in each position along its subdiagonal and 0 in all other positions. The subdiagonal of N consists of k strings of ones of nonincreasing lengths

$$p_1 - 1 \geq p_2 - 1 \geq \cdots \geq p_k - 1,$$

with a single zero between consecutive strings. If any $p_j = 1$, that and all subsequent strings of ones are of length zero, and the subdiagonal of N then ends in a string of zeros. All other entries of N are zero. Finally, we note that $\mathbf{T}$ uniquely determines N. One and only one matrix in this form can represent $\mathbf{T}$, so the form is *canonical* for similarity of n-by-n nilpotent matrices. For example, if $n = 8$, $k = 4$, $p_1 = 3$, $p_2 = 2$, $p_3 = 2$, and $p_4 = 1$, the matrix N has 1, 1, 0, 1, 0, 1, 0 along its subdiagonal and zero elsewhere.

EXERCISES 7.5

1. Write a nilpotent matrix N in the form described in this section and satisfying the given data:

    (i) $n = 6$, $k = 2$, $p_1 = 5$, $p_2 = 1$.

    (ii) $n = 6$, $k = 2$, $p_1 = 3$, $p_2 = 3$.

    (iii) $n = 6$, $k = 3$, $p_1 = 3$, $p_2 = 2$, $p_3 = 1$.

    (iv) $n = 6$, $k = 3$, $p_1 = 2$, $p_2 = 2$, $p_3 = 2$.

    (v) $n = 6$, $k = 3$, $p_1 = 4$, $p_2 = 1$, $p_3 = 1$.

2. In the proof of Theorem 7.15, show that the $\mathbf{T}$-cyclic set C is a basis for $\mathscr{R}(\mathbf{T}) + \mathscr{N}(\mathbf{T})$.

3. Prove that if a matrix A is similar to a nilpotent matrix, then A is nilpotent with the same index of nilpotency.

4. Let $\mathbf{T}$ be the nilpotent linear transformation on $\mathbb{R}^3$ that is represented relative to the standard basis by the matrix

$$A = \begin{pmatrix} 0 & 0 & 0 \\ 2 & 0 & 0 \\ 1 & 3 & 0 \end{pmatrix}.$$

(i) Determine the rank, nullity, and index of nilpotency of **T**.

(ii) Let $\alpha_1 = (1, -1, 0)$. Show that $[(\alpha_1)_T] = \mathbb{R}^3$, and write the vectors of the **T**-cyclic basis $\{(\alpha_1)_T\}$.

(iii) Compute the matrix N that represents **T** relative to the **T**-cyclic basis $\{(\alpha_1)_T\}$, and use that matrix to verify your answers to (i).

5. Let N be the n-by-n matrix that has 1 in each subdiagonal position and 0 elsewhere, and let $A = (a_{ij})$ be any n-by-n matrix.

(i) Prove that A and N commute if and only if A is a "striped" matrix with a_{11} in each diagonal position, a_{21} in each subdiagonal position, and so on for each stripe below and parallel to the diagonal, and 0 in each position above the main diagonal.

(ii) If A is as described in (i) and if $a_{21} \neq 0$, show that the characteristic vectors of A span a one-dimensional space.

(iii) For $k \geq 2$, suppose that $a_{21} = a_{31} = \cdots = a_{k1} = 0$ but that $a_{k+1, 1} \neq 0$, where A is described in (i). Show that the characteristic vectors of A span a k-dimensional space.

6. Consider the vectors in the **T**-cyclic basis of Theorem 7.15 written in the following array in which the rows and columns are both of nonincreasing length:

$$\mathbf{T}^{p_1-1}\alpha_1, \ \mathbf{T}^{p_1-2}\alpha_1, \ldots \ldots , \alpha_1$$
$$\mathbf{T}^{p_2-1}\alpha_2, \ \mathbf{T}^{p_2-2}\alpha_2, \ldots \ldots , \alpha_2$$
$$\cdot$$
$$\cdot$$
$$\cdot$$
$$\mathbf{T}^{p_k-1}\alpha_k, \ \ldots , \alpha_k .$$

Deduce the following information.

(i) The first column forms a basis for $\mathcal{N}(\mathbf{T})$, so $k = n(\mathbf{T})$.

(ii) The first two columns form a basis for $\mathcal{N}(\mathbf{T}^2)$, and so the length of the second column is $\mathcal{N}(\mathbf{T}^2) - \mathcal{N}(\mathbf{T})$.

(iii) The length of each column is determined by **T**.

(iv) The length of the first row is determined by **T**.

(v) The total number of entries is n.

(vi) The numbers $n, k, p_1, \ldots, p_k$ are uniquely determined by **T**, and hence the subdiagonal matrix form N is canonical.

7. Prove that if **T** is nilpotent of index p on $\mathscr{V}_n$, then $\mathscr{R}(\mathbf{T}^{p-k}) \subseteq \mathscr{N}(\mathbf{T}^k)$ for $k = 1, \ldots, p-1$, with equality holding either for all values of k or for no values of k.

7.6   JORDAN CANONICAL FORM

At long last we are prepared to state and prove a theorem that will establish a canonical form for similarity.

Theorem 7.16   Let **T** be a linear transformation on $\mathscr{V}_n$ and let the characteristic polynomial of **T** be

$$p(x) = (-1)^n (x - \lambda_1)^{s_1} \cdots (x - \lambda_t)^{s_t},$$

where $\lambda_1, \ldots, \lambda_t$ are the distinct characteristic values of **T** and $s_1 + \cdots + s_t = n$. There exist t subspaces $\mathscr{S}_1, \ldots, \mathscr{S}_t$ of $\mathscr{V}_n$ such that

(a) each $\mathscr{S}_i$ is **T**-invariant,
(b) $\mathscr{V}_n = \mathscr{S}_1 \oplus \mathscr{S}_2 \oplus \cdots \oplus \mathscr{S}_t$,
(c) $\dim \mathscr{S}_i = s_i$,
(d) $\mathbf{T}_{\mathscr{S}_i} = \lambda_i \mathbf{I}(s_i) + \mathbf{N}_i$, where $\mathbf{N}_i$ is nilpotent.

Proof   Let $\mathbf{T}_1$ be the linear transformation on $\mathscr{V}_n$ defined by

$$\mathbf{T}_1 = \mathbf{T} - \lambda_1 \mathbf{I}.$$

By Theorem 7.8 there exist $\mathbf{T}_1$-invariant subspaces $\mathscr{S}_1$ and $\mathscr{R}_1$ such that

$$\mathscr{V}_n = \mathscr{S}_1 \oplus \mathscr{R}_1,$$

where $\mathbf{T}_1$ restricted to $\mathscr{S}_1$ is nilpotent and $\mathbf{T}_1$ restricted to $\mathscr{R}_1$ is nonsingular. Then $\mathscr{S}_1$ and $\mathscr{R}_1$ are also **T**-invariant because $\mathbf{T} = \mathbf{T}_1 + \lambda_1 \mathbf{I}$. Choose any basis for $\mathscr{S}_1$ and any basis for $\mathscr{R}_1$; their union is a basis for $\mathscr{V}_n$, relative to which **T** is represented by a block diagonal matrix

$$A = \left(\begin{array}{c|c} B_1 & Z \\ \hline Z & C_1 \end{array} \right),$$

where B_1 has dim $\mathscr{S}_1$ rows and columns. Hence $\det(A - \lambda_1 I) = \det(B_1 - \lambda_1 I)\det(C_1 - \lambda_1 I) = 0$. Because T_1 is nonsingular on $\mathscr{R}_1$, $\det(C_1 - \lambda_1 I) \neq 0$. Hence λ_1 is not a characteristic value of C_1. But the characteristic polynomial of A (and hence of T) is the product of the characteristic polynomials of B_1 and C_1. Then $(x - \lambda_1)^{s_1}$ must be a factor of $\det(B_1 - \lambda_1 I)$, so $s_1 \leq$ dim $\mathscr{S}_1$. Because T_1 is nilpotent on $\mathscr{S}_1$, a possible basis for $\mathscr{S}_1$ has the cyclic form described by Theorem 7.15. With that choice of basis, T_1 restricted to $\mathscr{S}_1$ is described by a matrix N_1 having dim $\mathscr{S}_1$ rows and columns and in the canonical subdiagonal form for nilpotent matrices,

$$
N_1 = \begin{pmatrix}
0 & & & & & \\
* & 0 & & & & \\
 & * & 0 & & & \\
 & & \cdot & \cdot & & \\
 & & & \cdot & \cdot & \\
 & & & & * & 0
\end{pmatrix},
$$

where each subdiagonal entry is 0 or 1. Then T restricted to $\mathscr{S}_1$ is represented by $B_1 = \lambda_1 I + N_1$:

$$
B_1 = \begin{pmatrix}
\lambda_1 & & & & \\
* & \lambda_1 & & & \\
 & * & \lambda_1 & & \\
 & & \cdot & \cdot & \\
 & & & \cdot & \cdot \\
 & & & & * & \lambda_1
\end{pmatrix}.
$$

Because this matrix has $(\lambda_1 - x)^{\dim \mathscr{S}_1}$ as its characteristic polynomial, dim $\mathscr{S}_1 \leq s_1$. Hence dim $\mathscr{S}_1 = s_1$, and all assertions of the theorem have been demonstrated for $i = 1$ without specializing the basis for $\mathscr{R}_1$ in any way. Hence we can now consider T restricted to the space $\mathscr{R}_1$; let $T_2 = T_{\mathscr{R}_1} - \lambda_2 I$, and repeat the argument. After t such steps the proof is complete.

To avoid cluttering the previous theorem with details that were not essential to its proof, we suppressed a precise description of the block matrices B_i by simply stating that each subdiagonal entry was 0 or 1. Now we fill in those details, all of which have been established in Theorems 7.15 and 7.16.

Jordan Canonical Form for Similarity   Any linear transformation **T** on $\mathcal{V}_n$ determines uniquely the following numbers:

$\lambda_1, \lambda_2, \ldots, \lambda_t$, the t distinct characteristic values of **T**,

$s_1, s_2, \ldots, s_t$, where s_i is the algebraic multiplicity of λ_i,

$p_{11} \geq p_{12} \geq \cdots \geq p_{1k(1)}$, where $p_{11} + \cdots + p_{1k(1)} = s_1$,

$$\begin{array}{ccc} \cdot & \cdot & \cdot \\ \cdot & \cdot & \cdot \\ \cdot & \cdot & \cdot \end{array}$$

$p_{t1} \geq p_{t2} \geq \cdots \geq p_{tk(t)}$, where $p_{t1} + \cdots + p_{tk(t)} = s_t$.

Relative to a suitable basis for $\mathcal{V}_n$, **T** is represented by an n-by-n diagonal block matrix J having t diagonal blocks $B_1, \ldots, B_t$. For each $i = 1, \ldots, t$, block B_i is an s_i-by-s_i matrix that is also a diagonal block matrix having $k(i)$ diagonal sub-blocks $S_{ij}, j = 1, \ldots, k(i)$. Each S_{ij} is a p_{ij}-by-p_{ij} matrix with λ_i in each diagonal position and 1 in each subdiagonal position. Every other subdiagonal entry of J is 0, and every entry that is not on the diagonal or subdiagonal of J is 0. For each ordering of the characteristic values of **T** there is one and only one matrix J in Jordan form, and in that sense the Jordan form is canonical for the relation of similarity. A change in the ordering of the λ_i simply produces a permutation of the major blocks in J.

As an illustration consider the 10-by-10 matrix shown below, which is in Jordan form.

$$J = \begin{pmatrix} 2 & & & & & & & & & \\ 1 & 2 & & & & & & & & \\ & 1 & 2 & & & & & & & \\ & & & 0 & 2 & & & & & \\ & & & & 1 & 2 & & & & \\ & & & & & & 0 & 1 & & \\ & & & & & & & 1 & 1 & \\ & & & & & & & & 1 & 1 \\ & & & & & & & & & 0 & 0 \\ & & & & & & & & & & 0 & 0 \end{pmatrix}$$

The diagonal entries are the characteristic values, each according to its algebraic multiplicity. Hence J represents a linear transformation whose characteristic polynomial is

$$p(x) = (x - 2)^5(x - 1)^3 x^2.$$

There are three distinct characteristic values, so $t = 3$. The major blocks along the diagonal are of dimensions 5, 3, and 2, each having one of the characteristic values in each diagonal position. Hence $s_1 = 5, s_2 = 3, s_3 = 2$. Within the first major block there are two sub-blocks, determined by each string of ones on the subdiagonal. The dimensions of those sub-blocks are $p_{11} = 3$ and $p_{12} = 2$. In the second major block there is only one sub-block, so $p_{21} = 3$. In the third major block there are two one-by-one sub-blocks, so $p_{31} = p_{32} = 1$. (If the bottom subdiagonal entry were 1 instead of 0, we would have $p_{31} = 2$ with only one sub-block corresponding to $\lambda_3 = 0$.)

All this information can be presented very concisely by listing the characteristic values in order and indicating the corresponding values of the numbers p_{ij}. For the above example we write

$$\lambda: \quad 2 \quad\quad 1 \quad\quad 0$$

$$\{(3, 2)\ (3)\ (1, 1)\}.$$

The symbol in the braces is called the *Segre characteristic* of J (or of any matrix similar to J). The sum of the numbers inside any set of parentheses is the size of that major block, whereas the numbers themselves specify the sizes of the sub-blocks. The number of sets of parentheses, of course, is the number t of distinct characteristic values. In general then, the Segre characteristic of $\mathbf{T}$ is

$$\lambda_1 \quad\quad\quad \lambda_2 \quad\quad \cdots \quad\quad \lambda_t$$

$$\{(p_{11}, \ldots, p_{1k(1)})\ (p_{21}, \ldots, p_{2k(2)}) \cdots (p_{t1}, \ldots, p_{tk(t)})\},$$

where the numbers p_{ij} describe the sizes of the sub-blocks of the Jordan form of a matrix that represents $\mathbf{T}$.

The Jordan canonical form is a powerful tool for proving theorems and analyzing linear transformations; it is not very helpful in computation because of the difficulty in practice of determining the characteristic values and the Segre characteristic of a given matrix A. Suppose we want to determine the Jordan form of A, and suppose we have the characteristic polynomial of A, completely factored into first-degree terms. Then we know $\lambda_1, \ldots, \lambda_t$ and $s_1, \ldots, s_t$, so for a fixed ordering of the λ_i we know the diagonal entries of J. To determine the subdiagonal entries of J we could (in theory) proceed by the method suggested by Exercise 7.5-6 and for each i determine the dimensions of the null spaces of $\mathbf{T}_i$, $\mathbf{T}_i^2$, ..., $\mathbf{T}_i^{p-1}$ on $\mathscr{S}_i$.

Fortunately when n is not too large, or more precisely when each s_i is small, we can obtain the needed information more readily. A string of m ones on the subdiagonal occurs only when some p_{ij} is $m + 1$, which implies that $s_i \geq m + 1$. Hence the only nonzero subdiagonal entries are associated with a characteristic value λ_i of algebraic multiplicity $s_i > 1$. If we look at the major block B_i in J that corresponds to λ_i, we have

$$B_i = \begin{pmatrix} \lambda_i & & & & \\ * & \lambda_i & & & \\ & * & \cdot & & \\ & & & \cdot & \\ & & & \cdot & \\ & & & * & \lambda_i \end{pmatrix}$$

where each $*$ is 0 or 1. The last column of B_i identifies a characteristic vector associated with λ_i. The column of each zero subdiagonal entry also identifies a characteristic vector associated with λ_i, and the set of all such characteristic vectors (one for each 0 on the subdiagonal and one for the last column) is linearly independent. Hence if there are m_i vectors in a maximal linearly independent subset of characteristic vectors associated with λ_i, then there are m_i sub-blocks in B_i; that is, $m_i = k(i)$. Also, the number of ones on the subdiagonal of the major block B_i is $s_i - m_i$. If $s_i - m_i = 1$, the subdiagonal of B_i is one followed by zeros. But if $s_i - m_i \geq 2$, the distribution of ones along the subdiagonal might remain in doubt. For example, let $s_i - m_i = 2$. The subdiagonal could be 1, 1, 0, 0, ... or 1, 0, 1, 0, ... In the former case there is a sub-block of size 3, in the latter there are two sub-blocks of size 2. This question can be decided by determining the minimal polynomial of A, which is the minimal polynomial of J. (See Exercise 7.4-6.) If $m(x)$ has a factor of $(x - \lambda_i)^3$, a three-by-three sub-block S_{ij} must exist because

$$\begin{pmatrix} 0 & 0 & 0 \\ 1 & 0 & 0 \\ 0 & 1 & 0 \end{pmatrix}$$

is nilpotent of index 3, whereas

$$\begin{pmatrix} 0 & 0 \\ 1 & 0 \end{pmatrix}$$

is nilpotent of index 2.

In summary, to find the Jordan form of A we first determine the characteristic values of A. For any characteristic value λ_i with algebraic multiplicity $s_i > 1$, we determine the maximal number $m_i \geq 1$ of linearly independent characteristic vectors associated with λ_i. Then B_i has m_i sub-blocks. If $s_i - m_i > 1$ for any i, we need more information, so we determine the minimal polynomial of A, which we shall prove in Section 7.8 to be

$$m(x) = (x - \lambda_1)^{p_{11}} \cdots (x - \lambda_t)^{p_{t1}}.$$

Then for each i we know the size s_i of the major block B_i. The number of sub-blocks of B_i is m_i, and the first and largest sub-block S_{i1} is of size p_{i1}. This information determines J uniquely whenever $s_i < 7$ for each i, but not for $s_i \geq 7$. (See Exercise 7.6-8.) For smaller values of s_i one can often find J by determining the characteristic and minimal polynomials of A without bothering to compute any characteristic vectors of A unless more information is needed.

A related problem, which we shall consider in the next section, is to determine a nonsingular matrix P such that $P^{-1}AP$ is in Jordan form.

Finally, we must remember that the theory of the Jordan form depends on a factorization of the characteristic polynomials into linear factors. This is always possible when the complex numbers are the scalars, but other canonical forms for similarity can be derived for real matrices when we are restricted to work only with real numbers.

EXERCISES 7.6

1. Given the matrix

$$A = \begin{pmatrix} 3 & -2 & -4 \\ 2 & 3 & 0 \\ -1 & 0 & 3 \end{pmatrix}$$

determine

    (i) the characteristic polynomial of A,

    (ii) the characteristic values of A,

    (iii) the minimal polynomial of A,

    (iv) a matrix J in Jordan form that is similar to A,

    (v) the Segre characteristic of A.

2. Carry out the instructions of Exercise 1 for each of the following matrices:

$$A_1 = \begin{pmatrix} 2 & 10 & 5 \\ -2 & -4 & -4 \\ 3 & 5 & 6 \end{pmatrix}, \qquad A_2 = \begin{pmatrix} 2 & -1 & 1 \\ 1 & 0 & 3 \\ 0 & 0 & 2 \end{pmatrix}.$$

Are A_1 and A_2 similar? Explain.

3. Carry out the instructions of Exercise 7.6-1 for the two matrices in Exercise 7.4-2(ii) and (iii). (You may use the answers to Exercise 7.4-2 and 7.4-3 to reduce computation.) Are these matrices similar? Explain.

4. Let A be a four-by-four matrix having $(x - 1)(x - 2)^2$ as its minimal polynomial.

(i) What can you deduce about the Jordan form J of A? Be as specific as possible.

(ii) What single additional bit of information will determine J uniquely?

(iii) Deduce from this example that matrices having the same minimal polynomial need not be similar. (Recall Exercise 7.4-6.)

5. Let **T** be a linear transformation on $\mathbb{R}^6$ that is nilpotent of index three and has three (but not four) linearly independent characteristic vectors. Determine all possible Jordan matrices that might represent **T**. Explain your reasoning.

6. A five-by-five matrix A has $(3 - x)^3(2 - x)^2$ as its characteristic polynomial. List all possible forms of the minimal polynomial of A, and for each write a matrix in Jordan form that is similar to A.

7. Let A and B be nilpotent n-by-n matrices.

(i) For $n = 3$ show that A and B are similar if and only if they have the same index p of nilpotency.

(ii) For $n = 4$ show that the assertion in (i) is false.

8. Show that the Jordan form of a seven-by-seven matrix A is not uniquely determined by the information $\lambda_1 = 0$, $s_1 = 7$, $m_1 = 3$, and $p_{11} = 3$.

9. Determine whether A and B are similar:

$$A = \begin{pmatrix} 1 & -1 \\ 4 & -3 \end{pmatrix}, \qquad B = \begin{pmatrix} -1 & 0 \\ 1 & -1 \end{pmatrix}.$$

10. A 12-by-12 Jordan matrix J has as diagonal entries in order 6 twos, 4

zeros, and 2 ones. Its subdiagonal entries in order are 1, 1, 1, 0, 1, 0, 1, 1, 1, 0, 0.

(i) Write J in block form showing all major blocks and all sub-blocks.

(ii) Write the characteristic polynomial of J.

(iii) Write the minimal polynomial of J.

(iv) Write the Segre characteristic of J.

(v) By examining the block form of J, show that $m(J) = Z$, where m is the polynomial you wrote in (iii).

11. Write a matrix whose only characteristic values are 2, 3, 4, 5 and whose corresponding Segre characteristic is $\{(2, 1, 1) (1, 1, 1) (3) (2, 1)\}$.

12. How many distinct six-by-six Jordan matrices have the characteristic polynomial $(x - 2)^6$? List all possible subdiagonals.

13. Given the companion matrix

$$A = \begin{pmatrix} 0 & 0 & 0 & 4 \\ 1 & 0 & 0 & -4 \\ 0 & 1 & 0 & -3 \\ 0 & 0 & 1 & 4 \end{pmatrix},$$

determine a Jordan matrix that is similar to A by finding the characteristic values of A and, if necessary, some of the characteristic vectors. Then check your conclusion by computing $P^{-1}AP$, where

$$P = \begin{pmatrix} -1 & 2 & 4 & -4 \\ 0 & -1 & 0 & 8 \\ 1 & -2 & -3 & -5 \\ 0 & 1 & 1 & 1 \end{pmatrix}, \quad P^{-1} = \frac{1}{18} \begin{pmatrix} 6 & 12 & 24 & 48 \\ -8 & -10 & -8 & 8 \\ 9 & 9 & 9 & 9 \\ -1 & 1 & -1 & 1 \end{pmatrix}.$$

7.7 REDUCTION TO JORDAN FORM

For some problems involving linear transformations we need to know only whether a given matrix A is diagonable. For other problems we need to find a matrix J in Jordan form that is similar to A. And sometimes we need also to find a nonsingular matrix P such that $P^{-1}AP = J$. This need arises when a question is posed in terms of a given coordinate system, leading to a matrix A. We can answer the question if we know the Jordan form of A; that is, we can answer the question in a new coordinate system. But we also need to translate that answer back to the original coordinate system to obtain a solution to the original problem. Examples of this type will arise in Chapter 11, and we now develop some computational techniques that are useful there and elsewhere.

We already know that A is diagonable if and only if there exists a basis $\{X_1, \ldots, X_n\}$ of characteristic vectors. Also if P has $X_1, \ldots, X_n$ as its column vectors, then $P^{-1}AP = J$ is diagonal. So we only need to consider the problem for matrices whose Jordan form contains at least one nonzero subdiagonal entry. We illustrate first with the four-by-four matrix A in Exercise 7.6-13. The characteristic polynomial of A is

$$p(x) = (x - 2)^2(x - 1)(x + 1)$$

so a Jordan matrix J similar to A is of the form

$$P^{-1}AP = J = \begin{pmatrix} 2 & 0 & 0 & 0 \\ * & 2 & 0 & 0 \\ 0 & 0 & 1 & 0 \\ 0 & 0 & 0 & -1 \end{pmatrix},$$

where only the $(2, 1)$ entry is in doubt. This question can be settled either by showing that the minimal polynomial is *not* $(x - 2)(x - 1)(x + 1)$ and therefore $m(x) = p(x)$, or by showing that the characteristic subspace $\mathscr{C}(2)$ has dimension one; that can be done by calculating the characteristic vectors associated with $\lambda_1 = 2$. Then the $(2, 1)$ entry of J is 1. To determine the matrix P we first calculate all the characteristic vectors:

$$\text{For } \lambda = -1, \quad X_4 = k \begin{pmatrix} -4 \\ 8 \\ -5 \\ 1 \end{pmatrix},$$

$$\text{For } \lambda = 1, \quad X_3 = k \begin{pmatrix} 4 \\ 0 \\ -3 \\ 1 \end{pmatrix},$$

$$\text{For } \lambda = 2, \quad X_2 = k \begin{pmatrix} 2 \\ -1 \\ -2 \\ 1 \end{pmatrix}.$$

Letting $k = 1$, we number these vectors with subscripts that correspond to the columns of J that identify those characteristic vectors in the new basis.

From the form of J, these vectors are linearly independent. To determine the fourth vector X_1 of the desired basis we see from the first column of J that it must satisfy the equation

$$AX_1 = 2X_1 + X_2,$$

$$(A - 2I)X_1 = X_2.$$

Because the entries of A and X_2 are known, this is a nonhomogeneous system of four linear equations in the four unknown components of X_1, and there is a unique solution,

$$X_1 = \begin{pmatrix} -1 \\ 0 \\ 1 \\ 0 \end{pmatrix}.$$

Then as shown in Exercise 7.6-13, if we let P be the matrix whose columns are X_1, X_2, X_3, X_4, then

$$P^{-1}AP = J.$$

What we have done, of course, is to determine a cyclic basis $\{X_1, X_2\}$ for the null space of $(A - 2I)^2$, because

$$(A - 2I)X_1 = X_2,$$

$$(A - 2I)X_2 = Z.$$

Now let us consider the problem generally. Given a linear transformation T, represented by a matrix A, we determine the characteristic polynomial,

$$p(x) = (-1)^n (x - \lambda_1)^{s_1} \cdots (x - \lambda_t)^{s_t}.$$

For each λ_i we determine a maximal linearly independent set of characteristic vectors $\{\xi_1, \ldots, \xi_m\}$, where $m = g_i$, the geometric multiplicity of λ_i. The number m is also the dimension of the characteristic subspace $\mathscr{C}(\lambda_i)$, and from Exercise 7.7-2, $s_i - g_i$ is the number of nonzero entries on the subdiagonal of the major block B_i of J:

$$B_i = \begin{pmatrix} \lambda_i & & & & \\ * & \lambda_i & & & \\ & * & \lambda_i & & \\ & & \ddots & \ddots & \\ & & & * & \lambda_i \end{pmatrix}.$$

At this stage we know only the number of subdiagonal ones but not their distribution in B_i. Because each sub-block S_{ij} of B_i has 1 in each subdiagonal position, there are m sub-blocks, and the last column of each sub-block identifies one of the characteristic vectors ξ_j, but we don't know which one. By computation we can determine the dimension p_{ij} of the sub-block S_{ij} associated with ξ_j. Denote the vectors of the basis associated with the sub-block S_{ij} by $\gamma_1, \gamma_2, \ldots, \gamma_r$, where $\gamma_1 = \xi_j$ but the other γ's are not yet known. But they must satisfy the equations

$$(\mathbf{T} - \lambda_i \mathbf{I})\gamma_1 = \theta,$$
$$(\mathbf{T} - \lambda_i \mathbf{I})\gamma_2 = \gamma_1,$$

(7.1)
$$\cdot$$
$$\cdot$$
$$\cdot$$

$$(\mathbf{T} - \lambda_i \mathbf{I})\gamma_r = \gamma_{r-1}.$$

Each of these equations represents a system of linear equations with the same coefficient matrix $(A - \lambda_i I)$ in the original coordinate system. The first system is homogeneous and a nonzero solution ξ_j is known to exist. Each of the subsequent systems is nonhomogeneous and might be inconsistent. Starting with $\gamma_1 = \xi_j$, we solve each successive system until we reach an inconsistent system,

$$(\mathbf{T} - \lambda_i \mathbf{I})\gamma_{r+1} = \gamma_r.$$

Then $r = p_{ij}$ is the size of the sub-block S_{ij} associated with the characteristic vector ξ_j. Because ξ_j corresponds to the last column of S_{ij}, we write the linearly independent set of γ's in the *reverse* order,

$$\mathscr{C}_j = \{\gamma_r, \ldots, \gamma_1\},$$

to correspond to the columns of S_{ij}.

After doing this for *each* ξ_j, we know the dimension p_{ij} of each sub-block in B_i. We renumber the ordered sets $\mathscr{C}_j$ so that

$$p_{i1} \geq p_{i2} \geq \cdots \geq p_{ij(i)}$$

and also renumber the vectors of those ordered sets to correspond to the corresponding columns of J. Then we do all of this again for each λ_i, finally obtaining an ordered basis for $\mathscr{V}_n$, relative to which $\mathbf{T}$ is represented by J. If that ordered basis for $\mathscr{V}_n$, represented as n-tuples relative to the original basis, is

$$\{X_1, \ldots, X_n\},$$

then we let P be the matrix having X_k in column k for $k = 1, \ldots, n$. Because the columns of P represent the new basis in terms of the old basis,

$$P^{-1}AP = J.$$

Observe carefully that each of the linear systems (7.1) has the same coefficient matrix. To avoid needless repetition of computation, when $s_i > 1$ it is prudent to solve the general system

(7.2) $(A - \lambda_i I)X = Y$

where Y has fixed but unspecified components. Gauss or Gauss-Jordan elimination will produce the consistency condition in terms of $y_1, \ldots, y_n$. Then let $Y = Z$ to obtain the characteristic vectors associated with λ_i, and select a maximal linearly independent subset. Pick any n-tuple C_1 in that subset, and then use your previous work with $Y = C_1$. If C_1 satisfies the consistency condition, a solution X will exist. Call that vector C_2 and repeat with $Y = C_2$. Eventually some C_r will fail to satisfy the consistency condition, and r is the size of the sub-block corresponding to C_1. The system (7.2) then is used for other characteristic vectors associated with λ_i. A similar system has to be solved for each value of λ_i for which $s_i > 1$.

Example   We wish to determine J and P such that $P^{-1}AP = J$, where

$$A = \begin{pmatrix} 9 & 7 & 3 \\ -9 & -7 & -4 \\ 4 & 4 & 4 \end{pmatrix}.$$

As in Exercise 7.2-1(iii) we determine the characteristic polynomial of A to be

$$p(x) = (2 - x)^3.$$

The system (7.2) is $(A - 2I)X = Y$. We reduce the augmented matrix

$$\begin{pmatrix} 7 & 7 & 3 & \vdots & y_1 \\ -9 & -9 & -4 & \vdots & y_2 \\ 4 & 4 & 2 & \vdots & y_3 \end{pmatrix}$$

to reduced echelon form:

$$\begin{pmatrix} 1 & 1 & 0 & \vdots & 4y_1 + 3y_2 \\ 0 & 0 & 1 & \vdots & -(9y_1 + 7y_2) \\ 0 & 0 & 0 & \vdots & 7(2y_1 + 2y_2 + y_3) \end{pmatrix}$$

The consistency condition is

$$2y_1 + 2y_2 + y_3 = 0,$$

and a solution of a consistent system is

(7.3)
$$x_3 = -(9y_1 + 7y_2)$$
$$x_2 = c \text{ (arbitrary)}$$
$$x_1 = 4y_1 + 3y_2 - c.$$

To find characteristic vectors, let each $y_i = 0$; the solution is

$$c \begin{pmatrix} -1 \\ 1 \\ 0 \end{pmatrix}.$$

Because $\lambda = 2$ has algebraic multiplicity three and geometric multiplicity one, we know that J has 2 ones on its subdiagonal, so

$$J = \begin{pmatrix} 2 & 0 & 0 \\ 1 & 2 & 0 \\ 0 & 1 & 2 \end{pmatrix}.$$

But to find a suitable P we choose $c = -1$ to obtain

$$C_1 = \begin{pmatrix} 1 \\ -1 \\ 0 \end{pmatrix}.$$

Letting $Y = C_1$ we note that the consistency condition is satisfied and use (7.3) with $c = -1$ to obtain

$$C_2 = \begin{pmatrix} 2 \\ -1 \\ -2 \end{pmatrix}.$$

Letting $Y = C_2$, we note again that the consistency condition is satisfied, so (7.3) yields

$$C_3 = \begin{pmatrix} 6 \\ -1 \\ -11 \end{pmatrix}.$$

Now for $Y = C_3$ the consistency condition is not satisfied, as expected.

Writing these n-tuples in reverse order as the columns of P, we obtain

$$P = \begin{pmatrix} 6 & 2 & 1 \\ -1 & -1 & -1 \\ -11 & -2 & 0 \end{pmatrix}.$$

Then

$$P^{-1} = \begin{pmatrix} -2 & -2 & -1 \\ 11 & 11 & 5 \\ -9 & -10 & -4 \end{pmatrix} \quad \text{and} \quad P^{-1}AP = J.$$

The matrix P depends on the particular characteristic vector used to start the process, and is not uniquely determined by A. The basis $\{(C_3)_{T_1}\}$ is a T_1-cyclic basis for $\mathbb{R}^3$, where $T_1 = T - 2I$.

EXERCISES 7.7

1. For each matrix A determine a matrix P such that $P^{-1}AP$ is in Jordan form, verifying your work by computation.

(i) $A = \begin{pmatrix} 5 & -6 & -6 \\ -1 & 4 & 2 \\ 3 & -6 & -4 \end{pmatrix}.$

(ii) $A = \begin{pmatrix} 3 & 1 & -1 \\ 2 & 2 & -1 \\ 2 & 2 & 0 \end{pmatrix}.$

(iii) $A = \begin{pmatrix} 3 & 1 & 1 \\ 0 & 4 & 0 \\ -1 & 1 & 5 \end{pmatrix}.$

(iv) $A = \begin{pmatrix} 4 & 2 & -1 & 4 \\ 0 & 2 & 0 & 0 \\ 1 & 3 & 2 & 1 \\ 0 & 0 & 0 & 2 \end{pmatrix}.$

(v) $A = \begin{pmatrix} 3 & 1 & 0 & 0 \\ 0 & 1 & 0 & 0 \\ 0 & 0 & 2 & 1 \\ 0 & 0 & 1 & 2 \end{pmatrix}.$

2. Show that the number of ones on the subdiagonal of the major block B_i of a Jordan matrix is $s_i - g_i$, the difference between the algebraic and geometric multiplicities of λ_i.

3. This exercise explores the following question: What is the largest number of m-by-m matrices that have the same characteristic polynomial but no two of which are similar? To simplify the problem, we assume that A has a single characteristic value, so that $p(x) = (-1)^m(x - \lambda)^m$ is the characteristic polynomial of A. Then $A - \lambda I$ either is Z or is nilpotent of index p for some $p = 2, 3, \ldots, m$. Let $N(m, p)$ denote the number of distinct m-by-m matrices that are nilpotent of index p and in canonical form (with 1 in the first subdiagonal position, 0 or 1 in every other subdiagonal position, and 0 elsewhere.) Then the number $N(m)$ of distinct m-by-m nilpotent matrices in canonical form is given by

$$N(m) = \sum_{p=2}^{m} N(m, p),$$

and $1 + N(m)$ is the number of distinct similarity classes of matrices that have characteristic polynomial $p(x) = (-x)^m$.

(i) Show that the values of $N(m)$ for $m = 1, 2, \ldots, 9$ are 0, 1, 2, 4, 6, 10, 14, 22, 30 respectively.

(ii) Use the data in (i) to conjecture a formula for $N(m)$.

(iii) Try to prove your conjecture.

4. Use Exercise 3 to show that there are more than 100 12-by-12 matrices, all of which have $x^5(x - 1)^4(x + 1)^3$ as characteristic polynomial but no two of which are similar.

7.8 AN APPLICATION OF
THE HAMILTON-CAYLEY THEOREM

We begin by using the Jordan form to sharpen Theorem 7.12 by proving the assertion made near the end of Section 7.6 that the minimal polynomial of A is

$$m(x) = (x - \lambda_1)^{p_{11}} \cdots (x - \lambda_t)^{p_{t1}}$$

Theorem 7.17   Let A be an n-by-n matrix with $\lambda_1, \ldots, \lambda_t$ as its distinct characteristic values and

$$\{(p_{11}, \ldots, p_{1k(1)}) \cdots (p_{t1}, \ldots, p_{tk(t)})\}$$

as its Segre characteristic. Then the minimal polynomial of A is

$$m(x) = (x - \lambda_1)^{p_{11}}(x - \lambda_2)^{p_{21}} \cdots (x - \lambda_t)^{p_{t1}}.$$

Proof Let $J = P^{-1}AP$ be the Jordan form of A for the given ordering of the λ_i. For any polynomial q, $q(J) = q(P^{-1}AP) = P^{-1}q(A)P$. Because P is nonsingular, $q(J) = Z$ if and only if $q(A) = Z$. We shall show that $m(J) = Z$, but $q(J) \neq Z$ if q has degree less than that of m, where m is defined in the statement of this theorem. Because J is a block diagonal matrix,

$$J = \text{diag } (B_1, \ldots, B_t),$$

the method of block multiplication shows that

$$J^k = \text{diag } (B_1^k, \ldots, B_t^k)$$

for any k. It follows that for any polynomial q

$$q(J) = \text{diag } (q(B_1), \ldots, q(B_t)).$$

Hence $q(J) = Z$ if and only if $q(B_i) = Z$ for $i = 1, 2, \ldots, t$. Let m denote the polynomial

$$m(x) = (x - \lambda_1)^{p_{11}}(x - \lambda_2)^{p_{21}} \cdots (x - \lambda_t)^{p_{t1}}.$$

We consider $m(B_i)$ where $B_i = \lambda_i I + N_i$ and where N_i is nilpotent of index p_{i1}:

$$m(B_i) = (B_i - \lambda_1 I)^{p_{11}}(B_i - \lambda_2 I)^{p_{21}} \cdots (B_i - \lambda_t I)^{p_{t1}}.$$

Because the factor $(B_i - \lambda_i I)^{p_{i1}} = N_i^{p_{i1}} = Z$, it follows that $m(B_i) = Z$ for $i = 1, \ldots, t$. Thus $m(B) = Z$. By Theorem 7.10 m is divisible by the minimal polynomial, and by Theorem 7.12 the minimal polynomial has the form

$$r(x) = (x - \lambda_1)^{r_1} \cdots (x - \lambda_t)^{r_t}, \ 1 \leq r_i \leq s_i.$$

Suppose $r_j < p_{j1}$ for at least one index j. If $j \neq i$, $(B_i - \lambda_j I)$ is nonsingular because its determinant is $(\lambda_i - \lambda_j)^{s_i} \neq 0$. If $j = i$, $(B_i - \lambda_i I)$ is nilpotent of index p_{i1} and therefore $(B_i - \lambda_i I)^{r_i} \neq Z$. Thus $r(B_i)$ has the same positive rank as $(B_i - \lambda_i I)^{r_i}$, so $r(B_i) \neq Z$. Hence for the minimal polynomial $r_i = p_{j1}$, as desired.

This theorem, of course, provides an immediate proof of the Hamilton-Cayley Theorem because $p_{i1} \leq s_i$ for each i. We now use the Hamilton-

Cayley Theorem to derive another method for calculating the inverse of a nonsingular matrix A. Let the characteristic polynomial of A be

$$p(x) = (-1)^n(x^n + c_1 x^{n-1} + \cdots + c_{n-1} x + c_n),$$
$$= (-1)^n(x - \lambda_1)(x - \lambda_2) \cdots (x - \lambda_n).$$

Then

$$A^n + c_1 A^{n-1} + \cdots + c_{n-1} A + c_n I = Z,$$

and from Exercise 7.1-9 we know that

$$c_n = (-1)^n \det A \neq 0$$

because A is nonsingular. Then

$$c_n I = -(A^n + c_1 A^{n-1} + \cdots + c_{n-1} A),$$

and if we multiply by $c_n^{-1} A^{-1}$ we obtain

$$A^{-1} = -c_n^{-1}(A^{n-1} + c_1 A^{n-2} + \cdots + c_{n-1} I).$$

Hence A^{-1} can be calculated as a linear combination of powers of A with the coefficients of the characteristic polynomial of A.

For large matrices the coefficients c_i might be hard to compute directly, but an alternate method can be used. In Exercise 7.1-10 the *trace* of A is defined to be the sum of the n characteristic values of A:

$$\text{tr } A = \sum_{i=1}^{n} \lambda_i.$$

In addition to the present application, the concept of trace is quite useful. We recall that similar matrices have equal characteristic values and therefore have equal traces. Thus the trace of a linear transformation **T** can be defined to be the trace of any matrix that represents **T**. We now prove that tr A is the sum of the diagonal elements of A.

Theorem 7.18

$$\text{tr } A = \sum_{i=1}^{n} a_{ii},$$

and hence if A and B are similar,

$$\sum_{i=1}^{n} a_{ii} = \sum_{i=1}^{n} b_{ii}.$$

Proof Consider the characteristic polynomial of A,

$$\det(A - \lambda I) = (-1)^n(\lambda^n + c_1 \lambda^{n-1} + \cdots + c_n)$$
$$= (-1)^n(\lambda - \lambda_1) \cdots (\lambda - \lambda_n).$$

From the determinant form, the coefficient of λ^{n-1} is seen to be

$$(-1)^{n-1} \sum_{i=1}^{n} a_{ii} = (-1)^n c_1,$$

while from the factored form we get

$$(-1)^{n+1} \sum_{i=1}^{n} \lambda_i = (-1)^n c_1.$$

Hence

$$\text{tr } A = \sum_{i=1}^{n} \lambda_i = \sum_{i=1}^{n} a_{ii}.$$

As a corollary we have the equation

$$c_1 = -\text{tr } A.$$

The other c_i can be determined similarly to give the following set of equations:

$$c_1 = -\text{tr } A$$
$$c_2 = -2^{-1}[c_1 \text{ tr } A + \text{tr } A^2]$$
$$c_3 = -3^{-1}[c_2 \text{ tr } A + c_1 \text{ tr } A^2 + \text{tr } A^3]$$
$$\cdot$$
$$\cdot$$
$$\cdot$$
$$c_n = -n^{-1}[c_{n-1} \text{ tr } A + c_{n-2} \text{ tr } A^2 + \cdots + c_1 \text{ tr } A^{n-1} + \text{tr } A^n].$$

These equations permit us to calculate A^{-1} by calculating A, $A^2, \ldots, A^{n-1}$ and the diagonal elements of A^n. From these matrices we can calculate the traces as sums of the diagonal elements, determine the c_i, and finally use the equation

$$A^{-1} = -c_n^{-1}(A^{n-1} + c_1 A^{n-2} + \cdots + c_{n-1} I).$$

This method for finding A^{-1} requires fewer than n^4 multiplications and is easily described in the language of high-speed computers. Also we obtain the characteristic polynomial of A as a by-product of the computation of A^{-1}.

However, if a computer is used to calculate A^{-1} and the characteristic polynomial p of A, the results probably will not be exact because of rounding errors in the computation process. Moreover, if the entries of A are obtained as a result of scientific measurements, those measurements provide an independent source of error. Finding solutions of a polynomial equation of fifth or higher degree is in itself quite difficult, and special approximation methods (discussed in books on numerical analysis or numerical linear algebra) are needed even to approximate the characteristic values of A.

EXERCISES 7.8

1. Use the method following Theorem 7.18 to calculate the inverse and the characteristic polynomial of the matrix of Exercise 4.4-1(ii).

2. Prove that the trace of A^k is the sum of the kth powers of the characteristic values of A.

3. Verify that

$$c_2 = -\tfrac{1}{2}(c_1 \text{ tr } A + \text{ tr } A^2).$$

4. Prove the following properties of the trace.

(i) If A is nilpotent, tr $A = 0$.

(ii) If A is idempotent, tr $A = r(A)$.

(iii) $\text{tr}(A + B) = \text{tr } A + \text{tr } B$.

(iv) $\text{tr}(kA) = k \text{ tr } A$.

(v) tr $A^t = $ tr A.

(vi) $\text{tr}(AB) = \text{tr}(BA)$.

5. Let A and B be two-by-two matrices for which det $A = $ det B and tr $A = $ tr B.

(i) Do A and B have the same characteristic values? Prove your answer.

(ii) Are A and B similar? Prove your answer.

(iii) Would your answers to (i) or (ii) be different if A and B were three-by-three matrices?

6. Let A be a real n-by-n matrix other than Z such that

$$a_{ik}a_{jk} = a_{kk}a_{ij} \text{ for all } i, j, k.$$

Prove the following statements:

   (i) $AA^t = (\text{tr } A)A$.

   (ii) $\text{tr } A \neq 0$.

   (iii) A is symmetric.

   (iv) The characteristic polynomial of A is $(-1)^n x^{n-1}(x - \text{tr } A)$.

7. Let H_p be the p-by-p matrix obtained by reflecting the identity matrix I across a horizontal line through its center.

   (i) Show that $H_p A$ is the matrix obtained by reflecting the p-by-p matrix A across a horizontal line through its center.

   (ii) Show that AH_p is the matrix obtained by reflecting A across a vertical line through its center.

   (iii) Describe $H_p^{-1} A H_p$.

8. Use the results of Exercise 7.8-7 and the Jordan form of A to prove that A and A^t are similar.

CHAPTER 8
INNER PRODUCT SPACES

8.1 INNER PRODUCTS

Thus far our study of vector spaces has been carried out without any dependence on the kinds of measurement that often are used to describe geometric objects, such as length, distance, angle, and orthogonality. An exception to this policy of benign neglect was allowed in Section 2.2 to facilitate the use of familiar geometric language to interpret the algebraic operations in $\mathbb{R}^n$, providing a model for the general definition of a vector space. You will recall that metric concepts in $\mathbb{R}^n$ were derived from a specific, real-valued function defined for each pair of vectors in $\mathbb{R}^n$. This function, called the dot product, was defined as follows:

If $X = (x_1, \ldots, x_n)$ and $Y = (y_1, \ldots, y_n)$,

then $X \cdot Y = x_1 y_1 + \cdots + x_n y_n$.

Furthermore, we showed that the dot product is bilinear, symmetric, and positive definite.

If we consider an arbitrary vector space $\mathscr{V}$ over a field $\mathscr{F}$, a counterpart of the dot product is a special type of scalar-valued function p defined for each pair of vectors in $\mathscr{V}$; in other words, a mapping from $\mathscr{V} \times \mathscr{V}$ into $\mathscr{F}$. If we expect to define the length $\|\xi\|$ of a nonzero vector ξ (as we did in $\mathbb{R}^n$) to be the positive real number

$$\|\xi\| = p(\xi, \xi)^{1/2},$$

then we must impose further conditions on p to assure that $p(\xi, \xi)$ has a positive square root in an ordered subfield of $\mathscr{F}$. Thus we also must restrict the nature of $\mathscr{F}$, and we do so by assuming that $\mathscr{F}$ is either $\mathbb{R}$ or $\mathbb{C}$. If you are not familiar with the algebra of complex numbers, including the complex conjugate operation, you should refer to Appendix A.5 at this time.

When $\mathscr{F} = \mathbb{R}$, metric concepts can be developed essentially as they were in Section 2.2, so we now consider the situation when $\mathscr{F} = \mathbb{C}$. Then $p(\xi, \eta)$ is a complex number, say

$$p(\xi, \eta) = a + ib,$$

and we want to be sure that $p(\xi, \xi)$ is a positive real number when $\xi \neq \theta$ so that it has a real, positive square root, which will be the length of ξ. But a complex number z is real if and only if $z = \bar{z}$. Thus we can guarantee that $p(\xi, \xi)$ is real by requiring p to have the property that

(8.1) $$p(\eta, \xi) = \overline{p(\xi, \eta)} = a - ib,$$

because if this condition is satisfied for all ξ and η, and if

$$p(\xi, \xi) = c + id,$$

then by interchanging the positions of the two equal vectors we obtain

$$p(\xi, \xi) = \overline{(c + id)} = c - id$$

so $d = 0$. Equation (8.1) is a modified form of the symmetry property of the dot product in $\mathbb{R}^n$. Its adoption requires that we also modify the assumption of bilinearity for the function p, because if p is linear in the first component, then

(8.2) $$p(\xi, k\eta) = \overline{p(k\eta, \xi)} = \bar{k}\,\overline{p(\eta, \xi)} = \bar{k}p(\xi, \eta).$$

Thus $p(\xi, k\eta) \neq kp(\xi, \eta)$ unless k is real. Hence we are led to the following definition.

Definition 8.1 Let $\mathscr{V}$ be a vector space over $\mathbb{C}$. A *complex inner product* on $\mathscr{V}$ is a function p with domain $\mathscr{V} \times \mathscr{V}$ and range $\mathbb{C}$ that satisfies the following conditions for all vectors in $\mathscr{V}$ and all scalars in $\mathbb{C}$:

(a) $p(c_1 \xi_1 + c_2 \xi_2, \eta) = c_1 p(\xi_1, \eta) + c_2 p(\xi_2, \eta)$;

(b) $p(\eta, \xi) = \overline{p(\xi, \eta)}$;

(c) $p(\xi, \xi)$ is real and positive for each $\xi \neq \theta$.

The inner product $p(\xi, \eta)$ is also denoted by $\langle \xi, \eta \rangle$.

Property (b) is called *conjugate symmetry* or *Hermitian symmetry*. When $\langle \xi, \eta \rangle$ is real, conjugate symmetry reduces to symmetry. Property (a) asserts that p is a linear function of its first component vector. Combining Properties (a) and (b), we have

$$\langle \xi, c_1 \eta_1 + c_2 \eta_2 \rangle = \overline{\langle c_1 \eta_1 + c_2 \eta_2, \xi \rangle}$$
$$= \overline{\overline{c_1} \langle \eta_1, \xi \rangle + \overline{c_2} \langle \eta_2, \xi \rangle}$$
$$= \overline{c_1} \langle \xi, \eta_1 \rangle + \overline{c_2} \langle \xi, \eta_2 \rangle.$$

This is expressed by saying that p is a *conjugate* linear function of its second component vector. Hence p is said to be *conjugate bilinear*, signifying linearity in the first component and conjugate linearity in the second component. Property (a) implies that $\langle \theta, \xi \rangle = 0$ for all ξ, so Property (c) can be stated as

$$\langle \xi, \xi \rangle \geq 0 \text{ in } \mathbb{R} \text{ for all } \xi, \text{ and}$$

$$\langle \xi, \xi \rangle = 0 \text{ if and only if } \xi = \theta.$$

That is, p is *positive definite*. In summary a complex inner product on $\mathscr{V}$ is a conjugate bilinear, conjugate symmetric, positive definite function from $\mathscr{V} \times \mathscr{V}$ onto $\mathbb{C}$.

The definition of a real inner product can be obtained from Definition 8.1 by noting that when p is real valued, conjugate bilinearity reduces to bilinearity, and conjugate symmetry reduces to symmetry.

Definition 8.2 Let $\mathscr{V}$ be a vector space over $\mathbb{R}$. A *real inner product* on $\mathscr{V}$ is a bilinear, symmetric, and positive definite function from $\mathscr{V} \times \mathscr{V}$ onto $\mathbb{R}$. A real vector space on which a real inner product is defined is called a *Euclidean space*. A complex vector space on which a complex inner product is defined is called a *unitary space*. Euclidean and unitary spaces collectively are called *inner product spaces*.

Unitary and Euclidean spaces can be investigated simultaneously by studying unitary spaces and noting any essential differences that arise in the Euclidean case, when the conjugate operator becomes the identity operator.

Examples of Inner Product Spaces

(A) Euclidean n-space: $\mathbb{R}^n$ with the dot product, as in Section 2.2.

(B) $\mathbb{R}^2$ with the inner product defined by

$$\langle(x_1, x_2), (y_1, y_2)\rangle = x_1 y_1 - 2x_1 y_2 - 2x_2 y_1 + 5x_2 y_2.$$

This function is easily seen to be bilinear and symmetric. It is positive definite because

$$\langle(x_1, x_2), (x_1, x_2)\rangle = x_1^2 - 4x_1 x_2 + 5x_2^2 = (x_1 - 2x_2)^2 + x_2^2.$$

(C) The infinite-dimensional space of all real valued functions continuous on the interval $0 \le t \le 1$, with the inner product defined by

$$\langle f, g \rangle = \int_0^1 f(t)g(t)dt.$$

(D) Unitary n-space: $\mathbb{C}^n$ with the inner product defined by

$$\langle(x_1, \ldots, x_n), (y_1, \ldots, y_n)\rangle = x_1 \bar{y}_1 + \cdots + x_n \bar{y}_n.$$

(E) The infinite-dimensional space of all complex valued functions continuous on the real interval $0 \le t \le 1$ with the inner product defined by

$$\langle f, g \rangle = \int_0^1 f(t)\overline{g(t)}\, dt.$$

We recall that in $\mathbb{R}^n$ the Schwarz inequality was the key to many aspects of Euclidean space. The same is true for any inner product space, as we now see. Although the proof of Theorem 2.1 can be adapted to prove the Schwarz inequality in any inner product space (Exercise 8.1-5), we shall give an alternative proof here.

Theorem 8.1 (Schwarz inequality)   In any inner product space $\mathscr{V}$,

$$|\langle \xi, \eta \rangle|^2 \le \langle \xi, \xi \rangle \langle \eta, \eta \rangle \text{ for all } \xi, \eta \in \mathscr{V}.$$

Proof   For any $\alpha \in \mathscr{V}$, $\langle \alpha, \alpha \rangle$ is real and nonnegative. Let $\alpha = a\xi + b\eta$ for any scalars a, b; then we have

$$0 \le \langle a\xi + b\eta, a\xi + b\eta \rangle = a\langle \xi, a\xi + b\eta \rangle + b\langle \eta, a\xi + b\eta \rangle$$

$$= a\bar{a}\langle \xi, \xi \rangle + a\bar{b}\langle \xi, \eta \rangle + \bar{a}b\langle \eta, \xi \rangle + b\bar{b}\langle \eta, \eta \rangle.$$

Now choose $b = \langle \xi, \xi \rangle$, and $a = -\langle \eta, \xi \rangle$. Then $b = \bar{b} \in \mathbb{R}$ and $\bar{a} = -\langle \xi, \eta \rangle$, so the previous inequality becomes

$$0 \le a\bar{a}b - ab\bar{a} - \bar{a}ba + bb\langle \eta, \eta \rangle$$

$$= b[b\langle \eta, \eta \rangle - a\bar{a}]$$

$$= b[\langle \xi, \xi \rangle \langle \eta, \eta \rangle - \langle \xi, \eta \rangle \overline{\langle \xi, \eta \rangle}].$$

If $b = 0$ then $\xi = \theta$ and the Schwarz inequality is valid with each side equal to zero. Otherwise $b > 0$, so the last bracket is nonnegative, which is the assertion of the Schwarz inequality because for any complex number c, $|c|^2 = c\bar{c}$.

We note that for nonzero vectors ξ and η the Schwarz inequality can be written in the form

$$\frac{|\langle \xi, \eta \rangle|^2}{\langle \xi, \xi \rangle \langle \eta, \eta \rangle} \le 1.$$

In the Euclidean case $\langle \xi, \eta \rangle$ is real, so this becomes

$$-1 \le \frac{\langle \xi, \eta \rangle}{\sqrt{\langle \xi, \xi \rangle \langle \eta, \eta \rangle}} \le 1.$$

That is, every pair of nonzero real vectors determines a unique real number between -1 and 1, and this real number is the cosine of a unique angle Ψ between 0 and π radians. That angle is defined to be the angle determined by (or between) the vectors ξ and η. But in the unitary case we cannot make such a definition because we only know that

$$\frac{\langle \xi, \eta \rangle}{\sqrt{\langle \xi, \xi \rangle \langle \eta, \eta \rangle}}$$

is a complex number whose magnitude does not exceed one. However this expression is real in the special case that $\langle \xi, \eta \rangle = 0$, and therefore it provides a definition of perpendicularity, even in a unitary space. And that is sufficient for many purposes, as we shall see in the next section.

EXERCISES 8.1

1. In $\mathbb{R}^2$ let $\xi = (x_1, x_2)$ and $\eta = (y_1, y_2)$, and let p be defined by

$$p(\xi, \eta) = x_1 y_1 + x_1 y_2 + x_2 y_1 + k x_2 y_2,$$

where $k \in \mathbb{R}$.

(i) Show that p is bilinear and symmetric for each $k \in \mathbb{R}$.

(ii) Determine all values of k for which p is an inner product.

2. Verify that Examples C and D actually satisfy the definition of an inner product.

3. If p is any inner product verify that

(i) $p(c\xi, d\eta) = c\bar{d}p(\xi, \eta)$,

(ii) $p(\theta, \eta) = p(\xi, \theta) = 0$.

4. Prove that if p is any inner product and if η_1 and η_2 are vectors such that $p(\xi, \eta_1) = p(\xi, \eta_2)$ for all ξ, then $\eta_1 = \eta_2$.

5. Carry out in detail the following outline of an alternative proof of the Schwarz inequality. $\mathrm{Rl}(z)$ denotes the real part of the complex number z.

(i) For all complex numbers x, y and all vectors ξ, η

$$0 \le \langle x\xi + y\eta, x\xi + y\eta \rangle = x\bar{x}\langle \xi, \xi \rangle + 2\mathrm{Rl}[x\bar{y}\langle \xi, \eta \rangle] + y\bar{y}\langle \eta, \eta \rangle.$$

(ii) Specify x to be real and choose $y = \langle \xi, \eta \rangle$ to obtain the real quadratic inequality, valid for all real x,

$$0 \le \langle \xi, \xi \rangle x^2 + 2|\langle \xi, \eta \rangle|^2 x + |\langle \xi, \eta \rangle|^2 \langle \eta, \eta \rangle.$$

(iii) Apply a criterion for a real quadratic function to be nonnegative, obtaining the Schwarz inequality.

6. Prove that equality holds in the Schwarz inequality if and only if the set $\{\xi, \eta\}$ is linearly dependent.

7. Show that the following general theorems are direct consequences of the Schwarz inequality:

(i) If $x_1, \ldots, x_n$ and $y_1, \ldots, y_n$ are any real numbers, then

$$\left(\sum_{i=1}^{n} x_i y_i \right)^2 \le \left(\sum_{i=1}^{n} x_i^2 \right) \left(\sum_{i=1}^{n} y_i^2 \right).$$

(ii) (Cauchy inequality) If $x_1, \ldots, x_n$ and $y_1, \ldots, y_n$ are any complex numbers, then

$$\left| \sum_{i=1}^{n} x_i \bar{y}_i \right|^2 \le \left(\sum_{i=1}^{n} |x_i|^2 \right) \left(\sum_{i=1}^{n} |y_i|^2 \right).$$

(iii) If f and g are real functions continuous on the interval $a \le x \le b$, then

$$\left(\int_a^b f(x)g(x)\, dx \right)^2 \le \int_a^b f^2(x)\, dx \int_a^b g^2(x)\, dx.$$

8. Let **T** be a nonsingular linear mapping from a vector space $\mathscr{V}(\mathscr{F})$ to an inner product space $\mathscr{W}(\mathscr{F})$, where $\mathscr{F}$ is $\mathbb{R}$ or $\mathbb{C}$ and where the inner product in $\mathscr{W}$ is denoted by $\langle \alpha, \beta \rangle$. Let p be the function from $\mathscr{V} \times \mathscr{V}$ to $\mathscr{F}$, defined by

$$p(\xi, \eta) = \langle \mathbf{T}\xi, \mathbf{T}\eta \rangle.$$

Verify that p defines an inner product on $\mathscr{V}$.

9. Let A and B be complex n-by-n matrices. Verify that the function f given by

$$f(A, B) = \operatorname{tr}(A\bar{B}^{\mathrm{t}}),$$

defines a complex inner product on the space of all complex n-by-n matrices, where tr C denotes the trace of C and each entry of $\bar{B}$ is the conjugate of the corresponding entry of B.

10. Another approach to metric concepts in a real vector space $\mathscr{V}$ is to define a function N, called a *norm*, from $\mathscr{V}$ to $\mathbb{R}$ having the following properties for all $\alpha, \beta \in \mathscr{V}$ and all $c \in \mathbb{R}$:

(a) $N(c\alpha) = |c| N(\alpha)$,

(b) $N(\alpha) > 0$ if $\alpha \neq \theta$,

(c) $N(\alpha + \beta) \leq N(\alpha) + N(\beta)$.

For example, on any Euclidean space $N(\alpha) = \|\alpha\|$ is a norm. Verify that each of the following definitions provides a norm for $\mathbb{R}^n$.

(i) $N(a_1, \ldots, a_n) = |a_1| + \cdots + |a_n|$.

(ii) $N(a_1, \ldots, a_n) = \max |a_i|, i = 1, \ldots, n$.

11. Referring to Exercise 10 for the definition of a norm, show that a norm is provided for the space of all real functions that are continuous on the interval $0 \leq t \leq 1$ by the definition

$$N(f) = \max |f(t)|, 0 \leq t \leq 1.$$

8.2   LENGTH, DISTANCE, ORTHOGONALITY

Let $\mathscr{V}$ be an inner product space that is either Euclidean or unitary. The inner product function p is used to define length, distance, and orthogonality; we begin with length.

Definition 8.3   The *length* $\|\xi\|$ of a vector ξ is defined by

$$\|\xi\| = \langle \xi, \xi \rangle^{1/2}.$$

Theorem 8.2   Length has the following properties:

(a) $\|c\xi\| = |c| \|\xi\|$.

(b) $\|\xi\| > 0$ if $\xi \neq \theta$, and $\|\theta\| = 0$.

(c) $\|\xi + \eta\| \leq \|\xi\| + \|\eta\|$.

Proof   Property (a) follows from Exercise 8.1-3, and Property (b) from part (c) of Definition 8.1. To prove Property (c) we first note that the Schwarz inequality can be written

$$|\langle \xi, \eta \rangle| = |\langle \eta, \xi \rangle| \leq \|\xi\| \|\eta\|.$$

Because $\langle \xi, \eta \rangle = \overline{\langle \eta, \xi \rangle}$, $\langle \xi, \eta \rangle + \langle \eta, \xi \rangle$ is real, and

$$|\langle \xi, \eta \rangle + \langle \eta, \xi \rangle| \leq |\langle \xi, \eta \rangle| + |\langle \eta, \xi \rangle| \leq 2\|\xi\| \|\eta\|.$$

Thus

$$\begin{aligned}
\|\xi + \eta\|^2 &= \langle \xi + \eta, \xi + \eta \rangle \\
&= \langle \xi, \xi \rangle + \langle \xi, \eta \rangle + \langle \eta, \xi \rangle + \langle \eta, \eta \rangle \\
&\leq \langle \xi, \xi \rangle + |\langle \xi, \eta \rangle + \langle \eta, \xi \rangle| + \langle \eta, \eta \rangle \\
&\leq \|\xi\|^2 + 2\|\xi\| \|\eta\| + \|\eta\|^2,
\end{aligned}$$

from which Property (c) follows immediately.

For the case of Euclidean n-space where p is the dot product, the length of $\xi = (x_1, \ldots, x_n)$ is simply the familiar form

$$\|\xi\| = \sqrt{x_1^2 + x_2^2 + \cdots + x_n^2}.$$

Property (c) is interpreted geometrically as the observation that the length of any side of a triangle does not exceed the sum of the lengths of the other two sides. Hence (c) is called the *triangle inequality*.

Because the points of n-space may be interpreted as n-tuples or vectors, the distance between vectors can be regarded as the distance between those points, or, equivalently, the length of the arrow from one point to the other.

Definition 8.4   The *distance* $d(\xi, \eta)$ between two vectors ξ and η is defined by

$$d(\xi, \eta) = \|\xi - \eta\|.$$

Theorem 8.3 Distance has the following properties:

(a) $d(\xi, \eta) = d(\eta, \xi)$,

(b) $d(\xi, \eta) > 0$ if $\xi \neq \eta$, and $d(\xi, \xi) = 0$,

(c) $d(\xi, \eta) \leq d(\xi, \zeta) + d(\zeta, \eta)$.

Proof Exercise.

Thus the distance which results from any inner product has the familiar properties of distance as defined by coordinates in analytic geometry: it is symmetric, positive for distinct points, and satisfies the triangle inequality. Any space for which a distance function is defined having these three properties is called a *metric space*.

As we noted at the end of the previous section, angle can be defined in Euclidean spaces, but the process does not apply in unitary spaces where $\langle \xi, \eta \rangle$ is generally a complex number. But the special case $\langle \xi, \eta \rangle = 0$ defines orthogonality in Euclidean spaces and makes sense also in unitary spaces.

Definition 8.5 In any inner product space, two vectors ξ and η are said to be *orthogonal* if and only if $\langle \xi, \eta \rangle = 0$. Two sets S and T of vectors are said to be orthogonal if and only if $\langle \sigma, \tau \rangle = 0$ for each $\sigma \in S$ and each $\tau \in T$.

Theorem 8.4 In any inner product space

(a) ξ is orthogonal to every $\eta \in \mathscr{V}$ if and only if $\xi = \theta$,

(b) if ξ is orthogonal to every vector of a set S, then ξ is orthogonal to the subspace spanned by S,

(c) any set of mutually orthogonal nonzero vectors is linearly independent.

Proof Exercise.

Our next objective is to demonstrate how a basis of mutually orthogonal vectors can be constructed for any finite dimensional inner product space. The main tool in this construction, known as the *Gram-Schmidt orthogonalization process*, was outlined in Exercise 2.2-4. Given any subspace $\mathscr{S}$ in an inner product space $\mathscr{V}$ and given any $\xi \in \mathscr{V}$ such that $\xi \notin \mathscr{S}$, we want to write ξ as the sum of two vectors, $\xi = \sigma + \tau$, such that σ is in $\mathscr{S}$ and τ is orthogonal to $\mathscr{S}$. That is, we want to split ξ into two components, one in $\mathscr{S}$

and one orthogonal to $\mathscr{S}$. To illustrate the simplest case we suppose that $\mathscr{S}$ is a one-dimensional subspace, say $\mathscr{S} = [\alpha]$. Let σ and τ be defined by

$$\sigma = \frac{\langle \xi, \alpha \rangle}{\langle \alpha, \alpha \rangle} \alpha,$$

$$\tau = \xi - \sigma.$$

Then $\sigma \in \mathscr{S}$ and $\xi = \sigma + \tau$, as desired. We must also verify that τ is orthogonal to $\mathscr{S}$:

$$\langle \tau, \alpha \rangle = \langle \xi - \sigma, \alpha \rangle = \langle \xi, \alpha \rangle - \langle \sigma, \alpha \rangle$$

$$= \langle \xi, \alpha \rangle - \frac{\langle \xi, \alpha \rangle}{\langle \alpha, \alpha \rangle} \langle \alpha, \alpha \rangle = 0.$$

Figure 8.1 illustrates the Euclidean case in which the angle Ψ between ξ and α is defined by

$$\cos \Psi = \frac{\langle \xi, \alpha \rangle}{\sqrt{\langle \xi, \xi \rangle \langle \alpha, \alpha \rangle}} = \frac{\langle \xi, \alpha \rangle}{\|\xi\| \, \|\alpha\|}.$$

The length of σ, the perpendicular projection of ξ on α, must satisfy

$$\|\sigma\| = \|\xi\| \, |\cos \Psi|,$$

and σ will have the same direction as α if $\cos \Psi > 0$ but the opposite direction if $\cos \Psi < 0$. Because $\|\alpha\|^{-1} \alpha$ is a vector of unit length in the direction of α, we have

$$\sigma = (\|\xi\| \, \cos \Psi)\|\alpha\|^{-1}\alpha$$

$$= \frac{\|\alpha\| \, \|\xi\| \, \cos \Psi}{\|\alpha\|^2} \alpha = \frac{\langle \xi, \alpha \rangle}{\langle \alpha, \alpha \rangle} \alpha,$$

as claimed previously. When $\mathscr{S}$ has dimension greater than one, the Gram-Schmidt process is similar except that the perpendicular projections of ξ on each vector of a mutually orthogonal set of vectors that spans $\mathscr{S}$. The details are shown in the proof of the next theorem.

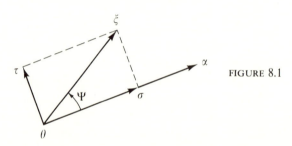

FIGURE 8.1

Theorem 8.5   In any finite-dimensional inner product space $\mathscr{V}$, there exists a basis consisting of mutually orthogonal vectors.

Proof   If $\mathscr{V}$ is one-dimensional, any nonzero vector forms such a basis. We proceed by induction, assuming the theorem for any space of dimension k. Let $\mathscr{V}$ be of dimension $k + 1$, and let $\mathscr{S}$ be any subspace of dimension k. By the induction hypothesis there exists an orthogonal basis $\{\alpha_1, \ldots, \alpha_k\}$ for $\mathscr{S}$. For $\xi \notin \mathscr{S}$, let

$$\sigma = \sum_{i=1}^{k} c_i \alpha_i, \text{ where } c_i = \frac{\langle \xi, \alpha_i \rangle}{\langle \alpha_i, \alpha_i \rangle},$$

and let

$$\tau = \xi - \sigma.$$

Clearly, $\sigma \in \mathscr{S}$, and we now show that τ is orthogonal to each α_i:

$$\langle \tau, \alpha_i \rangle = \langle \xi - \sigma, \alpha_i \rangle$$

$$= \langle \xi, \alpha_i \rangle - \left\langle \sum_{j=1}^{k} c_j \alpha_j, \alpha_i \right\rangle$$

$$= \langle \xi, \alpha_i \rangle - \sum_{j=1}^{k} c_j \langle \alpha_j, \alpha_i \rangle$$

$$= \langle \xi, \alpha_i \rangle - c_i \langle \alpha_i, \alpha_i \rangle$$

$$= \langle \xi, \alpha_i \rangle - \frac{\langle \xi, \alpha_i \rangle}{\langle \alpha_i, \alpha_i \rangle} \langle \alpha_i, \alpha_i \rangle = 0.$$

By Property (c) of Theorem 8.4 the set $\{\alpha_1, \ldots, \alpha_k, \tau\}$ is an orthogonal basis for $\mathscr{V}$.

The vector σ in the preceding proof is called the orthogonal projection of ξ on $\mathscr{S}$, and the numbers c_i are called *direction numbers* of σ in $\mathscr{S}$. In the Euclidean case if each vector α_i has unit length, we have

$$c_i = \langle \xi, \alpha_i \rangle = \langle \sigma, \alpha_i \rangle = \|\sigma\| \cos \Psi_i,$$

where Ψ_i is the angle between α_i and σ. In an inner product space of functions c_i is called the *Fourier coefficient* of σ relative to α_i.

Definition 8.6   In any inner product space a vector of unit length is called *normal*. A set of mutually orthogonal, normal vectors is called a *normal orthogonal* (or *orthonormal*) set.

Because $\|\alpha\|^{-1}\alpha$ is normal whenever $\alpha \neq \theta$, it is easy to convert any orthogonal basis into a normal orthogonal basis. Thus to construct a normal orthogonal basis for any finite-dimensional inner product space, we can begin with any nonzero vector α_1 and any $\xi_2 \notin [\alpha_1]$ and use the Gram-Schmidt process to obtain the component α_2 of ξ_2 that is orthogonal to $[\alpha_1]$. Then choose $\xi_3 \notin [\alpha_1, \alpha_2]$ and again compute the component α_3 of ξ_3 that is orthogonal to $[\alpha_1, \alpha_2]$, continuing in this fashion until an orthogonal basis is obtained. Then replace each α_1 by $\beta_i = \|\alpha_i\|^{-1}\alpha_i$ to obtain a normal orthogonal basis. As an exercise you may carry out this construction for the space $\mathscr{P}_2$ of real polynomials of degree not exceeding two.

Theorem 8.6   Let $\mathscr{V}_n$ be any n-dimensional inner product space, let $\{\alpha_1, \ldots, \alpha_n\}$ be a normal orthogonal basis for $\mathscr{V}_n$, and let

$$\xi = \sum_{i=1}^{n} x_i\alpha_i \text{ and } \eta = \sum_{j=1}^{n} y_j\alpha_j.$$

Then

$$\langle \xi, \eta \rangle = x_1\bar{y}_1 + x_2\bar{y}_2 + \cdots + x_n\bar{y}_n.$$

Proof $$\langle \xi, \eta \rangle = \left\langle \sum_{i=1}^{n} x_i\alpha_i, \sum_{j=1}^{n} y_j\alpha_j \right\rangle$$

$$= \sum_{i=1}^{n} x_i\left\langle \alpha_i, \sum_{j=1}^{n} y_j\alpha_j \right\rangle$$

$$= \sum_{i=1}^{n} \sum_{j=1}^{n} x_i\bar{y}_j\langle \alpha_i, \alpha_j \rangle$$

$$= \sum_{i=1}^{n} x_i\bar{y}_i\langle \alpha_i, \alpha_i \rangle = \sum_{i=1}^{n} x_i\bar{y}_i,$$

where the last equality holds because each α_i is normal, and the previous equality holds because the α_i are orthogonal.

This theorem shows that relative to a normal orthogonal basis for $\mathscr{V}_n$ any inner product assumes the algebraic form of the standard dot product, real or complex. In the Euclidean case, this means that length, distance, and angle are computed by the familiar formulas (2.1), (2.2), and (2.3) when a normal orthogonal basis is used for $\mathscr{E}^n$.

EXERCISES 8.2

1. Prove Theorem 8.3.

2. Prove Theorem 8.4.

3. Prove the assertion made in the final sentence of Section 8.2.

4. Prove that in every Euclidean space

$$\langle \xi + \eta, \xi - \eta \rangle = 0 \text{ if and only if } \|\xi\| = \|\eta\|.$$

Is the same result valid in every unitary space? What theorem of plane Euclidean geometry does this result assert?

5. Let α, β be any vectors in an inner product space.

   (i) Show that

$$\|\alpha + \beta\|^2 = \|\alpha\|^2 + 2 \operatorname{Rl}\langle \alpha, \beta \rangle + \|\beta\|^2,$$

where $\operatorname{Rl}(z)$ denotes the real part of the complex number z.

   (ii) Show that

$$\|\alpha + \beta\|^2 - \|\alpha - \beta\|^2 = 4 \operatorname{Rl}\langle \alpha, \beta \rangle.$$

   (iii) Explain how the values of a real inner product on $\mathscr{V}$ can be obtained if the length of each vector in $\mathscr{V}$ is known.

6. In $\mathscr{E}^4$ decompose the vector $\xi = (3, -1, 5, 1)$ into a component σ along the vector $\alpha = (-1, 2, 3, 2)$ and a component τ orthogonal to α. Use the standard dot product.

7. In $\mathscr{E}^4$ with the standard basis and inner product, let

$$\alpha_1 = (2, 1, -5, 0),$$
$$\alpha_2 = (3, -1, 1, 0).$$

Use the Gram-Schmidt process to construct an orthogonal basis $\{\alpha_1, \alpha_2, \alpha_3, \alpha_4\}$ and then convert that basis to a normal orthogonal basis.

8. Consider the polynomial space $\mathscr{P}_2$ with inner product defined by

$$\langle p, q \rangle = \int_0^1 p(x)q(x)\, dx.$$

(i) Is the usual basis $B = \{1, x, x^2\}$ for $\mathscr{P}_2$ a normal orthogonal basis?

(ii) Use the Gram-Schmidt process to convert B to the normal orthogonal basis

$$N = \{1, \sqrt{3}(2x - 1), \sqrt{5}(6x^2 - 6x + 1)\}$$

(iii) Illustrate Theorem 8.6 by expressing each of the two polynomials

$$p(x) = x^2$$
$$q(x) = 2x - 6x^2$$

as a linear combination of the vectors of N and then comparing the value of $\langle p, q \rangle$ with the dot product of those two linear combinations.

9. Let $\{\alpha_1, \ldots, \alpha_n\}$ be a normal orthogonal basis in Euclidean n-space. Prove

(i) *Bessel's inequality*: $\sum_{i=1}^{k} \langle \xi, \alpha_i \rangle^2 \leq \|\xi\|^2$ for each $k \leq n$,

(ii) *Parseval's identity*: $\langle \xi, \eta \rangle = \sum_{i=1}^{n} \langle \xi, \alpha_i \rangle \langle \alpha_i, \eta \rangle$.

10. Let $\mathscr{S}$ be a subspace of Euclidean n-space. Prove that

(i) any linear combination of vectors each of which is orthogonal to $\mathscr{S}$ is also orthogonal to $\mathscr{S}$;

(i) the only vector of $\mathscr{S}$ that is orthogonal to $\mathscr{S}$ is θ;

(iii) each vector ξ has a unique decomposition

$$\xi = \sigma + \tau$$

where $\sigma \in \mathscr{S}$ and τ is orthogonal to $\mathscr{S}$.

11. Let $\mathscr{S}$ and $\mathscr{T}$ be subspaces of any Euclidean space $\mathscr{E}$, and let $\mathscr{S}^{\perp} = \{\xi \in \mathscr{E} \mid \langle \xi, \sigma \rangle = 0 \text{ for all } \sigma \in \mathscr{S}\}$. The set $\mathscr{S}^{\perp}$ is called the orthogonal complement of $\mathscr{S}$. Prove the following statements:

(i) $\mathscr{S}^{\perp}$ is a subspace of $\mathscr{E}$.

(ii) If $\mathscr{E}$ is finite dimensional, $\mathscr{E} = \mathscr{S} \oplus \mathscr{S}^{\perp}$.

(iii) $\mathscr{S} \subseteq (\mathscr{S}^{\perp})^{\perp}$, and equality holds if $\mathscr{E}$ is finite-dimensional.

(iv) $(\mathscr{S} + \mathscr{T})^{\perp} = \mathscr{S}^{\perp} \cap \mathscr{T}^{\perp}$.

(v) $(\mathscr{S} \cap \mathscr{T})^{\perp} \supseteq \mathscr{S}^{\perp} + \mathscr{T}^{\perp}$, and equality holds if $\mathscr{E}$ is finite-dimensional.

8.3 ISOMETRIES

Many important physical problems involve transformations that preserve distance, called *rigid motions*. Intuitively it is clear that a succession of rigid motions is a rigid motion; also the identity transformation is a rigid motion, and the reversal of a rigid motion is a rigid motion. In other words, the set of all rigid motions of a space forms a group; for a Euclidean space this group is called the *Euclidean group*, and for a unitary space the group of rigid motions is called the *unitary group*.

Not all rigid motions are linear, however. For example, a translation

$$\mathbf{T}\xi = \xi + \alpha, \text{ for fixed } \alpha,$$

moves each point of space the same distance $\|\alpha\|$ along the direction of α. Because $\mathbf{T}\theta = \alpha$, a nonzero translation is never a linear transformation. However, any rigid motion $\mathbf{R}$ is the composition of a linear rigid motion $\mathbf{S}$ and a translation $\mathbf{T}$, where $\mathbf{T}\xi = \xi + \mathbf{R}\theta$. That is, $\mathbf{T}$ is a translation by the vector through which $\mathbf{R}$ moves the origin. We now propose to study linear rigid motions, those in which the origin remains fixed. Familiar examples in $\mathscr{E}^2$ are reflections and rotations as described in Example D in Section 3.1.

> **Definition 8.7** A linear transformation $\mathbf{T}$ on an inner product space $\mathscr{V}$ is called an *isometry* if and only if
>
> $$\|\mathbf{T}\xi\| = \|\xi\| \text{ for every } \xi \in \mathscr{V}.$$

An isometry on a Euclidean space is called an *orthogonal* transformation, whereas an isometry on a unitary space is called a *unitary* transformation.

By definition, an isometry $\mathbf{T}$ preserves the length of each vector. Because distance is defined in terms of length, we have

$$d(\xi, \eta) = \|\xi - \eta\| = \|\mathbf{T}(\xi - \eta)\| = \|\mathbf{T}\xi - \mathbf{T}\eta\| = d(\mathbf{T}\xi, \mathbf{T}\eta),$$

so $\mathbf{T}$ preserves distance. The following theorem shows that $\mathbf{T}$ also preserves the inner product, and hence orthogonality (and angle in the Euclidean case).

> **Theorem 8.7** A linear transformation $\mathbf{T}$ on an inner product space $\mathscr{V}$ is an isometry if and only if
>
> $$\langle \mathbf{T}\xi, \mathbf{T}\eta \rangle = \langle \xi, \eta \rangle.$$

Proof   We first compute $\|\xi + \eta\|^2$ and $\|\xi + i\eta\|^2$

$$\|\xi + \eta\|^2 = \langle \xi + \eta, \xi + \eta \rangle$$
$$= \langle \xi, \xi \rangle + \langle \xi, \eta \rangle + \langle \eta, \xi \rangle + \langle \eta, \eta \rangle$$
$$= \|\xi\|^2 + \langle \xi, \eta \rangle + \overline{\langle \xi, \eta \rangle} + \|\eta\|^2.$$

Similarly,

$$\|\xi + i\eta\|^2 = \|\xi\|^2 - i\langle \xi, \eta \rangle + i\overline{\langle \xi, \eta \rangle} + \|\eta\|^2.$$

Now let $\langle \xi, \eta \rangle = a + ib$, so that $\overline{\langle \xi, \eta \rangle} = a - ib$.

Hence $\langle \xi, \eta \rangle + \overline{\langle \xi, \eta \rangle} = 2a = 2 \, \mathrm{Rl}\langle \xi, \eta \rangle,$

$\langle \xi, \eta \rangle - \overline{\langle \xi, \eta \rangle} = 2ib = 2i \, \mathrm{Im}\langle \xi, \eta \rangle.$

Hence $\mathrm{Rl}\langle \xi, \eta \rangle = \frac{1}{2}(\|\xi + \eta\|^2 - \|\xi\|^2 - \|\eta\|^2),$

$\mathrm{Im}\langle \xi, \eta \rangle = \frac{1}{2}(\|\xi + i\eta\|^2 - \|\xi\|^2 - \|\eta\|^2).$

Because the length of every vector is preserved by an isometry **T**, both the real and the imaginary parts of $\langle \xi, \eta \rangle$ are also preserved by **T**, so $\langle \mathbf{T}\xi, \mathbf{T}\eta \rangle = \langle \xi, \eta \rangle$ if **T** is an isometry. Conversely, if **T** preserves inner products, it must also preserve lengths, because $\|\xi\| = \langle \xi, \xi \rangle^{1/2}$.

To represent an isometry on $\mathscr{V}_n$ by a matrix, it is convenient to choose a normal orthogonal basis $\{\alpha_1, \ldots, \alpha_n\}$. Then $\|\alpha_i\| = 1 = \|\mathbf{T}\alpha_i\|$ for $i = 1, \ldots, n$; also $\langle \alpha_i, \alpha_j \rangle = \langle \mathbf{T}\alpha_i, \mathbf{T}\alpha_j \rangle = \delta_{ij}$. Hence if A denotes the matrix that represents **T** relative to the α-basis, each column vector of A is of unit length, and distinct column vectors are orthogonal. By Theorem 8.6 the inner product in $\mathscr{V}_n$ assumes the form of the dot product, so we adopt the following terminology.

Definition 8.8   A real n-by-n matrix $A = (a_{ij})$ is said to be *orthogonal* if and only if

$$\sum_{k=1}^{n} a_{ki} a_{kj} = \delta_{ij}.$$

A complex n-by-n matrix $A = (a_{ij})$ is said to be *unitary* if and only if

$$\sum_{k=1}^{n} a_{ki} \bar{a}_{kj} = \delta_{ij}.$$

If $A = (a_{ij})$ is an *m*-by-*n* matrix of complex numbers, then the *conjugate* $\bar{A}$ of A denotes the *m*-by-*n* matrix that has $\bar{a}_{ij}$ in the (i, j) position for each i and each j. See Exercise 8.3-1.

Theorem 8.8 Relative to a normal orthogonal basis, a real *n*-by-*n* matrix A represents an orthogonal linear transformation **T** if and only if A is orthogonal; a complex *n*-by-*n* matrix A represents a unitary linear transformation **T** if and only if A is unitary.

Proof Our previous remarks show that an orthogonal linear transformation is represented by an orthogonal matrix relative to a normal orthogonal basis. Conversely, the columns of a matrix represent the **T**-image of the basis vectors; the linear transformation that corresponds to an orthogonal matrix relative to a normal orthogonal basis preserves inner product and thus is orthogonal. The same argument is valid for the unitary case.

Theorem 8.9 Any orthogonal or unitary matrix is nonsingular.
Proof The column vectors are mutually orthogonal and hence linearly independent by Theorem 8.4(c).

Theorem 8.10 A real *n*-by-*n* matrix is orthogonal if and only if $A^{-1} = A^t$. A complex *n*-by-*n* matrix is unitary if and only if $A^{-1} = \bar{A}^t$.
Proof By Definition 8.8, A is orthogonal if and only if

$$\sum_{k=1}^{n} a_{ki} a_{kj} = \delta_{ij},$$

which is the (i, j) entry of $A^t A$. Hence $A^t A = I$. Also A is unitary if and only if

$$\sum a_{ki} \bar{a}_{kj} = \delta_{ij},$$

so $A^t \bar{A} = I = \bar{I} = \bar{A}^t A$.

Hence the inverse of an orthogonal or a unitary matrix is very easy to compute. In the unitary case the transpose of the conjugate of A is the same as the conjugate of the transpose of A; that is

$$\bar{A}^t = \overline{A^t}.$$

It is convenient to adopt special terminology and notation for this matrix. We call $\bar{A}^t$ the *adjoint* of A and denote it simply by A^*.

Theorem 8.11   If A is orthogonal or unitary, $|\det A| = 1$.
Proof   Exercise.

As a final observation we note that if two matrices A and B represent the same linear transformation **T** relative to two bases and if one basis is carried into the other by an isometry, then we have

$$B = P^{-1}AP, \qquad P \text{ an isometry,}$$
$$= P^t AP \qquad \text{if } P \text{ is orthogonal,}$$
$$= P^* AP \qquad \text{if } P \text{ is unitary.}$$

In problems with physical significance, one frequently wants to use only isometric changes of coordinates to preserve the geometry of the problem. We shall examine these ideas further in the next chapter.

EXERCISES 8.3

1. Verify the following properties of conjugation for complex matrices, where it is assumed that the matrices are of proper dimensions for the indicated operations to be defined.

 (i) $\overline{\overline{A}} = A$.

 (ii) $\overline{A + B} = \overline{A} + \overline{B}$.

 (iii) $\overline{A^t} = \overline{A}^t$.

 (iv) $\overline{AB} = \overline{A}\,\overline{B}$.

 (v) $\overline{A} = A$ if and only if each entry of A is real.

2. Let $\mathbf{R}(\Psi)$ be the linear transformation on $\mathscr{E}^2$ that rotates each point counterclockwise about the origin through a fixed angle Ψ.

 (i) Write a matrix A that represents $\mathbf{R}(\Psi)$.

 (ii) Show that A is orthogonal by computing dot products of the column vectors and also by showing that $AA^t = I$.

3. (i) Find a matrix representation B for the rigid motion of $\mathscr{E}^3$ that reflects each vector across the line

$$x_3 = 0$$
$$x_1 = x_2 .$$

(ii) Prove that B is orthogonal because it satisfies Definition 8.8.

(iii) Find B^{-1} and describe the linear transformation it represents.

4. Show that if A is orthogonal, then so is A^t.

5. Does the following matrix represent a rigid motion in $\mathscr{E}^3$?

$$A = \frac{1}{6}\begin{pmatrix} 1 & -5 & \sqrt{10} \\ -5 & 1 & \sqrt{10} \\ \sqrt{10} & \sqrt{10} & 4 \end{pmatrix}.$$

6. Prove Theorem 8.11.

7. Reason as follows to show that any linear rigid motion of the real plane is either a rotation or a rotation followed by a reflection across an axis.

(i) Write four or more quadratic conditions on the elements a, b, c, d of a real two-by-two matrix A that are necessary and sufficient that A be orthogonal.

(ii) Show that the only such matrices are

$$\begin{pmatrix} a & b \\ -b & a \end{pmatrix} \quad \text{and} \quad \begin{pmatrix} a & b \\ b & -a \end{pmatrix},$$

where $a^2 + b^2 = 1$.

(iii) Show that the first matrix in (ii) describes a rotation through the angle $\cos^{-1} a$ and the second describes a rotation followed by a reflection across an axis.

8. Prove that each characteristic value of an isometry satisfies $|\lambda| = 1$ and that any two characteristic vectors that are associated with distinct characteristic values are orthogonal. What can you deduce about the Jordan form of any matrix that represents an isometry?

9. Prove that if K is a real skew matrix and if $I + K$ is nonsingular, then

$$(I - K)(I + K)^{-1}$$

is orthogonal.

10. Let α_i denote column i of a real n-by-n matrix A; let $AA^t = C = (c_{ij})$. Prove the following:

(i) $c_{ij} = p(\alpha_i, \alpha_j)$, p an inner product;

(ii) if the α_i are mutually orthogonal,

$$\det C = [\|\alpha_1\| \, \|\alpha_2\| \, \cdots \, \|\alpha_n\|]^2;$$

(iii) if $|a_{ij}| \leq k$ for $i, j = 1, 2, \ldots, n$ and some fixed number k, then $\|\alpha_i\| \leq k\sqrt{n}$;

(iv) if $|a_{ij}| \leq k$, and if the α_i are mutually orthogonal, then

$$|\det A| = \|\alpha_1\| \, \|\alpha_2\| \, \cdots \, \|\alpha_n\| \leq k^n n^{n/2}.$$

CHAPTER 9
SCALAR-VALUED
FUNCTIONS

9.1 LINEAR FUNCTIONALS

Introductory calculus is concerned principally with the properties of functions from $\mathbb{R}$ to $\mathbb{R}$. That study is extended in intermediate calculus and elementary analysis to include a wider variety of functions. For example, a plane curve can be regarded in parametric form as the graph of a function from $\mathbb{R}$ to $\mathbb{R}^2$. A real-valued function of three variables is a function from $\mathbb{R}^3$ to $\mathbb{R}$, and a vector-valued function of several real variables is a function from $\mathbb{R}^n$ to $\mathbb{R}^m$. To make significant statements about functions, it is necessary to restrict the type of function considered by assuming some additional property, such as continuity, integrability, boundedness, or differentiability. Typically the restrictions imposed on the functions under study are kept as light as possible to obtain theorems that apply to a large class of functions.

Although most functions that we need to study are nonlinear, linearity plays a very important role in their study. For example, the tangent line to a curve at a point (or the tangent plane to a surface at a point) provides a local linear approximation to the curve (or the surface). Linearity also occurs in the context of numbers that are associated with certain functions. For example, the derivative at c is linear because

$$(af + bg)'(c) = af'(c) + bg'(c).$$

Likewise, the definite integral is linear because

$$\int_c^d (af(x) + bg(x))\, dx = a \int_c^d f(x)\, dx + b \int_c^d g(x)\, dx.$$

Each of these examples illustrates a linear mapping from a vector space over $\mathbb{R}$ to the field $\mathbb{R}$. We have seen many other examples of linear mappings from one vector space to another vector space and of linear transformations from a vector space to the same vector space.

Other forms of linearity also have occurred in our study. For example, the cross-product $\alpha \times \beta$ of vectors in $\mathbb{R}^3$ is a bilinear function from $\mathbb{R}^3 \times \mathbb{R}^3$ to $\mathbb{R}^3$. A real inner product on $\mathscr{V}$ is a bilinear function from $\mathscr{V} \times \mathscr{V}$ to $\mathbb{R}$. A complex inner product on $\mathscr{V}$ is a conjugate bilinear function from $\mathscr{V} \times \mathscr{V}$ to $\mathbb{C}$. Furthermore the determinant function on $\mathscr{V}_n(\mathscr{F})$ is a function from $\mathscr{V} \times \mathscr{V} \times \cdots \times \mathscr{V}$ (n times) to $\mathscr{F}$ that is linear in each of its n components. It is an example of an n-linear function, and a general study of such functions is called *multilinear algebra*.

In this section we consider linear functions from a vector space $\mathscr{V}(\mathscr{F})$ to the field $\mathscr{F}$. Because $\mathscr{F}$ itself can be regarded as a vector space of dimension one over $\mathscr{F}$, we know from Sections 3.1 and 4.1 that the set $\mathscr{L}(\mathscr{V}, \mathscr{F})$ forms a linear space over $\mathscr{F}$, and if $\mathscr{V}$ is n-dimensional then $\mathscr{L}(\mathscr{V}, \mathscr{F})$ is also n-dimensional. We also know that any two n-dimensional vector spaces over $\mathscr{F}$ are isomorphic, so $\mathscr{L}(\mathscr{V}_n, \mathscr{F})$ is isomorphic to $\mathscr{V}_n$. However, $\mathscr{L}(\mathscr{V}, \mathscr{F})$ and $\mathscr{V}$ are not isomorphic when $\mathscr{V}$ is infinite-dimensional, as we illustrate in Exercise 9.1-9.

Definition 9.1 Let $\mathscr{V}(\mathscr{F})$ be a vector space. A linear mapping $\mathbf{f}$ from $\mathscr{V}$ to $\mathscr{F}$ is called a *linear functional*. The vector space $\mathscr{L}(\mathscr{V}, \mathscr{F})$ is called the *dual space* of $\mathscr{V}$ and is denoted by $\mathscr{V}'$.

Two examples of linear functionals on function spaces are provided by the derivative and integral mappings cited above. For an example of a linear functional on $\mathscr{V}_n$, let $\{\alpha_1, \ldots, \alpha_n\}$ be any basis and let $\gamma = c_1 \alpha_1 + \cdots + c_n \alpha_n$ be any fixed vector. We define $\mathbf{f}_\gamma$ by this rule:

$$\text{if } \xi = x_1 \alpha_1 + \cdots + x_n \alpha_n \quad \text{then } \mathbf{f}_\gamma(\xi) = x_1 c_1 + \cdots + x_n c_n.$$

In words, $\mathbf{f}_\gamma$ attaches to each vector ξ the scalar whose value is the dot produced of the n-tuples that represent ξ and γ. Properties of the dot product show that $\mathbf{f}_\gamma$ is a linear functional in $\mathscr{V}'_n$ for each $\gamma \in \mathscr{V}_n$. Furthermore, if $\gamma \neq \delta$ then from Exercise 2.2-3(iii) we know that $\mathbf{f}_\gamma \neq \mathbf{f}_\delta$. Hence the correspondence $\gamma \to \mathbf{f}_\gamma$ is one-to-one from $\mathscr{V}_n$ into $\mathscr{V}'_n$. It is not difficult to show that the correspondence is an isomorphism of $\mathscr{V}_n$ onto $\mathscr{V}'_n$.

We now use the correspondence $\gamma \to \mathbf{f}_\gamma$ and the chosen basis for $\mathscr{V}_n$ to obtain a basis for the dual space $\mathscr{V}'_n$. For each $i = 1, \ldots, n$ we choose $\gamma = \alpha_i$, thus obtaining the subset $\{\mathbf{f}_{\alpha_1}, \ldots, \mathbf{f}_{\alpha_n}\}$ of $\mathscr{V}'_n$. From the dot product description of $\mathbf{f}_\gamma$ it is clear that

$$\mathbf{f}_{\alpha_i}(\alpha_j) = \delta_{ij}.$$

That is, $\mathbf{f}_{\alpha_i}$ maps α_i into 1 and α_j into 0 for $j \neq i$. To show that $\{\mathbf{f}_{\alpha_1}, \ldots, \mathbf{f}_{\alpha_n}\}$ is a basis for $\mathscr{V}'_n$, it suffices to show that these linear functionals span $\mathscr{V}'_n$ because $\dim(\mathscr{V}'_n) = n$. Let $\mathbf{f} \in \mathscr{V}'_n$, and let $a_i = \mathbf{f}(\alpha_i)$, $i = 1, \ldots, n$. Then

$$\mathbf{f}(\alpha_i) = a_i = \sum_{j=1}^{n} a_j \delta_{ij} = \sum_{j=1}^{n} a_j \mathbf{f}_{\alpha_j}(\alpha_i) = \left(\sum_{j=1}^{n} a_j \mathbf{f}_{\alpha_j}\right)(\alpha_i).$$

Hence $\mathbf{f}$ and $a_1 \mathbf{f}_{\alpha_1} + \cdots + a_n \mathbf{f}_{\alpha_n}$ are linear functionals whose values coincide at each vector in a basis for $\mathscr{V}_n$, and by linearity coincide on all of $\mathscr{V}_n$, so

$$\mathbf{f} = \sum_{j=1}^{n} a_j \mathbf{f}_{\alpha_j}.$$

The basis $\{\mathbf{f}_{\alpha_1}, \ldots, \mathbf{f}_{\alpha_n}\}$ for $\mathscr{V}'_n$ is called the *dual basis* of the basis $\{\alpha_1, \ldots, \alpha_n\}$ for $\mathscr{V}_n$. We have proved the following statement.

Theorem 9.1 Let $\{\alpha_1, \ldots, \alpha_n\}$ be any basis of $\mathscr{V}_n$, and let $\mathbf{f}_i \in \mathscr{V}'_n$ be defined for $i = 1, \ldots, n$ by

$$\mathbf{f}_i(\alpha_j) = \delta_{ij}.$$

Then $\{\mathbf{f}_1, \ldots, \mathbf{f}_n\}$ is a basis for $\mathscr{V}'_n$.

We now explore further the relationship between a vector space $\mathscr{V}$ and its dual space $\mathscr{V}'$. If S is a nonvoid set of vectors in $\mathscr{V}$ we denote by S^0 the set of all linear functionals in $\mathscr{V}'$ that map each vector of S into 0. That is,

$$S^0 = \{\mathbf{f} \in \mathscr{V}' \,|\, \mathbf{f}\sigma = 0 \text{ for each } \sigma \in S\}.$$

S^0 is called the *annihilator* of S.

Theorem 9.2   Let $\mathscr{V}$ be a vector space and $\mathscr{V}'$ its dual space.

(a) If $S \subseteq T$ in $\mathscr{V}$, then $T^0 \subseteq S^0$ in $\mathscr{V}'$.

(b) If $\mathscr{M}$ is a subspace of $\mathscr{V}$, $\mathscr{M}^0$ is a subspace of $\mathscr{V}'$.

(c) If $\mathscr{M}$ is an m-dimensional subspace of $\mathscr{V}_n$, then $\mathscr{M}^0$ is an $n - m$ dimensional subspace of $\mathscr{V}'_n$.

(d) If $\mathscr{M}$ and $\mathscr{N}$ are subspaces of $\mathscr{V}_n$, then

$$(\mathscr{M} \cap \mathscr{N})^0 = \mathscr{M}^0 + \mathscr{N}^0,$$

$$(\mathscr{M} + \mathscr{N})^0 = \mathscr{M}^0 \cap \mathscr{N}^0.$$

Proof   You may prove (a) and (d) as exercises. To prove (b) we let $\mathscr{M}$ be a subspace of $\mathscr{V}$, let $\mathbf{f}, \mathbf{g} \in \mathscr{M}^0$ and $a, b \in \mathscr{F}$. We want to show that $a\mathbf{f} + b\mathbf{g}$ is in $\mathscr{M}^0$; that is, that $a\mathbf{f} + b\mathbf{g}$ maps each $\mu \in \mathscr{M}$ into 0:

$$(a\mathbf{f} + b\mathbf{g})\mu = a\mathbf{f}\mu + b\mathbf{g}\mu = a0 + b0 = 0.$$

(Observe that the hypothesis that $\mathscr{M}$ is a subspace of $\mathscr{V}$ is not needed. We have shown that if S is a sub*set* of $\mathscr{V}$, then S^0 is a sub*space* of $\mathscr{V}'$, and $S^0 = [S]^0$; that is S^0 annihilates the entire subspace spanned by S.) To prove (c), let $\{\alpha_1, \ldots, \alpha_m\}$ be a basis for $\mathscr{M}$; extend this to a basis $\{\alpha_1, \ldots, \alpha_n\}$ for $\mathscr{V}$, and let $\{\mathbf{f}_1, \ldots, \mathbf{f}_n\}$ be the corresponding dual basis for $\mathscr{V}'_n$. Then $\mathbf{f}_i \in \mathscr{M}^0$ for $i = m + 1, \ldots, n$, so $[\mathbf{f}_{m+1}, \ldots, \mathbf{f}_n] \subseteq \mathscr{M}^0$. Let $\mathbf{f} = \sum_{j=1}^{n} c_j \mathbf{f}_j \in \mathscr{M}^0$. Then for $i = 1, \ldots, m$

$$0 = \mathbf{f}\alpha_i = \sum_{j=1}^{n} c_j \mathbf{f}_j \alpha_i = \sum_{j=1}^{n} c_j \delta_{ij} = c_i,$$

so $\mathbf{f} \in [\mathbf{f}_{m+1}, \ldots, \mathbf{f}_n]$. Thus dim $\mathscr{M}^0 = n - m$.

Be sure you understand the geometric meaning of the information in Theorem 9.2. Any nonvoid set of vectors in $\mathscr{V}$ determines its annihilator, a subspace of $\mathscr{V}'$. A set S and the subspace $[S]$ that it spans have the same annihilator. But each inclusion relation between subsets of $\mathscr{V}$ is reversed for the corresponding annihilator subspaces in $\mathscr{V}'$. In the process of passing from subspaces to their annihilators, intersections are turned into sums, and sums are turned into intersections. The annihilator of $\mathscr{V}$ is the zero subspace of $\mathscr{V}'$, and the annihilator of $[\theta]$ is the entire space $\mathscr{V}'$. For n-dimensional spaces, the annihilator of a one-dimensional subspace (a line) in $\mathscr{V}$ is an $(n - 1)$-dimensional subspace (a hyperplane) in $\mathscr{V}'$. These observations form the basis for the duality theory of projective geometry.

EXERCISES 9.1

1. Let **f** be a nonzero linear functional on $\mathscr{V}$, and let c be a nonzero scalar. Show that there exists a nonzero vector ξ in $\mathscr{V}$ such that $\mathbf{f}\xi = c$.

2. Let $\alpha = (a_1, a_2, a_3)$ in $\mathscr{E}^3$ relative to the standard basis, let **f** be a linear functional on $\mathscr{E}^3$, and let $\beta = (\mathbf{f}\varepsilon_1, \mathbf{f}\varepsilon_2, \mathbf{f}\varepsilon_3)$. Show that $\mathbf{f}\alpha = \alpha \cdot \beta$.

3. In $\mathbb{R}^3$ the vectors $\{\alpha_1, \alpha_2, \alpha_3\}$ form a basis, where

$$\alpha_1 = (0, 1, 1),$$
$$\alpha_2 = (1, 0, 1),$$
$$\alpha_3 = (1, 1, 0).$$

Determine the dual basis $\{\mathbf{f}_1, \mathbf{f}_2, \mathbf{f}_3\}$ for the dual space of $\mathbb{R}^3$, and compute $\mathbf{f}_i(a, b, c)$ for $i = 1, 2, 3$.

4. Let $\mathscr{C}$ denote the space of real valued functions that are continuous on the interval $c \leq x \leq d$. For each $g \in \mathscr{C}$ let φ_g be defined by

$$\varphi_g f = \int_c^d f(x)g(x)\, dx.$$

Show that $\varphi_g \in \mathscr{C}'$.

5. If ξ is any nonzero vector in $\mathscr{V}_n$, show that there exists a linear functional **f** in $\mathscr{V}_n'$ such that $\mathbf{f}\xi \neq 0$.

6. Show that the mapping $\gamma \to \mathbf{f}_\gamma$ is an isomorphism from $\mathscr{V}_n$ onto $\mathscr{V}_n'$, where $\mathbf{f}_\gamma$ is defined in the text following Definition 9.1.

7. Prove Statements (a) and (d) of Theorem 9.2.

8. Let **f** be a fixed nonzero linear functional on $\mathscr{V}_n(\mathscr{F})$, and let $\mathscr{H} = \{\xi \in \mathscr{V}_n \mid \mathbf{f}\xi = 0\}$. Show that $\mathscr{H}$ is a subspace of $\mathscr{V}_n$ and has dimension $n - 1$.

9. Let $\mathscr{P}$ denote the infinite-dimensional space of all polynomials with real coefficients, and let $\mathscr{S}$ denote the infinite-dimensional space of all infinite sequences of real numbers. Carry out the method suggested below to show that $\mathscr{S}$ and $\mathscr{P}'$ are isomorphic.

(i) With each real sequence $\sigma = \{c_i \,|\, i = 0, 1, \ldots\}$ we associate the mapping $\mathbf{f}_\sigma$ from $\mathscr{P}$ to $\mathbb{R}$ defined by specifying that $\mathbf{f}_\sigma$ is linear and that $\mathbf{f}_\sigma(x^k) = c_k$ for $k = 0, 1, 2, \ldots$ Write the value of $\mathbf{f}_\sigma(p)$, where

$$p(x) = a_0 + a_1 x + a_2 x^2 \in \mathscr{P}.$$

(ii) Verify that $\mathbf{f}_\sigma \in \mathscr{P}'$ for each $\sigma \in \mathscr{S}$.

(iii) Show that the mapping $\sigma \to \mathbf{f}_\sigma$ from $\mathscr{S}$ to $\mathscr{P}'$ is one-to-one and onto $\mathscr{P}'$.

(iv) Complete the proof that $\mathscr{S}$ and $\mathscr{P}'$ are isomorphic. (The methods of Exercises 2.3-3 and 2.4-8 show that $\mathscr{P}$ is isomorphic to a proper subspace $\mathscr{T}$ of $\mathscr{S}$. Because $\mathscr{T}$ and $\mathscr{S}$ are not isomorphic, this example shows that the conclusion that $\mathscr{V}$ and $\mathscr{V}'$ are isomorphic when $\mathscr{V}$ is finite-dimensional is not valid when $\mathscr{V}$ is infinite-dimensional.)

9.2   TRANSPOSE
OF A LINEAR MAPPING

Each linear mapping $\mathbf{T}$ from a vector space $\mathscr{V}$ to a vector space $\mathscr{W}$ can be associated in a natural way with a linear mapping S from the dual space $\mathscr{W}'$ to the dual space $\mathscr{V}'$. (We again observe that the process of passing to the dual spaces manages to turn things around.) In this construction each element of $\mathscr{W}'$ will be regarded in two ways—as a mapping $\mathbf{g}$ from $\mathscr{W}$ to $\mathscr{F}$ and as a "vector" g in $\mathscr{W}'$ that is mapped by S into $S(g)$ in $\mathscr{V}'$. Boldface type denotes $\mathbf{g}$ as a mapping from $\mathscr{W}$ to $\mathscr{F}$, and italic type denotes g as a vector in $\mathscr{W}'$.

A linear mapping S from $\mathscr{W}'$ to $\mathscr{V}'$ carries each linear functional g in $\mathscr{W}'$ into a linear functional $S(g)$ in $\mathscr{V}'$. Because g is a linear functional in $\mathscr{W}'$, the mapping $\mathbf{g}$ carries each vector $\eta \in \mathscr{W}$ into the scalar $\mathbf{g}\eta \in \mathscr{F}$. Similarly the functional $S(g)$ carries each vector $\xi \in \mathscr{V}$ into the scalar $S(g)(\xi) \in \mathscr{F}$. Let η denote $\mathbf{T}\xi$. Then

$$\mathbf{g}\eta = \mathbf{g}(\mathbf{T}\xi) \text{ in } \mathscr{F},$$

and we define S by the rule that for each $g \in \mathscr{W}'$

$$S(g)(\xi) = \mathbf{g}\mathbf{T}(\xi) \text{ for each } \xi \in \mathscr{V},$$

or simply

$$S(g) = \mathbf{g}\mathbf{T} \text{ for each } g \in \mathscr{W}'.$$

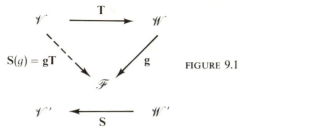

FIGURE 9.1

But **T** is a linear mapping from $\mathscr{V}$ to $\mathscr{W}$, and **g** is a linear mapping from $\mathscr{W}$ to $\mathscr{F}$, so the composite mapping **gT** is linear from $\mathscr{V}$ to $\mathscr{F}$. Hence **gT**, and therefore **S**(g) is in $\mathscr{V}'$ for each $g \in \mathscr{W}'$. As an exercise you may show that **S** is linear. Figure 9.1 illustrates the various mappings involved in the definition of **S**.

Theorem 9.3   Let **T** be a linear mapping of rank k from $\mathscr{V}_n$ to $\mathscr{W}_m$, and let **S** be the linear mapping from $\mathscr{W}'_m$ to $\mathscr{V}'_n$ defined by

$$\mathbf{S}(g) = \mathbf{gT} \text{ for each } \mathbf{g} \in \mathscr{W}'_m.$$

Then **S** has rank k.

Proof   By hypothesis $\mathscr{R}(\mathbf{T})$ is a subspace of $\mathscr{W}_m$ having dimension k. By Theorem 9.2 the annihilator $\mathscr{R}(\mathbf{T})^0$ of $\mathscr{R}(\mathbf{T})$ is a subspace of $\mathscr{W}'_m$ having dimension $m - k$. We note that

$$f \in \mathscr{R}(\mathbf{T})^0 \text{ if and only if } \mathbf{f}(\eta) = 0 \text{ for each } \eta \in \mathscr{R}(\mathbf{T}),$$

$$\text{if and only if } \mathbf{f}(\mathbf{T}\xi) = 0 \text{ for each } \xi \in \mathscr{V}_n,$$

$$\text{if and only if } \mathbf{fT} = \theta \text{ in } \mathscr{V}'_n,$$

$$\text{if and only if } \mathbf{S}(f) = \theta \text{ in } \mathscr{V}'_n,$$

$$\text{if and only if } f \in \mathscr{N}(\mathbf{S}).$$

Hence $\mathscr{N}(\mathbf{S}) = \mathscr{R}(\mathbf{T})^0$ and $n(\mathbf{S}) = m - k$, so $r(\mathbf{S}) = k$.

We will comment on the significance of this result after we derive a matrix representation for **S**.

Theorem 9.4   Let **T** and **S** be as in the previous theorem. Let **T** be represented by the matrix $^{\beta}A_{\alpha}$, relative to the basis $\{\alpha_1, \ldots, \alpha_n\}$ for $\mathscr{V}_n$ and the basis $\{\beta_1, \ldots, \beta_m\}$ for $\mathscr{W}_m$. Let $\{f_1, \ldots, f_n\}$ and $\{g_1, \ldots, g_m\}$ be the corresponding dual bases for $\mathscr{V}'_n$ and $\mathscr{W}'_m$, and let **S** be represented by the matrix $^{f}B_{g}$. Then

$$^{f}B_{g} = (^{\beta}A_{\alpha})^{t}.$$

Proof   If $^f B_g$ represents $\mathbf{S}$ as described, then

$$\mathbf{S}(g_j) = \sum_{k=1}^{n} b_{kj} f_k = g_j \mathbf{T}, \, j = 1, \ldots, m.$$

Hence for $i = 1, \ldots, n$,

$$g_j \mathbf{T} \alpha_i = \left(\sum_{k=1}^{n} b_{kj} \mathbf{f}_k \right) \alpha_i = \sum_{k=1}^{n} b_{kj} (\mathbf{f}_k \alpha_i)$$

$$= \sum_{k=1}^{n} b_{kj} \delta_{ki} = b_{ij},$$

for $i = 1, \ldots, n$ and $j = 1, \ldots, m$.

But because $^\beta A_\alpha$ represents $\mathbf{T}$ we also have

$$g_j \mathbf{T} \alpha_i = g_j \left(\sum_{k=1}^{m} a_{ki} \beta_k \right) = \sum_{k=1}^{m} a_{ki} g_j \beta_k$$

$$= \sum_{k=1}^{m} a_{ki} \delta_{jk} = a_{ji}.$$

Thus $b_{ij} = a_{ji}$ for $i = 1, \ldots, n$ and $j = 1, \ldots, m$, so $^f B_g$ is the transpose of $^\beta A_\alpha$.

Definition 9.2   Let $\mathbf{T}$ be a linear mapping from $\mathscr{V}_n$ to $\mathscr{W}_m$. The *transpose* $\mathbf{T}^t$ of $\mathbf{T}$ is the linear mapping from $\mathscr{W}'_m$ to $\mathscr{V}'_n$ defined by

$$\mathbf{T}^t(g) = g\mathbf{T} \text{ for every } g \in \mathscr{W}'_m.$$

Hence the mapping $\mathbf{S}$ described in Figure 9.1 is called the transpose of $\mathbf{T}$. We deferred naming $\mathbf{S}$ until now because the justification for that terminology is the relation established in Theorem 9.4. We also see that Theorem 9.3 simply states that a linear mapping and its transpose have the same rank, which was independently established for matrices in Theorem 4.15. It is natural to expect the algebra of transposes to be the same for linear mappings as for matrices; that is,

$$(c\mathbf{T})^t = c\mathbf{T}^t,$$

$$(\mathbf{T} + \mathbf{S})^t = \mathbf{T}^t + \mathbf{S}^t,$$

$$\mathbf{I}^t = \mathbf{I},$$

$$(\mathbf{RT})^t = \mathbf{T}^t \mathbf{R}^t,$$

when the indicated mappings are defined. You may derive these results as exercises.

Now we turn to a further question involving the dual space $\mathscr{V}'$ of $\mathscr{V}$. Because $\mathscr{V}'$ is a vector space, it too has a dual space $(\mathscr{V}')'$, denoted by $\mathscr{V}''$ and called the *bidual* space of $\mathscr{V}$. Of course, if $\mathscr{V}$ is n-dimensional, so are $\mathscr{V}'$ and $\mathscr{V}''$, and all these spaces are isomorphic. But in general all we can say is that $\mathscr{V}$ is isomorphic to a subspace of its bidual. We now see how to establish a natural correspondence between the elements of $\mathscr{V}$ and certain of the elements of $\mathscr{V}''$, and we can define this correspondence without reference to any basis.

To establish the notation, we use italic Greek letters for elements of $\mathscr{V}$, boldface Greek letters for elements of $\mathscr{V}''$ and italic or boldface Latin letters for elements of $\mathscr{V}'$. The elements of $\mathscr{V}'$ again lead a double life, sometimes as mappings that carry elements of $\mathscr{V}$ into $\mathscr{F}$ and sometimes as the objects that are mapped into $\mathscr{F}$ by elements of $\mathscr{V}''$. Now let ξ be a fixed vector in $\mathscr{V}$. For each $\mathbf{f}$ in $\mathscr{V}'$, $\mathbf{f}\xi \in \mathscr{F}$. Hence ξ can be regarded as attaching a scalar value $\mathbf{f}\xi$ to each $\mathbf{f} \in \mathscr{V}'$. Hence ξ itself determines a mapping from $\mathscr{V}'$ to $\mathscr{F}$. Furthermore, that mapping is linear, because

$$(a\mathbf{f} + b\mathbf{g})\xi = a\mathbf{f}\xi + b\mathbf{g}\xi.$$

Thus ξ determines a linear functional $\boldsymbol{\xi}$ on $\mathscr{V}'$. With each vector $\xi \in \mathscr{V}$ we therefore associate the linear functional $\boldsymbol{\xi} \in \mathscr{V}''$, defined by

$$\boldsymbol{\xi}(f) = \mathbf{f}(\xi) \quad \text{for each } f \in \mathscr{V}'.$$

To see that the mapping $\xi \to \boldsymbol{\xi}$ from $\mathscr{V}$ to $\mathscr{V}''$ is linear, let $\xi, \eta \in \mathscr{V}$ and a, $b \in \mathscr{F}$. Then for all $f \in \mathscr{V}'$

$$(a\boldsymbol{\xi} + b\boldsymbol{\eta})(f) = \mathbf{f}(a\xi + b\eta)$$

$$= a\mathbf{f}(\xi) + b\mathbf{f}(\eta)$$

$$= a\boldsymbol{\xi}(f) + b\boldsymbol{\eta}(f).$$

To see that the mapping $\xi \to \boldsymbol{\xi}$ from $\mathscr{V}$ to $\mathscr{V}''$ is one-to-one, suppose that $\boldsymbol{\xi} = \boldsymbol{\eta}$. Then $\boldsymbol{\xi}(f) = \boldsymbol{\eta}(f)$ for every $f \in \mathscr{V}'$ and $\mathbf{f}(\xi) = \mathbf{f}(\eta)$ for every $\mathbf{f} \in \mathscr{V}'$. Then $\mathbf{f}(\xi - \eta) = 0$ for every $\mathbf{f} \in \mathscr{V}'$. If $\mathscr{V}$ is finite-dimensional, Exercise 9.1-5 shows that $\xi - \eta = 0$; because Exercise 9.1-5 can be established for any vector space, we conclude that $\mathscr{V}$ is isomorphic to a subspace of $\mathscr{V}''$.

We can now use this isomorphism to interpret the equation $(\mathbf{T}^t)^t = \mathbf{T}$, which we might expect to be valid because of the corresponding equation for matrices. But $\mathbf{T}$ is a linear mapping from $\mathscr{V}$ to $\mathscr{W}$, and $(\mathbf{T}^t)^t$ is a linear mapping from $\mathscr{V}''$ to $\mathscr{W}''$. However, $\mathscr{V}$ is isomorphic to a subspace $\mathscr{S}$ of $\mathscr{V}''$, $\mathscr{W}$ is isomorphic to a subspace $\mathscr{T}$ of $\mathscr{W}''$, and $(\mathbf{T}^t)^t$ as a mapping from $\mathscr{S}$ to $\mathscr{T}$ is a carbon copy of the mapping $\mathbf{T}$ from $\mathscr{V}$ to $\mathscr{W}$.

EXERCISES 9.2

1. Prove that the transpose of a linear mapping is linear.

2. Prove the following properties of transposes of linear mappings.

   (i) $\mathbf{I}^t = \mathbf{I}$.

   (ii) $(c\mathbf{T})^t = c\mathbf{T}^t$.

   (iii) $(\mathbf{S} + \mathbf{T})^t = \mathbf{S}^t + \mathbf{T}^t$

   (iv) $(\mathbf{RT})^t = \mathbf{T}^t\mathbf{R}^t$.

3. If $\mathbf{S}$ is a nonsingular mapping from $\mathscr{V}$ onto $\mathscr{W}$, show that $\mathbf{S}^t$ is nonsingular and that $(\mathbf{S}^t)^{-1} = (\mathbf{S}^{-1})^t$.

4. Let $\mathbf{T}$ be a linear mapping from $\mathscr{V}_n$ to $\mathscr{W}_m$. Interpret the meaning of the equation $(\mathbf{T}^t)^t = \mathbf{T}$, and then prove that it is valid in the sense of that interpretation.

9.3 BILINEAR FUNCTIONS

A real inner product on $\mathscr{V}(\mathbb{R})$ is a particular type of scalar valued function, defined for each ordered *pair* of vectors of $\mathscr{V}$, and therefore is a mapping p from the Cartesian product $\mathscr{V} \times \mathscr{V}$ into $\mathbb{R}$, where p is specified to be bilinear, symmetric, and positive definite. Similarly, a complex inner product on $\mathscr{V}(\mathbb{C})$ is a mapping p from $\mathscr{V} \times \mathscr{V}$ to $\mathbb{C}$, where p is specified to be conjugate bilinear, conjugate symmetric, and positive definite.

In this section we shall begin, somewhat more generally, with two vector spaces $\mathscr{V}$ and $\mathscr{W}$ over the same field $\mathscr{F}$. Again we shall assume that $\mathscr{F}$ is either $\mathbb{R}$ or $\mathbb{C}$. When $\mathscr{F} = \mathbb{R}$ we shall study *bilinear* functions from $\mathscr{V} \times \mathscr{W}$ to $\mathbb{R}$, and when $\mathscr{F} = \mathbb{C}$ we shall study *conjugate bilinear* functions from $\mathscr{V} \times \mathscr{W}$ to $\mathbb{C}$. Because conjugate bilinearity reduces to bilinearity when the scalars are real, we can develop the two theories in parallel by considering the case for $\mathscr{F} = \mathbb{C}$, extracting the corresponding results for $\mathscr{F} = \mathbb{R}$.

Definition 9.3 Let $\mathscr{V}_n$ and $\mathscr{W}_m$ be vector spaces over $\mathscr{F}$ (real or complex). A *conjugate bilinear function* f is a function that assigns to each pair of vectors $(\xi, \eta) \in \mathscr{V}_n \times \mathscr{W}_m$ a scalar value $f(\xi, \eta)$ in such a way that the following properties hold:

(a) $f(a\xi_1 + b\xi_2, \eta) = af(\xi_1, \eta) + bf(\xi_2, \eta)$.

(b) $f(\xi, a\eta_1 + b\eta_2) = \bar{a}f(\xi, \eta_1) + \bar{b}f(\xi, \eta_2)$.

Condition (a) states that f is a linear function of its first variable (which is a vector); Condition (b) is a modified form of linearity in the second variable:

$$f(\xi, \eta_1 + \eta_2) = f(\xi, \eta_1) + f(\xi, \eta_2),$$
$$f(\xi, b\eta) = \bar{b}f(\xi, \eta).$$

If the scalar field is real, then $\bar{b} = b$, in which case f is called simply a *bilinear function*.

First, we see how conjugate bilinear functions can be represented by matrices; the work we do here will bear a striking similarity to the corresponding matrix representation of linear transformations. Let $\{\alpha_1, \ldots, \alpha_n\}$ be a basis for $\mathscr{V}_n$ and $\{\beta_1, \ldots, \beta_m\}$ a basis for $\mathscr{W}_m$. Let $\xi = \sum_{s=1}^{n} x_s \alpha_s \in \mathscr{V}_n$, and $\eta = \sum_{r=1}^{m} y_r \beta_r \in \mathscr{W}_m$. Then

$$f(\xi, \eta) = f\left(\sum_{s=1}^{n} x_s \alpha_s, \eta\right) = \sum_{s=1}^{n} x_s f(\alpha_s, \eta)$$

$$= \sum_{s=1}^{n} x_s f\left(\alpha_s, \sum_{r=1}^{m} y_r \beta_r\right) = \sum_{s=1}^{n} x_s \left(\sum_{r=1}^{m} \bar{y}_r f(\alpha_s, \beta_r)\right)$$

$$= \sum_{s=1}^{n} \sum_{r=1}^{m} x_s \bar{y}_r f(\alpha_s, \beta_r).$$

The mn scalars $f(\alpha_s, \beta_r)$ therefore completely determine the value of the function f. Now consider the m-by-n matrix

$$A = (a_{ij}), \text{ where } a_{ij} = f(\alpha_j, \beta_i),$$

which is uniquely determined by f relative to the α- and β-bases. Then

$$f(\xi, \eta) = \sum_{r=1}^{m} \bar{y}_r \left(\sum_{s=1}^{n} a_{rs} x_s\right) = \bar{Y}^t A X = Y^* A X,$$

where X is the n-component column vector that represents ξ in the α-basis for $\mathscr{V}_n$, Y is the m-component column vector that represents η in the β-basis for $\mathscr{W}_m$, and $Y^* = \bar{Y}^t$ is the *adjoint* (conjugate transpose) of Y.

Observe carefully that we have adopted the convention whereby the (i, j) entry of the matrix A that represents f relative to the α-basis for $\mathscr{V}_n$ and the β-basis for $\mathscr{W}_m$ is $f(\alpha_j, \beta_i)$. It might appear to be more natural not to interchange the order of subscripts in this way, but with this convention subsequent relations work out in a convenient form.

Definition 9.4   An expression of the type

$$\sum_{i=1}^{m} \sum_{j=1}^{n} \bar{y}_i a_{ij} x_j$$

is called a *conjugate bilinear form* in the $(m + n)$ variables $x_1, \ldots, x_n$, $y_1, \ldots, y_m$.

Thus each conjugate bilinear function on $\mathscr{V}_n \times \mathscr{W}_m$ determines a conjugate bilinear form relative to a choice of bases for $\mathscr{V}_n$ and $\mathscr{W}_m$. Conversely, the coefficient matrix $A = (a_{ij})$ of a conjugate bilinear form determines a conjugate bilinear function f on $\mathscr{V}_n \times \mathscr{W}_m$ whose values on the pairs of basis vectors are specified by the rule

$$a_{ij} = f(\alpha_j, \beta_i),$$

for $i = 1, \ldots, m$ and $j = 1, \ldots, n$.

Now let us work out the details for a change of basis. Given a conjugate bilinear function f on $\mathscr{V}_n \times \mathscr{W}_m$, which is represented relative to an α-basis for $\mathscr{V}_n$ and a β-basis for $\mathscr{W}_m$ by the m-by-n matrix $A = (f(\alpha_j, \beta_i))$, let $\{\gamma_1, \ldots, \gamma_n\}$ be a new basis for $\mathscr{V}_n$ and $\{\delta_1, \ldots, \delta_m\}$ a new basis for $\mathscr{W}_m$. As in Theorem 6.3 we let

$$\gamma_j = \sum_{s=1}^{n} q_{sj}\alpha_s, \ j = 1, \ldots, n,$$

$$\delta_i = \sum_{r=1}^{m} p_{ri}\beta_r$$

The matrix $C = (c_{ij})$ that represents f relative to the new bases is defined by

$$c_{ij} = f(\gamma_j, \delta_i) = f\left(\sum_{s=1}^{n} q_{sj}\alpha_s, \sum_{r=1}^{m} p_{ri}\beta_r\right)$$

$$= \sum_{s=1}^{n} \sum_{r=1}^{m} q_{sj}\bar{p}_{ri} f(\alpha_s, \beta_r)$$

$$= \sum_{r=1}^{m} \bar{p}_{ri}\left(\sum_{s=1}^{n} a_{rs}q_{sj}\right).$$

The expression in parentheses is the (r, j) entry of AQ, and $\bar{p}_{ri}$ is the (i, r) entry of $\bar{P}^t = P^*$. Hence we have

$$C = P^*AQ,$$

where P and Q are nonsingular, because Q represents a change of basis in $\mathscr{V}_n$ and P represents a change of basis in $\mathscr{W}_m$.

Theorem 9.5   A conjugate bilinear function f from $\mathscr{V}_n(\mathbb{C}) \times \mathscr{W}_m(\mathbb{C})$ to $\mathbb{C}$ is represented uniquely relative to a pair α, β of bases by the m-by-n matrix

$$A = (a_{ij}), \text{ where } a_{ij} = f(\alpha_j, \beta_i).$$

If $\xi \in \mathscr{V}_n$ and $\eta \in \mathscr{W}_m$ are represented by X and Y, respectively, then

$$f(\xi, \eta) = Y^*AX.$$

An m-by-n matrix C also represents f relative to a suitable choice of bases if and only if

$$C = P^*AQ$$

for suitable nonsingular matrices P and Q (that is, if and only if C is equivalent to A.)

Proof   Our previous discussion has established all these assertions except the claim that if $C = P^*AQ$ then C also represents f. But in that case we can interpret Q and P as defining changes of bases in $\mathscr{V}_n$ and $\mathscr{W}_m$ and retrace our calculations to conclude that C and A represent the same conjugate bilinear function.

Theorem 9.6   A bilinear function from $\mathscr{V}_n(\mathbb{R}) \times \mathscr{W}_m(\mathbb{R})$ to $\mathbb{R}$ satisfies the assertions of Theorem 9.5, with Y^* replaced by Y^t and P^* replaced by P^t.

Proof   When all scalars are real, the conjugate operation is the identity mapping on $\mathbb{R}$. Hence the previous argument applies without further change.

Thus equivalence of m-by-n matrices plays the same role in the representation of conjugate bilinear functions as it does in the representation of linear mappings. But from Theorem 6.8 we know that two m-by-n matrices are equivalent if and only if they have the same rank, and from Theorem 6.7 we know that any m-by-n matrix of rank r is equivalent to the block matrix

$$\left(\begin{array}{c|c} I_r & Z \\ \hline Z & Z \end{array}\right).$$

Hence each conjugate bilinear function and each real bilinear function can be assigned a rank.

Definition 9.5   The *rank* of a complex conjugate bilinear function (or of a real bilinear function) is the rank of any matrix that represents such a function.

Theorem 9.7   Let f denote either a conjugate bilinear function from $\mathscr{V}_n(\mathbb{C}) \times \mathscr{W}_m(\mathbb{C})$ to $\mathbb{C}$ or a bilinear function from $\mathscr{V}_n(\mathbb{R}) \times \mathscr{W}_m(\mathbb{R})$ to $\mathbb{R}$. There exist a basis $\{\gamma_1, \ldots, \gamma_n\}$ for $\mathscr{V}_n$ and a basis $\{\delta_1, \ldots, \delta_m\}$ for $\mathscr{W}_m$ such that if $\xi = \sum_{i=1}^{n} x_i \gamma_i$ and $\eta = \sum_{j=1}^{m} y_j \delta_j$, then

$$f(\xi, \eta) = x_1 \bar{y}_1 + \cdots + x_r \bar{y}_r,$$

where r is the rank of f.

Proof   Relative to any choice of bases f is represented by a matrix A of rank r. Also A is equivalent to the canonical matrix

$$C = \left(\begin{array}{c|c} I_r & Z \\ \hline Z & Z \end{array} \right) = RAQ$$

for some nonsingular matrices R and Q. Let $P^* = R$. The changes of bases represented by P and Q establish the assertion, because

$$f(\xi, \eta) = Y^*CX.$$

In many important applications of conjugate bilinear functions we are interested in the special case in which $\mathscr{W}_m = \mathscr{V}_n$. Because nothing in our previous work precluded these two spaces from being the same space, we obtain the following special form of Theorem 9.5.

Theorem 9.8   A conjugate bilinear function f on $\mathscr{V}_n(\mathbb{C})$ is represented relative to a basis $\{\alpha_1, \ldots, \alpha_n\}$ for $\mathscr{V}_n$ by the n-by-n matrix

$$A = (a_{ij}), \text{ where } a_{ij} = f(\alpha_j, \alpha_i).$$

If ξ and η are represented by X and Y, then

$$f(\xi, \eta) = Y^*AX.$$

An n-by-n matrix C represents f relative to a suitable basis for $\mathscr{V}_n$ if and only if

$$C = Q^*AQ$$

for some nonsingular matrix Q.

In the same manner a special form of Theorem 9.6 holds for a bilinear function on $\mathscr{V}_n(\mathbb{R})$. In the next section we shall examine two equivalence relations on $\mathscr{M}_{n \times n}$ that follow from Theorem 9.8 and its real analogue.

EXERCISES 9.3

1. Let P_{ij}, $R_{i,\,i+cj}$, and $M_i(c)$ denote the elementary matrices over the complex field $\mathbb{C}$.

(i) Compute the conjugate and the adjoint of each.

(ii) Describe the effect of the conjugate and the adjoint of each elementary matrix as an operation on the rows of a square matrix A.

(iii) Describe the effect of the conjugate and the adjoint of each elementary matrix as an operation on the columns of A.

2. Using Definition 9.4 as a guide, write a full definition of a real bilinear form.

3. Let $\xi = x_1 \varepsilon_1 + x_2 \varepsilon_2 + x_3 \varepsilon_3$ in $\mathbb{R}^3$ and $\eta = y_1 \varepsilon_1 + y_2 \varepsilon_2$ in $\mathbb{R}^2$. For each of the following bilinear forms write a matrix A such that $f(\xi, \eta) = Y^t A X$ is the given form.

(i) $4x_1 y_1 + 4x_1 y_2 + 2x_2 y_1 + 4x_2 y_2 - x_3 y_3$,

(ii) $2x_1 y_1 + 8x_1 y_2 + x_2 y_1 + 3x_2 y_2 + x_3 y_1 + 6x_3 y_2$.

4. Let f be a bilinear function on $\mathbb{R}^3 \times \mathbb{R}^2$ that is represented relative to bases $\{\alpha_1, \alpha_2, \alpha_3\}$ for $\mathbb{R}^3$ and $\{\beta_1, \beta_2\}$ for $\mathbb{R}^2$ by the matrix

$$A = \begin{pmatrix} 1 & 3 & 4 \\ 2 & -1 & 0 \end{pmatrix}.$$

Write the values of $f(\alpha_2, \beta_1), f(\alpha_1, \beta_2)$, and $f(\alpha_2, \beta_2)$.

5. Prove the following properties of the adjoint of complex matrices:

(i) $(A^*)^* = A$.

(ii) $r(A^*) = r(A)$.

(iii) $(A + B)^* = A^* + B^*$.

(iv) $(AB)^* = B^* A^*$.

(v) If A is nonsingular, $(A^{-1})^* = (A^*)^{-1}$.

6. Show in detail how Theorem 9.8 follows from Theorem 9.5.

7. State in full detail a theorem derived from Theorem 9.6 by specializing to the case $\mathcal{W}_m = \mathcal{V}_n$.

8. State in full detail the special form of Theorem 9.7 that applies when f is a bilinear function from $\mathcal{V}_n(\mathbb{R}) \times \mathcal{W}_m(\mathbb{R})$ to $\mathbb{R}$.

9. Fill in the details of the argument outlined in the proof of Theorem 9.5 to show that if $C = P^*AQ$, then C and A represent the same conjugate bilinear function.

9.4   CONJUNCTIVITY AND CONGRUENCE

Previously we have studied several equivalence relations on families of matrices; such relations occur when a given mathematical entity can be represented by different matrices relative to different coordinate systems. We now examine two equivalence relations on n-by-n matrices that arise from the matrix representation of conjugate bilinear functions on $\mathcal{V}_n(\mathbb{C})$ and bilinear functions on $\mathcal{V}_n(\mathscr{F})$ for any $\mathscr{F}$. The following definition is motivated by Theorem 9.8.

Definition 9.6

(a) Let A and C be n-by-n matrices of complex numbers. C is said to be *conjunctive* to A over $\mathbb{C}$ if and only if

$$C = Q^*AQ, \text{ where } Q^* = \bar{Q}^t,$$

for some nonsingular complex matrix Q.

(b) Let A and C be n-by-n matrices over $\mathscr{F}$. C is said to be *congruent* to A over $\mathscr{F}$ if and only if

$$C = Q^tAQ$$

for some nonsingular matrix Q over $\mathscr{F}$.

It is easy to verify that conjunctivity is an equivalence relation on the set of all n-by-n complex matrices and that congruence is an equivalence relation on the set of all n-by-n matrices with entries in any field $\mathscr{F}$. Furthermore, each of these equivalence relations is distinct from those that we have studied previously: row equivalence, equivalence, and similarity.

Congruence is easily stated in terms of elementary row and column operations. Because Q^t is nonsingular, it is a product of elementary matrices,

$$Q^t = E_k \cdots E_1,$$

$$Q = E_1^t \cdots E_k^t.$$

Hence C is congruent to A if and only if C can be obtained from A by a finite sequence of changes, each change being an elementary row operation followed by the same operation on the corresponding columns. (Recall Theorem 4.7 and Exercise 4.3-7.)

Conjunctivity can also be stated in terms of elementary row and column operations, but a slight modification is needed. Because Q^* is nonsingular, it corresponds to a sequence of elementary row operations. Then Q corresponds to a corresponding sequence of conjugate column operations. From Exercise 9.3-1, $P_{ij}^* = P_{ij}$, $(R_{i,\,i+cj})^* = R_{j,\,j+\bar{c}i}$, $(M_i(c))^* = M_i(\bar{c})$. Thus C is conjunctive to A if and only if C can be obtained from A by a finite sequence of changes, each change being an elementary row operation followed by the corresponding *conjugate* operation on the corresponding columns.

In Section 4.2 we saw that any n-by-n matrix A can be written uniquely as the sum of a symmetric matrix S and a skew-symmetric matrix K,

$$A = S + K,$$

$$S = \tfrac{1}{2}(A + A^t),$$

$$K = \tfrac{1}{2}(A - A^t).$$

This relation remains valid, of course, for complex matrices, but the conjugate operation in C permits a variation of this decomposition that turns out to be quite useful by replacing A^t by A^*. Thus any n-by-n complex A can be written uniquely as the sum

$$A = H + K, \text{ where } H^* = H \text{ and } K^* = -K.$$

Definition 9.7   An n-by-n complex matrix B is said to be *Hermitian* if and only if $B^* = B$. B is said to be *skew-Hermitian* if and only if $B^* = -B$.

Because the matrix B^* is called the adjoint of B, a Hermitian matrix is *self-adjoint* and any self-adjoint matrix is Hermitian. Sometimes the term *Hermitian congruence* is used in place of conjunctivity. Separate canonical forms relative to conjunctivity can be derived for Hermitian and skew-Hermitian matrices. We shall focus attention on Hermitian matrices, outlining the skew-Hermitian case as an exercise.

Theorem 9.9   Every complex Hermitian matrix of rank r is conjunctive over $\mathbb{C}$ to a diagonal matrix with nonzero real numbers in the first r diagonal positions and zeros elsewhere.
Proof   Let A be Hermitian and let P be nonsingular. Then $(P^*AP)^* = P^*A^*P^{**} = P^*AP$, so that a Hermitian matrix of the same rank is obtained from A by applying an elementary row operation followed by the corresponding conjugate column operation. Every dia-

gonal element of a Hermitian matrix is real. We show that if $A \neq Z$, A is conjunctive to a matrix in which some diagonal element is nonzero. Suppose that all diagonal elements of A are zero, and let $a_{ij} \neq 0$. If the real part of a_{ij} is zero, then $M_i(-i)AM_i(i)$ is a matrix, conjunctive to A, in which the real part of the (i, j) entry is nonzero. Hence we may assume that the real part of a_{ij} is nonzero. Then we can obtain a matrix B in which $b_{jj} \neq 0$ by replacing row j of A by $R_j + R_i$ and replacing column j by $C_j + C_i$; that is,

$$B = R_{j, j+i} A R_{i, i+j} = (R_{i, i+j})^* A R_{i, i+j}.$$

Thus A is conjunctive to a matrix B in which b_{jj} is real and nonzero;

$$b_{jj} = a_{ij} + a_{ji} = a_{ij} + \bar{a}_{ij} \neq 0.$$

By permuting row j and row 1, then column j and column 1, we obtain a matrix C conjunctive to A and having $c_{11} = b_{jj}$. For $i > 1$ if $c_{i1} \neq 0$, we replace row i by $R_i - c_{11}^{-1} c_{i1} R_1$ and replace column i by $C_i - c_{11}^{-1} \bar{c}_{i1} C_1$ to produce zeros in the $(i, 1)$ and $(1, i)$ positions. The resulting matrix is

$$P_1^* A P_1 = \begin{pmatrix} \begin{array}{c|ccc} c_{11} & 0 & \cdots & 0 \\ \hline 0 & & & \\ \cdot & & & \\ \cdot & & A_1 & \\ \cdot & & & \\ 0 & & & \end{array} \end{pmatrix}.$$

If $A_1 = Z$, we are through; otherwise the process may be repeated on A_1. Because conjunctive matrices have the same rank, after r steps we are through.

Theorem 9.10   Every symmetric matrix of rank r over $\mathscr{F}$ is congruent over $\mathscr{F}$ to a diagonal matrix with nonzero elements in the first r diagonal positions and zeros elsewhere, provided $1 + 1 \neq 0$ in $\mathscr{F}$.
Proof   The proof of Theorem 9.9, modified by using transposes instead of conjugate transposes, is effective.

Theorem 9.11   Every complex Hermitian matrix of rank r is conjunctive over $\mathbb{C}$ to a diagonal matrix with 1 in the first p diagonal positions, -1 in the next $r - p$ diagonal positions, and zeros elsewhere.

Proof Let a Hermitian matrix A be conjunctive to a matrix D in the diagonal form of Theorem 9.9:

$$D = \begin{pmatrix} d_1 & & & & & & \\ & \ddots & & & & & \\ & & d_r & & & & \\ & & & 0 & & & \\ & & & & \ddots & & \\ & & & & & 0 \end{pmatrix}.$$

Let p denote the number of positive elements of D. Then $P_{ij}^* D P_{ij}$ coincides with D except that d_i and d_j have been interchanged. Hence A is conjunctive to a diagonal matrix E in which the positive diagonal elements come first—say $e_1, \ldots, e_p$; the negative diagonal elements can be expressed as $-e_{p+1}, \ldots, -e_r$, and the remaining elements are zero. Thus $e_i^{-1/2}$ is a real number for $i = 1, \ldots, r$, and

$$P^* E P$$

is in the form stated in the theorem, where

$$P = M_r(e_r^{-1/2}) \cdots M_1(e_1^{-1/2}).$$

Theorem 9.12 Every real symmetric matrix of rank r is congruent over $\mathbb{R}$ to a diagonal matrix with 1 in the first p diagonal positions, -1 in the next $r - p$ diagonal positions, and zeros elsewhere.

Proof The proof of Theorem 9.11, modified for congruence instead of conjunctivity, is effective. Notice that the argument will not be valid in any field that fails to contain a positive square root of each of its positive elements.

It is an important fact that the form stated in Theorem 9.11 is canonical for Hermitian matrices under conjunctivity. Similarly, the form stated in Theorem 9.12 is canonical for real symmetric matrices under congruence. This means, of course, that each Hermitian matrix determines a unique value of r and of p, and similarly for each real symmetric matrix.

To prove uniqueness, let A be an n-by-n Hermitian matrix. Relative to an arbitrary basis $\{\alpha_1, \ldots, \alpha_n\}$ for $\mathscr{V}_n$, A determines a conjugate bilinear function f from $\mathcal{V}_n \times \mathcal{V}_n$ to $\mathbb{C}$: $f(\alpha_j, \alpha_i) = a_{ij}$. Because A is Hermitian, Theorem 9.11 guarantees that there exists a basis $\{\beta_1, \ldots, \beta_n\}$ relative to which f is represented by the matrix $B = P^*AP$ of the block form

$$B = \left(\begin{array}{c|c|c} I_p & & \\ \hline & -I_{r-p} & \\ \hline & & Z \end{array} \right)$$

for some r and p. Suppose there exists another basis $\{\gamma_1, \ldots, \gamma_n\}$, relative to which f is represented by $C = Q^*AQ$, where

$$C = \left(\begin{array}{c|c|c} I_q & & \\ \hline & -I_{s-q} & \\ \hline & & Z \end{array} \right).$$

Because conjunctive matrices have the same rank,

$$r = r(B) = r(A) = r(C) = s.$$

Let $\xi = \sum_{i=1}^{n} x_i \beta_i = \sum_{i=1}^{n} u_i \gamma_i$ and $\eta = \sum_{i=1}^{n} y_i \beta_i = \sum_{i=1}^{n} v_i \gamma_i$. Then

$$f(\xi, \eta) = Y^*BX = x_1 \bar{y}_1 + \cdots + x_p \bar{y}_p - x_{p+1}\bar{y}_{p+1} - \cdots - x_r \bar{y}_r,$$

$$= V^*CU = u_1 \bar{v}_1 + \cdots + u_q \bar{v}_q - u_{q+1}\bar{v}_{q+1} - \cdots - u_r \bar{v}_r.$$

Furthermore let $\mathcal{M} = [\beta_{p+1}, \ldots, \beta_r]$ and $\mathcal{N} = [\gamma_1, \ldots, \gamma_q, \gamma_{r+1}, \ldots, \gamma_n]$. Then $\dim \mathcal{M} + \dim \mathcal{N} = (r - p) + (n - r + q) = n + (q - p)$. Suppose that $q > p$; then $\mathcal{M} \cap \mathcal{N} \neq [\theta]$. For any nonzero $\xi \in \mathcal{M} \cap \mathcal{N}$ we have

$$f(\xi, \xi) = -x_{p+1}\bar{x}_{p+1} - \cdots - x_r \bar{x}_r < 0,$$

$$= u_1 \bar{u}_1 + \cdots + u_q \bar{u}_q \geq 0,$$

a contradiction. Hence $q \leq p$. By reversing the roles of p and q, the reverse inequality follows, so $p = q$.

    Hence any n-by-n Hermitian matrix A uniquely determines two non-negative numbers p and r such that A is conjunctive to one and only one matrix B in the form given above. Instead of specifying r and p to identify the conjunctivity class to which A belongs, it is customary to specify r and s, where $s = 2p - r$; clearly s is uniquely determined by A.

Definition 9.8   The *signature* s of a Hermitian matrix A is defined by

$$s = p - (r - p) = 2p - r,$$

which is the number of diagonal ones diminished by the number of diagonal minus ones in the canonical form of A relative to conjunctivity over $\mathbb{C}$. Similarly, the *signature* of a real symmetric matrix A is $2p - r$, where p and r are determined from the canonical form of A relative to congruence over $\mathbb{R}$.

The second part of this definition anticipates that p and r are uniquely determined for each real symmetric matrix, which may be proved by a straightforward modification of the argument given for the Hermitian case preceding Definition 9.8.

Theorem 9.13   Two *n-by-n* Hermitian matrices are conjunctive over $\mathbb{C}$ if and only if they have the same rank and the same signature.

Proof   If A and B have the same rank and signature, each is conjunctive over $\mathbb{C}$ to the same matrix in canonical form and hence they are conjunctive to each other. Conversely, if A and B are conjunctive, so are their respective canonical forms, which by the uniqueness argument must be equal. Hence they have the same rank and the same signature.

Theorem 9.14   Two *n-by-n* real symmetric matrices are congruent over $\mathbb{R}$ if and only if they have the same rank and the same signature.

Proof   Exercise. First you must establish the uniqueness of the canonical form for congruence over $\mathbb{R}$ of a real symmetric matrix.

Finally we observe that for complex symmetric matrices the diagonalization process described for the real case by Theorem 9.12 can be extended to obtain a particularly simple standard form for congruence over $\mathbb{C}$.

Theorem 9.15   Every complex symmetric matrix of rank r is congruent over $\mathbb{C}$ to the matrix

$$\left(\begin{array}{c|c} I_r & Z \\ \hline Z & Z \end{array}\right).$$

Proof   Exercise.

EXERCISES 9.4

1. Do the two real bilinear forms given in Exercise 9.3-3(i) and (ii) represent the same bilinear function relative to different bases for $\mathbb{R}^2$?

2. Show that conjunctivity over $\mathbb{C}$ is an equivalence relation on the set of *n-by-n* matrices over $\mathbb{C}$.

3. Show that congruence over $\mathscr{F}$ is an equivalence relation on the set of *n*-by-*n* matrices over $\mathscr{F}$.

4. Using the results of Exercise 9.3-1, write a full proof of the text assertion that *C* is conjunctive to *A* if and only if *C* can be obtained from *A* by a finite sequence of changes, each change being an elementary row operation followed by the corresponding conjugate column operation.

5. Let *A* be a skew-symmetric matrix over $\mathbb{C}$. Show that $a_{ii} = 0$ for each *i* and that any matrix that is congruent to *A* is skew-symmetric.

6. Prove the following properties of the adjoint of complex matrices. (See also Exercise 9.3-5.)

(i) AA^* and A^*A are Hermitian.

(ii) If *A* is Hermitian, then a_{ii} is real for each *i*; if *A* is real and symmetric, *A* is Hermitian.

(iii) If *A* is skew-Hermitian, then a_{ii} is pure imaginary for each *i*; if *A* is real and skew, then *A* is skew-Hermitian.

(iv) Let *B* be conjunctive to *A*. If *A* is Hermitian, so is *B*; if *A* is skew-Hermitian, so is *B*.

7. Prove that any *n*-by-*n* complex matrix *A* has a unique representation $A = H + K$, where *H* is Hermitian and *K* is skew-Hermitian.

8. (i) Prove that *H* is Hermitian if and only if *iH* is skew-Hermitian, where $i^2 = -1$.

(ii) Combine (i) and Theorem 9.11 to deduce a canonical form for conjunctivity of skew-Hermitian matrices.

9. Let *A* be an *n*-by-*n* skew-symmetric matrix over a field $\mathscr{F}$ in which $1 + 1 \neq 0$. Recalling the results of Exercise 5, use appropriate row and column operations to show that *A* is congruent to a matrix of the diagonal block form

$$\begin{pmatrix} A_1 & & & & \\ & \cdot & & & \\ & & \cdot & & \\ & & & \cdot & \\ & & & A_t & \\ & & & & Z \end{pmatrix}, \quad \text{where } A_i = \begin{pmatrix} 0 & 1 \\ -1 & 0 \end{pmatrix}$$

for $i = 1, \ldots, t$. Deduce that *A* has an even rank and that this form is canonical relative to congruence over $\mathscr{F}$ for skew-symmetric matrices; that is, two *n*-by-*n* skew-symmetric matrices are congruent if and only if they have the same rank.

10. Perform row and column operations to reduce the following skew matrix to a congruent matrix in the canonical form stated in Exercise 9:

$$\begin{pmatrix} 0 & 3 & -2 & 1 & 0 \\ -3 & 0 & 1 & -4 & 1 \\ 2 & -1 & 0 & 0 & -2 \\ -1 & 4 & 0 & 0 & 1 \\ 0 & -1 & 2 & -1 & 0 \end{pmatrix}.$$

11. Given

$$A = \begin{pmatrix} -10 & 5 & 2 \\ 5 & 0 & 3 \\ 2 & 3 & 6 \end{pmatrix}.$$

(i) Find a matrix congruent to A over the rational field and in the form of Theorem 9.10.

(ii) Find a matrix congruent to A over $\mathbb{R}$ and in the form of Theorem 9.12. Determine the rank and signature of A.

(iii) Find a matrix congruent to A over $\mathbb{C}$ and in the form of Theorem 9.15.

(iv) Illustrate that Theorem 9.10 does not describe a canonical form for congruence over $\mathscr{F}$.

12. Determine the canonical form relative to conjunctivity of the following Hermitian matrix and determine its rank and signature:

$$\begin{pmatrix} 1 & i & 1+i \\ -i & 0 & 1 \\ 1-i & 1 & 2 \end{pmatrix}, \quad i^2 = -1.$$

13. Prove Theorem 9.15 and deduce that two symmetric complex matrices are congruent over $\mathbb{C}$ if and only if they have the same rank.

14. Show that any nonsingular Hermitian matrix and its inverse are conjunctive.

9.5   HERMITIAN AND QUADRATIC FUNCTIONS

We begin by considering a complex vector space $\mathscr{V}_n$ and a conjugate bilinear function f from $\mathscr{V}_n \times \mathscr{V}_n$ to $\mathbb{C}$. A function h from $\mathscr{V}_n$ to $\mathbb{C}$ can be obtained from f by the rule

$$h(\xi) = f(\xi, \xi) \text{ for each } \xi \in \mathscr{V}_n.$$

If f is conjugate symmetric then $h(\xi) = \overline{h(\xi)}$, so h then is a function from $\mathscr{V}_n$ to $\mathbb{R}$. Indeed it is precisely this property that makes conjugate symmetry and conjugate bilinearity natural conditions to impose for an inner product function on a complex vector space. Hence we now assume that f is conjugate symmetric, and $h(\xi) = f(\xi, \xi)$. Relative to any basis $\{\alpha_1, \ldots, \alpha_n\}$ for $\mathscr{V}_n$, h assumes the form

$$h(\xi) = X^*AX = \sum_{i=1}^{n} \sum_{j=1}^{n} f(\alpha_j, \alpha_i)\bar{x}_i x_j,$$

where $a_{ij} = f(\alpha_j, \alpha_i)$ and $\xi = x_1\alpha_1 + \cdots + x_n\alpha_n$. Thus h is represented by the matrix A, which is Hermitian because

$$a_{ij}^* = \bar{a}_{ji} = \overline{f(\alpha_i, \alpha_j)} = f(\alpha_j, \alpha_i) = a_{ij}.$$

By Theorem 9.8 another matrix B represents h relative to a suitable basis if and only if A and B are conjunctive.

Now we consider the corresponding situation when $\mathscr{V}_n$ is real and f is a symmetric, bilinear function from $\mathscr{V}_n \times \mathscr{V}_n$ to $\mathbb{R}$. Then the function q from $\mathscr{V}_n$ to $\mathbb{R}$ is defined by

$$q(\xi) = f(\xi, \xi) \text{ for each } \xi \in \mathscr{V}_n.$$

Relative to an α-basis q takes the form

$$q(\xi) = X^tAX = \sum_{i=1}^{n} \sum_{j=1}^{n} f(\alpha_i, \alpha_j)x_i x_j,$$

where $a_{ij} = f(\alpha_j, \alpha_i) = f(\alpha_i, \alpha_j)$. Hence A is real and symmetric. Another matrix B represents q relative to a suitable basis if and only if A and B are congruent.

This discussion is summarized in the following definition and theorems.

Definition 9.9

(a) Let f be a conjugate symmetric, conjugate bilinear function from $\mathscr{V}_n(\mathbb{C}) \times \mathscr{V}_n(\mathbb{C})$ to $\mathbb{C}$ and let h be defined from $\mathscr{V}_n$ to $\mathbb{R}$ by

$$h(\xi) = f(\xi, \xi) \text{ for all } \xi \in \mathscr{V}_n.$$

Then h is called a *Hermitian function* on $\mathscr{V}_n$. If A is any complex n-by-n matrix, the expression

$$X*AX = \sum_{i=1}^{n} \sum_{j=1}^{n} a_{ij}\bar{x}_i x_j$$

is called a *Hermitian form*.

(b) Let f be a symmetric bilinear function from $\mathscr{V}_n(\mathbb{R}) \times \mathscr{V}_n(\mathbb{R})$ to $\mathbb{R}$ and let q be defined from $\mathscr{V}_n$ to $\mathbb{R}$ by

$$q(\xi) = f(\xi, \xi) \text{ for all } \xi \in \mathscr{V}_n.$$

Then q is called a *quadratic function* on $\mathscr{V}_n$. If A is any real n-by-n matrix, the expression

$$X^t AX = \sum_{i=1}^{n} \sum_{j=1}^{n} a_{ij} x_i x_j$$

is called a *quadratic form*.

Theorem 9.16   Let $\mathscr{V}_n$ be a complex vector space, and let h be a Hermitian function on $\mathscr{V}_n$. Relative to a basis $\{\alpha_1, \ldots, \alpha_n\}$ for $\mathscr{V}_n$, h is represented by a uniquely determined Hermitian matrix A, where if $\xi = \sum_{i=1}^{n} x_i \alpha_i$, then $h(\xi) = X*AX$. Relative to a basis $\{\beta_1, \ldots, \beta_n\}$, h is represented by the Hermitian matrix B if and only if A and B are conjunctive over $\mathbb{C}$.

Theorem 9.17   Let $\mathscr{V}_n$ be a real vector space, and let q be a quadratic function on $\mathscr{V}_n$. Relative to a basis $\{\alpha_1, \ldots, \alpha_n\}$ for $\mathscr{V}_n$, q is represented by a uniquely determined real symmetric matrix A, where if $\xi = \sum_{i=1}^{n} x_i \alpha_i$, then $q(\xi) = X^t AX$. Relative to a basis $\{\beta_1, \ldots, \beta_n\}$, q is represented by the real symmetric matrix B if and only if A and B are congruent over $\mathbb{R}$.

Each quadratic (or Hermitian) function on a real (or complex) vector space $\mathscr{V}_n$ determines a real symmetric (or Hermitian) matrix relative to a given basis, which in turn determines a quadratic (or Hermitian) form. We now consider the problem in reverse: starting with a quadratic (or Hermitian) form, can we find a quadratic (or Hermitian) function whose values are given by that form? Notice that a quadratic form can be represented by various matrices. For example,

$$3x_1^2 + 4x_1 x_2 - x_2^2$$

can be written as

$$(x_1 \ x_2)\begin{pmatrix} 3 & 0 \\ 4 & -1 \end{pmatrix}\begin{pmatrix} x_1 \\ x_2 \end{pmatrix},$$

$$(x_1 \ x_2)\begin{pmatrix} 3 & 4 \\ 0 & -1 \end{pmatrix}\begin{pmatrix} x_1 \\ x_2 \end{pmatrix},$$

$$(x_1 \ x_2)\begin{pmatrix} 3 & 2 \\ 2 & -1 \end{pmatrix}\begin{pmatrix} x_1 \\ x_2 \end{pmatrix},$$

and so on. The variations occur from different decompositions of the coefficient of $x_i x_j$ into the two numbers a_{ij} and a_{ji}. An easy solution of the ambiguity appears if we decompose A as the sum of its symmetric and skew-symmetric parts:

$$A = S + K.$$

Then

$$X^t A X = X^t(S + K)X = X^t S X + X^t K X;$$

but a simple calculation verifies that

$$X^t K X = 0.$$

Hence only the symmetric component of A contributes to the value of $q(\xi)$, and we lose nothing by insisting that the real form

$$\sum_{i=1}^{n} \sum_{j=1}^{n} a_{ij} x_i x_j$$

be represented by the real symmetric matrix $B = (b_{ij})$, where $b_{ii} = a_{ii}$ for $i = 1, \ldots, n$ and $b_{ij} = b_{ji} = \frac{1}{2}(a_{ij} + a_{ji})$ when $i \neq j$.

Correspondingly, if for given complex numbers a_{ij} the form

$$\sum_{i=1}^{n} \sum_{j=1}^{n} a_{ij} \bar{x}_i x_j$$

is real for all complex vectors $(x_1, \ldots, x_n)$, then the form can be represented by the *Hermitian* matrix $B = (b_{ij})$, where $b_{ii} = a_{ii}$ for $i = 1, \ldots, n$ and $b_{ij} = \bar{b}_{ji} = \frac{1}{2}(a_{ij} + \bar{a}_{ji})$. (See Exercise 9.5-4.)

Henceforth we shall assume that the forms we consider have been appropriately symmetrized:

$$X^* A X, \quad \text{where } a_{ij} = \bar{a}_{ji}.$$

This condition reduces in the real case to

$$X^t AX, \quad \text{where } a_{ij} = a_{ji}.$$

Then for a given basis $\{\alpha_1, \ldots, \alpha_n\}$ for a complex vector space, the function $h(\xi) = X^*AX$, where $\xi = \sum_{i=1}^{n} x_i \alpha_i$, is a Hermitian function whose values are specified by the given form. Similarly, in the real case, $q(\xi) = X^t AX$ is a quadratic function.

Theorems 9.13 and 9.16 show that two Hermitian matrices represent the same Hermitian function if and only if they have the same rank and signature. Similarly, two real symmetric matrices represent the same quadratic function if and only if they have the same rank and signature.

Definition 9.10

   (a) The *rank* of a Hermitian function or of a Hermitian form is the rank of any Hermitian matrix that represents that function or form.

   (b) The *rank* of a quadratic function or of a quadratic form is the rank of any real symmetric matrix that represents that function or form.

   (c) The *signature* of a Hermitian function or form is the signature of any Hermitian matrix that represents that function or form.

   (d) The *signature* of a quadratic function or form is the signature of any real symmetric matrix that represents that function or form.

   (e) A Hermitian (quadratic) function is said to be *positive definite* if and only if

$$h(\xi) > 0 \text{ for all } \xi \neq \theta,$$

$$(q(\xi) > 0 \text{ for all } \xi \neq \theta).$$

Theorem 9.18   A Hermitian function h on a complex vector space $\mathscr{V}_n$ is positive definite if and only if it has rank n and signature n.
Proof   According to Theorem 9.11, h is represented relative to a suitable basis $\{\alpha_1, \ldots, \alpha_n\}$ by the canonical form under conjunctivity, so

$$h(\xi) = x_1 \bar{x}_1 + \cdots + x_p \bar{x}_p - x_{p+1} \bar{x}_{p+1} - \cdots - x_r \bar{x}_r,$$

where $s = 2p - r = p - (r - p)$. If $r = n = s$, then $p = n$ and $h(\xi) > 0$ for all $\xi \neq \theta$. If $r < n$, then $h(\alpha_n) = 0$, and h is not positive definite. If $s < n$, then $h(\alpha_n) = -1$, and again h is not positive definite.

Theorem 9.19   A quadratic function q on a real vector space $\mathcal{V}_n$ is positive definite if and only if it has rank n and signature n.
Proof   Exercise.

Theorem 9.20   A Hermitian matrix A represents a positive definite Hermitian function if and only if

$$A = P^*P$$

for some nonsingular complex matrix P.
Proof   A positive definite Hermitian function is represented in canonical form relative to conjunctivity by the identity matrix. Hence A represents that form if and only if

$$A = P^*IP$$

for some nonsingular complex matrix P.

Theorem 9.21   A real symmetric matrix A represents a positive definite quadratic function if and only if

$$A = P^tP$$

for some nonsingular real matrix P.
Proof   Exercise.

Examples of Real Symmetric Quadratic Forms

(A) The expression for the fundamental metric (element of arc length) in three-dimensional space is

$$ds^2 = \begin{cases} dx^2 + dy^2 + dz^2 & \text{in rectangular coordinates,} \\ dr^2 + r^2\, d\theta^2 + dz^2 & \text{in cylindrical coordinates,} \\ d\rho^2 + \rho^2 \sin^2 \psi\, d\theta^2 + \rho^2\, d\psi^2 & \text{in spherical coordinates.} \end{cases}$$

In the study of differential geometry the expression for the fundamental metric of a surface is of basic importance.

(B) In classical mechanics the kinetic energy of a particle of mass m and having n degrees of freedom is given by

$$\text{KE} = \frac{m}{2} \sum_{i=1}^{n} \left(\frac{dx_i}{dt}\right)^2,$$

where the x_i are the position coordinates of the particle. In more general systems the kinetic energy is represented by more complicated quadratic forms.

(C) The equation of a central quadric surface is

$$ax^2 + bxy + cy^2 + dxz + eyz + fz^2 = k,$$

the left-hand member being a quadratic form. The numbers b, d, and e are zero only if the axes of the quadric surface coincide with the coordinate axes; if these are not zero we are interested in finding an orthogonal change of coordinates that will simplify the form to a sum of squares.

In a Hermitian or quadratic function, it is natural to search for a coordinate system relative to which the form that represents that function is as simple as possible. The usual situation is that we are given a Hermitian or quadratic form, as in Example C, and we wish to change coordinates in such a way that the form is reduced to a sum of squares. With physical and geometric applications in mind, we are particularly interested in using only rigid motions (isometries) to change coordinates. Our investigation of Hermitian and quadratic forms reduces this problem to a study of Hermitian and real symmetric matrices; we shall see that the characteristic values and vectors of such matrices possess remarkable properties.

Theorem 9.22 Let A be a matrix that is either Hermitian or real and symmetric. Every characteristic value of A is real.
Proof By either hypothesis, $A = A^*$. If X is a characteristic vector associated with the characteristic value λ, then

$$AX = \lambda X$$

$$X^*AX = \lambda X^*X.$$

Clearly X^*X is real and positive; X^*AX is also real because A is either Hermitian or real and symmetric. Hence λ is the quotient of two real numbers.

Theorem 9.22 implies that the characteristic polynomial (and therefore the minimal polynomial) of A can be factored as a product of real linear factors.

Theorem 9.23   Let A be a matrix that is either Hermitian or real and symmetric. If X_1 and X_2 are characteristic vectors associated with distinct characteristic values λ_1 and λ_2, then $X_1^* X_2 = 0$.

Proof   Let $AX_i = \lambda_i X_i$, $i = 1, 2$.

Then

$$X_1^*(AX_2) = \lambda_2 X_1^* X_2,$$

$$(X_1^* A)X_2 = (AX_1)^* X_2 = (\lambda_1 X_1)^* X_2 = \lambda_1 X_1^* X_2.$$

Hence $(\lambda_1 - \lambda_2)X_1^* X_2 = 0$. Because $\lambda_1 \neq \lambda_2$, we conclude that $X_1^* X_2 = 0$.

Geometrically, this proves that in a Euclidean space the characteristic vectors of a real symmetric matrix are orthogonal whenever they are associated with distinct characteristic values, because $X_1^t X_2 = \langle \xi_1, \xi_2 \rangle$ relative to a normal orthogonal basis. Similarly, in a unitary space two characteristic vectors of a Hermitian matrix are orthogonal whenever they are associated with distinct characteristic values.

The next result is the matrix form of the Principal Axes Theorem, which asserts that any quadratic form can be reduced to the sum of squares by an appropriate orthogonal transformation. In particular, the axes of any quadric surface are orthogonal, and a rigid motion of the axes of any rectangular coordinate system aligns the new coordinate axes with those of the quadric. The corresponding result for Hermitian forms also is valid.

Theorem 9.24 (*Principal Axes Theorem*)   Any real symmetric matrix A is simultaneously similar to and congruent to a diagonal matrix D; that is, there exists an orthogonal matrix P, such that

$$D = P^{-1}AP = P^t AP$$

is diagonal, with the characteristic values of A along the diagonal.

Proof   Let $\{\alpha_1, \ldots, \alpha_n\}$ be a normal orthogonal basis for the Euclidean space $\mathscr{V}$, and let $\mathbf{T}$ be the linear transformation represented by A in that basis. If ξ is any characteristic vector of $\mathbf{T}$, then $\|\xi\|^{-1}\xi$ is of unit length, characteristic, and associated with the same characteristic value as ξ. Let $\beta_1 = \|\xi\|^{-1}\xi$, where ξ is characteristic and associated with λ_1. Extend to a normal orthogonal basis $\{\beta_1, \ldots, \beta_n\}$ for $\mathscr{V}$. Relative to the new basis, $\mathbf{T}$ is represented by the matrix

$$B = R^{-1}AR = R^t AR,$$

where R represents an orthogonal change of basis and is therefore orthogonal. Thus B is real and symmetric, of the form

$$B = \left(\begin{array}{c|c} \lambda_1 & Z \\ \hline Z & B_1 \end{array} \right),$$

where B_1 is a real symmetric square matrix of dimension $n - 1$. We repeat the argument, selecting γ_2 as a normal characteristic vector in the **T**-invariant space $[\beta_2, \ldots, \beta_n]$, letting $\gamma_1 = \beta_1$. Because β_1 is orthogonal to $[\beta_2, \ldots, \beta_n]$, γ_1 and γ_2 form a normal orthogonal set that can be extended to a normal orthogonal basis $\{\gamma_1, \gamma_2, \ldots, \gamma_n\}$ for $\mathscr{V}$. Then **T** is represented by C, where

$$C = S^{-1}BS = S^tBS, \quad S \text{ orthogonal,}$$

$$C = \left(\begin{array}{cc|c} \lambda_1 & 0 & \\ & & Z \\ 0 & \lambda_2 & \\ \hline & Z & C_1 \end{array} \right).$$

Because R and S are orthogonal, so is RS, and

$$C = S^t(R^tAR)S = (RS)^tA(RS).$$

Hence A is simultaneously similar and congruent to C; the theorem follows after n steps.

Theorem 9.25 (*Principal Axes Theorem*) Any Hermitian matrix A is simultaneously similar to and conjunctive to a real diagonal matrix D with the characteristic values of A as the diagonal elements:

$$D = P^{-1}AP = P^*AP$$

for some unitary matrix P.
Proof Exercise.

Theorems 9.24 and 9.25 are strengthened versions of Theorems 9.10 and 9.3, respectively. The reduction of a Hermitian (or real symmetric) matrix to diagonal form can be accomplished by unitary (or orthogonal) transformations. Indeed, it is clear that in selecting a basis in the proof of Theorem 9.25 (or 9.24) we could begin with a characteristic vector corresponding to a *positive* characteristic value (if one exists) and continue as long as such values remain; then we could select characteristic vectors associated with negative characteristic values, as long as they remain, obtaining the diagonal matrix

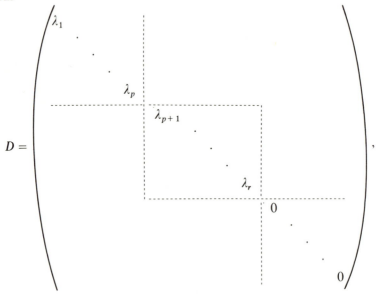

$$D = \begin{pmatrix} \lambda_1 & & & & & & & & \\ & \ddots & & & & & & & \\ & & \ddots & & & & & & \\ & & & \lambda_p & & & & & \\ & & & & \lambda_{p+1} & & & & \\ & & & & & \ddots & & & \\ & & & & & & \lambda_r & & \\ & & & & & & & 0 & \\ & & & & & & & & \ddots \\ & & & & & & & & & 0 \end{pmatrix},$$

where $\lambda_1, \ldots, \lambda_p$ are positive and $\lambda_{p+1}, \ldots, \lambda_r$ are negative. Clearly r is the rank and $2p - r$ is the signature of A and of the Hermitian (or quadratic) form that A represents. D is similar *and* conjunctive (or congruent) to A. But this is as far as unitary (or orthogonal) transformations can be used in reducing A to canonical form relative to conjunctivity (or congruence) as described by Theorem 9.9 (or Theorem 9.10). Indeed, D is the Jordan form of A, the canonical form of A relative to similarity. The further changes needed to produce 1 and -1 in the nonzero diagonal positions correspond to a change of scale along each of the principal axes already obtained.

Furthermore, we have obtained an alternative description of the rank and signature of a Hermitian (or real symmetric) matrix. As we already know, the rank of any matrix is the number of nonzero characteristic values, each counted according to its multiplicity as a zero of the characteristic polynomial. The signature of a Hermitian (or real symmetric) matrix is the number of positive characteristic values minus the number of negative characteristic values, again each counted according to its algebraic multiplicity.

The actual process of reducing a quadratic form to a sum and difference of squares can be accomplished by the classical process of successively completing squares. However, the Principal Axes Theorem provides an alternative method, which we describe for quadratic forms although it is valid for

Hermitian forms as well. In a quadratic form we write the real symmetric matrix A that form defines. The characteristic values of A are real and allow us to write immediately a diagonal matrix that represents the form after an orthogonal change of coordinates. If we also want to describe this change of coordinates explicitly, we obtain a complete set of mutually orthogonal characteristic vectors ξ_1, ξ_2, ..., ξ_n, normalized so that each is of unit length, expressed in terms of the original coordinates. The matrix P, whose column vectors are $\xi_1, \ldots, \xi_n$, is then orthogonal, and $P^t A P = D$. The change of coordinates can be obtained explicitly from P. Of course, calculating characteristic values involves finding the solutions of a polynomial of degree n; this is often difficult. An alternate method of reducing A to diagonal form is to use row and column operations as in Theorem 9.9; however, the change of coordinates represented by such operations is not orthogonal in general.

As a final result, which is important in such applications as the solution of vibration problems in dynamics, we show that, given any two Hermitian (or quadratic) forms, one of which is positive definite, there exists a single change of coordinates that diagonalizes both forms. This result is easily explained in terms of two central quadric surfaces in three-dimensional Euclidean space. Each surch surface determines a quadratic form, and positive definite forms correspond to ellipsoids. Given a central ellipsoid and a second quadratic surface, we can rotate axes to align the new coordinate axes with the axes of the ellipse. This transformation is distance-preserving. Then we change scale along the axes of the ellipsoid, deforming it into a sphere for which any direction is a principal axis. We again rotate axes to align the coordinate axes with the principal axes of the second quadric surface. Because the two surfaces have a common center, the new equation for each will be of the form

$$a_1 x_1^2 + a_2 x_2^2 + a_3 x_3^2 = a_4,$$

where a_1, a_2, and a_3 are ± 1.

Theorem 9.26   Let A and B be n-by-n Hermitian matrices. If A is positive definite, there exists a nonsingular complex matrix P such that

$$P^*AP = I,$$

$$P^*BP = D, \quad \text{where } D \text{ is diagonal.}$$

Proof   Because A is positive definite, there exists a nonsingular matrix Q such that

$$Q^*AQ = I.$$

Then for *any* unitary matrix R,

$$R^*Q^*AQR = R^*IR = R^{-1}R = I.$$

Because B is Hermitian, Q^*BQ is also Hermitian; so by Theorem 9.25 there exists a unitary matrix R such that

$$R^*Q^*BQR = D,$$

where D is diagonal. Let $P = QR$. Then $P^*AP = I$, and $P^*BP = D$.

EXERCISES 9.5

1. Given the real quadratic form $ax_1^2 + 2bx_1 x_2 + cx_2^2$:

(i) Prove that the form is positive definite if and only if $a > 0$ and $b^2 - ac < 0$.

(ii) Show that the central conic $ax^2 + 2bxy + cy^2 = 1$ is an ellipse or hyperbola, depending on whether the quadratic form on the left has $r = 2$ and $s = 2$ or $r = 2$ and $s = 0$.

2. Represent each of the following quadratic forms by a real symmetric matrix, and determine the rank and signature of each.

(i) $x_1^2 - 2x_1 x_3 + 2x_2^2 + 4x_2 x_3 + 6x_3^2$.

(ii) $16x_1 x_2 - x_3^2$.

(iii) $3x_1^2 + 4x_1 x_2 + 8x_1 x_3 + 4x_2 x_3 + 3x_3^2$.

(iv) $x_1 x_2 + 2x_1 x_3 + x_3^2$.

(v) $3x_1^2 + 8x_1 x_2 - 3x_2^2 - 5x_3^2$.

(vi) $2x_1 x_2 + 8x_1 x_3 + x_2^2 + 4x_2 x_3 + 6x_3^2$.

(vii) $4x_1^2 - 10x_1 x_2 - 4x_1 x_3 + 4x_2^2 - 4x_2 x_3 - 8x_3^2$.

3. Refer to the quadratic forms in Exercise 2.

(i) Which of the forms, if any, represent the same quadratic function relative to different bases for $\mathbb{R}^3$?

(ii) Which of the forms, if any, could be used to define an inner product on $\mathbb{R}^3$?

4. Let A be a complex n-by-n matrix with the property that X^*AX is real for every complex column vector X. Let $A = H + K$ be the decomposition of A as the sum of a Hermitian matrix H and a skew-Hermitian matrix K. Prove that $X^*AX = X^*HX$ and that $X^*KX = 0$.

5. Prove Theorem 9.19.

6. Prove Theorem 9.21.

7. Prove Theorem 9.25.

8. Make a table that summarizes the various equivalence relations on matrices that we have studied. The first column should list Systems of Linear Equations, Linear Mappings, Linear Transformations, Conjugate Bilinear Functions, Hermitian Functions, and Quadratic Functions. The second and third columns should give the name and matrix description of the corresponding relation that must hold between matrices which represent the same object of the type listed in the first column. The fourth column should name and concisely describe the corresponding canonical form.

9. The Taylor expansion of a function f of two variables at (a, b) is expressed in terms of the partial derivatives of f by

$$f(a + h, b + k) = f(a, b) + hf_x(a, b) + kf_y(a, b)$$
$$+ \tfrac{1}{2}[h^2 f_{xx}(a, b) + 2hk f_{xy}(a, b) + k^2 f_{yy}(a, b)] + \cdots,$$

provided that

$$f_{xy} = \frac{\partial}{\partial y}\left(\frac{\partial f}{\partial x}\right) = \frac{\partial}{\partial x}\left(\frac{\partial f}{\partial y}\right) = f_{yx}.$$

If $f_x(a, b) = 0 = f_y(a, b)$ then (a, b) is a critical point for maximum or minimum. The term in brackets is a quadratic form in h and k; if the form has rank two, it determines whether f has a local maximum or minimum or neither at (a, b), because for h and k sufficiently small the remainder term is negligible. Assuming $r = 2$, show that

(i) $f(a, b)$ is a relative maximum if $s = -2$,

(ii) $f(a, b)$ is a relative minimum if $s = 2$,

(iii) $f(a, b)$ is neither maximum nor minimum otherwise.

10. As we know, any quadratic form in three variables can be reduced by an orthogonal change of coordinates to the form

$$\lambda_1 x_1^2 + \lambda_2 x_2^2 + \lambda_3 x_3^2,$$

where the λ_i are real. Hence any centrally symmetric quadric surface has an equation of the form

$$\lambda_1 x_1^2 + \lambda_2 x_2^2 + \lambda_3 x_3^2 = 1$$

in a suitable rectangular coordinate system. Use the rank and signature of the quadratic form to classify all possible types of centrally symmetric quadric surfaces and identify each type by means of a sketch.

11. Use the following chain of reasoning to prove Hadamard's inequality: If A is a real n-by-n matrix such that $|a_{ij}| \leq k$ for all i, j, then

$$|\det A| \leq k^n n^{n/2}.$$

(A special case of this result was derived in Exercise 8.3-9.)

    (i) Let $B_0 = A^t A$; then B_0 is real and symmetric.

    (ii) If A is nonsingular, B_0 is positive definite.

    (iii) If B is any n-by-n real, symmetric, positive definite matrix, if $x_1, \ldots, x_n$ are any nonzero numbers, and if $c_{ij} = b_{ij} x_i x_j$, then C is symmetric and positive definite.

    (iv) If C is any n-by-n real, symmetric, positive definite matrix, then

$$\det C \leq \left[\frac{\operatorname{tr} C}{n} \right]^n.$$

    (v) Let C be defined as in (iii), using $B = B_0$ and $x_i = b_{ii}^{-1/2}$. Show that $\operatorname{tr} C = n$, and $\det B = \det C \cdot \Pi_{i=1}^n b_{ii}$.

    (vi) Complete the proof of Hadamard's inequality.

12. Use Hadamard's inequality (Exercise 11) to show that if A is a real n-by-n matrix such that $|a_{ij}| \leq k$ for all i, j, and if the characteristic polynomial of A is

$$(-1)^n[\lambda^n + c_1 \lambda^{n-1} + \cdots + c_n],$$

then

$$|c_j| \leq \binom{n}{j} j^{j/2} k^j, \text{ for } j = 1, 2, \ldots, n.$$

9.6 ISOMETRIC DIAGONABILITY

In the previous section we saw that each Hermitian matrix and each real symmetric matrix is diagonable and that the required change of coordinates can be performed by an isometry, a linear rigid motion of the coordinate axes. When we first discussed diagonability generally, no inner product had been defined on the underlying vector space, so we were not concerned with the type of transformation used to diagonalize a matrix except that it had to be nonsingular. In this section we shall reconsider diagonability, regarding a given matrix as representing a linear transformation on an inner product space and restricting our attention to isometric transformations.

We recall that an isometry **T** is called unitary or orthogonal, according to whether the underlying vector space is complex or real. Relative to a normal orthogonal basis, **T** is represented by a matrix P, which is unitary or orthogonal, respectively. A unitary matrix is characterized by the equation $P^* = P^{-1}$ and an orthogonal matrix by $P^t = P^{-1}$. Because the former condition reduces to the latter when A is real, we shall use the term *isometric* to describe a matrix that is either unitary or orthogonal; that is, any matrix P for which $P^* = P^{-1}$. We shall say that A is *isometrically diagonable* if and only if $P^{-1}AP$ is diagonal for some isometric matrix P.

If A is isometrically diagonable, then for some isometric matrix P and some diagonal matrix D, we have

$$D = P^{-1}AP = P^*AP,$$

$$D^* = \bar{D} = P^*A^*P = P^{-1}A^*P,$$

$$DD^* = (P^{-1}AP)(P^{-1}A^*P) = P^{-1}AA^*P,$$

$$D^*D = (P^{-1}A^*P)(P^{-1}AP) = P^{-1}A^*AP.$$

Because diagonal matrices commute and P is nonsingular, we conclude that

$$AA^* = A^*A.$$

Hence, a necessary condition that A be isometrically diagonable is that A and A^* commute. Our next objective is to prove that this condition is also sufficient.

Definition 9.11   A matrix A of complex numbers is said to be *normal* if and only if

$$AA^* = A^*A.$$

Because any real number is also a complex number, this definition also applies to a real matrix, in which case $AA^t = A^tA$.

Theorem 9.27   Let A be any n-by-n matrix and P any n-by-n isometric matrix. Then A is normal if and only if P^*AP is normal.
Proof   Exercise.

The fact that normality is preserved under isometric changes of bases suggests a way to proceed. We first show that any matrix can be reduced to upper triangular form by means of an isometry. Then we show that any upper triangular matrix which is normal must be diagonal.

Theorem 9.28   If A is any n-by-n matrix, there exists a unitary matrix U such that U^*AU is upper triangular.

Proof   Choose any normal orthogonal basis for complex $\mathcal{V}_n$, and let $\mathbf{T}$ be the linear transformation represented by A relative to that basis. Let λ_1 be a characteristic value of $\mathbf{T}$, and γ_1 a corresponding characteristic vector of unit length. Extend to a normal orthogonal basis $\{\gamma_1, \beta_2, \ldots, \beta_n\}$ for $\mathcal{V}_n$. Then $\mathbf{T}$ is represented relative to the new basis by a matrix $B = P_1^* A P_1$, where P_1 is unitary and where B has the block form

$$B = \left(\begin{array}{c|c} \lambda_1 & B_2 \\ \hline Z & B_4 \end{array} \right).$$

If $n = 1$ or 2, B is upper triangular, and the theorem is proved. For $n > 2$, we proceed by induction, assuming that the theorem is valid for all $(n - 1)$-by-$(n - 1)$ matrices. Thus there exists a unitary matrix Q such that $Q^*B_4 Q$ is upper triangular. Let P_2 be defined by the block form

$$P_2 = \left(\begin{array}{c|c} 1 & Z \\ \hline Z & Q \end{array} \right).$$

Because Q is unitary, the column vectors of P_2 are orthogonal and of unit length. Hence P_2 is unitary, and by direct calculation

$$P_2^* B P_2 = \left(\begin{array}{c|c} \lambda_1 & B_2 Q \\ \hline Z & Q^* B_4 Q \end{array} \right),$$

which is upper triangular. Let $U = P_1 P_2$. Then U is unitary, and U^*AU is upper triangular.

The reason for using unitary transformations and complex $\mathcal{V}_n$ in Theorem 9.28 was necessity rather than convenience: a real matrix need not have any real characteristic values. If a real matrix A is orthogonally similar over the real numbers to an upper triangular matrix B, the diagonal elements of B are real, and these are the characteristic values of B and of A. Conversely, if all the characteristic values of A are real, then the proof of Theorem 9.28 can be adapted to show that P^tAP is upper triangular for some real orthogonal matrix P.

Theorem 9.29   An *n*-by-*n* matrix A is normal if and only if there exists a unitary matrix U such that U^*AU is diagonal.

Proof   The argument that precedes Definition 9.11 proves the "if" part. Suppose A is normal. Then by Theorems 9.27 and 9.28, for some unitary matrix U, U^*AU is upper triangular and normal:

$$B = U^*AU = \begin{vmatrix} b_{11} & b_{12} & \cdots & b_{1n} \\ 0 & b_{22} & \cdots & b_{2n} \\ \cdot & \cdot & & \cdot \\ \cdot & \cdot & & \cdot \\ \cdot & \cdot & & \cdot \\ 0 & 0 & \cdots & b_{nn} \end{vmatrix}.$$

Equating the expressions for the (i, j) elements of BB^* and B^*B, we have

$$\sum_{k=1}^{n} b_{ik}\bar{b}_{jk} = \sum_{k=1}^{n} \bar{b}_{ki}b_{kj}.$$

But $b_{rs} = 0$ whenever $r > s$. For $i = j = 1$, we have

$$b_{11}\bar{b}_{11} + b_{12}\bar{b}_{12} + \cdots + b_{1n}\bar{b}_{1n} = \bar{b}_{11}b_{11}.$$

Hence $b_{1s} = 0$ whenever $s > 1$. Then for $i = j = 2$, we have

$$b_{22}\bar{b}_{22} + b_{23}\bar{b}_{23} + \cdots + b_{2n}\bar{b}_{2n} = \bar{b}_{22}b_{22}.$$

Hence $b_{2s} = 0$ whenever $s > 2$. Continuing in this way, we have $b_{rs} = 0$ whenever $r < s$. Hence B is diagonal.

As a further consideration let us investigate what A^* means when interpreted as a linear transformation on an inner product space. That is, given a linear transformation **T** on a finite-dimensional inner product space $\mathscr{V}$, **T** is represented relative to a normal orthogonal basis by a matrix A, for which A^* is easily calculated. Relative to the same basis, A^* represents a linear transformation on $\mathscr{V}$, say **T***. How are **T** and **T*** related?

In the following computation each index of summation runs from 1 to *n*. We have

$$\mathbf{T}\alpha_i = \sum_r a_{ri}\alpha_r \quad \text{and} \quad \mathbf{T}^*\alpha_j = \sum_s \bar{a}_{js}\alpha_s.$$

Let

$$\xi = \sum_i x_i\alpha_i \quad \text{and} \quad \eta = \sum_k y_k\alpha_k.$$

Then we have

$$\langle \mathbf{T}\xi, \eta \rangle = \sum_i \sum_k x_i \bar{y}_k \langle \mathbf{T}\alpha_i, \alpha_k \rangle$$

$$= \sum_i \sum_k \sum_r x_i a_{ri} \bar{y}_k \langle \alpha_r, \alpha_k \rangle$$

$$= \sum_i \sum_k x_i a_{ki} \bar{y}_k$$

$$= \sum_i \sum_k a_{ki} \bar{y}_k \left\langle \sum_t x_t \alpha_t, \alpha_i \right\rangle$$

$$= \sum_k \bar{y}_k \left\langle \sum_t x_t \alpha_t, \sum_i \bar{a}_{ki} \alpha_i \right\rangle$$

$$= \sum_k \bar{y}_k \langle \xi, \mathbf{T}^*\alpha_k \rangle$$

$$= \left\langle \xi, \sum_k y_k \mathbf{T}^*\alpha_k \right\rangle$$

$$= \langle \xi, \mathbf{T}^*\eta \rangle.$$

Hence for all $\xi, \eta \in \mathscr{V}$, $\mathbf{T}^*$ satisfies the property

$$\langle \mathbf{T}\xi, \eta \rangle = \langle \xi, \mathbf{T}^*\eta \rangle$$

But from Exercise 8.1-4 we know that a vector is uniquely determined by the value of its inner product with each vector of the space. Thus $\mathbf{T}^*$ is uniquely defined on $\mathscr{V}$ by $\mathbf{T}$ and the given inner product.

Definition 9.12   The *adjoint* of a linear transformation $\mathbf{T}$ on an inner product space $\mathscr{V}$ is the mapping $\mathbf{T}^*$ of $\mathscr{V}$ into $\mathscr{V}$ which is defined by the equation

$$\langle \xi, \mathbf{T}^*\eta \rangle = \langle \mathbf{T}\xi, \eta \rangle \text{ for all } \xi, \eta \in \mathscr{V}.$$

As an exercise you may verify that $\mathbf{T}^*$ is linear; that is, for all ξ, η_1, $\eta_2 \in \mathscr{V}$ and all scalars a, b,

$$\langle \xi, \mathbf{T}^*(a\eta_1 + b\eta_2) \rangle = \langle \xi, a(\mathbf{T}^*\eta_1) + b(\mathbf{T}^*\eta_2) \rangle.$$

Theorem 9.30   Relative to a given normal orthogonal basis, if $\mathbf{T}$ is represented by a matrix A, then $\mathbf{T}^*$ is represented by A^*.
Proof Exercise.

Following the terminology introduced for matrices, a *normal* transformation is one that commutes with its adjoint, $TT^* = T^*T$. This class of transformations is of particular interest because Theorem 9.29 guarantees that a normal transformation decomposes the underlying space into a direct sum of the characteristic subspaces of **T** that furthermore are mutually orthogonal. This implies that if **T** is normal and if ξ_1 and ξ_2 are characteristic vectors associated with distinct characteristic values, then $\langle \xi_1, \xi_2 \rangle = 0$. An important subclass of normal transformations are *self-adjoint* transformations, $T = T^*$. Because the characteristic values of **T*** are the complex conjugates of the characteristic values of **T**, the characteristic values of a self-adjoint transformation are real. Self-adjoint transformations are called *Hermitian* in the complex case, *symmetric* in the real case.

If we consider the real case, another question comes to mind. If **T** is represented by A, then **T***, a transformation on $\mathscr{V}$, is represented by A^t. But in Theorem 9.4 we saw that A^t represents a transformation T^t on the dual space $\mathscr{V}'$, relative to the dual basis. How are **T*** and **T^t** related? Even for complex inner product spaces it makes sense to ask whether there is a natural relation between the adjoint **T*** and the transpose **T^t** of **T**, so we shall investigate the question in that form.

Let $\mathscr{V}$ be a finite-dimensional inner product space over the complex numbers, $\mathscr{V}'$ the dual space of all linear mappings from $\mathscr{V}$ to $\mathbb{C}$. With each $f \in \mathscr{V}'$ we associate the vector $\varphi(f) \in \mathscr{V}$, defined by

$$\langle \xi, \varphi(f) \rangle = \mathbf{f}(\xi) \text{ for every } \xi \in \mathscr{V}.$$

As we observed previously $\varphi(f)$ is uniquely determined because the value of its inner product with each vector has been specified. This mapping φ from $\mathscr{V}'$ to $\mathscr{V}$ is one-to-one, for if $\varphi(f) = \varphi(g)$ then $\mathbf{f}(\xi) = \mathbf{g}(\xi)$ for all $\xi \in \mathscr{V}$, so $\mathbf{f} = \mathbf{g}$. Conversely, to each $\eta \in \mathscr{V}$ we can associate the function $h(\eta)$ from $\mathscr{V}$ into $\mathbb{C}$, defined by

$$h(\eta)(\xi) = \langle \xi, \eta \rangle, \text{ for each } \xi \in \mathscr{V}.$$

Then $h(\eta)$ is linear because the inner product is linear in its first component. Thus $h(\eta) \in \mathscr{V}'$, so h is a mapping from $\mathscr{V}$ to $\mathscr{V}'$. Furthermore, for each $\xi \in \mathscr{V}$

$$\langle \xi, \varphi(h(\eta)) \rangle = h(\eta)(\xi) = \langle \xi, \eta \rangle \text{ for all } \eta \in \mathscr{V}.$$

Hence $\varphi(h(\eta)) = \eta$, and φ is onto $\mathscr{V}$. Clearly $h = \varphi^{-1}$.

It is a matter of direct computation to verify that for all $f, g \in \mathscr{V}'$

$$\varphi(f + g) = \varphi(f) + \varphi(g),$$

$$\varphi(af) = \bar{a}\varphi(f) \text{ for all } a \in \mathbb{C},$$

because the inner product is conjugate linear in the second component. Hence if $\mathscr{V}$ is a real vector space, φ is a vector space isomorphism from $\mathscr{V}'$ onto $\mathscr{V}$; but if $\mathscr{V}$ is complex, φ is one-to-one onto $\mathscr{V}$ but not an isomorphism.

Now we are ready to compare $\mathbf{T}^t$ as a transformation of $\mathscr{V}'$ with $\mathbf{T}^*$ as a transformation of $\mathscr{V}$. These mappings were defined by the respective equations

$$(\mathbf{T}^t f)(\xi) = \mathbf{f}(\mathbf{T}\xi) \text{ for all } \xi \in \mathscr{V}, f \in \mathscr{V}',$$

$$\langle \xi, \mathbf{T}^* \rangle = \langle \mathbf{T}\xi, \eta \rangle \text{ for all } \xi, \eta \in \mathscr{V}.$$

We will show that these two mappings preserve the one-to-one correspondence φ, which exists between the space $\mathscr{V}$ and $\mathscr{V}'$: $\mathbf{T}^*\varphi(f) = \varphi(\mathbf{T}^t f)$ all $f \in \mathscr{V}'$. To see this we let $\xi \in \mathscr{V}$ and $f \in \mathscr{V}'$; from the defining equations for φ, $\mathbf{T}$, and $\mathbf{T}^*$ we have

$$(\mathbf{T}^t f)(\xi) = \langle \xi, \varphi(\mathbf{T}^t f) \rangle$$

$$= \mathbf{f}(\mathbf{T}\xi)$$

$$= \langle \mathbf{T}\xi, \varphi(f) \rangle$$

$$= \langle \xi, \mathbf{T}^*\varphi(f) \rangle$$

Again using the fact that a vector is uniquely determined by values of its inner product with every vector of the space implies that $\mathbf{T}^*\varphi(f) = \varphi(\mathbf{T}^t f)$. Hence for a finite-dimensional real inner product space the mapping φ between $\mathscr{V}$ and its dual space $\mathscr{V}'$ permits us to regard $\mathbf{T}^t$ either as a linear transformation on $\mathscr{V}'$ or as a linear transformation on $\mathscr{V}$, with the assurance that each pair of corresponding vectors in the two spaces is mapped by $\mathbf{T}^t$ into a pair of corresponding vectors. See Figure 9.2.

As a variation of Exercise 8.1-9 you may show in Exercise 9.6-12 that the mapping φ from $\mathscr{V}'$ to $\mathscr{V}$ can be used to define an inner product p on $\mathscr{V}'$ in terms of the inner product on $\mathscr{V}$ by defining for all $f, g \in \mathscr{V}'$

$$p(f, g) = \langle \varphi(g), \varphi(f) \rangle.$$

$$\phi(f) \in \mathscr{V} \xrightarrow{\quad \mathbf{T}^* \quad} \mathbf{T}^*\phi(f) = \phi(\mathbf{T}^t f) \in \mathscr{V}$$

$$\Big\uparrow \phi \qquad\qquad \Big\uparrow \phi \qquad\qquad \text{FIGURE 9.2}$$

$$f \in \mathscr{V}' \xrightarrow{\quad \mathbf{T}^t \quad} \mathbf{T}^t f \in \mathscr{V}'$$

The reversal of the order of components is needed in the complex case to compensate for the fact that φ narrowly falls short of being linear because

$$\varphi(af) = \bar{a}\varphi(f).$$

EXERCISES 9.6

1. Prove Theorem 9.27.

2. Let A be a real matrix for which every characteristic value is real. Deduce that A is orthogonally diagonable if and only if A is normal.

3. Prove that an n-by-n matrix A is normal if and only if there exists a set of n mutually orthogonal characteristic vectors.

4. To illustrate Exercises 2 and 3, refer to the matrix A given below. Show that the characteristic values of A are real, that A is diagonable, but that A is not orthogonally diagonable:

$$A = \begin{pmatrix} 2 & 0 & 0 \\ 0 & 1 & 0 \\ 0 & 2 & 2 \end{pmatrix}.$$

5. Prove that $\mathbf{T}^*$ is linear.

6. Prove Theorem 9.30.

7. Prove the following assertions about a normal transformation $\mathbf{T}$ and its adjoint.

(i) $\|\mathbf{T}\xi\| = \|\mathbf{T}^*\xi\|$ for every $\xi \in \mathcal{V}$.

(ii) λ is a characteristic value of $\mathbf{T}$ if and only if $\bar{\lambda}$ is a characteristic value of $\mathbf{T}^*$.

(iii) ξ is a characteristic vector of $\mathbf{T}$ if and only if ξ is a characteristic vector of $\mathbf{T}^*$.

(iv) If ξ_1 and ξ_2 are characteristic vectors of $\mathbf{T}$ associated with distinct characteristic values, then ξ_1 and ξ_2 are orthogonal.

8. Which, if any, of the following matrices are normal?

$$\text{(i)}\ A = \begin{pmatrix} 2 & 1 & 0 \\ 1 & 3 & 2 \\ 0 & -2 & 4 \end{pmatrix}.$$

$$\text{(ii) } A = \begin{pmatrix} 2 & 0 & 1 \\ 0 & 2 & 0 \\ 1 & 0 & 2 \end{pmatrix}.$$

$$\text{(iii) } A = \begin{pmatrix} 1 & 2 & i \\ 2 & 3 & -1 \\ -i & -1 & 0 \end{pmatrix}.$$

9. Let A be a normal matrix. Prove:

(i) If $A^p = Z$, then $A = Z$.

(ii) If A is idempotent, A is Hermitian.

(iii) If $A^3 = A^2$, A is idempotent.

10. Show that the following properties hold for any linear transformation $\mathbf{T}$ on a unitary space $\mathscr{V}_n$.

(i) If c_1 and c_2 are complex numbers such that $|c_1| = |c_2|$, then $c_1 \mathbf{T} + c_2 \mathbf{T}^*$ is normal.

(ii) If $\langle \mathbf{T}\xi, \xi \rangle$ is real for each $\xi \in \mathscr{V}_n$, then $\mathbf{T}$ is Hermitian.

(iii) If $\|\mathbf{T}\xi\| = \|\mathbf{T}^*\|$ for each $\xi \in \mathscr{V}_n$, then $\mathbf{T}$ is normal.

11. Prove the following statements, which suggest that Hermitian matrices that commute have an algebraic relation to normal matrices that is analogous to the algebraic relation of real numbers to complex numbers.

(i) If H_1 and H_2 are Hermitian matrices such that $H_1 H_2 = H_2 H_1$, then $H_1 + iH_2$ is normal, where $i^2 = -1$.

(ii) Conversely, if A is normal, then there exist Hermitian matrices H_1 and H_2 such that $H_1 H_2 = H_2 H_1$ and $A = H_1 + iH_2$.

12. Let $\mathscr{V}$ be a finite-dimensional complex inner product space, $\mathscr{V}'$ its dual space, and φ the one-to-one mapping of $\mathscr{V}'$ onto $\mathscr{V}$ defined in the text. Let p be the mapping on $\mathscr{V}' \times \mathscr{V}'$ defined by

$$p(f, g) = \langle \varphi(g), \varphi(f) \rangle.$$

Prove that p is an inner product on $\mathscr{V}'$.

13. Carry out the details of the following derivation of the *polar decomposition* of a nonsingular matrix A as the product of an isometric matrix Q and a positive definite Hermitian matrix H.

(i) A^*A is Hermitian.

(ii) For some isometry P, $P^*A^*AP = D$ is diagonal with a positive real number in each diagonal position.

(iii) There exists a diagonal matrix E for which $E^2 = D$ and $e_{ii} > 0$ for every i.

(iv) Let $H = PEP^*$; then H is Hermitian and positive definite.

(v) Let $Q = AH^{-1}$; then Q is an isometry.

(vi) $A = QH$.

CHAPTER 10 APPLICATION: LINEAR PROGRAMMING

Most of the material we have considered until now is central in a study of linear algebra. This is not to claim that we have explored every important idea, nor that all previous topics are needed for various applications of matrix theory. But the material in Chapters 1–9 will be included in almost any serious introduction to linear algebra, perhaps in different sequence and with different emphasis.

The interests of the student should determine which topics should be studied next. For prospective mathematicians there are important applications to geometry, analysis, probability, and algebra. Generalizations to infinite-dimensional spaces is of special interest and importance. Physicists might be more interested in applications to Newtonian mechanics, quantum mechanics, or relativity, the chemist in crystal structure or spectroscopy. Engineers will find applications to elasticity, electrical networks, wave propagation, and aircraft design. Economists might prefer to apply matrix theory to input-output analysis and various problems in optimization. Biologists can relate matrix theory to genetics and ecology, psychologists to the theory of learning, sociologists to group relations.

Such applications of linear algebra are of genuine interest, not only because matrices are such versatile tools in modeling problems of social and scientific importance, but also because such applications stimulate the development of new knowledge about matrices. Thus there is a strong temptation to discuss a wide spectrum of applications in the remainder of this book. However, we have chosen to concentrate on only two areas of application: linear programming and linear differential equations.

This chapter presents the basic methods of linear programming. The approach is informal and at an introductory level, so that it can be considered after Chapter 4 in a first-quarter or first-semester course in linear algebra as an application of special significance for students of economics and management science. Chapter 11 applies the material of Chapter 7 to use matrix methods in solving systems of first order linear differential equations, a topic of interest to students of mathematics, physical science, and engineering.

10.1   LINEAR INEQUALITIES

Linear programming is an application of linear algebra that has been developed within the last thirty years as a technique for economic planning and decision making. Stated concisely, a central problem of management is to utilize available resources for the production of goods and the provision of services in such a way that specified objectives are achieved in the best possible manner. The available resources can include raw materials, labor supply, energy sources, plant capacity, and distribution methods among others. Goods and services can include manufactured products, agricultural output, transportation and communication, as well as health care and other needs of society. A specific objective of management might be to maximize employment or profit or production or to minimize costs or delivery time or fuel consumption. Executing such activity typically is subject to various quantitative constraints, such as a limited supply of raw materials, skilled workers, or machine capacity. Constraints also can be imposed by contractual agreements and standards of quality in the finished products.

We introduce the two types of linear programming problems through two specific examples. Although these examples are stated in simplistic terms, it should be evident that Example A illustrates the general problem of utilizing limited resources to maximize a well-defined objective, whereas Example B illustrates the general problem of providing required goods or services at minimal cost.

Example A To raise money for camp a scout troop is planning to sell two different mixtures of nuts. The standard mix consists of 60 percent peanuts, 20 percent filberts, and 20 percent cashews and sells for $2 a pound. The fancy mix consists of 20 percent peanuts, 20 percent filberts, and 60 percent cashews, and sells for $3 a pound. The troop has 60 pounds of peanuts, 24 pounds of filberts, and 52 pounds of cashews to combine into standard mix and fancy mix. How many pounds of each mix will maximize the return the scouts receive for the present supply of nuts?

Let x_1 denote the number of pounds of standard mix and x_2 the number of pounds of fancy mix to be prepared. Then the number of pounds of peanuts used will be $0.6x_1 + 0.2x_2$, which cannot exceed 60. Writing similar expressions for filberts and cashews, we obtain

$$0.6x_1 + 0.2x_2 \leq 60 \qquad \text{(peanut supply)}$$

$$0.2x_1 + 0.2x_2 \leq 24 \qquad \text{(filbert supply)}$$

$$0.2x_1 + 0.6x_2 \leq 52 \qquad \text{(cashew supply)}$$

We seek values of $x_1 \geq 0$ and $x_2 \geq 0$ that maximize the receipts,

$$2x_1 + 3x_2,$$

subject to those inequalities.

A geometric solution can be obtained for this example. By graphing the inequalities, as in Figure 10.1, we see that the points (x_1, x_2) that satisfy the inequalities are precisely those of the shaded region F. Hence the problem will be solved when we determine the point of that region at which the function $2x_1 + 3x_2$ assumes its largest value. The total receipts $2x_1 + 3x_2$ have a constant value along the points on the dashed line and another constant value along any line parallel to that line. As we shall see, for points of F the maximum value of $2x_1 + 3x_2$ is 310, occurring at the point (50, 70). Hence the scouts will maximize their receipts at $310 by selling 50 pounds of standard mix and 70 pounds of fancy mix. This mixture will exactly use up the supply of filberts and cashews, but will leave 16 pounds of peanuts unsold.

Example B Now suppose that an oil company operates two refineries from which it supplies three of its distribution centers. Refinery R_1 has a daily capacity of 4,000 barrels and Refinery R_2 has a daily capacity of 7,000 barrels. The distribution centers D_1, D_2, and D_3 have daily needs of 2,000, 3,000, and 5,000 barrels, respectively. Shipping costs in cents

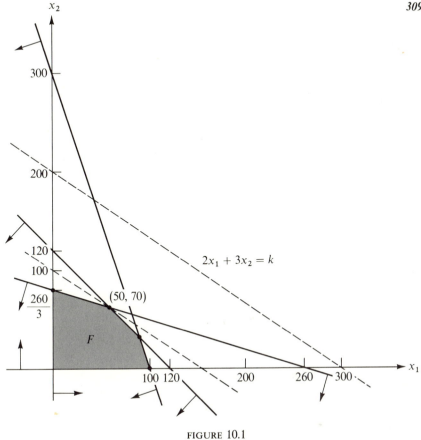

FIGURE 10.1

per barrel from the refineries to the distribution centers are given in the following table.

	D_1	D_2	D_3
R_1	1	2	3
R_2	2	4	6

Plan a shipping schedule that will meet all demands from the available supplies at minimum shipping cost.

Let y_{ij} denote the number of barrels shipped daily from refinery R_i to distribution center D_j. The limited outputs of the refineries impose two constraints

$$y_{11} + y_{12} + y_{13} \le 4,000,$$

$$y_{21} + y_{22} + y_{23} \le 7,000.$$

The demands of the three distribution centers impose three more constraints

$$y_{11} + y_{21} \geq 2,000,$$

$$y_{12} + y_{22} \geq 3,000,$$

$$y_{13} + y_{23} \geq 5,000.$$

The objective is to *minimize* the total shipping cost,

$$C = 1y_{11} + 2y_{12} + 3y_{13} + 2y_{21} + 4y_{22} + 6y_{23},$$

by choosing *nonnegative* values for the six variables y_{ij}, where $i = 1, 2$ and $j = 1, 2, 3$, in such a way that each of the five constraint inequalities is satisfied. The solution of this problem, which we are not yet prepared to derive, is to let $y_{11} = 0 = y_{12}, y_{13} = 4,000, y_{21} = 2,000, y_{22} = 3,000,$ $y_{23} = 1,000$ to achieve a minimal shipping cost of \$340. Surprisingly perhaps, this optimum choice does not use the least expensive route (from R_1 to D_1) but uses the most expensive (from R_2 to D_3).

These two examples illustrate the nature of linear programming problems, which we shall describe more formally in the next section. The essential ingredients of such problems are a set of linear constraints and a linear objective function. The constraints can be expressed by a system of m linear inequalities in n variables,

(10.1)
$$
\begin{aligned}
a_{11}x_1 + a_{12}x_2 + \cdots + a_{1n}x_n &\leq c_1, \\
a_{21}x_1 + a_{22}x_2 + \cdots + a_{2n}x_n &\leq c_2, \\
&\ \ \vdots \\
a_{m1}x_1 + a_{m2}x_2 + \cdots + a_{mn}x_n &\leq c_m,
\end{aligned}
$$

and the objective function is a linear function,

$$f(x_1, x_2, \ldots, x_n) = b_1 x_1 + b_2 x_2 + \cdots + b_n x_n.$$

The problem is to choose values for $x_1, \ldots, x_n$ that satisfy all the inequalities and also produce an extreme value of the objective function. Typically each variable represents a physical quantity and hence must be *nonnegative*; $x_j \geq 0$ for $j = 1, \ldots, n$. In all, therefore, the n variables must satisfy a total of $m + n$ linear inequalities.

To prepare for a general study of linear programming problems, we now explore the solution of the system (10.1) of m linear inequalities in n variables. Using matrix notation we can write (10.1) as

$$AX \le C,$$

where $X \ge Z$ and

$$A = \begin{pmatrix} a_{11} & a_{12} & \cdots & a_{1n} \\ a_{21} & a_{22} & \cdots & a_{2n} \\ \cdot & \cdot & & \cdot \\ \cdot & \cdot & & \cdot \\ \cdot & \cdot & & \cdot \\ a_{m1} & a_{m2} & \cdots & a_{mn} \end{pmatrix}, \quad X = \begin{pmatrix} x_1 \\ x_2 \\ \cdot \\ \cdot \\ \cdot \\ x_n \end{pmatrix}, \quad C = \begin{pmatrix} c_1 \\ c_2 \\ \cdot \\ \cdot \\ \cdot \\ c_m \end{pmatrix}.$$

In this notation AX is a column vector with m components, as is C. The *vector inequality* $AX \le C$ is defined to mean that

$$\sum_{j=1}^{n} a_{ij}x_j \le c_i \quad \text{for} \quad i = 1, \ldots, m;$$

that is, each component of C is as large as the corresponding component of AX.

Now let i be any fixed index from 1 to n. Geometrically each vector X represents a point of $\mathscr{E}^n$, and for each X in $\mathscr{E}^n$ the number

$$\sum_{j=1}^{n} a_{ij}x_j$$

is either less than c_i, equal to c_i, or greater than c_i. Thus each linear inequality separates the points of $\mathscr{E}^n$ into three disjoint sets according to those three possibilities. Furthermore, each of the three sets is *convex*, which means that whenever X_1 and X_2 are both in the same set, then *every* point on the line segment joining X_1 and X_2 also must be in that set. (See Figure 10.2 and Exercise 10.1-5.)

The points X that satisfy $\sum_{j=1}^{n} a_{ij}x_j = c_i$ form a hyperplane in $\mathscr{E}^n$; all the points on one side of that hyperplane satisfy

$$\sum_{j=1}^{n} a_{ij}x_j < c_i,$$

and the points on the other side satisfy the opposite inequality. Hence the solution of each of the m inequalities of (10.1) is a *half space* of $\mathscr{E}^n$, consisting of all points lying on or to one side of a hyperplane. The solution to the system (10.1) is the set of all points that lie in each of the m half spaces

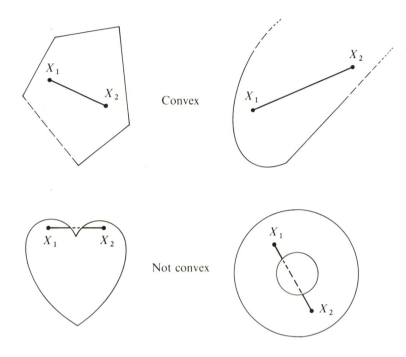

Convex

Not convex

FIGURE 10.2

determined by the m linear inequalities. That solution set might be empty, but in any case it too will be convex. (See Exercise 10.1-6.)

Next we consider the behavior of a linear function

$$f(X) = b_1 x_1 + \cdots + b_n x_n$$

along any line segment in $\mathscr{E}^n$. Suppose that X_1 and X_2 are in $\mathscr{E}^n$. Then each point Y on the line segment from X_1 to X_2 can be written

$$Y = X_1 + t(X_2 - X_1) = (1 - t)X_1 + tX_2,$$

for $0 \le t \le 1$. Because f is linear,

$$f(Y) = (1 - t)f(X_1) + tf(X_2).$$

In particular, if $f(X_1) = f(X_2)$, then $f(Y)$ has the constant value $f(X_1)$ for all Y on the line segment from X_1 to X_2. Because f is continuous, this shows that the values of $f(Y)$ vary monotonically (either constant or steadily increasing or steadily decreasing) along each line segment within $\mathscr{E}^n$. Furthermore, the value of f at any point Y on the line segment from X_1 to X_2 must lie between the numbers $f(X_1)$ and $f(X_2)$. Hence, *among the values that a linear function assumes on the points of a bounded convex set, the greatest and least values, if any, must occur at points on the boundary of that set.*

This observation is especially significant in the context of linear programming. The simultaneous linear constraints define a convex set F in $\mathscr{E}^n$, and the problem is to locate a point at which a linear objective function assumes an extreme value on F. We now know that if such a point exists, it must lie on the boundary of F. Indeed, because the boundary of F consists of hyperplanes, we can extend the previous argument to show that if the objective function has an extreme value on F, that extreme value must occur at a corner point of F. With this information you are now urged to restudy Example A to see how that problem can be solved simply by locating the five corner points of F and by evaluating the objective function $2x_1 + 3x_2$ at each of those five points.

EXERCISES 10.1

1. (i) On one set of axes for $\mathscr{E}^2$ sketch the graphs of each of the following inequalities, shading the region F that is common to all:

$$3x - 2y \geq 6,$$

$$x + 4y \leq 2,$$

$$x + y \geq 0.$$

   (ii) Determine the maximum and minimum values of the linear function $5x - 3y$ on the region F determined in (i).

2. Sketch the solution F of the following system of linear inequalities and determine the extreme values of the linear function $4x + 3y$ on F.

$$x + 2y \geq 8,$$

$$3x + y \geq 6,$$

$$x - y \leq 2,$$

$$x + y \geq 1,$$

$$x \geq 0,$$

$$y \geq 0.$$

3. Sketch the graphs of each of the following systems of linear inequalities, indicating the solution set.

(i) $\quad x_1 -\quad 2x_2 \geq 4,$
$\quad\quad x_1 +\quad x_2 \leq 7,$
$\quad\quad\quad\quad x_2 \geq 0.$

(ii) $\quad x_1 -\quad 2x_2 \geq 4,$
$\quad\quad x_1 +\quad x_2 \geq 7,$
$\quad\quad\quad x_1 \geq 0,$
$\quad\quad\quad\quad x_2 \geq 0.$

(iii) $\quad x_1 -\quad 2x_2 \leq 4,$
$\quad\quad x_1 +\quad x_2 \leq 7,$
$\quad\quad 2x_1 -\quad x_2 \leq -8,$
$\quad\quad\quad x_1 \geq 0,$
$\quad\quad\quad\quad x_2 \geq 0.$

4. Determine the maximum and minimum of the function $3x_1 + 2x_2$ on each of the solution sets of Exercise 3.

5. Let $U = (u_1, \ldots, u_n)$ and $V = (v_1, \ldots, v_n)$ be points of $\mathscr{E}^n$ that satisfy the linear inequality

$$a_1 x_1 + a_2 x_2 + \cdots + a_n x_n < c,$$

and let Y be a point on the line segment from U to V.

(i) Show that $Y = (1 - t)U + tV$ for some value of t in $[0, 1]$.

(ii) Show that $a_1 y_1 + a_2 y_2 + \cdots + a_n y_n < c$.

6. Let S and T be convex subsets of $\mathscr{E}^n$.

(i) Show that $S \cap T$ is convex.

(ii) Use the results of Exercises 5 and 6(i) to show that the solution to the system (10.1) is a convex set.

7. Classify each of the sets defined below as convex or not convex.

(i) All points in the second quadrant of $\mathscr{E}^2$.

(ii) All points of $\mathscr{E}^2$ that lie on the line $2y + 3x = 7$.

(iii) All points of $\mathscr{E}^2$ that lie beneath the graph of $y = x^2$.

(iv) All points of $\mathscr{E}^2$ that lie above the graph of $y = x^2$.

(v) All points of $\mathscr{E}^2$ that lie on the graph of $y = x^2$.

(vi) All points of $\mathscr{E}^2$.

(vii) A set consisting of a single point.

(viii) The empty set.

(ix) A set consisting of two distinct points in $\mathscr{E}^2$.

8. A farmer has two orchards, each of which produces apples of three levels of quality. One hour of picking apples in orchard A costs \$20 for labor and equipment and produces 6 bushels of grade I apples, 2 bushels of grade II, and 4 bushels of grade III. An hour of picking in orchard B costs \$16 and produces 2 bushels of grade I, 2 bushels of grade II, and 12 bushels of grade III. The farmer contracts to furnish a market each day with 12 bushels of grade I apples, 8 bushels of grade II, and 24 bushels of grade III. The farmer wants to fulfill his contract by minimizing his total cost for labor and equipment. Suppose that he decides to pick y_1 hours daily in orchard A and y_2 hours daily in orchard B. Clearly $y_1 \geq 0$ and $y_2 \geq 0$. Write a complete system of inequalities that express the constraints on y_1 and y_2 arising from the contractual commitments of the farmer, and also write a function that expresses the cost of labor and equipment.

9. Solve the linear program described in Exercise 8 by graphing the system of inequalities, determining the corner points of the associated convex region, and evaluating the cost function at each corner point.

10. Given a linear form f defined for each $X = (x_1, \ldots, x_n)$ in $\mathscr{E}^n$ by

$$f(X) = c_1 x_1 + c_2 x_2 + \cdots + c_n x_n.$$

(i) Show that f is a monotone function along any line segment in $\mathscr{E}^n$.

(ii) Let C be a closed and bounded convex set in $\mathscr{E}^n$. Deduce that the extreme value of f on C must occur on the boundary of C.

10.2   DUAL LINEAR PROGRAMS

In Examples A and B of the preceding section we have seen two types of linear programs, each involving a system of linear inequalities and a linear function for which we seek an extreme value on the convex set F of points that satisfy all of the inequalities. In Example A the inequalities represented production ceilings imposed by limited supplies of raw materials, and the objective function represented a financial return that we sought to *maximize*. In Example B the inequalities expressed production floors imposed by market demands for delivered materials, and the objective function represented a cost of delivery that we sought to *minimize*. We now state the maximum problem in general form, using matrix notation for conciseness.

Maximum Problem   Let $A = (a_{ij})$ be an m-by-n matrix of real numbers. Let B be an n-component row vector, and let C be an m-component column vector. Determine an n-component column vector X such that

$$AX \leq C,$$

(10.2) $$X \geq Z, \text{ and}$$

$$BX \text{ is maximized.}$$

It might be helpful to restate this problem in component form in terms that interpret a typical economic situation in which the problem arises naturally. A company manufactures n different products, $P_1, P_2, \ldots, P_n$, from m different raw materials, $M_1, M_2, \ldots, M_m$. Each unit of product P_j requires a_{ij} units of raw material M_i. The total monthly supply of M_i is c_i units, $i = 1, \ldots, m$. Let x_j denote the number of units of P_j produced each month. The total amount of material M_i used in this production schedule is

$$a_{i1}x_1 + a_{i2}x_2 + \cdots + a_{in}x_n,$$

which cannot exceed the monthly supply c_i, $i = 1, 2, \ldots, m$. Hence we have these constraints on the numbers $x_1, \ldots, x_n$:

$$a_{11}x_1 + a_{12}x_2 + \cdots + a_{1n}x_n \leq c_1,$$
$$a_{21}x_1 + a_{22}x_2 + \cdots + a_{2n}x_n \leq c_2,$$
$$\cdot$$
$$\cdot$$
$$\cdot$$
$$a_{m1}x_1 + a_{m2}x_2 + \cdots + a_{mn}x_n \leq c_m,$$
$$x_1 \geq 0, x_2 \geq 0, \ldots, x_n \geq 0.$$

We want to choose the numbers x_j to satisfy each of these $m + n$ inequalities and to maximize the company's profits:

$$\text{maximize } (b_1x_1 + b_2x_2 + \cdots + b_nx_n),$$

where b_j is the profit obtained from each unit of P_j that is produced. It should now be apparent that (10.2) expresses this problem in matrix form.

The given matrix A and the two given vectors B and C can be used to formulate another linear programming problem that is closely related to the problem stated in (10.2). For the related problem the economic significance of the numbers b_j, a_{ij}, and c_i must be given new interpretations, but the mathematical statement of the related problem is readily formulated.

Minimum Problem   Let $A = (a_{ij})$ be an m-by-n matrix of real numbers. Let B be an n-component row vector, and let C be an m-component column vector. Determine an m-component row vector Y such that

$$YA \geq B,$$

(10.3) $$Y \geq Z, \text{ and}$$

$$YC \text{ is minimized.}$$

This problem is easily restated in component form:

$$a_{11}y_1 + a_{21}y_2 + \cdots + a_{m1}y_m \geq b_1,$$
$$a_{12}y_1 + a_{22}y_2 + \cdots + a_{m2}y_m \geq b_2,$$
$$.$$
$$.$$
$$.$$
$$a_{1n}y_1 + a_{2n}y_2 + \cdots + a_{mn}y_m \geq b_n,$$
$$y_1 \geq 0, \, y_2 \geq 0, \, \ldots, \, y_m \geq 0.$$
$$\text{Minimize } (c_1 y_1 + c_2 y_2 + \cdots + c_m y_m).$$

Hence, given an m-by-n matrix A, a one-by-n matrix B and an m-by-one matrix C, we can state two linear programming problems, one of each type as expressed by (10.2) and (10.3). These two associated problems are said to be *dual* to each other.

A vector X that satisfies all of the inequalities of (10.2) is said to be *feasible* for the *maximum* problem (or *maximum feasible*). A maximum feasible vector X_0 such that $BX_0 \geq BX$ for all maximum feasible vectors X is said to be *maximum optimal*.

A vector Y that satisfies all of the inequalities of (10.3) is said to be *feasible* for the *minimum* problem (or *minimum feasible*). A minimum feasible vector Y_0 such that $Y_0 C \leq YC$ for all minimum feasible vectors Y is said to be *minimum optimal*.

Now for a pair of dual problems let X denote any maximum feasible vector and let Y denote any minimum feasible vector. We have

$$X \geq Z \text{ and } AX \leq C,$$

$$Y \geq Z \text{ and } YA \geq B.$$

Because Y is an m-component row vector and AX and C are m-component column vectors, YAX and YC are scalars (one-by-one matrices). Furthermore, because for each $i = 1, \ldots, m$, we have

$$\sum_{j=1}^{n} a_{ij}x_j \leq c_i,$$

$$y_i \geq 0,$$

it follows that

$$\sum_{j=1}^{n} y_i a_{ij} x_j \leq y_i c_i,$$

$$YAX = \sum_{i=1}^{m} \sum_{j=1}^{n} y_i a_{ij} x_j \leq \sum_{i=1}^{m} y_i c_i = YC.$$

Similarly, because $X \geq Z$ and $YA \geq B$, we have

$$YAX \geq BX.$$

Hence

$$YC \geq YAX \geq BX,$$

whenever X and Y are feasible vectors. Now fix X temporarily. Because $YC \geq BX$ for all feasible X, the *minimum* value of YC for all feasible Y must be at least as large as BX:

$$\min YC \geq BX.$$

Because this inequality holds for all feasible X, the *maximum* value of BX for all feasible X cannot exceed min YC. In summary,

(10.4) $\min YC \geq \max BX.$

That is, *the optimum value of the objective function for a maximum problem cannot exceed the optimum value of the objective function of the dual minimum problem.* This means that if we can find any dual pair X_1 and Y_1 of feasible vectors such that

$$BX_1 \geq Y_1 C,$$

then both X_1 and Y_1 are optimal, and the optimum values of the two objective functions must coincide.

Of course, such a dual pair may or may not exist. But the first of the next two theorems, which we state without proof, describes precisely when a solution exists. The second theorem describes the basic relationship between a linear programming problem and its dual problem when a solution does exist.

Fundamental Existence Theorem   A necessary and sufficient condition that *either* a maximum linear programming problem or its dual minimum problem has an optimal vector (a solution) is that *both* problems have feasible vectors.

Fundamental Duality Theorem   If an optimal vector exists for a given linear programming problem (maximum or minimum), then an optimal vector also exists for the dual problem, and the optimum values of the two objective functions are *equal*.

In the remaining sections of this chapter we shall use these relationships between a linear program and its dual to justify and interpret a computational algorithm that is guaranteed to produce simultaneously a solution to a linear program and its dual or else show that no solution exists.

As a final consideration we reinterpret a linear program and its dual in vector space notation in order to show how the terms "dual" and "duality" are related to the dual space as developed in Sections 9.1 and 9.2. (If you haven't studied those sections, you should either do so now or omit the rest of this section.)

From systems (10.2) and (10.3) we see that a linear program and its dual are described by an m-by-n matrix A, a row vector B with n components, and a column vector C having m components. The maximum problem requires the determination of a column vector X with n components, while the minimum problem requires the determination of a row vector Y with m components, both satisfying certain properties. We interpret the given matrix A as a linear mapping $\mathbf{T}$ from $\mathscr{V}_n$ to $\mathscr{W}_m$, the given column vector C as a vector γ in $\mathscr{W}_m$, and the unknown column vector X as a vector ξ in $\mathscr{V}_n$. We can interpret the given row vector B as a linear functional g in the dual space $\mathscr{V}'_n$ and interpret the unknown vector Y as a linear functional f in the dual space $\mathscr{W}'_m$. We rewrite the condition $YA \geq B$ in the form $A^t Y^t \geq B^t$. Then the inequalities and objective functions that define the dual pair of programming problems are reformulated as follows:

$$AX \leq C \text{ translates into } \quad \mathbf{T}\xi \leq \gamma \text{ in } \mathscr{W}_m,$$

$$X \geq Z \text{ translates into } \quad \xi \geq \theta \text{ in } \mathscr{V}_n,$$

$$A^t Y^t \geq B^t \text{ translates into } \mathbf{T}^t f \geq g \text{ in } \mathscr{V}'_n,$$

$$Y \geq Z \text{ translates into } \quad f \geq z \text{ in } \mathscr{W}'_m.$$

(See Figure 10.3.) The objective function BX of the maximum problem is the scalar $g(\xi)$, and the objective function YC of the minimal problem is the

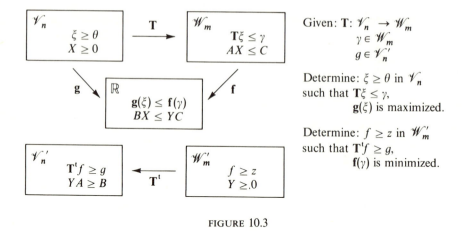

FIGURE 10.3

scalar $f(\gamma)$. The mapping $\mathbf{T}$ and vectors $\gamma \in \mathscr{W}_m$ and $g \in \mathscr{V}_n'$ are given, whereas $\xi \in \mathscr{V}_n$ and $f \in \mathscr{W}_m'$ are to be determined, subject to the conditions listed above, to maximize $g(\xi)$ and to minimize $f(\gamma)$.

The feasible vectors of the maximum problem are those $\xi \geq \theta$ in $\mathscr{V}_n$ for which $\mathbf{T}\xi \leq \gamma$ in $\mathscr{W}_m$, while the feasible vectors of the minimum problem are those linear functionals $f \geq z$ in $\mathscr{W}_m'$ such that $\mathbf{T}f \geq g$ in $\mathscr{V}_n'$. For feasible vectors ξ and f we have

$$\mathbf{g}(\xi) \leq (\mathbf{T}f)(\xi) = \mathbf{f}(\mathbf{T}\xi) \leq \mathbf{f}(\gamma),$$

so the translated form of (10.4) becomes

$$\max_{\xi} \mathbf{g}(\xi) \leq \min_{f} \mathbf{f}(\gamma).$$

The duality theorem shows that equality holds in this relationship if and only if both ξ and f are optimal.

EXERCISES 10.2

1. Write in component form the dual linear programming problem of Example A in Section 10.1.

2. Write in component form the dual linear programming problem of Example B in Section 10.1.

3. Write in component form the dual linear programming problem of the orchard problem from Exercise 10.1-8. Find any feasible vector

$X = (x_1, x_2, x_3)$ for the dual problem, and evaluate the objective function f for the dual problem at X. Then find any feasible vector Y for the orchard problem, evaluate the cost function g at Y, and confirm that $f(X) \leq g(Y)$.

4. The following example illustrates the *diet problem*, one of the earliest applications of linear programming. Foods A and B each provide three nutrients essential to health. The number of units of each nutrient provided by one ounce of each food is given in the table below. Suppose that the minimal daily requirement of each nutrient is given in the last column of the table and that the cost of each food, in cents per ounce, is given in the bottom row.

Nutrient	Food A	Food B	Minimum Daily Requirement
N_1	0.15	0.10	1
N_2	75.0	150.0	750
N_3	1.2	1.2	10
Cost	4	3	

The problem is to determine the number of ounces y_1 of Food A and y_2 of Food B that will provide the minimal daily requirements of these nutrients at minimum cost.

(i) Express this diet problem in component form.

(ii) Solve this problem by graphical methods in $\mathscr{E}^2$.

(iii) Write in component form the dual of the diet problem, using x_1, x_2, and x_3 as variables.

5. A metal processor wishes to produce an alloy of tin and lead containing at least 60 percent lead and at least 35 percent tin. He can purchase four different alloys with the percentage compositions and hundredweight prices shown in the table below.

Alloys

	A_1	A_2	A_3	A_4	Desired
Lead	40	60	80	70	≥ 60
Tin	60	40	20	30	≥ 35
Costs	240	180	160	210	Minimum

Carry out the following steps to determine how the processor should blend the alloys to minimize his costs.

(i) Write the dual of this program.

(ii) Solve the dual program by graphical methods in $\mathscr{E}^2$, and show that the maximum value of the dual objective function $60x_1 + 35x_2$ is 168.

(iii) Show that at the optimal point determined in (ii) the value of the *modified* objective function $65x_1 + 35x_2$ is 175. (This modification recognizes that the numbers in the first two lines of the table are *percentages*.)

(iv) Find by trial and error a feasible vector for the original problem at which the cost function has the value 175.

(v) Explain why the feasible vector found in (iv) is optimal for the original problem.

6. (i) Use graphical methods in $\mathscr{E}^2$ to sketch the set F of feasible vectors that satisfy the following system of inequalities:

$$2y_1 + y_2 \ge 12,$$
$$5y_1 + 8y_2 \ge 74,$$
$$y_1 + 6y_2 \ge 24,$$
$$y_1 \ge 0,$$
$$y_2 \ge 0.$$

(ii) Determine the minimum value at points of F for each of the following objective functions:

$$y_1 + y_2,$$
$$3y_1 + y_2,$$
$$y_1 + 3y_2,$$
$$3y_1 - y_2.$$

7. (i) Use graphical methods in $\mathscr{E}^2$ to sketch the set F of feasible vectors that satisfy the following system of inequalities:

$$x_1 + 2x_2 \leq 600,$$

$$5x_1 + 4x_2 \leq 1600,$$

$$2x_1 + x_2 \leq 600,$$

$$x_1 \geq 0,$$

$$x_2 \geq 0.$$

(ii) Determine the maximum value at points of F for each of the following objective functions:

$$40x_1 + 50x_2,$$

$$40x_1 + 70x_2,$$

$$40x_1 + 20x_2.$$

10.3 COMPUTATIONAL TECHNIQUES

Although vector notation provides a concise and convenient means for expressing a dual pair of linear programs, as in (10.2) and (10.3), constructing a numerical solution requires computation with the numerical entries of the matrix A and vectors B and C. All the numbers and variables needed to describe both problems of a dual pair can be presented in a single tableau as shown in tableau 10.5.

(10.5)

		x_1	$\cdots$	x_j	$\cdots$	x_n	$\leq$	
	y_1	a_{11}	$\cdots$	a_{1j}	$\cdots$	a_{1n}		c_1
	.	.		.		.		.
	.	.		.		.		.
	.	.		.		.		.
	y_i	a_{i1}		a_{ij}		a_{in}		c_i
	.	.		.		.		.
	.	.		.		.		.
	.	.		.		.		.
	y_m	a_{m1}		a_{mj}		a_{mn}		c_m
	$\geq$							
		b_1	$\cdots$	b_j	$\cdots$	b_n		

The information of the maximum problem is presented by the rows: the variables x_j, listed as labels across the top of the tableau, are to be assigned nonnegative values in such a way that for each $i = 1, 2, \ldots, m$

$$a_{i1} x_1 + \cdots + a_{ij} x_j + \cdots + a_{in} x_n \leq c_i,$$

and in such a way that the value of the objective function

$$b_1 x_1 + \cdots + b_j x_j + \cdots + b_n x_n$$

is *maximized*. Similarly, the information of the dual of the maximum problem is presented by the columns: the variables y_i, listed as labels along the left, are to be assigned nonnegative values in such a way that for each $j = 1, 2, \ldots, n$

$$a_{1j} y_1 + \cdots + a_{ij} y_i + \cdots + a_{mj} y_m \geq b_j,$$

and in such a way that the value of the objective function

$$c_1 y_1 + \cdots + c_i y_i + \cdots + c_m y_m$$

is *minimized*.

Our first step is to convert the system $AX \leq C$ of m linear inequalities in n variables into a system of m linear *equations* in $m + n$ variables. To do this, we define the m-component vector V by the equation

$$V = C - AX,$$

or

$$v_i = c_i - (a_{i1} x_1 + \cdots + a_{in} x_n), \qquad i = 1, \ldots, m.$$

Then X is feasible if and only if $V \geq Z$. Indeed, when X is feasible, v_i describes the amount by which the value $a_{i1} x_1 + \cdots + a_{in} x_n$ is less than the constraint c_i. Hence each v_i is called a *slack variable*. In particular, we note that if each $c_i \geq 0$, then $X = Z$ is feasible, and for that choice of X, $V = C \geq Z$. To represent the variables v_i in the tableau, it is convenient first to write the system of linear equations in the equivalent form $AX - C = -V$, or

$$a_{11} x_1 + \cdots + a_{12} x_2 + \cdots + a_{1n} x_n - c_1 = -v_1,$$

$$a_{i1} x_1 + \cdots + a_{ij} x_j + \cdots + a_{in} x_n - c_i = -v_i,$$

$$a_{m1} x_1 + \cdots + a_{mj} x_j + \cdots + a_{mn} x_n - c_m = -v_m.$$

Then we modify tableau (10.5) by replacing the inequality $\leq$ above the vertical dashed line by the label -1 above the last column, and by adding the label $-v_i$ on the right-hand side of row i, $i = 1, 2, \ldots, m$.

(10.6)

	x_1	$\cdots$	x_j	$\cdots$	x_n	-1	$=$
y_1	a_{11}	$\cdots$	a_{1j}	$\cdots$	a_{1n}	c_1	$-v_1$
y_i	a_{i1}	$\cdots$	a_{ij}	$\cdots$	a_{in}	c_i	$-v_i$
y_m	a_{m1}	$\cdots$	a_{mj}	$\cdots$	a_{mn}	c_m	$-v_m$
-1	b_1	$\cdots$	b_j	$\cdots$	b_n		
$=$	u_1	$\cdots$	u_j	$\cdots$	u_n		

The new labels u_j that appear below the tableau (10.6) are obtained by applying to the inequalities $YA \geq B$ of the dual minimum problem the same sort of reasoning that produced the slack variables $-v_j$ for the maximum problem. We define the n-component vector U by the equation

$$U = YA - B.$$

Then Y is feasible if and only if $U \geq Z$. When Y is feasible each u_j, $j = 1, \ldots, n$, represents the amount by which $a_{1j}y_1 + \cdots + a_{mj}y_m$ exceeds the constraint b_j. Hence each u_j is called a *surplus variable*. We note that if each $b_j \leq 0$, then $Y = Z$ is feasible, and for that choice of Y, $U = -B \geq Z$.

By introducing slack and surplus variables, we have converted the m inequalities in n variables of the maximum linear program into a system of m equations in $m + n$ variables, and the n inequalities in m variables of the dual minimum program into a system of n equations in $m + n$ variables. Furthermore, in tableau (10.6) if each entry c_i in the last column is *nonnegative* and if each entry b_j in the last row is *nonpositive*, then we can obtain a dual pair X, Y of feasible vectors by letting each $x_j = 0$ and each $y_i = 0$, letting $v_i = c_i$ for each i, and letting $u_j = -b_j$ for each j. By the Fundamental Existence Theorem the existence of any dual pair of feasible vectors guarantees the existence of a dual pair of feasible vectors that produce the common extreme value of the two objective functions.

Tableau (10.6), therefore, represents two systems of linear equations (in two sets of $m + n$ variables). If each $c_i \geq 0$, a solution of one system can be obtained by assigning the value zero to each variable whose label appears in the top margin of the tableau and by assigning to each variable whose label appears in the right-hand margin the number directly beside it in the last column. Such a solution is called a *basic solution* of the system; the variables whose labels appear along the right-hand side are called *basic variables* and those whose labels appear across the top are called *nonbasic variables*. Similarly, a solution of the dual system of linear equations can be obtained by assigning the value zero to each variable whose label appears in the left-hand margin (nonbasic variables) and assigning to each variable whose label appears in the bottom margin (basic variables) the *negative* of the number directly above it in the bottom row of the tableau. In this description of basic and nonbasic variables we have carefully avoided using the labels themselves but have referred only to where those labels appear. Such care is necessary because the simplex method for solving a linear program uses pivot operations to replace a given tableau by another tableau that represents the same dual pair of linear programs. In this process two pairs of labels of variables are interchanged. Whether a given variable is considered to be basic or nonbasic depends entirely on where its label appears in the current tableau presentation of the system.

Basic and *nonbasic* variables can also be understood from another point of view. Each nonbasic variable of a system of linear equations can be assigned any value, but then the value of each basic variable is uniquely determined by the values assigned to the nonbasic variables. In a system of m linear equations in $m + n$ variables, we can designate any m of the variables as basic and write the system in a form that expresses each basic variable as a linear combination of the other n variables. Hence the designation basic and nonbasic is not at all intrinsic but depends only on the algebraic form in which the system is written.

Until now we have made no reference to the relation of slack and surplus variables to the objective functions, BX for the maximum problem and YC for the dual minimum problem. For any X and Y, with their associated slack and surplus vectors V and U, we have

$$AX + V = C \qquad \text{and} \quad YA - U = B,$$

$$YC = Y(AX + V) \quad \text{and} \quad BX = (YA - U)X,$$

$$YC - BX = YV + UX = \sum_{i=1}^{m} y_i v_i + \sum_{j=1}^{n} u_j x_j.$$

Fundamental Duality Equation For any dual pair of linear programs as described by tableau (10.6) let X and Y be any vectors and let V and U be the associated slack and surplus vectors. The *difference* $YC - BX$ between the value of the objective function of the minimum program at Y and the value of the objective function of the maximum program at X is $UX + VY$,

$$u_1 x_1 + \cdots + u_n x_n + v_1 y_1 + \cdots + v_m y_m.$$

This expression is simply the dot product of the vector $(X; V)$ of the variables for the maximum problem with the vector $(U; Y)$ of the variables for the dual minimum problem. When Y and X are both feasible, the components of X, V, Y, and U are nonnegative. Then $YC - BX \geq 0$ as we observed in the previous section. By the Fundamental Duality Theorem if X and Y are optimal, then $YC - BX = 0$. But also $UX \geq 0$ and $YV \geq 0$, so we conclude that $UX = 0$ and $YV = 0$ when X and Y are optimal. Geometrically, with each X in $\mathscr{E}^n$ we have associated a slack vector V in $\mathscr{E}^m$, and with each Y in $\mathscr{E}^m$ we have associated a surplus vector U in $\mathscr{E}^n$:

> Feasible vectors X and Y are an optimal dual pair if and only if X and U are orthogonal in $\mathscr{E}^n$ and Y and V are orthogonal in $\mathscr{E}^m$.

In many applications of linear programming all the entries of A, B, and C are nonnegative. When $C \geq Z$, $X = Z$ is feasible for the maximum problem, but a feasible vector for the dual minimum problem is not so easily identified. The strategy that we shall use is to rewrite the tableau for the programs by interchanging the role of a basic variable with that of a nonbasic variable for each of the two programs, obtaining a new tableau presentation of the original problem. By a sequence of such interchanges, we hope to arrive at a tableau, given by A_1, B_1, and C_1, in which $B_1 \leq Z$ and $C_1 \geq Z$, from which we can read off optimal solutions for both problems. Such interchanges of variables are performed by *pivot operations*, which we now describe. Procedures for choosing pivot operations that will lead systematically to optimal solutions will be considered in the next section.

To perform a pivot operation on a system of linear equations as described by the tableau (10.6), we apply the Gaussian procedure of solving one equation for one variable in terms of the other variables, substituting that expression into all the other equations of the system. Written fully, the equations of the maximum problem represented by (10.6) become

$$a_{11}x_1 + \cdots + a_{1j}x_j + \cdots + a_{1n}x_n - c_1 = -v_1,$$

$$a_{i1}x_1 + \cdots + a_{ij}x_j + \cdots + a_{in}x_n - c_i = -v_i,$$

$$a_{m1}x_1 + \cdots + a_{mj}x_j + \cdots + a_{mn}x_n - c_m = -v_m.$$

If $a_{ij} \neq 0$ we can "pivot" on a_{ij} by solving equation i for x_j,

$$x_j = -\left[\frac{a_{i1}}{a_{ij}}x_1 + \cdots + \frac{1}{a_{ij}}v_i + \cdots + \frac{a_{in}}{a_{ij}}x_n - \frac{c_i}{a_{ij}}\right],$$

where the basic variable v_i is written in the position previously occupied by x_j. When this value of x_j is substituted into the other equations, the system has the following form:

$$a'_{11}x_1 + \cdots + a'_{ij}v_i + \cdots + a'_{1n}x_n - c'_1 = -v_1,$$

$$a'_{i1}x_1 + \cdots + a'_{ij}v_i + \cdots + a'_{in}x_n - c'_i = -x_j,$$

$$a'_{m1}x_1 + \cdots + a'_{mj}v_i + \cdots + a'_{mn}x_n - c'_m = -v_m,$$

where the new coefficients are computed as follows.

(1) $a'_{ij} = \dfrac{1}{a_{ij}}$. (Replace the pivot a_{ij} by its reciprocal.)

(2) $a'_{is} = \dfrac{a_{is}}{a_{ij}}$ if $s \neq j$, and $c'_i = \dfrac{c_i}{a_{ij}}$. (Divide every nonpivot entry

in the *row* of the pivot by the pivot.)

(3) $a'_{rj} = -\dfrac{a_{rj}}{a_{ij}}$ if $r \neq i$. (Divide every nonpivot entry in the *column*

of the pivot by the *negative* of the pivot.)

(4) $a'_{rs} = a_{rs} - \dfrac{a_{rj}a_{is}}{a_{ij}}$ if $r \neq i$ and $s \neq j$, and $c'_r = c_r - \dfrac{a_{rj}c_i}{a_{ij}}$ if $r \neq i$.

This last rule is more easily understood by considering the numbers in the corners of the "rectangle" determined in the tableau by the pivot p and any entry q in the (r, s) position in neither the row nor the column of the pivot p.

Rule 4 states that q should be replaced by $q - tw/p$.

Thus a pivot on the entry a_{ij} in the tableau (10.6) produces the following new tableau.

		x_1		$\circled{v_i}$		x_n	-1	$=$
y_1		$\left(a_{11} - \dfrac{a_{1j}a_{i1}}{a_{ij}}\right)$	$\cdots$	$-\dfrac{a_{1j}}{a_{ij}}$	$\cdots$	$\left(a_{1n} - \dfrac{a_{1j}a_{in}}{a_{ij}}\right)$	$\left(c_1 - \dfrac{a_{1j}c_i}{a_{ij}}\right)$	$-v_1$
$\circled{u_j}$		$\dfrac{a_{i1}}{a_{ij}}$	$\cdots$	$\dfrac{1}{a_{ij}}$	$\cdots$	$\dfrac{a_{in}}{a_{ij}}$	$\dfrac{c_i}{a_{ij}}$	$-\circled{x_j}$
row i								
y_m		$\left(a_{m1} - \dfrac{a_{mj}a_{i1}}{a_{ij}}\right)$	$\cdots$	$-\dfrac{a_{mj}}{a_{ij}}$	$\cdots$	$\left(a_{mn} - \dfrac{a_{mj}a_{in}}{a_{ij}}\right)$	$\left(c_m - \dfrac{a_{mj}c_i}{a_{ij}}\right)$	$-v_m$
-1		$\left(b_1 - \dfrac{b_j a_{i1}}{a_{ij}}\right)$	$\cdots$	$-\dfrac{b_j}{a_{ij}}$	$\cdots$	$\left(b_n - \dfrac{b_j a_{in}}{a_{ij}}\right)$		
$=$		u_1	$\cdots$	$\circled{y_i}$	$\cdots$	u_n		

column j (above table, over the pivot column)

Observe that the variable names that label the pivot row at the right and the pivot column at the top have been interchanged. Also the variable labels at the left of the pivot row and the bottom of the pivot column have been interchanged. To see that this latter interchange is justified, we write the system of linear equations that express the minimum problem described by tableau (10.6), solve equation j for y_i in terms of the other y's and u_j. Equation j of that system is

$$a_{1j}y_1 + \cdots + a_{ij}y_i + \cdots + a_{mj}y_m - b_j = u_j,$$

so we have

$$-\frac{a_{1j}}{a_{ij}}y_1 - \cdots + \frac{1}{a_{ij}}u_j - \cdots - \frac{a_{mj}}{a_{ij}}y_m + \frac{b_j}{a_{ij}} = y_i,$$

which agrees with the information in column j of the new tableau and shows that a pivot on a_{ij} interchanges the nonbasic variable of row i with the basic variable of column j, with labels along the left and bottom margins of the tableau.

Example   We illustrate the pivot procedure with Example A of Section 10.1. The tableau form of the program of that example is shown below.

Nonbasic variables

		x_1	x_2	-1	$=$	
	y_1	0.6	0.2	60	$-v_1$	
Nonbasic	y_2	0.2	0.2	24	$-v_2$	Basic variables
variables	y_3	0.2	0.6*	52	$-v_3$	
	-1	2	3			
	$=$	u_1	u_2			

Basic variables

A pivot on the entry 0.6 in the (3, 2) position produces a new tableau.

Nonbasic variables

		x_1	v_3	-1	$=$	
	y_1	8/15	$-1/3$	128/3	$-v_1$	
Nonbasic	y_2	2*/15	$-1/3$	20/3	$-v_2$	Basic variables
variables	u_2	1/3	5/3	260/3	$-x_2$	
	-1	1	-5			
	$=$	u_1	y_3			

Basic variables

Another pivot, this time on the entry $2/15$ in the $(2, 1)$ position produces the following new tableau.

Nonbasic variables

		v_2	v_3	-1	$=$
	y_1	-4	1	16	$-v_1$
Nonbasic	u_1	$15/2$	$-5/2$	50	$-x_1$
variables	u_2	$-5/2$	$5/2$	70	$-x_2$
	-1	$-15/2$	$-5/2$		
	$=$	y_2	y_3		

(Basic variables labels: $-v_1$, $-x_1$, $-x_2$ with "Basic variables" to the right)

Basic variables

Observe that each entry in the last column in this tableau is nonnegative and each entry in the bottom row is nonpositive. Under those conditions we can read an optimal solution to the maximum problem directly from the tableau. Let each nonbasic variable be zero; $v_2 = 0 = v_3$. Let the basic variables v_1, x_1, and x_2 have the values appearing in the last column: $v_1 = 16$, $x_1 = 50$, $x_2 = 70$. Then the vector $(X; V)$ is $(50, 70; 16, 0, 0)$. The corresponding value of the objective function BX is $2x_1 + 3x_2 = 310$.

For the dual minimum problem, we also let each nonbasic variable equal zero. Let each basic variable equal the *negative* of the number above it in the bottom row: $y_2 = 15/2$, $y_3 = 5/2$. Then the vector $(U; Y)$ is $(0, 0; 0, 15/2, 5/2)$. The corresponding value of the objective function YC is $60y_1 + 24y_2 + 52y_3 = 310$. Because Y and X are feasible and $BX = YC = 310$, this solution is optimal. As we discovered earlier by geometric methods, in Example A of Section 10.1 the scouts should prepare 50 pounds of standard mix and 70 pounds of fancy mix. The value 16 of the slack variable v_1 is the number of pounds of peanuts that are unused in the optimal solution. We also verify our previous observation that $UX = 0 = YV$.

EXERCISES 10.3

1. Write a complete tableau in the form (10.6) for each of the following linear programming problems and its dual problem. (Use labels $x_1, \ldots, x_n$ across the top of the tableau for the variables of the *maximum* problem of the dual pair.)

(i) The orchard problem of Exercise 10.1-8.

(ii) The diet problem of Exercise 10.2-4.

(iii) Example B of Section 10.1

2. The following tableau defines a dual pair of linear programs.

	x_1	x_2	x_3	-1	$=$
y_1	1	1	1	100	$-v_1$
y_2	3	2	4*	210	$-v_2$
y_3	3	2	0	150	$-v_3$
-1	5	4	6		
$=$	u_1	u_2	u_3		

Pivot on the (2, 3) entry and write the resulting tableau, fully labeled.

3. The tableau below represents a dual pair of linear programs.

	x_1	x_2	x_3	-1	$=$
y_1	1	2	1	2	$-v_1$
y_2	3	6	-2	5	$-v_2$
y_3	3*	-3	1	3	$-v_3$
-1	25	10	15		
$=$	u_1	u_2	u_3		

(i) Verify that a pivot on the (3, 1) entry produces the following tableau.

	v_3	x_2	x_3	-1	$=$
y_1	$-1/3$	3	$2/3$	1	$-v_1$
y_2	-1	9*	-3	2	$-v_2$
u_1	$1/3$	-1	$1/3$	1	$-x_1$
-1	$-25/3$	35	$20/3$		
$=$	y_3	u_2	u_3		

(ii) Compute the tableau that occurs from a pivot on the $(2, 2)$ entry of the tableau in (i).

(iii) Compute the tableau that results from a pivot on the $(1, 3)$ entry of the tableau in (ii).

(iv) The tableau of (iii) shows that $X = (11/9, 13/45, 1/5)$ and $V = (0, 0, 0)$ is a feasible solution of the linear system and that $Y = (11, 2/9, 40/9)$ and $U = (0, 0, 0)$ is a feasible solution of the dual system. Compute the values of the objective functions of the two systems for this choice of X and Y. What do you conclude from comparing these two values?

4. A boat manufacturer makes rowboats and canoes. Those made in spring and summer bring profits of $20 per rowboat and $18 per canoe. Those made in fall and winter sell during the boating season and bring profits of $40 per rowboat and $35 per canoe. Each rowboat requires 5 hours of carpentry and 3 hours of finishing, whereas each canoe requires 6 hours of carpentry and 1 hour of finishing. During each six-month period there are 12,000 available hours for carpentry and 15,000 available hours for finishing. The supply of materials allow for no more than 3,000 rowboats and 3,000 canoes to be built each year. The manufacturer wants to maximize his annual profit.

(i) Write in component form the complete set of inequalities that express the constraints of this linear programming problem.

(ii) Write in component form the objective function.

(iii) Write a fully labeled tableau that presents this maximum linear programming problem and its dual.

10.4 SIMPLEX METHOD

The simplex method is a computational procedure for solving linear programming problems. The method is extremely efficient because it makes use of fundamental relationships between a linear program and its dual program, actually producing optimal vectors (solutions) for *both* programs or else demonstrating that no optimal vector exists for *either* program.

We begin with a dual pair of linear programs, presented in the tableau form (10.6) in which slack and surplus variables have been introduced.

Nonbasic variables	x_1	$\cdots$	x_j	$\cdots$	x_n	-1	$=$
y_1	a_{11}	$\cdots$	a_{1j}	$\cdots$	a_{1n}	c_1	$-v_1$
$\cdot$	$\cdot$		$\cdot$		$\cdot$	$\cdot$	$\cdot$
$\cdot$	$\cdot$		$\cdot$		$\cdot$	$\cdot$	$\cdot$
$\cdot$	$\cdot$		$\cdot$		$\cdot$	$\cdot$	$\cdot$
y_i	a_{i1}	$\cdots$	a_{ij}	$\cdots$	a_{in}	c_i	$-v_i$
$\cdot$	$\cdot$		$\cdot$		$\cdot$	$\cdot$	$\cdot$
$\cdot$	$\cdot$		$\cdot$		$\cdot$	$\cdot$	$\cdot$
$\cdot$	$\cdot$		$\cdot$		$\cdot$	$\cdot$	$\cdot$
y_m	a_{m1}	$\cdots$	a_{mj}	$\cdots$	a_{mn}	c_m	$-v_m$
-1	b_1	$\cdots$	b_j	$\cdots$	b_n		
$=$	u_1	$\cdots$	u_j	$\cdots$	u_n		Basic variables

The information of the maximum program is read from the rows of the tableau: determine a nonnegative vector $(X; V)$ with $n + m$ components such that

$$(1)\ a_{i1}x_1 + \cdots + a_{in}x_n - c_i = -v_i \text{ for } i = 1, \ldots, m,$$

$$(2)\ b_1x_1 + \cdots + b_nx_n \text{ is maximized.}$$

Similarly the information of the dual minimum problem is read from the columns: determine a nonnegative vector $(U; Y)$ such that

$$(1')\ a_{1j}y_1 + \cdots + a_{mj}y_m - b_j = u_j \text{ for } j = 1, 2, \ldots, n,$$

$$(2')\ c_1 y_1 + \cdots + c_m y_m \text{ is minimized.}$$

To explain the simplex method we shall first consider a special case that frequently occurs in applications in which each number c_i in the last column is *positive*. This special condition allows us to start with a feasible vector $(X; V)$ for the maximal program, obtained by letting

$$x_j = 0 \quad \text{for } j = 1, \ldots, n,$$

$$v_i = c_i \quad \text{for } i = 1, \ldots, m.$$

The corresponding value of the objective function BX is zero. Similarly in the minimum program we arbitrarily assign the value zero to each nonbasic variable y_i, and use (1') to evaluate $u_j = -b_j$, $j = 1, \ldots, n$. The resulting vector $(U; Y)$ is orthogonal to $(X; V)$, as desired for an optimal solution, and the value of the objective function YC coincides with the value of BX, as desired for an optimal solution. However, the vector $(U; Y)$ is *not feasible* for the minimum problem unless each $b_j \leq 0$. Thus if any entry in the bottom row of the tableau is positive, the value BX is not maximal.

The simplex algorithm can now be applied. In essence it is a set of rules for selecting a pivot entry a_{ij} of the tableau. A pivot on that entry exchanges a nonbasic variable x_j and a basic variable v_i (and similarly exchanges y_i and u_j). This produces a new tableau, and the rules of selection are designed so that each entry of the last column of the new tableau remains nonnegative. Hence by letting each currently nonbasic variable of the maximum problem be zero and letting each currently basic variable of the maximum problem have the value in the last column adjacent to its label, we obtain a new feasible vector $(X; V)$ of the maximum problem. Furthermore, the new value of the objective function BX is as large or larger than it was at the previous step. For the minimum problem we also let each currently nonbasic variable be zero and let each currently basic variable have the negative of the value in the last row of the new tableau. The rules also guarantee that the new value YC of the objective function of the minimum problem again equals the new value BX of the objective function of the maximum problem. Hence, if the new vector $(U; Y)$ is feasible, the dual pair of problems is solved. Otherwise, we again apply the simplex rules to choose another pivot and repeat the process. After each pivot step the new tableau has a nonnegative entry in each position of the last column; also the new tableau must satisfy exactly one of the following three conditions:

(1) each entry of the bottom row is nonpositive,

(2) some entry of the bottom row is positive and there is at least one positive entry in the column above it,

(3) some entry of the bottom row is positive, and every entry in the column above it is negative.

If (1) occurs, an optimal solution can be read from that tableau. If (2) occurs, the simplex algorithm can be continued. If (3) occurs, the tableau shows that the value of BX is unbounded on the set of all feasible vectors for the maximum problem, in which case no optimal solution exists.

Now we are ready to state the rules of the *simplex method* for selecting a pivot element, *assuming that each entry of the last column is nonnegative* and that some entry of the bottom row is positive.

Step 1   Locate, say in column $j \le n$, any *positive entry* in the bottom row. (Usually the largest positive entry is chosen.) The pivot entry will be in column j.

Step 2   Locate all the *positive* entries a_{kj} in column j, and compute the *test ratio* c_k/a_{kj} for each of them, where c_k is the entry in the last column of row k. (Ignore all nonpositive a_{kj}.)Let i be the value of k for which the test ratio is *smallest*.

Step 3   Pivot on a_{ij}, obtaining a new tableau.

Step 4   Interchange both pairs of variable labels of the pivot row and column.

Step 5   Examine the new tableau to see which of the three conditions (1), (2), or (3) applies. If (2) holds, repeat Steps 1-5 with the new tableau. Otherwise stop, obtaining a solution (or proof of no solution) from the new tableau.

Example   Maximize $24x_1 + 25x_2$ subject to the constraints

$$x_1 + 5x_2 \le 1,$$
$$4x_1 + x_2 \le 3,$$
$$x_1 \ge 0,$$
$$x_2 \ge 0.$$

The tableau for this problem follows.

Nonbasic variables	x_1	x_2	-1	$=$
y_1	1	5*	1	$-v_1$
y_2	4	1	3	$-v_2$
-1	24	25	0	
$=$	u_1	u_2		Basic variables

Observe that 0 (the *negative* of the value of the objective function $24x_1 + 25x_2$ when $x_1 = 0 = x_2$) has been entered in the lower right corner. Because $25 > 24$, we choose the second column for the first pivot. The corresponding test ratios are $1/5$ and $3/1$, and we choose 5 as pivot, corresponding to the smaller test ratio. The pivot operation then produces a new tableau with interchanged labels for the pivot variables.

	x_1	v_1	-1	$=$
u_2	$1/5$	$1/5$	$1/5$	$-x_2$
y_2	$19/5*$	$-1/5$	$14/5$	$-v_2$
-1	19	-5	-5	
$=$	u_1	y_1		

You should verify these computations: the pivot is replaced by its reciprocal, every other entry in the pivot row is divided by the pivot to obtain the new entry for that position, every other entry in the pivot column is divided by the negative of the pivot to obtain the new entry for that position, and each entry q, not in the row or column of the pivot p, is replaced according to the "rectangle scheme"

by $q - tw/p$. Notice that the value -5 in the lower right corner, obtained by the rectangle scheme in pivoting on 5, is also the *negative* of the value of the objective function $24x_1 + 25x_2$ when the nonbasic variables x_1 and v_1 equal zero and the basic variables are assigned the values in the last column, $x_2 = 1/5$ and $v_2 = 14/5$.

In the new tableau we see that one positive entry remains in the bottom row. Hence we pivot again, this time on an entry in the first column. The test ratios are 1 for the first row and $14/19$ for the second row. So we pivot on $19/5$ to obtain the following tableau.

	v_2	v_1	-1	$=$
u_2			$1/19$	$-\dot{x}_2$
u_1	·		$14/19$	$-x_1$
-1	-5	-4	-19	
$=$	y_2	y_1		

As indicated here, it is good practice when pivoting to first calculate the entries of the bottom row and last column because these are the numbers needed for a solution. Because the last row is now nonpositive and the last column is nonnegative (ignoring the lower right-hand corner), we now have an optimal solution, and we don't need to compute the other tableau entries.

If we let

$$(X; V) = (14/19, 1/19; 0, 0),$$

$$(U; Y) = (0, 0; 4, 5),$$

we see that the objective function for maximum problem has the value

$$24\left(\frac{14}{19}\right) + 25\left(\frac{1}{19}\right) = \frac{336 + 25}{19} = 19,$$

as indicated by the entry in the lower right-hand corner of the last tableau. The dual minimum problem has the objective function

$$y_1 + 3y_2,$$

which also assumes the value 19 when $y_1 = 4$ and $y_2 = 5$. Because $X = (14/19, 1/19)$ and $Y = (4, 5)$ are *feasible* vectors for which the values of the objective functions coincide, each represents an optimal solution to the corresponding problem.

Now let us reexamine the first two steps of the simplex algorithm to see why the method works. We begin with each $x_i = 0$ and $BX = 0$. BX can be increased by increasing any x_j for which $b_j > 0$. Too large an increase in x_j could violate one or more constraints of the maximum problem. If the coefficient of x_j in a given constraint is negative, an increase in x_j will not violate that constraint, so we consider only the positive a_{kj} in Step 2. Constraint k has a slack of c_k when all nonbasic variables are zero, so x_j can be increased by any amount up to c_k/a_{kj} without violating that constraint. This explains the significance of the test ratios. The choice of pivot then guarantees that the prospective change in x_j will not violate any constraint. (The new entries in the last column will be nonnegative, and can be used to find a new feasible vector for the maximal problem.) In Step 1 when $b_j > 0$ for several values of j, it is desirable (but not necessary) to choose a pivot that produces the largest increase in BX. To do so, for each j such that $b_j > 0$ compute the test ratio r_j and then choose j so that $b_j r_j$ is maximized, because $b_j r_j$ is the amount by which BX is increased by a pivot on the specified entry of column j. The remarkable computational efficiency of the simplex algorithm results from the choice of pivot to indicate a direction in which the objective function increases from corner to corner along the edges of the convex polyhedron defined by the feasible vectors of the maximal problem.

Finally we consider how to apply the simplex algorithm if some entry of the last column of the initial tableau is negative. The simplex method then specifies a pivot designed to produce an equivalent tableau in which the number of nonnegative entries in the last column is increased. When all such elements are nonnegative, the algorithm returns to the steps described previously. Consider, then, a tableau in which *some element of the last column is negative*.

Step A   Let $c_r < 0$ be such that $c_t \geq 0$ for all $t > r$. If $a_{rs} \geq 0$ for all s, then no assignment of nonnegative values to the variables x_j and u_i can satisfy the constraint represented by row r. In that case no feasible vector exists, so there is no optimal solution. If $a_{rj} < 0$ for some j, choose any such value of j as the pivot column.

Step B   To choose the pivot row, examine the test ratios c_r/a_{rj} and c_t/a_{tj} for all $t > r$ for which $a_{tj} > 0$. Each such test ratio is positive, and we let $i \geq r$ denote the row that produces the smallest test ratio. Then a pivot on a_{ij} will produce a tableau in which the last column has nonnegative entries in at least row r and all subsequent rows.

Step C   Repeat this procedure, if necessary, to obtain a tableau in which the last column is nonnegative. The five-step simplex algorithm as previously described can then be applied to complete the solution.

Example   A student organization has \$120 to spend on Christmas toys for underprivileged children. They can buy sleds at \$3 each, dolls at \$2 each, and wagons at \$5 each. The number of wagons and dolls combined must not exceed 18, and the number of dolls and sleds combined must be at least 20. Determine the largest number of toys that can be purchased.

Let s, d, and w denote the number of toys of each respective type that are purchased. Then we have $d \geq 0$, $s \geq 0$, $w \geq 0$,

$$d + \quad w \leq 18,$$

$$s + d \qquad \geq 20,$$

$$3s + 2d + 5w \leq 120,$$

and we want to maximize $s + d + w$. First multiply the second displayed inequality by -1. The resulting tableau is shown below.

	s	d	w	-1	$=$
y_1	0	1	1	18	$-v_1$
y_2	-1^*	-1	0	-20	$-v_2$
y_3	3	2	5	120	$-v_3$
-1	1	1	1	0	
$=$	u_1	u_2	u_3		

To produce positive entries in the last column we can pivot on -1 in either the first or second column. Choosing the first column we obtain the test ratios $-20/-1 < 120/3$, so we pivot on the indicated entry.

	v_2	d	w	-1	$=$
y_1	0	1	1	18	$-v_1$
u_1	-1	1	0	20	$-s$
y_3	3^*	-1	5	60	$-v_3$
-1	1	0	1	-20	
$=$	y_2	u_2	u_3		

For the next pivot we could choose the first or third column because the two positive entries in the bottom row are equal. Arbitrarily we choose the first column; then 3 is the only possible choice of pivot in that column.

	v_3	d	w	-1	$=$
y_1	0	1^*	1	18	$-v_1$
u_1	1/3	2/3	5/3	40	$-s$
y_2	1/3	$-1/3$	5/3	20	$-v_2$
-1	$-1/3$	1/3	$-2/3$	-40	
$=$	y_3	u_2	u_3		

The test ratios in the second column are $18 < 60$, so we pivot on 1.

	v_3	v_1	w	-1	$=$
u_2				18	$-d$
u_1				28	$-s$
y_2				26	$-v_2$
-1	$-1/3$	$-1/3$	-1	-46	
$=$	y_3	y_1	u_3		

Because each entry in the last column is positive (except the lower corner entry) and each entry in the last row is negative, we can write an optimal solution,

$$(s, d, w; v_1, v_2, v_3) = (28, 18, 0; 0, 26, 0)$$

with a maximum value of 46 for the total number of toys. No wagons are purchased; all the money is spent, because $v_3 = 0$. There are exactly 18 dolls and wagons, because $v_1 = 0$. There are 46 sleds and dolls purchased, 26 more than the constraint, because $v_2 = 26$. For the dual minimum problem the objective function is $18y_1 - 20y_2 + 120y_3$ from the first tableau. An optimal solution is

$$(u_1, u_2, u_3; y_1, y_2, y_3) = (0, 0, 1; 1/3, 0, 1/3),$$

and the minimum value of YC is 46.

We have not discussed some variations and complications that can arise in the solution of linear programs by the simplex method, nor have we formally proved that the simplex method either will produce an optimal solution or will show that none exists. After becoming familiar with these techniques by solving the following exercises, you might want to consult a more extensive treatment of linear programming in which such matters are examined in detail.

EXERCISES 10.4

1. Use the simplex method to solve each of the following dual pairs of linear programs. In each case write an optimal vector $(X; V)$ for the maximum program, an optimal vector $(U; Y)$ for the minimum program, and the common extreme value of the two objective functions.

(i) The orchard problem (Exercises 10.1-8 and 10.3-1(i)).

(ii) The diet problem (Exercises 10.2-4 and 10.3-1(ii)).

(iii) The transportation problem (Example B in Section 10.1 and Exercise 10.3-1(iii)).

(iv) The boat manufacturer problem (Exercise 10.3-4).

(v) Exercise 10.3-2.

2. Machines 1, 2, and 3 each turn out products A, B, C, and D. The profit from each item of products A, B, C, D are \$9, \$7, \$6, \$4, respectively. The numbers of hours for Machine 1 to turn out one item of each product are 10, 5, 2, 1, respectively. Corresponding hours for Machine 2 are 6, 6, 2, 2; for Machine 3 the times are 9/2, 18, 3/2, 6. The numbers of hours available on each machine are, in order, 50, 36, and 81. Develop a plan that specifies how many items of each product should be produced by each machine to maximize profits.

3. (i) Use the simplex method to maximize $3x_1 + 2x_2$ subject to the following constraints:

$$x_1 + x_2 \leq 2,$$

$$x_1 + x_2 \geq 1,$$

$$x_1 + 2x_2 \geq 2,$$

$$6x_1 + 5x_2 \leq 10,$$

$$x_1 \qquad \leq 1,$$

$$x_1 \qquad \geq 0,$$

$$x_2 \geq 0.$$

(ii) Solve the same problem graphically in $\mathscr{E}^2$.

4. Use the simplex method to solve each of the dual pairs of linear programming problems given by the following tableaus. Write your solutions as described in Exercise 1.

(i)

	x_1	x_2	x_3	-1	$=$
y_1	1	1	3	25	$-v_1$
y_2	1	-1	0	-10	$-v_2$
y_3	0	1	2	15	$-v_3$
-1	3	2	1		
$=$	u_1	u_2	u_3		

(ii)

	x_1	x_2	-1	$=$
y_1	1	2	16	$-v_1$
y_2	3	1	23	$-v_2$
y_3	-1	0	-7	$-v_3$
-1	20	30		
$=$	u_1	u_2		

5. Six types of bonds have the following yields.

Bond	A_1	A_2	B_1	B_2	C_1	C_2
Yield	0.06	0.05	0.07	0.08	0.10	0.09

The trustees of a fund want to invest in these bonds but are restricted in their actions by several conditions:

(a) At least 40 percent of the total investment must be in bonds of type A_1 and A_2;

(b) not more than 35 percent of the total investment may be in bonds of type B_1 or B_2,

(c) not more than 35 percent of the total investment may be in bonds of type C_1 or C_2.

Determine an investment plan subject to these constraints that will produce maximum yield on the capital invested.

CHAPTER 11
APPLICATION:
LINEAR
DIFFERENTAL
EQUATIONS

11.1 SEQUENCES AND SERIES OF MATRICES

Previously, we observed that each scalar polynomial

$$p(x) = \sum_{k=0}^{m} a_k x^k$$

determines a corresponding matrix polynomial

$$p(X) = \sum_{k=0}^{m} a_k X^k,$$

where X is an n-by-n matrix. The matrix $p(X)$ is defined for any positive integer n, and for $n = 1$ the matrix $p(X)$ is simply the scalar $p(x)$. Thus matrix polynomials with scalar coefficients are a generalized form of scalar polynomials. Furthermore this is a significant generalization in the sense that the algebra of matrix polynomials with scalar coefficients is different from that of scalar polynomials. For example, any factorization of $p(x)$ induces a like factorization of $p(X)$, but the scalar equation

$$(x - a)(x - b) = 0$$

implies that $x = a$ or $x = b$, whereas the corresponding matrix equation

$$(X - aI)(X - bI) = Z$$

does not imply that $X = aI$ or $X = bI$.

A matrix polynomial with scalar coefficients is a particular type of function in which the variable is a square matrix. In this chapter we shall see that a wider class of matrix functions can be defined, and then we shall apply that information to solve systems of linear differential equations. To do so, we first summarize some information about scalar polynomials and, more generally, about power series in a scalar variable.

Let f be a scalar-valued function that has the property that its first m derivatives all exist in an open interval of the x-axis around zero. Then the *Taylor polynomial* T_m of order m at zero is defined by

$$T_m(x) = \frac{f(0)}{0!} + \frac{f'(0)}{1!} x + \frac{f''(0)}{2!} x^2 + \cdots + \frac{f^{(m)}(0)}{m!} x^m,$$

where $0! = 1$ and $k! = (k - 1)!k$, and where $f^{(k)}(0)$ denotes the value at zero of the kth derivative of f. It is easily verified that

$$T_m(0) = f(0), \ T'_m(0) = f'(0), \ \ldots, \ T_m^{(m)}(0) = f^{(m)}(0).$$

Thus the values of T_m and its first m derivatives coincide at zero with the corresponding values of f and its first m derivatives. Near zero, therefore, the polynomial T_m is a good approximation to the function f. At each $x \neq 0$, $T_m(x)$ and $f(x)$ may or may not coincide; the *error* or *remainder* function R_m is defined by

$$R_m(x) = f(x) - T_m(x).$$

Various expressions for $R_m(x)$ are known, although they do not concern us here.

Now suppose that f has a derivative of every order at zero. Then $T_m(x)$ is defined for each $m = 0, 1, 2, \ldots$. In that case the *Taylor series* of f at zero is the expression

$$T(x) = \sum_{k=0}^{\infty} \frac{f^{(k)}(0)x^k}{k!}.$$

To interpret this expression as a function it is necessary to be familiar with the concept of convergence of a series of functions, which is described in many calculus textbooks. We now list some needed information about power series.

(1) Any power series $\sum_{k=0}^{\infty} a_k x^k$ converges for $x = 0$.

(2) If a power series converges at $x = r$, it also converges for every value x such that $|x| < |r|$.

(3) The *largest* real number r such that a power series converges for all x such that $|x| < r$ is called the *radius of convergence* of that power series. If no such largest number exists, the series is said to have an infinite radius of convergence, in which case the series converges for all x.

(4) If r is the radius of convergence of a power series and if $|x_0| = r$, the series might or might not converge at x_0.

The preceding statements probably are familiar to you for the case in which x and each a_i are real. The same statements are valid when x is a complex variable and each a_i is a complex number. In that case they assert that any complex power series has a radius r of convergence (perhaps infinite), and that the series converges at every point of the complex plane interior to the circle C of radius r with center at the origin. The series fails to converge at every point exterior to C, and it might or might not converge at any given point on C. When x is a real variable and each a_i is real, the intersection of C with the real axis determines the interval of convergence of the real power series. The series might converge at both, one, or neither of the end points of that interval, but it converges for every point interior to that interval and fails to converge at each point exterior to that interval.

A power series defines a function g for each x for which the series converges; furthermore if the power series is the Taylor series at zero of a function f, then $f(x) = g(x)$ for each x interior to the circle of convergence. For example, we can write

$$e^x = \sum_{k=0}^{\infty} \frac{x^k}{k!} \qquad \text{for all } x \in \mathbb{C},$$

$$\sin x = \sum_{k=0}^{\infty} \frac{(-1)^k x^{2k+1}}{(2k+1)!} \qquad \text{for all } x \in \mathbb{C},$$

$$\cos x = \sum_{k=0}^{\infty} \frac{(-1)^k x^{2k}}{(2k)!} \qquad \text{for all } x \in \mathbb{C},$$

$$\log(1 + x) = \sum_{k=1}^{\infty} \frac{(-1)^{k+1} x^k}{k} \qquad \text{for all } x \in \mathbb{C} \text{ such that } |x| < 1.$$

Because scalar polynomials extend readily to matrix polynomials with scalar coefficients, it is reasonable to ask whether scalar power series can be extended to power series of a square matrix, the coefficients of the series being scalars. To answer this query we must first investigate sequences and series of matrices.

Definition 11.1   For each positive integer k let $A^{(k)}$ denote an m-by-n matrix whose (i, j) entry is denoted $a_{ij}^{(k)}$. The sequence given by $\{A^{(k)}\} = \{A^{(0)}, A^{(1)}, \dots, A^{(k)}, \dots\}$ is said to *converge* to the matrix $A = (a_{ij})$ if and only if for each $i = 1, \dots, m$ and each $j = 1, \dots, n$ the number sequence $\{a_{ij}^{(k)}\} = \{a_{ij}^{(0)}, \dots, a_{ij}^{(k)}, \dots\}$ converges to the number a_{ij}.

According to this definition the convergence of a sequence of m-by-n matrices requires the convergence of the sequence of numbers in position (i, j) of the given matrices for each i and each j. For example, given

$$A^{(k)} = \begin{pmatrix} \dfrac{1}{k+2} & 3 \\ \dfrac{1-2k}{k+1} & \dfrac{k}{k+1} \end{pmatrix}, \quad B^{(k)} = \begin{pmatrix} 1 & 0 \\ 0 & k \end{pmatrix},$$

the sequence $\{A^{(k)}\}$ converges to

$$A = \begin{pmatrix} 0 & 3 \\ -2 & 1 \end{pmatrix},$$

but the sequence $\{B^{(k)}\}$ fails to converge because the number sequence $\{k\}$ fails to converge.

Theorem 11.1   Let the sequence $\{A^{(k)}\}$ of m-by-n matrices converge to A, and let P and Q be any p-by-m and n-by-q matrices, respectively. Then the matrix sequence $\{PA^{(k)}Q\}$ converges to PAQ.
Proof   Let $b_{ij}^{(k)}$ denote the (i, j) entry of $PA^{(k)}Q$. Then

$$b_{ij}^{(k)} = \sum_{r=1}^{m} \sum_{s=1}^{n} p_{ir} a_{rs}^{(k)} q_{sj}.$$

For each $r = 1, \dots, m$ and each $s = 1, \dots, n$ the sequence $\{a_{rs}^{(k)}\}$ converges to a_{rs}. Hence $\{p_{ir} a_{rs}^{(k)} q_{sj}\}$ converges to $p_{ir} a_{rs} q_{sq}$. Because the sum of convergent number sequences is convergent, $\{b_{ij}^{(k)}\}$ converges to

$$\sum_{r=1}^{m} \sum_{s=1}^{n} p_{ir} a_{rs} q_{sj},$$

which is the (i, j) entry of PAQ.

Theorem 11.1 is significant because it guarantees that convergence is preserved under the various equivalence relations that we have considered. In particular, when $m = n = p = q$ and $P = Q^{-1}$, we see that convergence of a sequence of n-by-n matrices is not affected by a change of coordinates in $\mathscr{V}_n$. Thus it is possible to define convergence of a sequence $\{\mathbf{T}^{(k)}\}$ of linear transformations on $\mathscr{V}_n$ in terms of the convergence of the sequence of matrices that represent those transformations in any coordinate system.

Recall that an infinite series $\sum_{k=0}^{\infty} a_k$ of numbers is said to converge to a number s if and only if the sequence $\{s_n\}$ of partial sums converges to s, where

$$s_n = a_0 + a_1 + \cdots + a_n.$$

We define convergence of an infinite series $\sum_{k=0}^{\infty} A^{(k)}$ of m-by-n matrices in terms of the convergence of the mn series $\sum_{k=0}^{\infty} a_{ij}^{(k)}$ of the numbers that occupy the (i, j) positions for each i and each j.

Definition 11.2 An infinite series $\sum_{k=0}^{\infty} A^{(k)} = A^{(0)} + A^{(1)} + \cdots$ of m-by-n matrices is said to *converge* to the matrix $S = (s_{ij})$ if and only if for each $i = 1, \ldots, m$ and each $j = 1, \ldots, n$ the number series $\sum_{k=0}^{\infty} a_{ij}^{(k)}$ converges to s_{ij}.

Theorem 11.2 Let the infinite series $\sum_{k=0}^{\infty} A^{(k)}$ of m-by-n matrices converge to the m-by-n matrix A. Let P be any p-by-m matrix, and let Q be any n-by-q matrix. Then the infinite series $\sum_{k=0}^{\infty} PA^{(k)}Q$ of p-by-q matrices converges to PAQ.
Proof Exercise.

In the next section we shall restrict our attention to power series $\sum_{k=0}^{\infty} a_k X^k$, where X is an n-by-n matrix, using the results obtained to give meaning to functions of matrices, such as $\cos X$ and e^x.

EXERCISES 11.1

1. For each matrix $A^{(k)}$ defined below, if the sequence $\{A^{(k)}\}$ converges, find the matrix to which the sequence converges. Otherwise state why the sequence does not converge.

(i) $A^{(k)} = \begin{pmatrix} \dfrac{(-1)^k}{k} & \dfrac{2^k}{3^k + 1} \\ \dfrac{k^2 - 1}{k^2 + 1} & \dfrac{1 - 3k}{k} \end{pmatrix}$, $k = 1, 2, \ldots.$

(ii) $A^{(k)} = \begin{pmatrix} k^{-2} & \sin \pi k \\ \cos \pi k & k^{-1} \end{pmatrix}, \qquad k = 1, 2, \ldots .$

2. Prove that the infinite series $\sum_{k=0}^{\infty} A^{(k)}$ of m-by-n matrices converges if and only if the infinite sequence $\{S^{(p)}\}$ of matrices converges, where $S^{(p)}$ denotes the partial sum,

$$S^{(p)} = A^{(0)} + A^{(1)} + \cdots + A^{(p)}.$$

3. Verify that the Taylor series at $x = 0$ for each of the following functions is correctly stated in the text:

(i) e^x,

(ii) $\log (1 + x)$.

4. Determine the Taylor series at $x = 0$ for the function $f(x) = (1 - x)^{-1}$. Determine also the radius of convergence of that series.

5. Determine the Taylor series at zero of the polynomial

$$p(x) = a_0 + a_1 x + \cdots + a_n x^n.$$

6. Prove Theorem 11.2.

11.2 MATRIX POWER SERIES

Let X denote an n-by-n matrix. For any nonnegative integer k and any scalar a_k, the expression $a_k X^k$ denotes an n-by-n matrix, and so does any finite sum of such terms. A denumerably infinite sum of such terms,

$$\sum_{k=0}^{\infty} a_k X^k,$$

is called a *power series* in the matrix X. A power series in X is a formal expression, but if X is a matrix for which the power series converges, that series determines the n-by-n matrix to which the matrix converges. Hence any power series can be used to define a function from a subset of $\mathcal{M}_{n \times n}$ into $\mathcal{M}_{n \times n}$ for any n. For $n = 1$ such a function is scalar-valued.

Definition 11.3 Let $\sum_{k=0}^{\infty} a_k x^k$ be a scalar power series, and let f denote the scalar-valued function defined by

$$f(x) = \sum_{k=0}^{\infty} a_k x^k$$

for all x for which the series converges. The *matrix-valued function f* is defined by

$$f(X) = \sum_{k=0}^{\infty} a_k X^k$$

for each square matrix X for which the series converges.

This definition is a natural extension of the correspondence between scalar polynomials and matrix polynomials with scalar coefficients, and it permits us to consider matrix functions other than polynomials. For example, we can now write e^X, $(I - X)^{-1}$, and $\cos X$ for at least some square matrices, meaning that

$$e^X = \sum_{k=0}^{\infty} \frac{X^k}{k!},$$

$$(I - X)^{-1} = \sum_{k=0}^{\infty} X^k,$$

$$\cos X = \sum_{k=0}^{\infty} \frac{(-1)^k X^{2k}}{(2k)!}.$$

Two questions arise immediately:

For what square matrices X does a power series converge?

If a power series in X converges, to what matrix does it converge? We recognize of course that the minimal polynomial of X makes it unnecessary to compute the higher powers of X. Consider the power series

$$f(X) = \sum_{k=0}^{\infty} a_k X^k$$

where X is a matrix whose minimal polynomial is

$$m(x) = b_0 + b_1 x + \cdots + x^m,$$

and where $m \le n$ by the Hamilton-Cayley Theorem. Then $m(X) = Z$, so

$$X^m = -(b_0 + b_1 X + \cdots + b_{m-1} X^{m-1}).$$

Then $f(X)$ can be written in the form

$$f(X) = \sum_{k=0}^{m-1} s_k X^k,$$

where each s_k is an infinite series of scalars. The question of convergence of $f(X)$ is then reduced to the question of convergence of the m scalar series s_k.

However, Theorem 11.2 provides a better attack on the question of convergence. Given $f(X)$ let $Y = P^{-1}XP$ for any nonsingular n-by-n matrix P. Then $f(X)$ converges if and only if $f(Y)$ converges, and if either converges, then $f(Y) = P^{-1}f(X)P$. We choose P such that Y is in Jordan form, having diagonal blocks of a simple form. We determine whether $f(Y)$ converges, and if so, the matrix to which it converges. Then we use P to extract corresponding answers about $f(X)$. The next few theorems establish the details of this method.

Theorem 11.3   Let Y be an n-by-n matrix in diagonal block form

$$Y = \begin{pmatrix} Y_1 & Z \\ Z & Y_2 \end{pmatrix}$$

where Y_1 and Y_2 are square matrices. Let f be defined by a power series with scalar coefficients. Then
   (a) $f(Y)$ converges if and only if $f(Y_1)$ and $f(Y_2)$ both converge,
   (b) if $f(Y)$ converges, then

$$f(Y) = \begin{pmatrix} f(Y_1) & Z \\ Z & f(Y_2) \end{pmatrix}.$$

Proof   Exercise. Observe that this theorem can be generalized to the case in which there are p diagonal blocks, for any $p \leq n$.

If Y is a Jordan matrix, $Y = \mathrm{diag}(Y_1, \ldots, Y_t)$, the previous theorem shows that we need only be concerned with the convergence of $f(Y_i)$, where

$$Y_i = (\lambda_i I + N_i)$$

and where N_i is nilpotent. In case $\lambda_i = 0$, the problem is quickly settled by the following theorem.

Theorem 11.4   If Y is nilpotent, $f(Y)$ converges.
Proof   Exercise.

In general, however, the question of convergence of $f(X)$ depends on the location of the characteristic values of X relative to the circle of convergence of the series that defines f in the complex plane.

Theorem 11.5   Let $f(x) = \sum_{k=0}^{\infty} a_k x^k$ converge for all complex x such that $|x| < r$. If $|\lambda_i| < r$ for every characteristic value λ_i of an n-by-n matrix X, then $f(X)$ converges.

Proof   Let $P^{-1}XP = \operatorname{diag}(J_1, J_2, \ldots, J_t)$ be a Jordan form of X. By Theorem 11.3 $f(X)$ converges if and only if $f(J_i)$ converges for each i. Furthermore it suffices to show that $f(J)$ converges for each sub-block J. Hence let $J = (\lambda I + N)$ where J denotes a p-by-p sub-block in which each diagonal entry is λ and each subdiagonal entry is one. We consider the partial sum of the series $f(J)$,

$$S_m(J) = \sum_{k=0}^{m} a_k(\lambda I + N)^k.$$

Because N is nilpotent of index p, there are no more than p powers of N in the expansion of $(\lambda I + N)^k$ even for large k. Let $m > p$. Using the binomial theorem we can write

$$
\begin{aligned}
S_m(J) = \quad & a_0 I \\
+ \quad & a_1 \lambda I + & a_1 N \\
+ \quad & a_2 \lambda^2 I + & 2a_2 \lambda N + a_2 N^2 \\
& \vdots & \vdots \\
+ \; & a_{m-1}\lambda^{m-1}I + C_1^{m-1}a_{m-1}\lambda^{m-2}N + \cdots + C_{p-1}^{m-1}a_{m-1}\lambda^{m-p}N^{p-1} \\
+ \; & a_m \lambda^m I + & C_1^m a_m \lambda^{m-1}N + \cdots + C_{p-1}^m a_m \lambda^{m-p+1}N^{p-1},
\end{aligned}
$$

where $C_s^m = \dfrac{m!}{s!(m-s)!}$ is the binomial coefficient.

Summing on like powers of N, we have

$$S_m(J) = \sum_{r=0}^{p-1} \left(\sum_{k=r}^{m} a_k C_r^k \lambda^{k-r} \right) N^r.$$

But

$$a_k C_r^k \lambda^{k-r} = \frac{a_k}{r!} \frac{k!}{(k-r)!} \lambda^{k-r} = \frac{a_k}{r!} \left. \frac{d^r}{dx^r}(x^k) \right]_{x=\lambda},$$

so

$$\sum_{k=r}^{m} a_k C_r^k \lambda^{k-r} = \frac{1}{r!} \sum_{k=r}^{m} a_k \left. \frac{d^r}{dx^r}(x^k) \right]_{x=\lambda} = \frac{1}{r!} S_m^{(r)}(\lambda),$$

where $S_m^{(r)}(\lambda)$ is the rth derivative of $S_m(x)$, evaluated at $x = \lambda$. Therefore $S_m(J)$ has the lower triangular form

$$
\begin{vmatrix}
S_m(\lambda) & 0 & 0 & \cdots & 0 & 0 \\
S_m^{(1)}(\lambda) & S_m(\lambda) & 0 & \cdots & 0 & 0 \\
\dfrac{S_m^{(2)}(\lambda)}{2!} & S_m^{(1)}(\lambda) & S_m(\lambda) & \cdots & 0 & 0 \\
\cdot & \cdot & \cdot & & \cdot & \cdot \\
\cdot & \cdot & \cdot & & \cdot & \cdot \\
\cdot & \cdot & \cdot & & \cdot & \cdot \\
\dfrac{S_m^{(p-1)}(\lambda)}{(p-1)!} & \dfrac{S_m^{(p-2)}(\lambda)}{(p-2)!} & \dfrac{S_m^{(p-3)}(\lambda)}{(p-3)!} & \cdots & S_m^{(1)}(\lambda) & S_m(\lambda)
\end{vmatrix}
$$

in which the (i, j) entry is zero if $i < j$, $S_m(\lambda)$ if $i = j$, and $S_m^{(q)}(\lambda)/q!$ if $i = j + q$. By hypothesis, $\sum_{k=0}^{\infty} a_k \lambda^k$ converges because $|\lambda| < r$. Thus the number sequence of partial sums, $S_m(\lambda)$ converges. But also if a power series $\sum_{k=0}^{\infty} a_k x^k$ converges for $|x| < r$, then the series obtained by differentiating the given series term by term also will converge for $|x| < r$. Hence the matrix sequence $S_m(J)$ of partial sums converges, which implies that $f(P^{-1}XP)$ converges and so does $f(X)$.

As a consequence of the previous theorem we know that such functions as e^X, $\sin X$, and $\cos X$ are defined for every square matrix X, whereas $\log(I + X)$ is defined for any square matrix having all its characteristic values in the interior of the unit circle. Caution must be excercised, however, in working with matrix functions because of the differences between the algebra of matrices and the algebra of scalars. For example even though e^X and e^Y are defined for all n-by-n matrices X and Y,

$$e^X e^Y \neq e^{X+Y}$$

unless X and Y happen to commute.

Theorem 11.6 If $f(X)$ converges and if λ is a characteristic value of X, then $f(\lambda)$ is a characteristic value of $f(X)$.

Proof Let $J = P^{-1}XP$ be a Jordan form of X. Then X and J have the same characteristic values, and $f(J) = P^{-1}f(X)P$. But from the proof of Theorem 11.5, $f(J)$ is a lower triangular matrix with $f(\lambda_i)$ as the diagonal entries, where $\lambda_1, \ldots, \lambda_t$ are the distinct characteristic values of X.

Theorem 11.7   For every n-by-n matrix X,

(a) $\det e^X = e^{\operatorname{tr} X}$, where the tr X denotes the trace of X,

(b) e^X is nonsingular,

(c) $(e^X)^{-1} = e^{-X}$.

Proof   Exercise.

Let us summarize these results by describing a method for calculating the matrix $f(A)$, where f is defined by a power series and A is an n-by-n matrix such that every characteristic value of A lies within the circle of convergence of the power series. To find $f(A)$ we first reduce A to Jordan form as described in Section 7.7:

$$P^{-1}AP = J = \operatorname{diag}(J_1, \ldots, J_s),$$

where each J_i is one of the *sub*-blocks of J with the corresponding characteristic value in each diagonal position and 1 in each subdiagonal position. By Theorems 11.3 and 11.2

$$f(J) = \operatorname{diag}(f(J_1), \ldots, f(J_s))$$
$$= f(P^{-1}AP) = P^{-1}f(A)P,$$

so $f(A) = P \operatorname{diag}(f(J_1), \ldots, f(J_s))P^{-1}$. Hence the problem will be solved if we determine $f(J_i)$ for each sub-block J_i of J. But from the proof of Theorem 11.5 it follows that $f(J_i)$ is lower triangular with

$f(\lambda_i)$ in each position on the diagonal,

$f'(\lambda_i)$ in each position on the first subdiagonal,

$f''(\lambda_i)/2!$ in each position on the second subdiagonal,

and so on, with $f^{(p-1)}(\lambda_i)/(p-1)!$ in the lower left-hand corner position, where J_i is a p-by-p sub-block.

Example   Let us calculate e^A, where A is the matrix given in the Example of Section 7.7. There it was shown that

$$P^{-1}AP = J = \begin{pmatrix} 2 & 0 & 0 \\ 1 & 2 & 0 \\ 0 & 1 & 2 \end{pmatrix},$$

so there is only one sub-block in J. Because $\lambda = 2$ and each derivative of e^x at $x = 2$ is e^2, we have

$$e^J = \begin{pmatrix} e^2 & 0 & 0 \\ e^2 & e^2 & 0 \\ e^2/2 & e^2 & e^2 \end{pmatrix} = \frac{e^2}{2} \begin{pmatrix} 2 & 0 & 0 \\ 2 & 2 & 0 \\ 1 & 2 & 2 \end{pmatrix}.$$

Then using P and P^{-1} as determined in Section 7.7, we have

$$e^A = Pe^J P^{-1} = \frac{e^2}{2} \begin{pmatrix} 14 & 12 & 5 \\ -16 & -14 & -7 \\ 8 & 8 & 6 \end{pmatrix}.$$

EXERCISES 11.2

1. Evaluate e^A for each of the following matrices.

(i) $A = \begin{pmatrix} -1 & 0 \\ 0 & 1 \end{pmatrix}$. Note that A is diagonal.

(ii) $A = \begin{pmatrix} 0 & 0 & 0 \\ 1 & 0 & 0 \\ 1 & 1 & 0 \end{pmatrix}$. Note that A is nilpotent.

(iii) $A = \begin{pmatrix} -1 & 3 \\ 1 & 1 \end{pmatrix}$. Note that A is diagonable.

(iv) $A = \begin{pmatrix} 2 & 0 & 0 \\ 0 & -1 & 3 \\ 0 & 1 & 1 \end{pmatrix}$. Note that A is in block diagonal form.

Use Theorem 11.3 and the results of (iii).

2. Using the results of Exercise 7.7(i) evaluate e^A, where

$$A = \begin{pmatrix} 5 & -6 & -6 \\ -1 & 4 & 2 \\ 3 & -6 & -4 \end{pmatrix}.$$

3. Using the results of Exercise 7.7(ii), evaluate e^A, where

$$A = \begin{pmatrix} 3 & 1 & -1 \\ 2 & 2 & -1 \\ 2 & 2 & 0 \end{pmatrix}.$$

4. Prove each of the following theorems:

    (i) Theorem 11.3.

    (ii) Theorem 11.4.

    (iii) Theorem 11.7.

5. Let $f(x) = \sum_{k=0}^{\infty} a_k x^k$ and let A be a matrix with distinct character-istic values $\lambda_1, \lambda_2, \ldots, \lambda_n$ for which $f(A)$ converges. It can be proved that

$$f(A) = D^{-1}[D_0 I + D_1 A + \cdots + D_{n-1} A^{n-1}],$$

where D is the Vandermonde determinant

$$\det \begin{vmatrix} 1 & 1 & \cdots & 1 \\ \lambda_1 & \lambda_2 & \cdots & \lambda_n \\ \lambda_1^2 & \lambda_2^2 & \cdots & \lambda_n^2 \\ \cdot & \cdot & & \cdot \\ \cdot & \cdot & & \cdot \\ \cdot & \cdot & & \cdot \\ \lambda_1^{n-1} & \lambda_2^{n-1} & \cdots & \lambda_n^{n-1} \end{vmatrix},$$

and D_k is the determinant that coincides with D except that row $(k+1)$ of D_k is $(f(\lambda_1), f(\lambda_2), \ldots, f(\lambda_n))$. (This formula is reminiscent of Cramer's rule.)

    (i) Apply the method described above to calculate $f(A)$, where

$$A = \begin{pmatrix} 3 & 1 & -2 \\ 2 & 4 & -4 \\ 2 & 1 & -1 \end{pmatrix}.$$

    (ii) Check your result in (i) by using it to compute A^2.

    (iii) For A as given in (i) find necessary and sufficient conditions on f that $f(A)$ be a scalar matrix.

    (iv) Calculate e^A.

6. Let $A = \begin{pmatrix} 1 & 0 \\ 0 & 2 \end{pmatrix}$ and $B = \begin{pmatrix} 0 & 0 \\ 1 & 0 \end{pmatrix}$. Compute e^A, e^B, $e^A e^B$, $e^B e^A$, and s^{A+B}.

7. By writing the power series for e^A and e^B and multiplying term by term, show that if $AB = BA$ then $e^A e^B = e^{A+B} = e^B e^A$.

11.3 MATRICES OF FUNCTIONS

We now consider matrices whose entries are scalar-valued functions of a real variable t. To be specific, let $X(t)$ denote an m-by-n matrix of the form

$$
X(t) = \begin{pmatrix}
x_{11}(t) & x_{12}(t) & \cdots & x_{1n}(t) \\
x_{21}(t) & x_{22}(t) & \cdots & x_{2n}(t) \\
\cdot & \cdot & & \cdot \\
\cdot & \cdot & & \cdot \\
\cdot & \cdot & & \cdot \\
x_{m1}(t) & x_{m2}(t) & \cdots & x_{mn}(t)
\end{pmatrix},
$$

where for each $i = 1, 2, \ldots, m$ and each $j = 1, 2, \ldots, n$ the entry $x_{ij}(t)$ is the value at t of a real-valued function x_{ij} whose domain includes a fixed interval $a \leq t \leq b$ of real numbers. Then X is a function, defined at least for $a \leq t \leq b$, whose value for each t in its domain is the m-by-n matrix $X(t)$ of real numbers.

In this section we shall indicate how the basic notions of calculus can be defined for such functions, developing only enough of the technique to enable us to apply the results to the study of differential equations. The ideas are quite straightforward because continuity, derivative, and integral can be defined component-by-component.

Definition 11.4 Let X be an m-by-n matrix of real-valued functions x_{ij}, $i = 1, 2, \ldots, m$ and $j = 1, 2, \ldots, n$, each of which is defined for all t in some interval $a \leq t \leq b$ of real numbers.

(a) X is said to be *continuous* at t_0 if and only if for each $i = 1, \ldots, m$ and each $j = 1, \ldots, n$ the function x_{ij} is continuous at t_0.

(b) The *derivative* $\mathscr{D}X(t_0)$ of X at t_0 is the m-by-n matrix whose (i, j) entry is

$$
\frac{d}{dt} x_{ij}(t_0),
$$

provided that for each $i = 1, \ldots, m$ and each $j = 1, \ldots, n$ the function x_{ij} has a derivative at t_0.

(c) The *definite integral* $\int_a^b X$ is the *m*-by-*n* matrix whose (i, j) entry is

$$\int_a^b x_{ij}(t)\, dt,$$

provided that for each $i = 1, \ldots, m$ and each $j = 1, \ldots, n$ the function x_{ij} is integrable on the interval $a \le t \le b$.

As one might anticipate from experience with the algebra of matrices, the calculus of matrices requires special care because of the noncommutativity of matrix multiplication. But the results asserted in the following theorems are generalizations of the corresponding results of the calculus of scalar functions in the sense that the formulas for calculus of matrices reduce to the familiar formulas when all the matrices involved commute with each other.

Theorem 11.8   If Y is a *p*-by-*m* matrix of scalar-valued functions, if X is an *m*-by-*n* matrix of scalar-valued functions and if $\mathscr{D}Y$ and $\mathscr{D}X$ exist, then

(a) $\mathscr{D}(YX) = (\mathscr{D}Y)X + Y(\mathscr{D}X)$,

(b) $\mathscr{D}(X^r) = (\mathscr{D}X)X^{r-1} + X(\mathscr{D}X)X^{r-2} + \cdots + X^{r-1}(\mathscr{D}X)$, for any positive integer r,

(c) $\mathscr{D}(X^{-1}) = X^{-1}(\mathscr{D}X)X^{-1}$ if X is nonsingular.

Proof   Exercise.

Theorem 11.9   If A is an *m*-by-*m* matrix of scalars, then

$$\mathscr{D}(e^{tA}) = Ae^{tA} = e^{tA}A.$$

Proof   By definition e^{tA} denotes the series

$$\sum_{k=0}^{\infty} \frac{(tA)^k}{k!} = I + tA + \frac{t^2 A^2}{2!} + \cdots + \frac{t^k A^k}{k!} + \cdots,$$

which converges for every *m*-by-*m* matrix A and every real t. Because a power series can be differentiated term by term within its circle of convergence, we have

$$\mathscr{D}(e^{tA}) = Z + A + tA^2 + \cdots + \frac{t^{k-1}A^k}{(k-1)!} + \cdots$$

$$= A\left(I + tA + \cdots + \frac{(tA)^{k-1}}{(k-1)!} + \cdots\right)$$

$$= Ae^{tA},$$

as claimed. Because all the matrices involved commute with each other, we can also factor A to the right to obtain $\mathscr{D}(e^{tA}) = e^{tA}A$.

The formula of Theorem 11.9 is reminiscent of the corresponding property of the scalar exponential function,

$$\frac{d}{dt} e^{at} = ae^{at},$$

which plays such an important role in the solution of linear differential equations with constant coefficients. For example, recall that the solution of the scalar differential equation

$$x'(t) = ax(t)$$

is the function

$$x(t) = x(t_0)e^{a(t-t_0)}.$$

A matrix analogue of that differential equation is

$$\mathscr{D}X(t) = AX(t),$$

where $X(t)$ is an m-by-n matrix of scalar valued functions and A is an m-by-m matrix of scalars.

Theorem 11.10   Given the matrix differential equation

$$\mathscr{D}X(t) = AX(t),$$

where $X(t)$ is an m-by-n matrix of differentiable scalar valued functions and A is an m-by-m matrix of scalars,

(a) for any t_0, the matrix

$$e^{(t-t_0)A}X(t_0)$$

satisfies the given differential equation,

(b) any matrix that satisfies the equation is of the form specified in (a) for some t_0,

(c) the rank of a given solution $X(t)$ is the same for all values of t.

Proof Conclusion (a) follows directly from Theorem 11.8(a) and Theorem 11.9, because $X(t_0)$ is a matrix of scalars. To prove (b) let $V(t)$ be any solution, and let $Y(t) = e^{-tA}V(t)$. Then

$$\mathcal{D}Y(t) = e^{-tA}(-A)V(t) + e^{-tA}\mathcal{D}V(t)$$
$$= e^{-tA}[-AV(t) + \mathcal{D}V(t)] = Z,$$

because $\mathcal{D}V(t) = AV(t)$. Hence each entry of Y is a scalar and Y is a constant matrix C. Thus for any value t_0 of t

$$Y(t_0) = C = e^{-t_0 A}V(t_0),$$

and

$$V(t) = e^{tA}Y(t) = e^{tA}e^{-t_0 A}V(t_0)$$
$$= e^{(t-t_0)A}V(t_0),$$

as claimed. Furthermore from Theorem 11.7(b), $e^{(t-t_0)A}$ is nonsingular, so for all t the matrix $V(t)$ has the same rank as the matrix $V(t_0)$, proving (c).

In the next section we shall consider only the special case of these results in which $n = 1$; in that case $X(t)$ is a column vector of m scalar functions, A is an m-by-m matrix of scalars, and we seek to solve the matrix differential equation

$$\mathcal{D}X(t) = AX(t).$$

EXERCISES 11.3

1. Prove that the derivative of a matrix of functions as given in Definition 11.8(b) could also be defined by

$$\mathcal{D}X(t_0) = \lim_{h \to 0} \frac{1}{h}[X(t_0 + h) - X(t_0)],$$

where the limit operation, applied to a matrix, is understood to be the limit operation applied to each entry of the matrix.

2. Prove Theorem 11.8.

3. Does an analogue of the fundamental theorem of calculus hold for matrix calculus? Explain your answer.

4. Does an analogue of the mean value theorem of the derivative hold for matrix calculus? Explain your answer.

5. Show that if $X(t)$ is an m-by-m matrix of scalar-valued functions, then

$$\frac{d}{dt} \det X(t) = \det\left(\frac{d}{dt} X_1, X_2, \ldots, X_m\right) + \det\left(X_1, \frac{d}{dt} X_2, \ldots, X_m\right)$$

$$+ \cdots + \det\left(X_1, X_2, \ldots, \frac{d}{dt} X_m\right),$$

where X_j denotes column j of X.

6. Derive an expression for the derivative of tr $X(t)$, where $X(t)$ is an m-by-m matrix of scalar-valued functions. Show that

$$\frac{d}{dt} \operatorname{tr} X(t) = \operatorname{tr} \mathscr{D}X(t).$$

11.4 SYSTEMS OF LINEAR DIFFERENTIAL EQUATIONS

We shall now apply Theorem 11.10 in the special case in which $n = 1$. Then X is a column vector of m functions $x_i(t)$ of a real variable t, and A is an m-by-m matrix of real numbers. The vector differential equation

$$\mathscr{D}X(t) = AX(t)$$

represents a system of m simultaneous differential equations in m variables with constant coefficients;

$$x_1'(t) = a_{11} x_1(t) + a_{12} x_2(t) + \cdots + a_{1m} x_m(t),$$

$$x_2'(t) = a_{21} x_1(t) + a_{22} x_2(t) + \cdots + a_{2m} x_m(t),$$

(11.5)

.

.

.

$$x_m'(t) = a_{m1} x_1(t) + a_{m2} x_2(t) + \cdots + a_{mm} x_m(t).$$

We also are given the vector $X(t_0)$ for some t_0 as an initial condition, and we seek to find $X(t)$ for all t in terms of the constant coefficients a_{ij} and the numbers $x_i(t_0)$, $i = 1, \ldots, m$. Theorem 11.10, of course, provides a solution in the form

$$X(t) = e^{(t-t_0)A}X(t_0),$$

so the problem is reduced to evaluating the matrix $e^{(t-t_0)A}$, which we considered briefly in Section 11.2. Suppose that the m-by-m matrix P reduces A to a Jordan matrix J,

$$P^{-1}AP = J.$$

Then

$$P^{-1}X(t) = (P^{-1}e^{(t-t_0)A}P)(P^{-1}X(t_0))$$

$$= e^{(t-t_0)P^{-1}AP}(P^{-1}X(t_0)) = e^{(t-t_0)J}P^{-1}X(t_0).$$

If we write $Y(t) = P^{-1}X(t)$, we have

$$Y(t) = e^{(t-t_0)J}Y(t_0).$$

In short, if we can determine P and J, then we can adapt the method described at the end of Section 11.2 to compute $e^{(t-t_0)J}$ and thus to determine $Y(t)$, and hence $X(t)$. We now illustrate this method.

Example   Solve the following system of linear differential equations, subject to the initial condition that $x_1 = 1$, $x_2 = 2$, and $x_3 = 3$ when $t = 0$:

$$x_1'(t) = \quad 9x_1(t) + 7x_2(t) + 3x_2(t),$$

$$x_2'(t) = -9x_1(t) - 7x_2(t) - 4x_3(t),$$

$$x_3'(t) = \quad 4x_1(t) + 4x_2(t) + 4x_3(t).$$

In matrix form $X'(t) = AX(t)$, where

$$A = \begin{pmatrix} 9 & 7 & 3 \\ -9 & -7 & -4 \\ 4 & 4 & 4 \end{pmatrix}, \quad X(t) = \begin{pmatrix} x_1(t) \\ x_2(t) \\ x_3(t) \end{pmatrix}, \quad X(0) = \begin{pmatrix} 1 \\ 2 \\ 3 \end{pmatrix}.$$

The matrix A was considered in the Examples of Sections 7.7 and 11.2, and we make use of the results obtained there. The Jordan form of A is

$$J = \begin{pmatrix} 2 & 0 & 0 \\ 1 & 2 & 0 \\ 0 & 1 & 2 \end{pmatrix},$$

and $P^{-1}AP = J$, where

$$P = \begin{pmatrix} 6 & 2 & 1 \\ -1 & -1 & -1 \\ -11 & -2 & 0 \end{pmatrix}, \text{ and } P^{-1} = \begin{pmatrix} -2 & -2 & -1 \\ 11 & 11 & 5 \\ -9 & -10 & -4 \end{pmatrix}.$$

Then because $t_0 = 0$ we calculate $Y(0)$ and e^{tJ}:

$$Y(0) = P^{-1}X(0) = \begin{pmatrix} -9 \\ 48 \\ -41 \end{pmatrix},$$

$$e^{tJ} = \begin{pmatrix} e^{2t} & 0 & 0 \\ te^{2t} & e^{2t} & 0 \\ \dfrac{t^2}{2}e^{2t} & te^{2t} & e^{2t} \end{pmatrix} = \frac{e^{2t}}{2}\begin{pmatrix} 2 & 0 & 0 \\ 2t & 2 & 0 \\ t^2 & 2t & 2 \end{pmatrix}.$$

In the calculation of e^{tJ} we let $f(x) = e^{tx}$. Then $f'(x) = te^{tx}$ and $f''(x) = t^2 e^{tx}$. In each diagonal position we write $f(\lambda)$, in each subdiagonal we write $f'(\lambda)$, and in each position on the second subdiagonal we write $f''(\lambda)/2$, where $\lambda = 2$ for this example. Compare this result with e^J as calculated in the Example of Section 11.2.

Next we compute $Y(t) = e^{tJ}Y(0)$:

$$Y(t) = \frac{e^{2t}}{2}\begin{pmatrix} 2 & 0 & 0 \\ 2t & 2 & 0 \\ t^2 & 2t & 2 \end{pmatrix}\begin{pmatrix} -9 \\ 48 \\ -41 \end{pmatrix} = \frac{e^{2t}}{2}\begin{pmatrix} -18 \\ -18t + 96 \\ -9t^2 + 96t - 82 \end{pmatrix}.$$

Finally, then, $X(t) = PY(t)$, so

$$X(t) = \frac{e^{2t}}{2}\begin{pmatrix} 2 + 60t - 9t^2 \\ 4 - 78t + 9t^2 \\ 6 + 36t \end{pmatrix}, \text{ or } \begin{matrix} x_1 = \frac{1}{2}(2 + 60t - 9t^2)e^{2t}, \\ x_2 = \frac{1}{2}(4 - 78t + 9t^2)e^{2t}, \\ x_3 = (3 + 18t)e^{2t}. \end{matrix}$$

Clearly $X(0)$ has the correct value, and only a small amount of additional computation is needed to verify that these expressions for x_1, x_2, and x_3 satisfy the original system of differential equations for all values of t.

Now let us return to the general system (11.5). Clearly a lot of computation is required for its solution, and we wonder whether all that work is necessary. We can now observe that the computation of P can be avoided by a substitute procedure that does not necessarily reduce the amount of computation. Assume that the Jordan form of A has been determined and that $e^{(t-t_0)J}$ is known. For each sub-block J_i of J, $e^{(t-t_0)J}$ will have a corresponding sub-block along its diagonal and of the form

$$e^{\lambda_i(t-t_0)}\begin{pmatrix} 1 & 0 & 0 & \cdots & 0 \\ (t-t_0) & 1 & 0 & \cdots & 0 \\ \dfrac{(t-t_0)^2}{2!} & (t-t_0) & 1 & \cdots & 0 \\ \cdot & \cdot & \cdot & & \cdot \\ \dfrac{(t-t_0)^{k_i-1}}{(k_i-1)!} & \cdot & \cdot & \cdots & 1 \end{pmatrix}$$

Thus each entry of $e^{(t-t_0)J}$ is of the form

$$c(t-t_0)^p e^{\lambda_i(t-t_0)}$$

for some scalar c, some nonnegative integer p, and some characteristic value λ_i of A. Furthermore, because P is a matrix of constants, each entry of $e^{(t-t_0)A}$ is a linear combination of the entries of $e^{(t-t_0)J}$. This means that each entry of the solution vector

$$X(t) = e^{(t-t_0)A}X(0)$$

is of the form

$$x_j = \sum_{i=1}^{s} q_i(t)e^{\lambda_i(t-t_0)}$$

where $\lambda_1, \ldots, \lambda_s$ are the distinct characteristic values of A and where $q_i(t)$ is a polynomial whose degree is less than the size of the largest sub-block associated with λ_i. The λ_i are known, but the coefficients of each of these polynomials are unknown. But writing a trial solution in terms of these unknown polynomials, we can use the equations themselves to determine those coefficients.

To illustrate, in the preceding example there is one characteristic value ($\lambda = 2$), and an associated sub-block of dimension 3. Hence we write a trial solution,

$$x_1(t) = (a_1 + b_1 t + c_1 t^2)e^{2t}$$
$$x_2(t) = (a_2 + b_2 t + c_2 t^2)e^{2t}$$
$$x_3(t) = (a_3 + b_3 t + c_3 t^2)e^{2t},$$

where the coefficients a_i, b_i, c_i are unknown. The initial condition $X(0)$ yields the values of the a_i:

$$x_1(0) = 1 = a_1,$$
$$x_2(0) = 2 = a_2,$$
$$x_3(0) = 3 = a_3.$$

Next we can calculate the derivative of the trial solution vector,

$$x_1'(t) = [(2 + b_1) + (2b_1 + 2c_1)t + 2c_1 t^2]e^{2t},$$
$$x_2'(t) = [(4 + b_2) + (2b_2 + 2c_2)t + 2c_2 t^2]e^{2t},$$
$$x_3'(t) = [(6 + b_3) + (2b_3 + 2c_3)t + 2c_3 t^2]e^{2t}.$$

Substitution in the given system then yields the following equations and two others obtained similarly from $x_2'(t)$ and $x_3'(t)$:

$$[(2 + b) + (2b_1 + c_1)t + 2c_1 t^2]e^{2t} = 9(1 + b_1 t + c_1 t^2)e^{2t}$$
$$+ 7(2 + b_2 t + c_2 t^2)e^{2t}$$
$$+ 3(3 + b_3 t + c_3 t^2)e^{2t}.$$

Because this equation is to hold for all t we can equate coefficients of like powers of t to obtain

$$2 + b_1 = 32,$$
$$2b_1 + c_1 = 9b_1 + 7b_2 + 3b_3,$$
$$2c_1 = 9c_1 + 7c_2 + 3c_3;$$

this system reduces to

$$b_1 = 30,$$
$$c_1 = 210 + 7b_2 + 3b_3,$$
$$7c_1 = -7c_2 - 3c_3.$$

Similar equations obtained from $x_2'(t)$ and $x_3'(t)$ permit us to determine each of the nine undetermined coefficients in the trial solution. We omit the algebraic details but observe that the values $a_1 = 1, a_2 = 2, a_3 = 3, b_1 = 30$

already agree with the coefficients obtained in our earlier solution of this same problem.

As a final application, we consider a single linear, homogeneous differential equation of order m with constant coefficients,

$$(11.6) \quad \frac{d^m}{dt^m} x(t) + a_1 \frac{d^{m-1}}{dt^{m-1}} x(t) + \cdots + a_{m-1} \frac{d}{dt} x(t) + a_m x(t) = 0,$$

and with the initial conditions specified by

$$x(t_0) = c_1, \quad x'(t_0) = c_2, \ldots, x^{(m-1)}(t_0) = c_m,$$

where $x^{(k)}(t)$ denotes the kth derivative of x at t.

We solve this differential equation by first reducing it to a special form of the system (11.5) of m linear first order differential equations. To do this we let

$$y_1 = x(t),$$
$$y_2 = \frac{d}{dt} x(t)$$
$$\cdot$$
$$\cdot$$
$$\cdot$$
$$y_m = \frac{d^{m-1}}{dt^{m-1}} x(t).$$

Observe that $y_{i+1} = y'_i (i = 1, \ldots, m-1)$ and $y'_m = d^m/dt^m x(t)$. Using (11.6) we obtain the system

$$y'_1(t) = x'(t) = y_2(t),$$
$$y'_2(t) = x''(t) = y_3(t),$$
$$\cdot$$
$$\cdot$$
$$\cdot$$
$$y'_m(t) = x^{(m)}(t) = -[a_m x(t) + a_{m-1} x'(t) + \cdots + a_1 x^{(m-1)}(t)]$$
$$= -[a_m y_1(t) + a_{m-1} y_2(t) + \cdots + a_1 y_m(t)].$$

Comparing this system with (11.5), we observe that the new system can be written in vector notation as

$$Y'(t) = A Y(t),$$

where

$$A = \begin{pmatrix} 0 & 1 & 0 & \cdots & 0 \\ 0 & 0 & 1 & \cdots & 0 \\ 0 & 0 & 0 & \cdots & 0 \\ \cdot & \cdot & \cdot & & \\ \cdot & \cdot & \cdot & & \\ \cdot & \cdot & \cdot & & \\ 0 & 0 & 0 & \cdots & 1 \\ -a_m & -a_{m-1} & -a_{m-2} & \cdots & -a_1 \end{pmatrix}, \quad Y(t) = \begin{pmatrix} y_1(t) \\ y_2(t) \\ \cdot \\ \cdot \\ \cdot \\ y_m(t) \end{pmatrix}.$$

The initial condition is

$$Y(t_0) = \begin{pmatrix} c_1 \\ c_2 \\ \cdot \\ \cdot \\ \cdot \\ c_m \end{pmatrix}.$$

Because a matrix and its transpose have the same characteristic polynomial and because A^t has the form of a companion matrix (Exercise 7.1-9), the characteristic polynomial of A is

$$p(x) = (-1)^m(x^m + a_1 x^{m-1} + \cdots + a_{m-1}x + a_m).$$

The methods previously described for solving (11.5) therefore can be used to determine a solution $Y(t)$ in terms of t, and then the solution $x(t)$ of (11.6) is the first component $y_1(t)$ of $Y(t)$.

Example Solve $x''' - x'' - x' + x = 0$ subject to the initial conditions that $x(0) = 2$, $x'(0) = 1$, $x''(0) = 4$.

Let $y_1 = x(t)$, $y_2 = x'(t)$, $y_3 = x''(t)$. Then the given equation of order three is replaced by three simultaneous first-order equations:

$$y_1' = y_2$$

$$y_2' = y_3$$

$$y_3' = -y_1 + y_2 + y_3,$$

or in vector form, $Y'(t) = AY(t)$, where

$$Y(t) = \begin{pmatrix} y_1(t) \\ y_2(t) \\ y_3(t) \end{pmatrix}, \quad A = \begin{pmatrix} 0 & 1 & 0 \\ 0 & 0 & 1 \\ -1 & 1 & 1 \end{pmatrix}, \quad Y(0) = \begin{pmatrix} 2 \\ 1 \\ 4 \end{pmatrix}.$$

The characteristic polynomial of A is

$$p(x) = (-1)^3(x^3 - x^2 - x + 1) = (-1)(x - 1)^2(x + 1).$$

The characteristic subspace associated with $\lambda = 1$ is one-dimensional, so a Jordan form of A is

$$J = \begin{pmatrix} 1 & 0 & 0 \\ 1 & 1 & 0 \\ 0 & 0 & -1 \end{pmatrix} = P^{-1}AP,$$

where

$$P = \begin{pmatrix} -1 & 1 & 1 \\ 0 & 1 & -1 \\ 1 & 1 & 1 \end{pmatrix} \quad \text{and} \quad P^{-1} = \frac{1}{4}\begin{pmatrix} -2 & 0 & 2 \\ 1 & 2 & 1 \\ 1 & -2 & 1 \end{pmatrix}.$$

Hence

$$e^{tJ} = \begin{pmatrix} e^t & 0 & 0 \\ te^t & e^t & 0 \\ 0 & 0 & e^{-t} \end{pmatrix},$$

and we compute $e^{tA} = Pe^{tJ}P^{-1}$ to be

$$e^{tA} = \frac{1}{4}\begin{pmatrix} (3 - 2t)e^t + e^{-t} & 2e^t - 2e^{-t} & (-1 + 2t)e^t + e^{-t} \\ (1 - 2t)e^t - e^{-t} & 2e^t + 2e^{-t} & (1 + 2t)e^t - e^{-t} \\ (-1 - 2t)e^t + e^{-t} & 2e^t - 2e^{-t} & (3 + 2t)e^t + e^{-t} \end{pmatrix}.$$

Finally, $x(t) = y_1(t)$, which is the first component of

$$Y(t) = e^{tA}Y(0),$$

$$x(t) = \frac{e^t}{4}[2(3 - 2t) + 1(2) + 4(-1 + 2t)] + \frac{e^{-t}}{4}[2(1) + 1(-2) + 4(1)]$$

$$= e^t[1 + t] + e^{-t}.$$

Observe that the general form of this solution can be anticipated from the Jordan form of A. Because $\lambda = 1$ is a repeated characteristic value corresponding to a sub-block of dimension two in J, one term of the solution is the product of e^t and a polynomial of degree one in t. Because $\lambda = -1$ is a characteristic value corresponding to a sub-block of dimension 1 in J, one term of the solution is the product of e^{-t} and a polynomial of degree 0 (a constant).

We conclude with some general observations about the solution of the system $Y' = AY$, where A is an m-by-m matrix. The m characteristic values of

A may or may not be distinct. If λ_i has algebraic multiplicity, p_i, then one term of each component of the solution vector will be the product of $e^{\lambda_i t}$ and a polynomial in t of degree less than p_i. Each component of the solution vector is the sum of one such product for each of the distinct characteristic values of A. Hence a basis for the space of all solutions will be

$$\{e^{\lambda_1 t}, te^{\lambda_1 t}, \ldots, t^{p_1-1}e^{\lambda_1 t}; \ldots; e^{\lambda_k t}, te^{\lambda_k t}, \ldots, t^{p_k-1}e^{\lambda_k t}\},$$

where $\lambda_1, \ldots, \lambda_k$ are the distinct characteristic values of A and where the algebraic multiplicity of λ_i is p_i.

EXERCISES 11.4

1. Solve the system of linear differential equations $X' = AX$ with initial condition $X(0)$ for each of the following cases.

(i) $A = \begin{pmatrix} 3 & 1 & 1 \\ 0 & 4 & 0 \\ -1 & 1 & 5 \end{pmatrix}$, $X(0) = \begin{pmatrix} 2 \\ 1 \\ 0 \end{pmatrix}$.

(ii) $A = \begin{pmatrix} 5 & 4 & 3 \\ -1 & 0 & -3 \\ 1 & -2 & 1 \end{pmatrix}$, $X(0) = \begin{pmatrix} 1 \\ 2 \\ 2 \end{pmatrix}$.

2. Solve the system of linear differential equations $X' = AX$, where

$$A = \begin{pmatrix} 2 & 10 & 5 \\ -2 & -4 & -4 \\ 3 & 5 & 6 \end{pmatrix},$$

for each of the following initial conditions, noting the effect of the initial conditions upon the form of the solution.

(i) $X(0) = \begin{pmatrix} 2 \\ 0 \\ 1 \end{pmatrix}$.

(ii) $X(0) = \begin{pmatrix} -1 \\ 2 \\ 3 \end{pmatrix}$.

(iii) $X(0) = \begin{pmatrix} 5 \\ -2 \\ 1 \end{pmatrix}$.

3. Use the method illustrated in the text to transform each of the following third-order linear differential equations to a system of three first-order linear differential equations. Then solve that system to obtain the solution of the original equation.

(i) $\dfrac{d^3}{dt^3} x(t) - 3 \dfrac{d^2}{dt^2} x(t) + 3 \dfrac{d}{dt} x(t) - x(t) = 0$,

with initial condition $x(0) = 2$, $x'(0) = -1$, and $x''(0) = 0$.

(ii) $\dfrac{d^3}{dt^3} x(t) - 2 \dfrac{d^2}{dt^2} x(t) - 4 \dfrac{d}{dt} x(t) + 8x(t) = 0$,

with initial condition $x(0) = 1$, $x'(0) = 4$, and $x''(0) = -4$.

(iii) $\dfrac{d^3}{dt^3} x(t) - \dfrac{d}{dt} x(t) = 0$,

with initial condition $x(0) = 5$, $x'(0) = 3$, and $x''(0) = 1$.

4. Solve each differential equation in Exercise 3 by the following method. First determine all values of λ such that $x = e^{\lambda t}$ is a solution. (These are the characteristic values of the differential equation.) If λ is a characteristic value of algebraic multiplicity m, then each of the functions $e^{\lambda t}$, $te^{\lambda t}$, ..., $t^{m-1}e^{\lambda t}$ is a basic solution, and every solution is a linear combination of the set of all basic solutions. A differential equation of the form (11.6) will have m basic solutions. Hence, using m undetermined constants, let $x(t)$ be a linear combination of the basic solutions. The values of the unknown constants can then be determined to satisfy the given initial condition, thus solving the equation.

5. Recalling from Section 5.3 the definition of linear independence of functions and the role of the Wronskian determinant in verifying linear independence, show that if $r \neq s$ then the set $\{t^r e^{at}, t^s e^{at}\}$ is linearly independent.

APPENDIX A

A.1 ALGEBRAIC SYSTEMS AND FIELDS

The underlying mathematical structure for the study of linear algebra is called a *vector space*. Because vector spaces occur in many different contexts (a few of which are cited in Section 2.3), it would be quite inefficient to derive the basic properties that must hold in any vector space each time a particular example of a vector space arises. Thus we choose to consider the general concept of a vector space. Consequently we carefully avoid specifying a particular form that "vectors" should assume; instead we specify a set of algebraic properties that are common to all vector spaces. Similarly, we avoid specifying the nature of the "scalars" of a vector space except to require that the scalars form a *field*. A field, like a vector space, is a well-defined type of algebraic system. Because examples of fields are familiar, we shall first illustrate the notion of an algebraic system informally for fields, using our observations to motivate a more general formulation of the notion of an *algebraic system*.

Two well-known examples of fields are the system $\mathbb{Q}$ of rational numbers and the system $\mathbb{R}$ of real numbers. A rational number is of the form a/b, where a and b are integers and $b \neq 0$. The *sum* $r + s$ of two rational numbers r and s is a rational number, and so is their *product* $r \cdot s$. Thus $\mathbb{Q}$ is said to be *closed* under the operations of addition and multiplication. Furthermore, addition has the following familiar algebraic properties:

A1. Additional is *associative*: for all r, s, t in $\mathbb{Q}$

$$(r + s) + t = r + (s + t).$$

A2. Addition is *commutative*: for all r, s in $\mathbb{Q}$

$$r + s = s + r.$$

A3. There exists in $\mathbb{Q}$ an *identity element*, 0 (called zero), relative to addition:

$$r + 0 = r = 0 + r \quad \text{for each } r \text{ in } \mathbb{Q}.$$

A4. For each r in $\mathbb{Q}$ there exists in $\mathbb{Q}$ an *inverse element*, $-r$, relative to addition:

$$r + (-r) = 0 = (-r) + r.$$

With one slight modification, multiplication possesses an analogous set of properties.

M1. Multiplication is *associative*; for all r, s, t in $\mathbb{Q}$

$$(r \cdot s) \cdot t = r \cdot (s \cdot t).$$

M2. Multiplication is *commutative*: for all r, s in $\mathbb{Q}$

$$r \cdot s = s \cdot r.$$

M3. There exists an *identity element*, 1 (called unity), relative to multiplication:

$$r \cdot 1 = r = 1 \cdot r \quad \text{for each } r \text{ in } \mathbb{Q}.$$

M4. For each *nonzero* r in $\mathbb{Q}$ there exists in $\mathbb{Q}$ an *inverse element*, r^{-1}, relative to multiplication:

$$r \cdot r^{-1} = 1 = r^{-1} \cdot r.$$

Finally, the operations of addition and multiplication are related by the property

D. Multiplication is *distributive* over addition: for all r, s, t in $\mathbb{Q}$

$$r \cdot (s + t) = r \cdot s + r \cdot t.$$

Now consider the system $\mathbb{R}$ of real numbers, which also is closed under the operations of addition and multiplication of real numbers. Again we note that the nine postulates A1 to A4, M1 to M4, and D are satisfied. Although $\mathbb{Q}$ and $\mathbb{R}$ are different sets of numbers, the arithmetic operations on both sets have the same basic properties. Hence any theorem that can be deduced from these nine properties will be valid in $\mathbb{Q}$ and in $\mathbb{R}$. Each of these systems is a specific example of a type of algebraic system, called a *field*, which we now describe more formally.

A *field* is an algebraic system that consists of a set F of two or more elements, together with two closed, binary operations, denoted $\oplus$ and $\odot$ and called "addition" and "multiplication," such that the following postulates are satisfied:

A. Addition is associative and commutative, there exists an identity element o relative to addition, and each element of F has an additive inverse element in F.

M. Multiplication is associative and commutative, there exists an identity element i relative to multiplication, and each element of F *other than o* has a multiplicative inverse in F.

D. Multiplication is distributive over addition.

The smallest possible field has two elements, and we can verify directly that the set $F = \{0, 1\}$ forms a field when the customary rules for addition and multiplication of 0 and 1 are modified only by the rule that $1 \oplus 1 = 0$. There exist many other fields, finite and infinite; in particular, in Appendix A.5 we describe in some detail the field $\mathbb{C}$ of complex numbers.

Turning our attention again to the subject of algebraic systems in general, we point out that a system need not have exactly two operations; it could have fewer or more. For example, a *commutative group* is a system that consists of a set G of elements and one closed binary operation $\oplus$ that satisfies postulate A as stated above. An example is the set I of integers, where $\oplus$ signifies addition of integers. A different example is provided by the set $G = \{1, -1\}$ with the operation defined to be numerical multiplication. A third example is the set of all nonzero elements of a field, which form a commutative group relative to the operation $\odot$.

With these examples to guide us, we are ready to describe an *algebraic system* in general terms. A system consists of a nonempty set S of *elements*, a set O of *operations* that are closed on S, a set R of *relations* (including equality) that are defined on S, and a set P of *postulates* that are satisfied by the elements in S, the operations in O, and the relations in R. The properties of the system are derived from the postulates in the form of *theorems*, and the language of the system is simplified by the use of *definitions*.

Referring to Definition 2.2, we see that a vector space $\mathscr{V}$ over a field $\mathscr{F}$ has *two* types of elements, a set $\mathscr{V}$ of vectors and a set $\mathscr{F}$ of scalars. There are two operations, $+$ and $\cdot$, on the set of scalars, and relative to those operations the scalars form a field, as stated by Postulate (a) of Definition 2.2. There is a single operation $\oplus$ on the set $\mathscr{V}$ of vectors, and relative to that operation $\mathscr{V}$ forms a commutative group, as stated by Postulate (b). Also there is a mixed operation $\odot$ that combines any scalar and any vector to produce a vector. The properties of $\odot$ are given by Postulate (c).

Many other types of algebraic systems exist, and the investigation of algebraic properties of such systems constitutes the broad area of mathematics called *abstract algebra*. In Appendix A.2 we shall briefly examine another example of an algebraic system, the *algebra of sets*; then we use the language and notation of sets to describe what is meant by a binary relation and a closed binary operation.

A.2   SETS AND SUBSETS

The word *set* is used to denote a *collection, class, family,* or *aggregate* of objects. Each such object is called an *element* or *member* of the set. Usually we denote a set by a capital letter and elements of that set by small letters. To indicate that a given object b is a member of a given set S we write

$$b \in S,$$

which can be read as "b is a member of S", "b belongs to S," "b is an element of S," "b is in S." If b is *not* an element of S we write

$$b \notin S.$$

There are two common ways of specifying a set:

(1) listing all its elements within braces,

(2) stating a characteristic property that determines whether or not any given object is an element of that set.

The notation that we adopt for the second method comprises two parts, separated by a vertical line, within braces. The first part tells us what type of elements are being considered, and the second part specifies the characteristic property. For example, let I denote the set of all integers, and let S denote the set of all integers whose square is less than seven. The two methods of writing S are

$$S = \{-2, -1, 0, 1, 2\},$$

where we attach no importance to the order in which the elements are listed, and

$$S = \{x \in I \mid x^2 < 7\},$$

which is read, "S is the set of all integers x such that x^2 is less than 7."

We use the symbol Φ to denote the *void* or *empty* set, which contains no elements at all. Though the void set at first may seem to be an artificial notion, its acceptance as a bona fide set is convenient.

Sets S and T are said to be *equal* if and only if they contain exactly the same elements. In terms of the membership relation, this is expressed as:

$$T = S \quad \text{means "} x \in T \text{ if and only if } x \in S.\text{"}$$

The next concept is that of a *subset*. If S and T are sets, T is said to be a subset of S, written $T \subseteq S$, if and only if every element of T is an element of S; that is

$$T \subseteq S \quad \text{means "if } x \in T, \text{ then } x \in S.\text{"}$$

The subset notation $T \subseteq S$ can also be written $S \supseteq T$. This notation is analogous to the notation for inequality of numbers: $a \le b$ means the same as $b \ge a$. It is verified readily that equality of sets can be expressed in terms of the subset notation as:

$$T = S \quad \text{means "} T \subseteq S \text{ and } S \subseteq T.\text{"}$$

It is clear from the definition of subset that any set is a subset of itself. Also, because the void set has no elements, the statement "If $x \in \Phi$, then $x \in S$" is logically valid for any set S. Thus we have

$$\Phi \subseteq S \quad \text{and} \quad S \subseteq S \quad \text{for every set } S.$$

T is said to be a *proper subset* of S if and only if every element of T is an element of S, but not every element of S is an element of T. This is written:

$$T \subset S \text{ if and only if } T \subseteq S \text{ and } T \ne S.$$

In practice it is sometimes useful to adopt geometric language for sets, calling an element a "point" even though the element may have no obvious

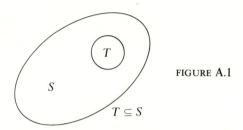

FIGURE A.1

geometric character. When we adopt such descriptive language we must bear in mind that this is done only for convenience, and we must carefully refrain from assuming any properties that might be suggested by our language but that are not otherwise legitimately established.

A geometric interpretation of sets suggests the use of sketches, called *Venn diagrams*, to represent sets and relations between sets. Thus the subset relation $T \subseteq S$ can be shown graphically as in Figure A.1. Such diagrams provide insight concerning sets, and often they suggest methods by which statements about sets can be proved or disproved. However, diagrams are not valid substitutes for formal proofs.

We now turn our attention to several ways in which sets can be combined to produce other sets. Let S be any set, and let $\mathcal{K}$ denote the collection of all subsets of S. (The elements of $\mathcal{K}$ are the subsets of S.) We shall define three operations on $\mathcal{K}$, called *union, intersection,* and *complementation*:

Union: $\qquad A \cup B = \{x \in S\} \,|\, x \in A \text{ or } x \in B\}$,

Intersection: $\qquad A \cap B = \{x \in S\} \,|\, x \in A \text{ and } x \in B\}$,

Complementation: $\qquad A' = \{x \in S\} \,|\, x \text{ is } not \text{ in } A\}$.

Thus if A and B are any elements of $\mathcal{K}$, the set "A union B" consists of all elements of S that belong to A, or to B, or to both. The set "A intersection B" consists of all elements of S that belong to both A and B. The "complement of A in S" consists of all elements of S that do not belong to A. If $A \cap B = \Phi$ (the void set), we say that A and B are *disjoint*. Venn diagrams that illustrate these operations are shown in Figure A.2. More generally, the union and intersection of any family $\mathcal{F}$ of subsets of S are defined in an analogous way:

$$\bigcup_{A \in \mathcal{F}} A = \{x \in S \,|\, x \in A \text{ for } some \ A \in \mathcal{F}\},$$

$$\bigcap_{A \in \mathcal{F}} A = \{x \in S \,|\, x \in A \text{ for } all \ A \in \mathcal{F}\}.$$

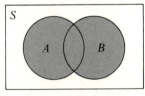
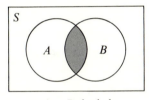

| $A \cup B$ shaded | $A \cap B$ shaded | A' shaded |

FIGURE A.2

Assuming an intuitive understanding of the meaning of the membership relation $x \in S$, we could now derive a number of theorems concerning the algebra of sets. For example, each of the operations of set union and set intersection is commutative and associative, and each is distributive over the other: for all $A, B, C \in \mathcal{K}$

$$A \cup B = B \cup A$$
and
$$A \cap B = B \cap A,$$

$$A \cup (B \cup C) = (A \cup B) \cup C$$
and
$$A \cap (B \cap C) = (A \cap B) \cap C,$$

$$A \cap (B \cup C) = (A \cap B) \cup (A \cap C)$$
and
$$A \cup (B \cap C) = (A \cup B) \cap (A \cup C).$$

Furthermore, S acts as an identity element for set intersection, and Φ acts as an identity element for set union, because for all $A \in \mathcal{K}$

$$A \cap S = A \quad \text{and} \quad A \cup \Phi = A.$$

Complementation satisfies the following properties:

$$A \cap A' = \Phi \quad \text{and} \quad A \cup A' = S,$$

$$(A \cup B)' = A' \cap B' \quad \text{and} \quad (A \cap B)' = A' \cup B'.$$

Many other results of this nature can be derived.

More formally we can consider an abstract algebraic system consisting of a nonvoid set $\mathcal{K}$, two binary operations $\cup$ and $\cap$ on the elements of $\mathcal{K}$, and one unary operation $'$ on the elements of $\mathcal{K}$. If we assume as postulates the properties described by the six pairs of equations in the preceding paragraph, the resulting system is called a *Boolean algebra*. Indeed, the associative laws and the last pair of equations (called the *de Morgan laws*) can be deduced from the other four pairs of equations, so those two pairs need not be included as postulates for a Boolean algebra.

A.3   RELATIONS AND
EQUIVALENCE RELATIONS

Set union and set intersection can be regarded as methods of combining given sets to form other sets, analogous to the familiar arithmetic operations of addition and multiplication as means of combining given numbers to form other numbers. We now define another very important construction of this nature for sets. Let A and B be sets, and let $A \times B$ denote the set of all *ordered pairs* (x, y) such that $x \in A$ and $y \in B$; in symbols,

$$A \times B = \{(x, y) \mid x \in A \text{ and } y \in B\}.$$

The set $A \times B$ is called the *cartesian product* of A and B, in recognition of the method by which René Descartes coordinatized the Euclidean plane by using ordered pairs of real numbers, one from each of two intersecting coordinate lines. The cartesian product construction can be used to formulate in set-theoretic terms concepts that arise in many different parts of mathematics. Here we are particularly interested in relations, functions, and operations.

A *binary relation* from a set A to a set B is defined to be a subset $\mathbf{R}$ of $A \times B$. Given any $\mathbf{R} \subseteq A \times B$, if $(a, b) \in \mathbf{R}$ we say that *a is related to b*, written $a \mathbf{R} b$. The *domain* of $\mathbf{R}$ is the set of all elements of A that are related by $\mathbf{R}$ to at least one element of B;

$$\text{dom } \mathbf{R} = \{a \in A \mid a \mathbf{R} y \text{ for some } y \in B\}.$$

The *range* of $\mathbf{R}$ is the set of all elements of B to which at least one element of A is related by $\mathbf{R}$;

$$\text{range } \mathbf{R} = \{b \in B \mid x \mathbf{R} b \text{ for some } x \in A\}.$$

A binary relation from A to A is called a *relation in A*.

It is sometimes convenient to think of a relation in geometric terms as an *association*, or many-valued correspondence, from A into B. Each element in dom $\mathbf{R}$ is associated by $\mathbf{R}$ with one or more elements in range $\mathbf{R}$. For example, let $A = \{a_1, a_2, a_3\}$, $B = \{b_1, b_2, b_3, b_4\}$; let $\mathbf{R}$ be the subset of $A \times B$, which consists of the pairs (a_1, b_1), (a_1, b_2), (a_2, b_1), (a_2, b_4), and (a_3, b_4). Then dom $\mathbf{R} = A$ and range $\mathbf{R} = (b_1, b_2, b_4\} \subset B$. $\mathbf{R}$ can be represented geometrically by the diagram in Figure A.3.

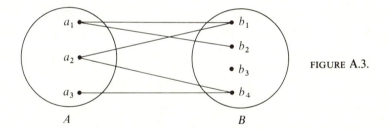

FIGURE A.3.

If $\mathscr{H}$ denotes the collection of all subsets of a set S, then the subset relation $\subseteq$ is a relation in $\mathscr{H}$. It is easy to verify that this relation has these properties:

Reflexivity: $A \subseteq A$ for every $A \in \mathscr{H}$.

Antisymmetry: If $A \subseteq B$ and $B \subseteq A$, then $A = B$.

Transitivity: If $A \subseteq B$ and $B \subseteq C$, then $A \subseteq C$.

Any relation having these three properties is called a *partial ordering*.

Another type of binary relation, called an *equivalence relation*, occurs throughout mathematics. An equivalence relation on a set A is a binary relation $\mathbf{E}$ in A which is reflexive, symmetric, and transitive:

Reflexive: $a \mathbf{E} a$ for each $a \in A$,

Symmetric: If $a \mathbf{E} b$, then $b \mathbf{E} A$,

Transitive: If $a \mathbf{E} b$ and $b \mathbf{E} c$, then $a \mathbf{E} c$.

We now consider the effect imposed by an equivalence relation $\mathbf{E}$ on a nonvoid set A. For each $a \in A$ let $[a]_{\mathbf{E}}$ denote the set of all elements that are related to a by $\mathbf{E}$:

$$[a]_{\mathbf{E}} = \{x \in A \,|\, x \mathbf{E} a\}.$$

The set $[a]_{\mathbf{E}}$ is called the *equivalence class determined by* a. Because $\mathbf{E}$ is reflexive, $a \mathbf{E} a$, so $a \in [a]_{\mathbf{E}}$. This implies that the set union of all the equivalence classes equals A. If $b \in [a]_{\mathbf{E}}$ then $b \mathbf{E} a$, and $a \mathbf{E} b$ because $\mathbf{E}$ is symmetric. Hence if $b \in [a]_{\mathbf{E}}$, then $a \in [b]_{\mathbf{E}}$. Suppose also that $y \in [b]_{\mathbf{E}}$. Then $y \mathbf{E} b$ and $b \mathbf{E} a$, so transitivity implies $y \mathbf{E} a$; hence $y \in [a]_{\mathbf{E}}$ and $[b]_{\mathbf{E}} \subseteq [a]_{\mathbf{E}}$. By reversing the roles of a and b we obtain $[a]_{\mathbf{E}} \subseteq [b]_{\mathbf{E}}$, so $[b]_{\mathbf{E}} = [a]_{\mathbf{E}}$ whenever $b \in [a]_{\mathbf{E}}$. This implies that two equivalence classes either are equal or have no elements in common.

Therefore, an equivalence relation **E** on a set A decomposes A into disjoint subsets, called equivalence classes. Such a decomposition of a set is called a partition.

Definition A.1   A *partition* of a set A is a collection $\mathscr{P}$ of subsets of A, called classes of the partition, such that

(a) if $x \in A$, then $x \in C$ for some $C \in \mathscr{P}$, and

(b) if $C, D \in \mathscr{P}$, then either $C = D$ or $C \cap D = \Phi$.

Now suppose we reverse the situation and start with any partition $\mathscr{P}$ of A. A relation **E** can be defined on A by writing a **E** b if and only if a and b are in the same class of $\mathscr{P}$. It can be proved as an exercise that **E** is an equivalence relation on A for which the equivalence classes are the classes of $\mathscr{P}$. Therefore, each equivalence relation on A determines a partition of A, and, conversely, each partition determines an equivalence relation.

A.4   FUNCTIONS AND OPERATIONS

It is assumed that you are familiar with numerical functions from your previous mathematical experience; our purpose here is to describe functions in a more general setting. Recall that each numerical function f has a *domain* of definition (a set D of numbers), and that to each $x \in D$ the function f assigns a unique number $f(x)$, which is called the *value* of f at x. As x varies over D, the numbers $f(x)$ form a set R, which is called the *range* of f.

A generalization from numerical functions to arbitrary functions is made very easily—we simply drop the requirement that D and R be sets of *numbers*, and consider them to be arbitrary sets.

Thus a function **F** from a set A into a set B consists of a nonvoid subset $D \subseteq A$, called the *domain* of **F**, and a correspondence that assigns to each $a \in D$ exactly one $b \in B$. The element b that is assigned by **F** to a is denoted **F**(a). The set of all $b \in B$ that are thus assigned to one or more $a \in D$ is called the *range* of **F**. If the range of **F** is the entire set B, we say that **F** is a function from A *onto* B.

Using the notion of a binary relation, we can state the definition of a function more concisely: a *function* **F** from A into B is a binary relation from A to B having the property that each element of the domain of **F** is related to *exactly one* element of the range of **F**. See Figure A.4.

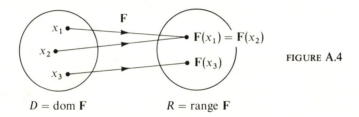

FIGURE A.4

$D = \text{dom } \mathbf{F}$ $R = \text{range } \mathbf{F}$

Often it is useful and convenient to use geometric language for functions, referring to the elements of the domain and range as points, and saying that $\mathbf{F}$ *maps* x into $\mathbf{F}(x)$. Indeed, a function is sometimes called a *mapping*, and $\mathbf{F}(x)$ is called the *image* of x under the mapping $\mathbf{F}$.

Notice that even though a mapping $\mathbf{F}$ assigns to each x in its domain a unique image $\mathbf{F}(x)$ in its range, two or more different points in the domain might have the same image, as indicated in Figure A.4. That is, from the equation $\mathbf{F}(x_1) = \mathbf{F}(x_2)$ we cannot deduce that $x_1 = x_2$ unless we know that $\mathbf{F}$ possesses the additional property that each point in its range is the image of *exactly* one point in its domain. In that event it is possible to define an *inverse mapping* $\mathbf{F}^*$, whose domain is the range of $\mathbf{F}$ and whose range is the domain of $\mathbf{F}$, such that for each y in the domain of $\mathbf{F}^*$

$$\mathbf{F}^*(y) = x \text{ if and only if } y = \mathbf{F}(x).$$

See Figure A.5.

In summary, we say that a mapping $\mathbf{F}$ is *one-to-one* (or *reversible*) if and only if $\mathbf{F}(x_1) = \mathbf{F}(x_2)$ implies $x_1 = x_2$. If $\mathbf{F}$ is one-to-one, then an inverse function $\mathbf{F}^*$ can be defined from range $\mathbf{F}$ to dom $\mathbf{F}$ such that for each $x \in \text{dom } \mathbf{F}$, $\mathbf{F}^*(\mathbf{F}(x)) = x$. Usually $\mathbf{F}^*$ is denoted by $\mathbf{F}^{-1}$.

The equation $\mathbf{F}^{-1}\mathbf{F}(x) = x$ states that if x is mapped first by $\mathbf{F}$ and if the image $\mathbf{F}(x)$ is then mapped by $\mathbf{F}^{-1}$, the resultant image is x itself. This is a special instance of the more general concept of *successive mappings*. Suppose that A, B, and C are sets, that $\mathbf{F}$ is a mapping from A into B, and that $\mathbf{G}$ is a

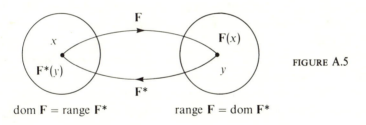

FIGURE A.5

dom $\mathbf{F}$ = range $\mathbf{F}^*$ range $\mathbf{F}$ = dom $\mathbf{F}^*$

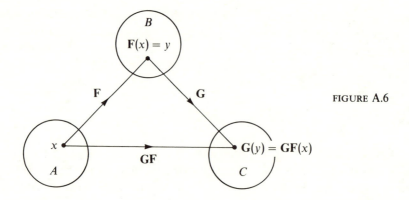

FIGURE A.6

mapping from B into C. Whenever the range of $\mathbf{F}$ is a subset of the domain of $\mathbf{G}$, we can define a direct mapping $\mathbf{GF}$ from A into C by the rule

$$\mathbf{GF}(x) = \mathbf{G}(\mathbf{F}(x)) \text{ for all } x \in \text{dom } \mathbf{F}.$$

See Figure A.6.

The concept of composite function is very important in linear algebra because matrix multiplication can be interpreted in terms of successive linear mappings of vector spaces. The notation that we have adopted for functions places the function symbol to the *left* of the symbol for the object that is mapped by that function, a convention that is time-honored and almost universally used in calculus and analysis. It has an awkward property in representing composite functions, however, because the successive mapping $\mathbf{F}$ followed by $\mathbf{G}$ is written $\mathbf{GF}$, reversing the order (as we read from left to right) in which the mappings take place. Largely for this reason some mathematicians prefer to write the function symbol to the *right* of the symbol for the object that it maps. You should be aware that some books on linear algebra use left-hand notation and others use right-hand notation for functions. Furthermore, any matrix that represents a linear mapping $\mathbf{F}$ in right-hand notation is the transpose of a matrix that represents $\mathbf{F}$ in left-hand notation, so many formulas of linear algebra appear in two different forms according to the notation used. When you browse among books on linear algebra, therefore, you must carefully determine in each book whether left-hand notation or right-hand notation is being used.

Using the notion of a function it is now quite easy to define what is meant by a *closed binary operation* on a set S. The idea is that each pair (x, y) of elements in S determines a unique element that is also in S. Hence a closed binary operation on S is a function from $S \times S$ to S; the domain of the function is the entire set $S \times S$. A *unary* operation on S is a function with domain S and range a subset of S. Special symbols, rather than function notation, typically are used to denote operations. For example, the operations previously listed for subsets of S are written $A \cup B$, $A \cap B$, and A'. Arithmetic operations in a field are written $a + b$, ab, $-a$, and a^{-1}.

A.5   COMPLEX NUMBERS

As an example of an algebraic system we briefly describe the field $\mathbb{C}$ of complex numbers. The elements of $\mathbb{C}$ are "numbers" z of the form

$$z = x + iy,$$

where x and y can be any real numbers. Equality in $\mathbb{C}$ is defined by $x_1 + iy_1 \ominus x_2 + iy_2$ if and only if $x_1 = x_2$ and $y_1 = y_2$ in $\mathbb{R}$. The two field operations, addition and multiplication, are defined in terms of the operations in $\mathbb{R}$ as follows:

$$(x_1 + iy_1) \oplus (x_2 + iy_2) \ominus (x_1 + x_2) + i(y_1 + y_2),$$

$$(x_1 + iy_1) \odot (x_2 + iy_2) \ominus (x_1 x_2 - y_1 y_2) + i(x_1 y_2 + x_2 y_1).$$

Using these definitions and the properties of real numbers one can verify directly that $\oplus$ and $\odot$ are closed operations on $\mathbb{C}$, that each is associative and commutative, and that $\odot$ is distributive over $\oplus$. The number $0 + i0$ is the identity of addition, and $1 + i0$ is the identity of multiplication. The additive inverse of $x + iy$ is $(-x) + i(-y)$, and if $x + iy \neq 0 + i0$, then the multiplicative inverse of $x + iy$ is $(x^2 + y^2)^{-1}x + i(x^2 + y^2)^{-1}(-y)$. Hence $\mathbb{C}$, together with the operations of $\oplus$ and $\odot$, forms a field.

By using ordered pairs (x, y) of real numbers one can define the complex number $x + iy$ without using either of the formal symbols $+$ and i, because those symbols only separate and identify the two components of the ordered pair (x, y) when it is written in the form $x + iy$. However, using the rule for multiplication we obtain the equation

$$(0 + i1) \odot (0 + i1) = -1 + 0i.$$

If we identify each complex number of the form $x + i0$ with the real number x, and if we write i in place of $0 + i1$, this equation becomes

$$i \odot i = -1,$$

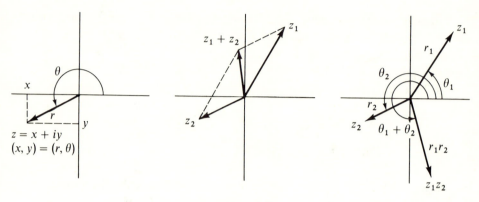

FIGURE A.7

or $i^2 = -1$ in $\mathbb{C}$. These conventions eliminate the need to use special symbols for addition and multiplication in $\mathbb{C}$.

The complex number $x + iy$ is represented geometrically as the point (x, y) of the Euclidean plane $\mathscr{E}^2$. In polar coordinates the point (x, y) can be represented as (r, θ) where $r = \sqrt{x^2 + y^2}$ and θ is the radian measure of the counterclockwise angle from the positive x-axis around to the line from the origin to (x, y). That is, $x = r \cos \theta$, $y = r \sin \theta$; or $z = r(\cos \theta + i \sin \theta)$.

Addition of complex numbers coincides with vector addition in $\mathscr{E}^2$. This is described geometrically, as in Figure A.7, by the parallelogram principle of vector addition. To interpret multiplication geometrically, we use the polar representation of complex numbers. If $z_1 = r_1(\cos \theta_1 + i \sin \theta_1)$ and $z_2 = r_2(\cos \theta_2 + i \sin \theta_2)$, then

$$z_1 z_2 = r_1 r_2 [(\cos \theta_1 \cos \theta_2 - \sin \theta_1 \sin \theta_2)$$
$$+ i(\cos \theta_1 \sin \theta_2 + \sin \theta_1 \cos \theta_2)]$$
$$= r_1 r_2 [\cos(\theta_1 + \theta_2) + i \sin(\theta_1 + \theta_2)].$$

The polar number r_1 is called the *magnitude* (or *modulus*) of the complex number z_1. (It is simply the length of the vector z_1.) The polar angle θ_1 is called the *angle* (or *amplitude*) of z_1. Hence the magnitude of the product $z_1 z_2$ is the product of the magnitudes of z_1 and z_2, whereas the angle of $z_1 z_2$ is the *sum* of the angles of z_1 and z_2.

The magnitude of z is denoted by $|z|$, also called the absolute value of z. Because

$$|x + iy| = \sqrt{x^2 + y^2},$$

$|z|$ is a nonnegative real number. Furthermore,

$$|z_1 z_2| = |z_1||z_2|,$$
$$|z_1 + z_2| \leq |z_1| + |z_2|.$$

The *conjugate* $\bar{z}$ of a complex number z is defined by the rule:

If $z = x + iy$, then $\bar{z} = x - iy$.

The following properties of the conjugate operation are easily verified:

If $z = z_1 + z_2$, then $\bar{z} = \bar{z}_1 + \bar{z}_2$,

If $z = z_1 z_2$, then $\bar{z} = \bar{z}_1 \bar{z}_2$,

$$\bar{\bar{z}} = z,$$

$z + \bar{z}$ is real,

$z\bar{z} = |z|^2$, real,

$$|\bar{z}| = |z|.$$

Geometrically, when z is represented by the point (x, y), $\bar{z}$ is represented by the point $(x, -y)$. That is, $\bar{z}$ is the reflection of z across the x-axis. In polar form if z is represented by (r, θ), then z is represented by $(r, -\theta)$.

SOLUTIONS FOR SELECTED EXERCISES

CHAPTER 1

Exercises 1.1

1. (i) $\begin{pmatrix} 2 + & c \\ 1 + & c \\ 0 - & 2c \end{pmatrix}$ for any c.

    (ii) $\begin{pmatrix} -1 \\ -2 \\ 4 \end{pmatrix}$.

    (iii) No solution exists.

    (iv) $\begin{pmatrix} -1 - 3c + & d \\ -1 - 3c + & 0d \\ 1 + & c - 2d \\ 0 + & c + 0d \\ 0 + 0c + & d \end{pmatrix}$.

2. (i) Various answers are correct, but the last equation of each is

$$0x_1 + 0x_2 + 0x_3 = b - 7.$$

(ii) A solution exists if and only if $b = 7$.

(iii) With $b = 7$ the solution is $\begin{pmatrix} 0 + c \\ 1 + c \\ 0 + c \end{pmatrix}$ for any c.

3. (i) If the n-tuple $(c_1, \ldots, c_n)$ is a solution of (1.1), it is also a solution of (1.2), and conversely, when $k \neq 0$. But if $k = 0$, the converse is not valid.

(ii) If the n-tuple $(c_1, \ldots, c_n)$ is a solution of (1.1), it is also a solution of (1.3). The converse is also valid.

(iii) The order in which the equations of a system are written does not affect the solution.

Exercises 1.2

1. Various correct answers are possible; several examples follow.

(i) $\begin{pmatrix} 1 & 2 & 3 \\ 0 & 1 & 2 \\ 0 & 0 & 1 \end{pmatrix}$.   (iv) $\begin{pmatrix} 1 & 1 & 1 & 0 \\ 0 & 1 & 3/2 & 1/2 \\ 0 & 0 & 1 & 4/3 \end{pmatrix}$.

2. (ii), (iii) $\begin{pmatrix} 1 & 0 & -1 \\ 0 & 1 & 1 \\ 0 & 0 & 0 \end{pmatrix}$.   (iv) $\begin{pmatrix} 1 & 0 & 0 & 1/6 \\ 0 & 1 & 0 & -3/2 \\ 0 & 0 & 1 & 4/3 \end{pmatrix}$.

3. (i) $\begin{pmatrix} 6 \\ 1 \\ -1 \end{pmatrix}$.   (ii) $\begin{pmatrix} 2 + c \\ 0 + c \\ 1/2 \end{pmatrix}$.

(iii) No solution.

Exercises 1.3

1. (i) The point $(6, 1, -1)$.

(ii) A line through the point $(2, 0, 1/2)$ parallel to the line through the origin and $(1, 1, 0)$.

(iv) The plane through the origin with equation

$$x_1 - 2x_2 + 2x_3 = 0.$$

4. A solution exists if and only if $3a - b - c = 0$. If that condition is satisfied, the solution set will be a plane; an echelon form for the matrix will have only two nonzero rows, so two of the four variables can be assigned arbitrary values.

CHAPTER 2

Exercises 2.1

1.

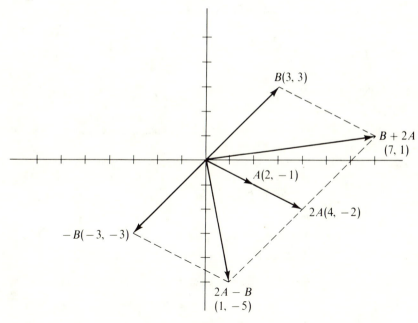

3. (i) tA is the line through the origin and A.

   (ii) $B + tA$ is the line through the B and $A + B$.

4. (i) L: $X = A + tB$, or $\begin{pmatrix} x_1 \\ x_2 \\ x_3 \end{pmatrix} = \begin{pmatrix} 2 \\ -1 \\ 6 \end{pmatrix} + t \begin{pmatrix} 4 \\ 1 \\ 3 \end{pmatrix}$.

   (ii) L: $\dfrac{x_1 - 2}{4} = \dfrac{x_2 + 1}{1} = \dfrac{x_3 - 6}{3}$.

6. $X = B + t(A - B)$ or $\begin{pmatrix} x_1 \\ x_2 \\ \cdot \\ \cdot \\ \cdot \\ x_n \end{pmatrix} = \begin{pmatrix} b_1 \\ b_2 \\ \cdot \\ \cdot \\ \cdot \\ b_n \end{pmatrix} + t \begin{pmatrix} a_1 - b_1 \\ a_2 - b_2 \\ \cdot \\ \cdot \\ \cdot \\ a_n - b_n \end{pmatrix}$.

Hence $\dfrac{x_1 - b_1}{a_1 - b_1} = \dfrac{x_2 - b_2}{a_2 - b_2} = \cdots = \dfrac{x_n - b_n}{a_n - b_n}$.

8. Property 9: $A = 1A = (1 + 0)A = 1A + 0A = A + 0A$.
   Add $-A$ to each side to obtain $Z = 0A$.

   Property 10: $Z = 0A = (1 + (-1))A = 1A + (-1)A = A + (-1)A$.
   Add $-A$ to each side to obtain $-A = (-1)A$.

   Property 11: $aZ = a(0A) = (a0)A = 0A = Z$.

Exercises 2.2

1. Properties (a), (b), and (c) are satisfied.

2. (a) $(dA + eB) \cdot C = (da_1 + eb_1, \ldots, da_n + eb_n) \cdot (c_1, \ldots, c_n)$
   $= (da_1 c_1 + eb_1 c_1) + \cdots + (da_n c_n + eb_n c_n)$
   $= dA \cdot C + eB \cdot C$.

   $A \cdot (dB + eC) = (a_1, \ldots, a_n) \cdot (db_1 + ec_1, \ldots, db_n + ec_n)$
   $= (da_1 b_1 + ea_1 c_1) + \cdots + (da_n b_n + ea_n c_n)$
   $= d\, A \cdot B + eA \cdot C$.

3. (i) If $A = Z$ then $A \cdot B = 0b_1 + \cdots + 0b_n = 0$ for every B.

   (ii) If $A \cdot B = 0$ for every B then $A \cdot A = 0$ so $A = Z$.

   (iii) If $A \cdot B = A \cdot C$ for every A, then

   $A \cdot B - A \cdot C = A \cdot (B - C) = 0$ for every A, so $B - C = Z$ by (ii).

4. (i) Let $c = \|A\|^{-1}$. Then $\|cA\| = c\|A\| = \|A\|^{-1}\|A\| = 1$.

   (ii) $\|B\| \cos \Psi(A, B) = A \cdot B/\|A\|$ by the formula preceding Theorem 2.1.
Also

$$\left(B - \frac{A \cdot B}{\|A\|} \frac{A}{\|A\|} \right) \cdot A = B \cdot A - A \cdot B = 0.$$

Observe that B is the sum of the vectors

$$\frac{A \cdot B}{\|A\|^2} A \text{ and } B - \frac{A \cdot B}{\|A\|^2} A.$$

The former is along A and the latter is perpendicular to A. Parts (iii) and (iv) are now demonstrated also.

6. Let Ψ denote the angle opposite the side AB. The sign of the number $(B - C) \cdot (A - C)$ indicates whether Ψ is acute $(+)$, right (0), or obtuse $(-)$. Analogous calculations can be made for each angle. The length of each side can be computed as an alternative method.

(i) Acute, not isosceles.

8. (i) Expand the left-hand side, using bilinearity; collect terms, using symmetry. In any parallelogram the sum of the squares of the lengths of the diagonals equals the sum of the squares of the lengths of the four sides.

9. (i) $(A + B) \cdot (A - B) = 0$ if and only if $\|A\| = \|B\|$.

(ii) Let the vertices of the quadrilateral be the origin, A, B, and C. Then the theorem becomes

$$\left\| \frac{B}{2} \right\| = \left\| \frac{A + B}{2} - \frac{A}{2} \right\| \text{ and } \left\| \frac{A}{2} - \frac{C}{2} \right\| = \left\| \frac{A + B}{2} - \frac{C + B}{2} \right\|,$$

and the proof is immediate.

Exercises 2.3

1. Each example except (iii) and (v) satisfies all properties of a vector space. But in (iii) and (v) a scalar multiple of a vector of the given form is not necessarily of that form.

2. Each example except (i) and (ii) satisfies all properties of a vector space. But in (i) scalar multiples of a function of the given form is not necessarily of that form. In (ii) the sum of two functions of the given form is not necessarily of that form.

3. Yes. All properties of a vector space are satisfied.

4. The zero polynomial is not of degree two, nor is the sum of $x^2 - x$ and $3x - x^2$ of degree two.

6. Let f and g be two solutions. Show that $f + g$ is a solution, and so is cf for every real number c. Also the zero function is a solution. All properties of a vector space are satisfied.

Exercises 2.4

1. The subsets defined in (ii), (iv), (v) are not subspaces. The zero function does not belong to (ii) or (iv). The set in (v) is not closed under scalar multiples, nor does each function have an additive inverse in the set.

2. The subsets defined in (i), (ii), (iv), and (v) are not subspaces, because (i), (iv), and (v) are not closed under vector sum, and (ii) is not closed under scalar multiples by negative scalars.

4. A possible choice of ξ is $(0, 0, 1)$. Any vector not in $\mathscr{S}$ will suffice.

5. Yes. $\mathscr{S}$ satisfies Theorem 2.7 as a subset of $\mathscr{V}$, and hence also as a subset of $\mathscr{T}$.

6.   (i) Let $\alpha \in (\mathscr{S} \cap \mathscr{U}) + (\mathscr{T} \cap \mathscr{U})$. Then $\alpha = \sigma + \tau$, where $\sigma \in \mathscr{S} \cap \mathscr{U}$ and $\tau \in \mathscr{T} \cap \mathscr{U}$. Thus $\sigma + \tau \in \mathscr{S} + \mathscr{T}$ and $\sigma + \tau \in \mathscr{U}$.

(ii) In $\mathbb{R}^2$ let $\mathscr{U} = [(1, 1)]$, $\mathscr{S} = [(1, 0)]$, $\mathscr{T} = [(0, 1)]$. Then $(\mathscr{S} \cap \mathscr{U}) + (\mathscr{T} \cap \mathscr{U}) = [\theta]$ but $(\mathscr{S} + \mathscr{T}) \cap \mathscr{U} = \mathscr{U}$.

(iii) Let $\beta \in (\mathscr{S} + \mathscr{T}) \cap \mathscr{U}$, where $\beta = \sigma + \tau$, $\sigma \in \mathscr{S}$, $\tau \in \mathscr{T}$. Show that $\beta \in (\mathscr{S} \cap \mathscr{U}) + (\mathscr{T} \cap \mathscr{U})$ because $\mathscr{S} \subseteq \mathscr{U}$. Then use (i).

9. $\mathscr{E}$ and $\mathscr{O}$ are subspaces of $\mathscr{V}(\mathbb{R})$ by Exercise 2.3-2 (iv) and (v). Let $f \in \mathscr{V}(\mathbb{R})$, and write $f(x) = \frac{1}{2}[f(x) + f(-x)] + \frac{1}{2}[f(x) - f(-x)]$. The two brackets on the right define an even and an odd function, so $\mathscr{V} = \mathscr{E} + \mathscr{O}$. Show that the zero function is the only function that is both even and odd, and conclude that $\mathscr{V} = \mathscr{E} \oplus \mathscr{O}$.

Exercises 2.5

1. (i) Determine scalars a, b, c, d (not all zero) such that $a\alpha + b\beta + c\alpha + d\delta = \theta$. This is a linear system of three equations in four variables, which can be solved by Gaussian elimination to obtain $a = -d/2, b = 5d/2, c = 3d/2$ for any value of d. In particular, let $d = 2$.

(ii) $\{\alpha, \beta, \gamma\}$ is linearly independent, as is any three-vector subset.

4. Let $T \subseteq S$. Any linear combination of vectors in T is a linear combination of vectors in S.

5. If $\sum_{i=1}^{n} a_i \alpha_i = \theta$, then $a_1 = 0$, $a_1 + a_2 = 0$, ..., $a_1 + a_2 + \cdots + a_n = 0$.

6. (i) If $a + b\sqrt{2} = 0$ with $a, b \in \mathbb{Q}$, then $\sqrt{2}$ is rational.

(ii) But $a + b\sqrt{2} = 0$ with $a, b \in \mathbb{R}$ if $b = -1$ and $a = \sqrt{2}$.

8. The vector equation $\sum_{i=1}^{m} c_i \alpha_i = \theta$ corresponds to a homogeneous linear system of n equations in the m variables c_i. Because $m > n$, a nontrivial solution exists, so the set of vectors is linearly dependent.

10. Any linear combination of vectors in $S \cup T$ is of the form $\sigma + \tau$, where $\sigma \in S$ and $\tau \in T$. If $\sigma + \tau = \theta$, then $\sigma = -\tau \in T$. Because $\mathscr{S} \cap \mathscr{T} = [\theta]$, $\sigma = \theta = \tau$; hence $S \cup T$ is independent.

12. If $\{f, f'\}$ is linearly dependent, $f' = kf$ for some $k \neq 0$. The solution of that differential equation consists of all functions of the form $f(x) = ae^{kx}$.

Exercises 2.6

1. (i) Two.

   (ii) Three.

3. (i) If $\sum_{i=1}^{n} b_i(c_i \alpha_i) = \theta$ then $b_i c_i = 0$ for each i because $\{\alpha_1, \ldots, \alpha_n\}$ is linearly independent. But $c_i \neq 0$, so $b_i = 0$ for each i, so $\{c_1 \alpha_1, \ldots, c_n \alpha_n\}$ is linearly independent and hence a basis for $\mathscr{V}_n$.

5. Let A be a linearly independent subset of $\mathscr{V}_n$. If $[A] = \mathscr{V}_n$, A is a basis. Otherwise, choose $\beta \in \mathscr{V}_n$ so that $\beta \notin [A]$, and let $B = A \cup \{\beta\}$. Then B is linearly independent, and the argument can be repeated. Because $\mathscr{V}_n$ is finite-dimensional, this process will produce a basis for $\mathscr{V}_n$.

7. (i) Clearly $\mathscr{V} = \mathscr{S} + \mathscr{T}$. If $\xi \in \mathscr{S} \cap \mathscr{T}$ then

$$\xi = \sum_{i=1}^{k} c_i \alpha_i = \sum_{i=k+1}^{n} c_i \alpha_i.$$

Then $c_1 \alpha_1 + \cdots + c_k \alpha_k - c_{k+1} \alpha_{k+1} - \cdots - c_n \alpha_n = 0$, so each $c_i = 0$ and $\xi = \theta$. Thus $\mathscr{S} \cap \mathscr{T} = [\theta]$ and $\mathscr{V} = \mathscr{S} \oplus \mathscr{T}$.

   (ii) Because $\mathscr{S} = [\alpha_1, \ldots, \alpha_k]$ and $\mathscr{T} = [\beta_1, \ldots, \beta_m]$, $\mathscr{V} = \mathscr{S} + \mathscr{T}$ is spanned by the set $\{\alpha_1, \ldots, \alpha_k, \beta_1, \ldots, \beta_m\}$. To see that this set is linearly independent, let $\sum_{i=1}^{k} a_i \alpha_i + \sum_{i=1}^{m} b_i \beta_i = \theta$. Then $\sum_{i=1}^{k} a_i \alpha_i$ is in $\mathscr{S} \cap \mathscr{T}$ so it equals θ. Then each $a_i = 0$. Similarly each $b_i = 0$.

10. (i) One. Any nonzero element of $\mathbb{Q}$ is a basis.

   (ii) Two. $\{1, i\}$ is a basis.

   (iii) One. Any nonzero element of $\mathbb{C}$ is a basis.

   (iv) The answer here is not obvious, but $\mathbb{R}$ is infinite-dimensional over $\mathbb{Q}$. If it were finite-dimensional, then $\mathbb{R}$ would be denumerable, which it isn't.

Exercises 2.7

1. (i) $H(\theta) = H(0\alpha) = 0H(\alpha) = \theta$.

   (ii) If $H(\alpha) = \theta$ for $\alpha \neq \theta$, H is not one-to-one because $H(\theta) = \theta$. Conversely, if $H(\alpha) \neq \theta$ when $\alpha \neq \theta$, then H is one-to-one, because if $H(\beta) = H(\gamma)$ then $H(\beta - \gamma) = \theta$, and $\beta = \gamma$.

2. (i) The composition JH maps $\mathscr{V}$ into $\mathscr{X}$. $JH(\alpha + \beta) = J(H\alpha + H\beta) = JH\alpha + JH\beta$ because J and H are linear. Also $JH(c\alpha) = J(cH\alpha) = cJH\alpha$, so JH is linear.

(ii) If $\mathbf{J}$ and $\mathbf{H}$ are both one-to-one, so is $\mathbf{JH}$, because if $\mathbf{JH}\alpha = \mathbf{JH}\beta$, then $\mathbf{H}\alpha = \mathbf{H}\beta$ and $\alpha = \beta$. Also $\mathbf{JH}$ maps $\mathscr{V}(\mathscr{F})$ onto $\mathscr{X}(\mathscr{F})$, so $\mathbf{JH}$ is an isomorphism from $\mathscr{V}(\mathscr{F})$ onto $\mathscr{X}(\mathscr{F})$.

5. Let $p(x) = a_0 + a_1 x + \cdots + a_n x^n$, $a_n \neq 0$ be a polynomial in $\mathscr{P}(\mathbb{Q})$. Let $\mathbf{H}p$ be the sequence $\{a_0, a_1, \ldots, a_n, 0, 0, \ldots\}$. Verify that $\mathbf{H}$ is one-to-one, onto $\mathscr{S}(\mathbb{Q})$, and linear.

6. $(\mathbf{H}_1 + \mathbf{H}_2)(a\alpha + b\beta) = \mathbf{H}_1(a\alpha + b\beta) + \mathbf{H}_2(a\alpha + b\beta)$
$$= a\mathbf{H}_1\alpha + b\mathbf{H}_1\beta + a\mathbf{H}_2\alpha + b\mathbf{H}_2\beta$$
$$= a(\mathbf{H}_1 + \mathbf{H}_2)\alpha + b(\mathbf{H}_1 + \mathbf{H}_2)\beta.$$

Hence $\mathbf{H}_1 + \mathbf{H}_2$ is linear. So is $c\mathbf{H}$ because

$$(c\mathbf{H})(a\alpha + b\beta) = c\mathbf{H}(a\alpha + b\beta) = ca\mathbf{H}\alpha + cb\mathbf{H}\beta$$
$$= a(c\mathbf{H})\alpha + b(c\mathbf{H})\beta.$$

The eight axioms of a vector space can be verified directly, so $\mathscr{H}$ is a vector space.

8. (i) The first $\mathbf{H}$ is linear; the second is not.

(ii) The first $\mathbf{H}$ is linear; the second is not.

(iii) The first $\mathbf{H}$ and second $\mathbf{H}$ are linear; the third is not.

CHAPTER 3

Exercises 3.1

1. $(\mathbf{S} + \mathbf{T})(a\alpha + b\beta) = \mathbf{S}(a\alpha + b\beta) + \mathbf{T}(a\alpha + b\beta)$ by the definition of $\mathbf{S} + \mathbf{T}$. Use the linearity of $\mathbf{S}$ and $\mathbf{T}$ to expand, and then regroup the resulting terms to obtain $a(\mathbf{S} + \mathbf{T})\alpha + b(\mathbf{S} + \mathbf{T})\beta$. Similarly,

$$(c\mathbf{T})(a\alpha + b\beta) = c[\mathbf{T}(a\alpha + b\beta)] = c[a\mathbf{T}\alpha + b\mathbf{T}\beta] = a(c\mathbf{T})\alpha + b(c\mathbf{T})\beta.$$

2. $\mathbf{T}[k_1(x_1, y_1) + k_2(x_2, y_2)] = \mathbf{T}(k_1 x_1 + k_2 x_2, k_1 y_1 + k_2 y_2)$
$$= (a(k_1 x_1 + k_2 x_2) + b(k_1 y_1 + k_2 y_2), c(k_1 x_1 + k_2 x_2) + d(k_1 y_1 + k_2 y_2))$$
$$= (k_1(ax_1 + by_1) + k_2(ax_2 + by_2), k_1(cx_1 + dy_1) + k_2(cx_2 + dy_2))$$
$$= (k_1(ax_1 + by_1), k_1(cx_1 + dy_1)) + (k_2(ax_2 + by_2), k_2(cx_2 + dy_2))$$
$$= k_1\mathbf{T}(x_1, y_1) + k_2\mathbf{T}(x_2, y_2).$$

3. Verify that the conjugate of the sum of two complex numbers is the sum of the two conjugates, and that the conjugate of any real multiple of a complex number is the same real multiple of the conjugate. If we write $x + iy$ as (x, y) and let $\mathbf{T}$ denote the conjugate mapping $\mathbf{T}(x, y) = (x, -y)$ this becomes a special case of Example $\mathbf{D}$ with $a = 1 = -c$ and $b = 0 = d$.

4. (i) Linear. A vertical projection onto the x-y plane in $\mathscr{E}^3$.

   (ii) Nonlinear. A vertical projection onto the plane $z = 1$.

   (iii) Linear. A reflection across the vertical plane $x = y$.

8. (i) $\mathbf{D}(p(x) + q(x)) = \mathbf{D}p(x) + \mathbf{D}q(x)$, and $\mathbf{D}(kp(x)) = k\mathbf{D}p(x)$.

   $\mathbf{M}(p(x) + q(x)) = x(p(x) + q(x)) = xp(x) + xq(x) = \mathbf{M}p(x) + \mathbf{M}q(x)$,

$\mathbf{M}(kp(x)) = xkp(x) = kxp(x) = k\mathbf{M}p(x)$.

   (ii) $\mathbf{D}$ is not nilpotent on the space of *all* polynomials because for each k, $\mathbf{D}^k p(x)$ is a nonzero polynomial if p has degree larger than k.

9. (i) $\mathbf{P}_{\mathscr{S}}(a\xi_1 + b\xi_2) = \mathbf{P}_{\mathscr{S}}(a(\sigma_1 + \tau_1) + b(\sigma_2 + \tau_2))$
   $$= \mathbf{P}_{\mathscr{S}}(a\sigma_1 + b\sigma_2 + a\tau_1 + b\tau_2)$$
   $$= a\sigma_1 + b\sigma_2 = a\mathbf{P}_{\mathscr{S}}(\xi_1) + b\mathbf{P}_{\mathscr{S}}(\xi_2).$$

Thus $\mathbf{P}_{\mathscr{S}}$ is linear and maps $\mathscr{V}$ onto $\mathscr{S}$. Similarly $\mathbf{P}_{\mathscr{T}}$ is linear and maps $\mathscr{V}$ onto $\mathscr{T}$.

   (ii) $\mathbf{P}_{\mathscr{S}}^2(\xi) = \mathbf{P}_{\mathscr{S}}(\mathbf{P}_{\mathscr{S}}\,\xi) = \mathbf{P}_{\mathscr{S}}(\sigma) = \sigma = \mathbf{P}_{\mathscr{S}}(\xi)$, so $\mathbf{P}_{\mathscr{S}}^2 = \mathbf{P}_{\mathscr{S}}$.

10. $\mathbf{R}(a\mathbf{S} + b\mathbf{T})\alpha = \mathbf{R}[a\mathbf{S}\alpha + b\mathbf{T}\alpha] = a\mathbf{RS}\alpha = b\mathbf{RT}\alpha = (a\mathbf{RS} + b\mathbf{RT})\alpha$.
   $(a\mathbf{R} + b\mathbf{S})\mathbf{T}\alpha = a\mathbf{RT}\alpha + b\mathbf{ST}\alpha = (a\mathbf{RT} + b\mathbf{ST})\alpha$.

12. (i) $(a_1 1 + a_2 i + a_3 j + a_4 k)(b_1 1 + b_2 i + b_3 j + b_4 k)$
   $$= (a_1 b_1 - a_2 b_2 - a_3 b_3 - a_4 b_4)1 + (a_1 b_2 + a_2 b_1 + a_3 b_4 - a_4 b_3)i$$
   $$+ (a_1 b_3 - a_2 b_4 + a_3 b_1 + a_4 b_2)j + (a_1 b_4 + a_2 b_3 - a_3 b_2 + a_4 b_1)k.$$

   (ii) $(a_1 1 + a_2 i + a_3 j + a_4 k)^{-1}$
   $$= (a_1^2 + a_2^2 + a_3^2 + a_4^2)^{-1}(a_1 1 - a_2 i - a_3 j - a_3 k),$$
   if $a_1^2 + a_2^2 + a_3^2 + a_4^2 \neq 0$.

   (iii) No; $ij = k = -ji$.

Exercises 3.2

1. $\mathscr{R}(\mathbf{T}) = \mathscr{E}^2$ and $\mathscr{N}(\mathbf{T}) = [\theta]$ in (i), (ii), and (iv), so $r(\mathbf{T}) = 2$ and $n(\mathbf{T}) = 0$. In (iii) $\mathscr{R}(\mathbf{T})$ is the y-axis and $\mathscr{N}(\mathbf{T})$ is the x-axis so $r(\mathbf{T}) = 1 = n(\mathbf{T})$.

3. $\mathscr{R}(\mathbf{D}) = \mathscr{P}$; $\mathscr{N}(\mathbf{D}) = \mathscr{P}_0$,
   where $\mathscr{P}_0 = \{p(x) \in \mathscr{P} \mid p(x) = c \text{ for some } c \in \mathbb{R}\}$.
   $\mathscr{R}(\mathbf{M}) = \mathscr{P}(0)$; $\mathscr{N}(\mathbf{M}) = [\theta]$, where $\mathscr{P}(0) = \{p(x) \in \mathscr{P} \mid p(0) = 0\}$.
   $\mathscr{R}(\mathbf{DM}) = \mathscr{P}$; $\mathscr{N}(\mathbf{DM}) = [\theta]$.
   $\mathscr{R}(\mathbf{MD}) = \mathscr{P}(0)$; $\mathscr{N}(\mathbf{DM}) = \mathscr{P}_0$.

4. If $r(\mathbf{T}) = 1$, let $\mathscr{R}(\mathbf{T}) = [\alpha]$. Then each $\xi \in \mathscr{V}$ determines some scalar $k(\xi)$ such that $\mathbf{T}\xi = k(\xi)\alpha \cdot \mathbf{T}^2\xi = k(\alpha)k(\xi)\alpha = k(\alpha)\mathbf{T}\xi$. Hence $\mathbf{T}^2 = c\mathbf{T}$ for $c = k(\alpha)$. The converse is false, however, because in $\mathbb{R}^3$ the projection $\mathbf{T}(a_1, a_2, a_3) = (a_1, a_2, 0)$ satisfies $\mathbf{T}^2 = \mathbf{T}$ and $r(\mathbf{T}) = 2$.

5. (i) Because $T^{p-1} \neq Z$, $T^{p-1}\xi = \theta$ for some $\xi \in \mathscr{V}$. But $T^p\xi = \theta$ because $T^p = Z$.

(ii) Suppose that $\sum_1^p c_i T^{i-1}\xi = \theta$, where not all c_i are zero. Let c_k be the first nonzero coefficient, where $k \leq p$. Then

$$c_k T^{k-1}\xi + \cdots + c_p T^{p-1}\xi = \theta,$$
$$T^{p-k}(c_k T^{k-1}\xi + \cdots + c_p T^{p-1}\xi) = \theta = c_k T^{p-1}\xi + \cdots + c_p T^p T^{p-k-1}\xi,$$
$$c_k T^{p-1}\xi = \theta.$$

Because $c_k \neq 0$, $T^{p-1}\xi = \theta$, contradicting the choice of ξ.

(iii) $\alpha = \sum_{i=1}^p b_i T^{i-1}\xi$, so $T\alpha = b_1 T\xi + \cdots + b_{p-1}T^{p-1}\xi + b_p T^p\xi$. Because $T^p\xi = \theta$, $T\alpha \in [B]$.

8. (i) If $\alpha \in \mathscr{R}(S + T)$, then $\alpha = (S + T)\xi = S\xi + T\xi$ for some $\xi \in \mathscr{V}_n$. But $S\xi \in \mathscr{R}(S)$ and $T\xi \in \mathscr{R}(T)$ so $\alpha \in \mathscr{R}(S) + \mathscr{R}(T)$.

(ii) $r(S + T) = \dim \mathscr{R}(S + T) \leq \dim (\mathscr{R}(S) + \mathscr{R}(T))$
$$\leq \dim \mathscr{R}(S) + \dim \mathscr{R}(T).$$

(iii) Use Theorem 3.5 and (ii).

9. (i) Let $\{\alpha_1, \ldots, \alpha_k\}$ be a basis for $\mathscr{N}(T)$, and extend it to a basis $\{\alpha_1, \ldots, \alpha_k, \alpha_{k+1}, \ldots, \alpha_m\}$ for $\mathscr{N}(ST)$. First show that $\{T\alpha_{k+1}, \ldots, T\alpha_m\}$ is a basis for the space $T\mathscr{N}(ST)$. Then show that $T\mathscr{N}(ST) \subseteq \mathscr{N}(S)$. It follows that $k = n(T)$, $m - k \leq n(S)$, so $m = n(ST) \leq n(T) + n(S)$. Then use Exercise 7.

(ii) Use the result of (i) in Theorem 3.5.

Exercises 3.3

1. The projection of $\mathscr{E}^2$ into the y-axis, $T(x, y) = (0, y)$, is singular because $n(T) = 1$. The other three examples are nonsingular because $n(T) = 0$.

4. Let T be nonsingular. Then $T^{-1}T$ is the identity mapping on $\mathscr{V}$ and TT^{-1} is the identity mapping on $\mathscr{R}(T)$. Thus T^{-1} is reversible and $(T^{-1})^{-1} = T$. If $c \neq 0$ $(cT)^{-1}(cT) = I$ and $c^{-1}T^{-1}(cT) = c^{-1}cT^{-1}T = I$, so $(cT)^{-1} = c^{-1}T^{-1}$.

5. If ST is nonsingular, there is a linear mapping R from $\mathscr{R}(ST)$ to $\mathscr{V}$ such that $R(ST) = I = (RS)T$, so $RS = T^{-1}$. Then $T(RS) = I = (TR)S$, so $TR = S^{-1}$. Conversely, let S and T be nonsingular; if $ST\alpha = ST\beta$, then $T\alpha = T\beta$, so $\alpha = \beta$. Hence ST is nonsingular.

6. Let $\{\alpha_1, \ldots, \alpha_k\}$ be any linearly independent set in $\mathscr{V}_n$; extend this set to a basis $\{\alpha_1, \ldots, \alpha_n\}$ for $\mathscr{V}_n$. If T is nonsingular then $\{T\alpha_1, \ldots, T\alpha_n\}$ is linearly independent, and so is $\{T\alpha_1, \ldots, T\alpha_k\}$. Conversely, if T preserves linear indpendence, T carries any basis for $\mathscr{V}_n$ into a basis for $\mathscr{R}(T)$, and thus is nonsingular.

7. (i) **J** and **JD** have domain $\mathscr{P}$ and range $\mathscr{P}_0$;
    D, DJ, and **D$_0$J** have domain $\mathscr{P}$ and range $\mathscr{P}$;
    D$_0$ has domain $\mathscr{P}_0$ and range $\mathscr{P}$;
    JD$_0$ has domain $\mathscr{P}_0$ and range $\mathscr{P}_0$.

   (ii) **DJ, D$_0$J, JD$_0$.**

   (iii) Only **D** and **JD** are singular.

Exercises 3.4

2. (i) $A = \begin{pmatrix} 1 & 1 \\ -1 & 1 \end{pmatrix}$.

   (ii) $T\alpha_1 = T\varepsilon_1 + 2T\varepsilon_2 = \alpha_1 - \alpha_2 = 0\varepsilon_1 + 3\varepsilon_2$
   $T\alpha_2 = T\varepsilon_1 - T\varepsilon_2 = \alpha_1 + \alpha_2 = 2\varepsilon_1 + \varepsilon_2$.

Solve these equations for $T\varepsilon_1$ and $T\varepsilon_2$ in terms of ε_1 and ε_2:

$$\begin{array}{l} 3T\varepsilon_1 = 4\varepsilon_1 + 5\varepsilon_2 \\ 3T\varepsilon_2 = -2\varepsilon_1 + 2\varepsilon_2 \end{array} \text{. Then } B = \begin{pmatrix} 4/3 & -2/3 \\ 5/3 & 2/3 \end{pmatrix}.$$

3. (i) $\begin{array}{ll} S(\varepsilon_1 + \varepsilon_2) = -2\varepsilon_1, & S\varepsilon_1 = -\varepsilon_1 - \varepsilon_2 \\ S\varepsilon_2 = -\varepsilon_1 + \varepsilon_2. & S\varepsilon_2 = -\varepsilon_1 + \varepsilon_2. \end{array}$

   (ii) $A = \begin{pmatrix} -1 & -1 \\ -1 & 1 \end{pmatrix}$.

   (iii) $S(a\varepsilon_1 + b\varepsilon_2) = a(-\varepsilon_1 - \varepsilon_2) + b(-\varepsilon_1 + \varepsilon_2) = -(a+b)\varepsilon_1 - (a-b)\varepsilon_2$.

   (iv) Let $c_1\beta_1 + c_2\beta_2 = \theta = c_1(\varepsilon_1 + \varepsilon_2) + c_2\varepsilon_2 = c_1\varepsilon_1 + (c_1 + c_2)\varepsilon_2$.
Because $\{\varepsilon_1, \varepsilon_2\}$ is linearly independent, $c_1 = 0 = c_1 + c_2$, so $c_1 = 0 = c_2$. Hence $\{\beta_1, \beta_2\}$ is a basis for $\mathbb{R}^2$.

   (v) Compute $S\beta_1$ and $S\beta_2$ in terms of β_1 and β_2. $B = \begin{pmatrix} -2 & -1 \\ 2 & 2 \end{pmatrix}$.

   (vi) Show that $\{S\varepsilon_1, S\varepsilon_2\}$ is linearly independent; S is nonsingular.

Then $\begin{array}{l} \varepsilon_1 = -S^{-1}\varepsilon_1 - S^{-1}\varepsilon_2 \\ \varepsilon_2 = -S^{-1}\varepsilon_1 + S^{-1}\varepsilon_2, \end{array}$ so $\begin{array}{l} -2\,S^{-1}\varepsilon_1 = \varepsilon_1 + \varepsilon_2 \\ -2\,S^{-1}\varepsilon_2 = \varepsilon_1 - \varepsilon_2, \end{array}$ and $C = \begin{pmatrix} -1/2 & -1/2 \\ -1/2 & 1/2 \end{pmatrix}$.

4. (i) If $\theta = a\alpha_1 + b\alpha_2 = (a-b)\varepsilon_1 + (2a+b)\varepsilon_2$, then $a - b = 0 = 2a + b$.
Hence $a = b = 0$, so $\{\alpha_1, \alpha_2\}$ is linearly independent.

   (ii) $\begin{array}{l} T\varepsilon_1 = \alpha_1 = \varepsilon_1 + 2\varepsilon_2 \\ T\varepsilon_2 = \alpha_2 = -\varepsilon_1 + \varepsilon_2 \end{array}$ so $D = \begin{pmatrix} 1 & -1 \\ 2 & 1 \end{pmatrix}$.

   (iii) $\begin{array}{l} T\alpha_1 = T\varepsilon_1 + 2T\varepsilon_2 = \alpha_1 + 2\alpha_2 \\ T\alpha_2 = -T\varepsilon_1 + T\varepsilon_2 = -\alpha_1 + \alpha_2 \end{array}$ so $E = \begin{pmatrix} 1 & -1 \\ 2 & 1 \end{pmatrix}$.

7. $ST\alpha_j = S \sum\limits_{k=1}^{m} a_{kj}\beta_k = \sum\limits_{k=1}^{m} a_{kj}(S\beta_k) = \sum\limits_{k=1}^{m} a_{kj} \sum\limits_{i=1}^{p} b_{ik}\gamma_i$

$\qquad = \sum\limits_{i=1}^{p} \sum\limits_{k=1}^{m} b_{ik} a_{kj}\gamma_i = \sum\limits_{i=1}^{p} c_{ij}\gamma_i.$

Hence $c_{ij} = \sum\limits_{k=1}^{m} b_{ik} a_{kj}$ for $i = 1, \ldots, p$ and $j = 1, \ldots, n.$

CHAPTER 4

Exercises 4.1

2. $AB = \begin{pmatrix} 2 & -4 & 3 \\ 2 & 9 & -1 \end{pmatrix}, \quad AC = \begin{pmatrix} -3 & 1 \\ 7 & 3 \end{pmatrix},$

$B^2 = \begin{pmatrix} 3 & 10 & 14 \\ 2 & -7 & 2 \\ 3 & 3 & -5 \end{pmatrix}, \quad BC = \begin{pmatrix} 13 & 8 \\ -6 & 0 \\ 0 & 1 \end{pmatrix}, \quad CA = \begin{pmatrix} 0 & 4 & 6 \\ -1 & 0 & 1 \\ 3 & 2 & 0 \end{pmatrix}.$

No other binary products are defined.

3. (i) A is a reflection across the x-axis: $(x, y) \to (x, -y).$
$\qquad$ B is a reflection across the line $y = x$: $(x, y) \to (y, x).$
$\qquad$ C is a vertical projection onto the line $y = 2x$:
$(x, y) \to (x, 2x).$

$\quad$ (ii) $AB = \begin{pmatrix} 0 & 1 \\ -1 & 0 \end{pmatrix}$; a clockwise rotation of $90°$: $(x, y) \to (y, -x).$

$\qquad$ $BA = \begin{pmatrix} 0 & -1 \\ 1 & 0 \end{pmatrix}$; a counterclockwise rotation of $90°$:
$(x, y) \to (-y, x).$
$\qquad$ $A^2 = I = B^2$; the identity mapping: $(x, y) \to (x, y).$
$\qquad$ $C^2 = C$; the vertical projection onto $y = 2x$: $(x, y) \to (x, 2x).$

$\quad$ 5. Define the function h from $\mathcal{M}_{1 \times 1}$ to $\mathcal{F}$ by $h((a)) = a.$ Then $h((a) + (b)) = h((a + b)) = a + b = h((a)) + h((b))$, and $h(c(a)) = h((ca)) = ca = ch(a).$ Hence h is a linear mapping. If $h((a)) = h((b))$, then $a = b$ and $(a) = (b)$, so h is one-to-one. Also for each $c \in \mathcal{F}$, $c = h((c))$, so h is an onto mapping. Thus h is an isomorphism. The vector space operations in $\mathcal{M}_{1 \times 1}$ correspond to the field operations in $\mathcal{F}$. Because $\mathcal{F}$ is a field and h is an isomorphism, $\mathcal{M}_{1 \times 1}$ is also a field.

$\quad$ 6. Let h map the two-by-two matrix

$$A = \begin{pmatrix} a & b \\ -b & a \end{pmatrix}$$ into the complex number $a + ib.$ Let $C = \begin{pmatrix} c & d \\ -d & c \end{pmatrix}.$

Show that h is one-to-one, onto $\mathbb{C}$, $h(A + C) = h(A) + h(C)$, $h(kA) = kh(A)$ for all $k \in \mathbb{C}$.

9. (i) $\sum_{i=1}^{3} b_i \beta_i = \theta$ if and only if

$$(b_1 + b_2)\alpha_1 + (-2b_1 + b_2 + b_3)\alpha_2 + (b_2 - b_3)\alpha_3 = \theta,$$

if and only if

$$-b_1 + b_2 = 0, \ -2b_1 + b_2 + b_3 = 0, \text{ and } b_2 - b_3 = 0;$$

if and only if $b_1 = b_2 = b_3 = 0$.

$$\alpha_1 = \tfrac{1}{2}(\beta_1 + \beta_2 + \beta_3),$$
$$\alpha_2 = \tfrac{1}{4}(-\beta_1 + \beta_2 + \beta_3),$$
$$\alpha_3 = \tfrac{1}{4}(-\beta_1 + \beta_2 - 3\beta_3).$$

(ii) $A = \begin{pmatrix} 1 & 1 & 0 \\ -2 & 1 & 1 \\ 0 & 1 & -1 \end{pmatrix}$, $\qquad B = \tfrac{1}{4} \begin{pmatrix} 2 & -1 & -1 \\ 2 & 1 & 1 \\ 2 & 1 & -3 \end{pmatrix}$.

(iv) Compute AB and BA.

10. (i) Relative to a fixed basis for $\mathscr{V}_n(\mathscr{F})$, let X and A represent linear transformations $\mathbf{S}$ and $\mathbf{T}$ respectively on $\mathscr{V}_n$. By Theorem 4.3 $XA = I$ if and only if $\mathbf{ST} = \mathbf{I}$ on $\mathscr{V}_n$. But then $\mathbf{TS} = \mathbf{I}$ on $\mathscr{V}_n$ by Theorem 3.8, and $AX = I$ by Theorem 4.3.

(ii)
$$a_{11}x_{11} + a_{21}x_{12} \qquad\qquad\qquad = 1,$$
$$a_{12}x_{11} + a_{22}x_{12} \qquad\qquad\qquad = 0,$$
$$a_{11}x_{21} + a_{21}x_{22} = 0,$$
$$a_{12}x_{21} + a_{22}x_{22} = 1.$$

$$a_{11}x_{11} \qquad\qquad + a_{12}x_{21} \qquad\qquad = 1,$$
$$a_{11}x_{12} \qquad\qquad + a_{12}x_{22} = 0,$$
$$a_{21}x_{11} \qquad\qquad + a_{22}x_{21} \qquad\qquad = 0,$$
$$a_{21}x_{12} \qquad\qquad + a_{22}x_{22} = 1.$$

11. Let $c_{ij} = \sum_{k=1}^{n} a_{ik} b_{kj}$. Because $0 \le a_{ik} \le 1$ and $0 \le b_{kj} \le 1$, $0 \le a_{ik} b_{kj} \le b_{kj}$ and $0 \le c_{ij} = \sum a_{ik} b_{kj} \le \sum b_{kj} = 1$. Also

$$\sum_{j=1}^{n} c_{ij} = \sum_{j=1}^{n} \left(\sum_{k=1}^{n} a_{ik} b_{kj} \right) = \sum_{k=1}^{n} \left(a_{ik} \sum_{j=1}^{n} b_{kj} \right) = \sum_{k=1}^{n} a_{ik} = 1.$$

Hence $C = AB$ is also a Markov matrix.

Exercises 4.2

1. The (r, s) entry of $U_{ij}A$ is $\sum_{k=1}^{n} u_{rk} a_{ks}$. Each $u_{rk} = 0$ if $r \ne i$ or $k \ne j$, and $u_{ij} = 1$. Hence the sum is zero if $r \ne i$ and is a_{js} if $r = i$. That is, $U_{ij}A$ has zero in every position in each row other than row i, and row i of $U_{ij}A$ is row j of A. A similar calculation can be used to describe AU_{ij}.

4. (i) Let $D = \text{diag}(d_1, \ldots, d_n)$ and $E = \text{diag}(e_1, \ldots, e_n)$. Then

$$DE = \text{diag}(d_1 e_1, \ldots, d_n e_n) = ED.$$

(ii) Suppose $DA = AD$ for all diagonal D. Then the (i, j) entries are $d_i a_{ij}$ and $a_{ij} d_j$ respectively. These will be equal for *all* D if and only if $a_{ij} = 0$ for $i \neq j$. That makes A diagonal.

7. Let

$$A = \begin{pmatrix} a & b \\ c & d \end{pmatrix}, \text{ where } A^2 = A.$$

By computation this matrix equation yields four scalar equations that yield the necessary condition

$$(a - d)(a + d - 1) = 0.$$

Consider separately the equations $a - d = 0$ and $a + d = 1$ to obtain the solutions I, Z, and all matrices of the following forms for any a and any nonzero b and c:

$$\begin{pmatrix} a & b \\ \dfrac{a(1-a)}{b} & 1-a \end{pmatrix} \quad \text{and} \quad \begin{pmatrix} a & \dfrac{a(1-a)}{c} \\ c & 1-a \end{pmatrix}.$$

9. If $A \neq Z$ but $A^2 = Z$, then a necessary condition is $(a + d)(a - d) = 0$. Show that the only such matrices are of the form

$$\begin{pmatrix} ab & b^2 \\ -a^2 & -ab \end{pmatrix}.$$

10. Let

$$\begin{pmatrix} a & b \\ c & d \end{pmatrix}\begin{pmatrix} x & y \\ u & v \end{pmatrix} = \begin{pmatrix} 1 & 0 \\ 0 & 1 \end{pmatrix}.$$

Write a linear system in four variables and reduce the corresponding matrix to echelon form to show that a necessary and sufficient condition for a solution is $ad - bc \neq 0$.

11. (a) If $AA^{-1} = I = A^{-1}A$, then A^{-1} is nonsingular, and A is *an* inverse of A^{-1}. But a nonsingular matrix B has a unique inverse because if $BC = BD$ then $C = B^{-1}(BC) = B^{-1}(BD) = D$. Hence $(A^{-1})^{-1} = A$.

(b) $(c^{-1}A^{-1})(cA) = c^{-1}cA^{-1}A = I$.

(c) $(B^{-1}A^{-1})(AB) = I$.

14. (i) $(AA^t)^t = (A^t)^t A^t = AA^t$ by Theorem 4.6.

(ii) $(A + A^t)^t = A^t + (A^t)^t = A^t + A = A + A^t$.

(iii) $(A - A^t)^t = A^t - (A^t)^t = -(A - A^t)$.

16.　(i) $(A + B)^t = A^t + B^t = A + B$.

　　(ii) $(AB)^t = B^t A^t = BA$. Hence $(AB)^t = AB$ if and only if $AB = BA$.

18. If $AA^{-1} = I$, then $(AA^{-1})^t = (A^{-1})^t A^t = I^t = I$. Hence A^t is nonsingular and $(A^t)^{-1} = (A^{-1})^t$.

19. AB is a one-by-one matrix $(\sum_{i=1}^{n} a_i b_i)$. But BA is an n-by-n matrix whose (i, j) entry is $b_i a_j$.

20.　(i) From Exercise 10, $L(v)$ is nonsingular if and only if

$$b^2\left(1 - \frac{v^2}{c^2}\right) \neq 0. \text{ Because } b^2 = \frac{c^2}{c^2 - v^2} \text{ and } v^2 < c^2, L(v) \text{ is nonsingular.}$$

　　(ii) Show that $L(v_1)L(v_2) = L(v')$ where

$$v' = \frac{(v_1 + v_2)c^2}{v_1 v_2 + c^2} \cdot$$

The product of matrices is associative, $L(0) = I$, and each $L(v)$ has an inverse for $|v| < c$. Hence the group axioms are satisfied.

Exercises 4.3

1.　(i) $\begin{pmatrix} 1 & 0 & 0 & \frac{1}{6} \\ 0 & 1 & 0 & -\frac{3}{2} \\ 0 & 0 & 1 & \frac{4}{3} \end{pmatrix}$; the solution is $X = \frac{1}{6}\begin{pmatrix} 1 \\ -9 \\ 8 \end{pmatrix}$.

2.　(i) $\begin{pmatrix} 1 & 0 & 1 \\ 0 & 1 & 1 \\ 0 & 0 & 0 \end{pmatrix}$; the solution is $X = \begin{pmatrix} 1 \\ 1 \end{pmatrix}$.

　　(ii) $\begin{pmatrix} 1 & 0 & -1 \\ 0 & 1 & 1 \\ 0 & 0 & 0 \end{pmatrix}$; the solution is $X = \begin{pmatrix} -1 \\ 1 \end{pmatrix}$.

　　(iii)　The reduced echelon form and solution are the same as in (ii), so systems (ii) and (iii) are equivalent.

3. The (r, s) entry of $U_{ih} U_{kj}$ is $\sum_{t=1}^{n} u_{rt} u_{ts}$, which is 0 unless $r = i$ and $s = j$. The (i, j) entry of $U_{ih} U_{kj}$ is nonzero only when t coincides with both i and j, and then it is 1. Hence $U_{ih} U_{kj} = U_{ij}$ if $h = k$, but $U_{ih} U_{kj} = Z$ if $h \neq k$. In particular $U_{ij}^2 = \delta_{ji} U_{ij}$, so $U_{ii}^2 = U_{ii}$ but for $j \neq i$ $U_{ij}^2 = Z$.

4. From Exercise 4.2-1 recall that $U_{ij} A$ has 0 in each position except in row i, and in row i the entry in column k is a_{jk}. Also AU_{ij} has 0 in each position except in column j, while in column j the entry in row r is a_{ri}.

　　(a) $M_i(c)A = [I + (c - 1)U_{ii}]A = A + (c - 1)U_{ii} A$. But row i of $U_{ii} A$ equals row i of A, and the other rows of $U_{ii} A$ are zero rows.

(c) $P_{ij} A = [I - U_{ii} + U_{ij} - U_{jj} + U_{ji}]A$

$\qquad = A - U_{ii} A + U_{ij} A - U_{jj} A + U_{ji} A.$

Starting with A, these operations successively replace row i by a zero row, then replace the zeros in row i by the entries of row j of A; then $-U_{jj} A$ replaces row j by a zero row and then $U_{ji} A$ replaces the zeros in row j by the entries of row i of A. The result coincides with A, except that rows i and j are interchanged.

(e) $AR_{i, i+cj} = A[I + cU_{ij}] = A + cAU_{ij}$. But AU_{ij} has 0 in each position except in column j, while column j coincides with column i of A. Hence $AR_{i, i+cj}$ coincides with A except that column j of $AR_{i, i+cj}$ is the sum of column j of A and c times column i of A.

6. $[M_i(c)]^t = M_i(c)$; $(R_{i, i+cj})^t = R_{j, j+ci}$; $P_{ij}^t = P_{ij}$.

7. Use the results of Exercise 6 and Theorem 4.7(d), (e), and (f). In effect the transpose interchanges the roles of i and j, which has no effect on $M_i(c)$ and P_{ij} because each is symmetric. Because $R_{i, i+cj}$ as a column operator reverses the roles of i and j, $(R_{i, i+cj})^t$ restores the original roles.

Exercises 4.4

1. (i) $A^{-1} = \frac{1}{8} \begin{pmatrix} -2 & 6 & 4 \\ 1 & -3 & 2 \\ 1 & 5 & 2 \end{pmatrix}$. (ii) $B^{-1} = \frac{1}{2} \begin{pmatrix} 1 & 1 & 1 & 0 \\ 0 & 1 & 0 & 1 \\ 1 & 1 & -1 & 0 \\ 0 & 1 & 0 & -1 \end{pmatrix}$.

2. The number of nonzero rows in any echelon form of a matrix is its rank,

(i) $r(A) = 3$.

(ii) $r(B) = 2$.

3. (i) $r(A) = 2$.

(ii) $r(B) = 3$; $B^{-1} = \begin{pmatrix} 2 & 2 & 1 \\ -3 & -3 & -1 \\ 1 & 2 & 2 \end{pmatrix}$.

4. Let A be an upper triangular n-by-n matrix. If some diagonal entry is zero, any echelon form of A will have at least one zero row, so A will be singular. But if every diagonal entry is nonzero, the n row vectors of A are linearly independent, so A is nonsingular.

5. Use Theorem 4.3 and Exercises 3.2-8 and 3.2-9 to conclude that

$$r(A + B) \le r(A) + r(B),$$
$$r(A) + r(B) - n \le r(AB) \le \min(r(A), r(B)).$$

Exercises 4.5

1. (i) $m = 4$, $n = 5$, $r(A) = 3 = r(A \mid Y)$.

$x_1 = -1 - 3x_4 + x_5$.

$x_2 = -1 - 3x_4$.

$x_3 = 1 + x_4 - 2x_5$.

x_4 and x_5 arbitrary.

The solution set is a translation by the vector $(-1, -1, 1, 0, 0)$ of the two-dimensional subspace $[(0, 0, 0, 1, 0), (0, 0, 0, 0, 1)]$ of $\mathbb{R}^5$, and $n - r(A) = 2$.

2. Let the m-by-n matrix A represent a linear mapping **T** from $\mathbb{R}^n$ to $\mathbb{R}^m$. Then $AX = Y$ has a solution for all Y if and only if **T** maps $\mathbb{R}^n$ onto $\mathbb{R}^m$. That is, $r(\mathbf{T}) = r(A) = m$.

4. If $m < n$ then $r(A) \leq m < n$. If $r(A) = r(A) \mid Y)$, the solution set is a translated subspace of $\mathbb{R}^n$ having dimension $n - r(A) > 0$. If $r(A) < r(A \mid Y)$, there are no solutions.

5. (i) A plane in $\mathbb{R}^3$, passing through the origin if and only if $y_i = 0$.

(ii) The intersection of two planes in $\mathbb{R}^3$. These two planes will be noninter-secting, yielding no solutions, when $r(A) < r(A \mid Y)$. But if $r(A) = r(A \mid Y)$, the two planes will intersect in a line.

(iii) Let $m = n = 3$. If $r(A) = 1$, there will be no solutions when $r(A) < r(A \mid Y)$, or infinitely many solutions (an entire plane) when $r(A) = r(A \mid Y)$. If $r(A) = 2$, there will be no solutions when $r(A) < r(A \mid Y)$, or infinitely many solutions (an entire line) when $r(A) = r(A \mid Y)$. If $r(A) = 3$, there will be a unique solution.

Exercises 4.6

1. The (i, j) entry of AB is $\sum_{s=1}^{n} a_{is} b_{sj}$. Let A_r denote columns $p(r) + 1, \ldots,$ $p(r) + t(r)$ of A and let B_r denote rows $p(r) + 1, \ldots, p(r) + t(r)$ of B. The (i, j) entry of $A_r B_r$ is $\sum_{s=1}^{t(r)} a_{i, \, p(r)+s} b_{p(r)+s, \, j}$. The (i, j) entry of $A_1 B_1 + \cdots + A_k B_k$ is the sum of the (i, j) entries of $A_r B_r$ as r ranges from one to k, and hence it is the sum $a_{i1} b_{ij} + \cdots + a_{in} b_{nj}$.

3. (i) For each $\xi \in [\alpha_1, \ldots, \alpha_k]$, $\mathbf{T}\xi \in [\alpha_1, \ldots, \alpha_k]$. That is, the space spanned by $\{\alpha_1, \ldots, \alpha_k\}$ is mapped by **T** into itself.

(ii) $AB = \left(\begin{array}{c|c} A_1 & A_2 \\ \hline Z & A_4 \end{array} \right) \left(\begin{array}{c|c} B_1 & B_2 \\ \hline Z & B_4 \end{array} \right) = \left(\begin{array}{c|c} A_1 B_1 & A_1 B_2 + A_2 B_4 \\ \hline Z & A_4 B_4 \end{array} \right)$.

(iii) Let **T** and **S** be the linear transformations represented by A and B. Because **T** and **S** each carries the subspace $[\alpha_1, \ldots, \alpha_k]$ into itself, so does the transformation **TS**. Thus the matrix representing **TS** relative to the given basis has a block of zeros in the last $n - k$ rows and the first k columns.

CHAPTER 5

Exercises 5.1

1. In Figure 5.1 let Ψ denote the angle at the origin of the parallelogram, and recall that $(\alpha \cdot \beta)^2 = \|\alpha\|^2 \|\beta\|^2 \cos \Psi$. Then

$$A^2 = \|\alpha\|^2 \|\beta\|^2 \sin^2 \Psi = \|\alpha\|^2 \|\beta\|^2 (1 - \cos^2 \Psi)$$
$$= \|\alpha\|^2 \|\beta\|^2 - (\alpha \cdot \beta)^2,$$

where $\alpha = (a_{11}, a_{21})$ and $\beta = (a_{12}, a_{22})$. Hence

$$A^2 = (a_{11}^2 + a_{21}^2)(a_{12}^2 + a_{22}^2) - (a_{11}a_{12} + a_{21}a_{22})^2$$
$$= (a_{11}a_{22} - a_{21}a_{12})^2.$$

2. (i) The given system is equivalent to the system

$$a_{11}x + a_{12}y = e,$$
$$(a_{11}a_{22} - a_{12}a_{21})y = a_{11}f - a_{21}e.$$

A solution will exist if and only if either (1) $a_{11}a_{22} - a_{12}a_{21} \neq 0$, or (2) $a_{11}a_{22} - a_{12}a_{21} = 0 = a_{11}f - a_{21}e$.

(ii) If a unique solution (x, y) exists, then

$$a_{11}a_{22} - a_{12}a_{21} = \det\begin{pmatrix} a_{11} & a_{12} \\ a_{21} & a_{22} \end{pmatrix} \neq 0.$$

In that case the solution is

$$x = \frac{\det\begin{pmatrix} e & a_{12} \\ f & a_{22} \end{pmatrix}}{\det\begin{pmatrix} a_{11} & a_{12} \\ a_{21} & a_{22} \end{pmatrix}}, \qquad y = \frac{\det\begin{pmatrix} a_{11} & e \\ a_{21} & f \end{pmatrix}}{\det\begin{pmatrix} a_{11} & a_{12} \\ a_{21} & a_{22} \end{pmatrix}}.$$

3. (i) $(\alpha \times \beta) \cdot \alpha = (a_2b_3 - a_3b_2)a_1 + (a_3b_1 - a_1b_3)a_2$
$$+ (a_1b_2 - a_2b_1)a_3 = 0.$$
$(\alpha \times \beta) \cdot \beta = (a_2b_3 - a_3b_2)b_1 + (a_3b_1 - a_1b_3)b_2$
$$+ (a_1b_2 - a_2b_1)b_3 = 0.$$

(ii) $\|\alpha \times \beta\|^2 = (a_2 b_3 - a_3 b_2)^2 + (a_3 b_1 - a_1 b_3)^2 + (a_1 b_2 - a_2 b_1)^2$

$\qquad = (a_1^2 + a_2^2 + a_3^2)(b_1^2 + b_2^2 + b_3^2)$

$\qquad \quad - (a_1 b_1 + a_2 b_2 + a_3 b_3)^2$

$\qquad = \|\alpha\|^2 \|\beta\|^2 - (\alpha \cdot \beta)^2 = \|\alpha\|^2 \|\beta\|^2 - \|\alpha\|^2 \|\beta\|^2 \, \cos^2 \Psi(\alpha, \beta)$

$\qquad = \|\alpha\|^2 \|\beta\|^2 (1 - \cos^2 \Psi) = \|\alpha\|^2 \|\beta\|^2 \sin^2 \Psi.$

5.   (i) $\det \begin{pmatrix} x & y & 1 \\ a & b & 1 \\ c & d & 1 \end{pmatrix} = (ad - bc) + (cy - xd) + (bx - ay) = 0$

is a linear equation in x and y, and hence represents a line in $\mathbb{R}^2$. The points (a, b) and (c, d) satisfy that equation, and therefore the line passes through those two points.

6.   (i) $\det A(x) = (3 - x)(4 - x)(-1 - x) - 4 - 8 + 4(4 - x) + 2(1 + x)$

$\qquad \qquad + 4(3 - x)$

$\qquad = -(3 - x)(4 - x)(1 + x) - 6x + 18$

$\qquad = -(3 - x)[(4 - x)(1 + x) - 6] = -(x - 3)(x - 2)(x - 1).$

(ii) $\det A(x) = 0$ if and only if $x = 3, 2,$ or 1.

(iii) $A(3) = \begin{pmatrix} 0 & 2 & 2 \\ 1 & 1 & 1 \\ -2 & -4 & -4 \end{pmatrix}.$

Two column vectors are equal, so the column vectors are linearly dependent, and $A(3)$ is singular. A similar argument holds for $A(2)$ and $A(1)$.

7. If the column vectors are linearly dependent, the parallelotope having those vectors as adjacent sides lies in some $(n - 1)$-dimensional subspace of $\mathbb{R}^n$; hence its n-dimensional volume is zero.

8.   (ii) $\det BA$

$= \det \begin{pmatrix} a_{11} b_{11} & a_{12} b_{11} + a_{22} b_{12} \\ a_{11} b_{21} & a_{12} b_{21} + a_{22} b_{22} \end{pmatrix} + \det \begin{pmatrix} a_{21} b_{12} & a_{12} b_{11} + a_{22} b_{12} \\ a_{21} b_{22} & a_{12} b_{21} + a_{22} b_{22} \end{pmatrix}$

$= \det \begin{pmatrix} a_{11} b_{11} & a_{12} b_{11} \\ a_{11} b_{21} & a_{12} b_{21} \end{pmatrix} + \det \begin{pmatrix} a_{11} b_{11} & a_{22} b_{12} \\ a_{11} b_{21} & a_{22} b_{22} \end{pmatrix} + \det \begin{pmatrix} a_{21} b_{12} & a_{12} b_{11} \\ a_{21} b_{22} & a_{12} b_{21} \end{pmatrix}$

$\quad + \det \begin{pmatrix} a_{21} b_{12} & a_{22} b_{12} \\ a_{21} b_{22} & a_{22} b_{22} \end{pmatrix}$

$= a_{11} a_{12} \det \begin{pmatrix} b_{11} & b_{11} \\ b_{21} & b_{21} \end{pmatrix} + a_{11} a_{22} \det \begin{pmatrix} b_{11} & b_{12} \\ b_{21} & b_{22} \end{pmatrix}$

$\quad + a_{21} a_{12} \det \begin{pmatrix} b_{12} & b_{11} \\ b_{22} & b_{21} \end{pmatrix} + a_{21} a_{22} \det \begin{pmatrix} b_{12} & b_{12} \\ b_{22} & b_{22} \end{pmatrix}$

$= 0 + a_{11} a_{22} \det B + a_{21} a_{12}(-\det B) + 0$

$= (a_{11} a_{22} - a_{21} a_{12}) \det B.$

Exercises 5.2

1. In formula (5.1) the only nonzero term in the sum, given that A is triangular, is $\pm a_{11} a_{22} \cdots a_{nn}$ because if p is not the identity permutation, there exist indices i and j such that $p(i) < i$ and $p(j) > j$. Then either $a_{p(i)i}$ or $a_{p(j)j}$ is zero because A is triangular. Also because the identity permutation is even,

$$\det A = a_{11} a_{22} \cdots a_{nn}.$$

3. If A is nonsingular $AA^{-1} = I$, so $(\det A)(\det A^{-1}) = \det I = 1$. Hence $\det A \neq 0$. It follows that $(\det A^{-1}) = (\det A)^{-1}$. Conversely, if A is singular, then a sequence of elementary column operations produces a matrix B having a zero column. Then $\det A = k \det B = 0$.

4. One method is to replace R_3 by $R_3 + 2R_4$, and apply the Laplace expansion to R_3. In the resulting four-by-four determinant replace C_1 by $C_1 + C_4$, and apply the Laplace expansion to C_1. Then apply the Laplace expansion to R_2 of the resulting three-by-three determinant. The result is

$$(-7)(2)(-1) \det\begin{pmatrix} -2 & 1 \\ 2 & 1 \end{pmatrix} = -56.$$

6. Let $r \neq t$; $\sum a_{rs} \operatorname{cof} a_{ts}$ is the Laplace expansion by the entries of row t of the determinant below, having two identical rows.

$$\det \begin{vmatrix} a_{11} & a_{12} & \cdots & a_{1n} \\ \cdot & \cdot & & \cdot \\ \cdot & \cdot & & \cdot \\ \cdot & \cdot & & \cdot \\ a_{r1} & a_{r2} & \cdots & a_{rn} \\ \cdot & \cdot & & \cdot \\ \cdot & \cdot & & \cdot \\ a_{r1} & a_{r2} & \cdots & a_{rn} \\ \cdot & \cdot & & \cdot \\ \cdot & \cdot & & \cdot \\ a_{n1} & a_{n2} & \cdots & a_{nn} \end{vmatrix} \begin{matrix} \\ \\ \\ \\ \text{row } r \\ \\ \\ \text{row } t \\ \\ \\ \\ \end{matrix}$$

8. In formula (5.1) the only nonzero products are those that include no entries from the zero block of A. Thus any nonzero product is of the form

$$\pm [b_{p(1)1} \cdots b_{p(k)k}] [d_{p(k+1), k+1} \cdots d_{p(n)n}],$$

where p permutes the sets $\{1, \ldots, k\}$ and $\{k+1, \ldots, n\}$ separately. But the fact that p permutes separately the two subsets of indices also guarantees that the signs also agree.

9. (ii) det $V(x_1, \ldots, x_n)$ is a polynomial of degree $n - 1$ in *each* of the variables $x_1, \ldots, x_n$. That polynomial has the value zero whenever $x_i = x_j$ for $i \neq j$ because two columns of the resulting determinant are identical. Hence $(x_j - x_i)$ is a factor for each $j > i$, and

$$\det V = k \prod_{1 \le i < j \le n} (x_j - x_i)$$

for some constant k, perhaps depending on n. For $n = 2$ we know that $k = 1$. For general n the coefficient of x_n^{n-1} is det $V(x_1, \ldots, x_{n-1})$. Hence by induction, $k = 1$ for all n.

Another form of proof uses elementary row operations to reduce det V to the form

$$\det \begin{pmatrix} 1 & 1 & \cdots & 1 \\ 0 & x_2 - x_1 & \cdots & x_n - x_1 \\ \cdot & \cdot & & \cdot \\ \cdot & \cdot & & \cdot \\ \cdot & \cdot & & \cdot \\ 0 & x_2^{n-2}(x_2 - x_1) & \cdots & x_n^{n-2}(x_n - x_1) \end{pmatrix}.$$

The argument is completed by using a Laplace expansion, a factorization of columns, and then induction.

Exercises 5.3

1. For example,

$$c_{21} = \text{cof } a_{21} = (-1)^3 \det \begin{pmatrix} 1 & 2 & 3 \\ 2 & 3 & 0 \\ 0 & 1 & 2 \end{pmatrix} = -(4 - 3) = -1.$$

By computing cof a_{ij} for $i, j = 1, 2, 3$, we obtain the matrix C of cofactors of A. Then com $A = C^t$ and det $A = 4$, so

$$A^{-1} = (\det A)^{-1} \text{ com } A = \tfrac{1}{4} C^t.$$

2. (i) Write the system as $AX = Y$, where

$$A = \begin{pmatrix} 2 & -1 & 3 \\ 0 & 1 & 0 \\ 2 & 1 & 1 \end{pmatrix}, \quad X = \begin{pmatrix} x_1 \\ x_2 \\ x_3 \end{pmatrix}, \quad Y = \begin{pmatrix} 3 \\ -2 \\ 1 \end{pmatrix}.$$

Then $X = A^{-1}Y$, where $A^{-1} = (\det A)^{-1} \text{ com } A$. But det $A = -4$, and

$$A^{-1}Y = -\tfrac{1}{4} \begin{pmatrix} 1 & 4 & -3 \\ 0 & -4 & 0 \\ -2 & -4 & 2 \end{pmatrix} \begin{pmatrix} 3 \\ -2 \\ 1 \end{pmatrix} = -\tfrac{1}{4} \begin{pmatrix} -8 \\ 8 \\ 4 \end{pmatrix} = \begin{pmatrix} 2 \\ -2 \\ -1 \end{pmatrix}.$$

(ii) $\det(A_1, A_2, A_3) = -4$; $\det(Y, A_2, A_3) = -8$;

$\det(A_1, Y, A_3) = 8$; $\det(A_1, A_2, Y) = 4$.

Hence $x_1 = \dfrac{-8}{-4}$; $x_2 = \dfrac{8}{-4}$; $x_3 = \dfrac{4}{-4}$.

(iii) Use the second equation to obtain

$$2x_1 + 3x_3 = 1,$$
$$2x_1 + x_3 = 3.$$

Subtract to obtain $x_3 = -1$; then $x_1 = 2$. We also have $x_2 = -2$.

5. $\det(A \text{ com } A) = \det[(\det A)I] = (\det A)^n \det I = (\det A)^n$,
$$= (\det A)(\det \text{com } A).$$

Hence det com $A = (\det A)^{n-1}$ if A is nonsingular. If A is singular, A com $A = Z$, so com A is singular, and det com $A = 0 = (\det A)^{n-1}$. Both A and com A are singular, or else both are nonsingular.

7. Let $x = r \cos \theta$, $y = r \sin \theta$. Then the Jacobian determinant is

$$J = \det \begin{pmatrix} \dfrac{\partial x}{\partial r} & \dfrac{\partial x}{\partial \theta} \\ \dfrac{\partial y}{\partial r} & \dfrac{\partial y}{\partial \theta} \end{pmatrix} = \det \begin{pmatrix} \cos \theta & -r \sin \theta \\ \sin \theta & r \cos \theta \end{pmatrix} = r;$$

from this result and (5.4) the stated result follows.

9. (i) $W(x) = \det \begin{pmatrix} e^{ax} & e^{bx} \\ ae^{ax} & be^{bx} \end{pmatrix} = (b-a)e^{ax}e^{bx} \neq 0$; independent.

(ii) $W(x) = \det \begin{pmatrix} \sin bx & \cos bx \\ b \cos bx & -b \cos bx \end{pmatrix} = -b$; independent if $b \neq 0$.

(iii) $W(x) = \det \begin{pmatrix} 2 & \sin^2 x & \cos 2x \\ 0 & \sin 2x & -2 \sin 2x \\ 0 & 2 \cos 2x & -4 \cos 2x \end{pmatrix} = 0$; dependent.

10. $y_1' = 2x$ and $y_2' = 2|x|$. Hence $W(x) = \det \begin{pmatrix} x^2 & x|x| \\ 2x & 2|x| \end{pmatrix} = 0$.

But suppose $ay_1 + by_2$ is identically zero on $[-1, 1]$. For $x > 0$ this implies that $a + b = 0$, and for $x < 0$ this implies that $a - b = 0$. For both these relations to hold we conclude that $a = b = 0$. Hence $\{y_1, y_2\}$ is linearly independent.

CHAPTER 6

Exercises 6.1

1. (i) Similarity of triangles is reflexive, symmetric, and transitive.

(ii) Parallelism of lines is symmetric; it is reflexive and transitive only if each line is considered to be parallel to itself.

2. Two triangles in $\mathscr{E}^2$ are similar if and only if the three angles of one triangle equal those of the other. If we agree to write the radian measure of the angles of each triangle as a number triple in nondecreasing order, we obtain a canonical form:

$$(x, y, z) \text{ where } 0 < x \le y \le z < x + y + z = \pi.$$

4. A and B are row equivalent if and only if a finite sequence of elementary row operations on B produces A; that is,

$$A = E_k \cdots E_2 E_1 B$$

for some set of elementary matrices E_i, each of which is nonsingular. Let $P = E_k \cdots E_2 E_1$. Then P is nonsingular. Conversely by Theorem 4.16 every nonsingular matrix is the product of elementary matrices. This proves Statement (a). Statement (b) follows from (a) and Theorem 4.13. To prove (c) let $m = n$. The reduced echelon form of a matrix of rank k has unit vectors in k columns. A is nonsingular if and only if $k = n$, and then the reduced echelon form of A is I_n.

5. Determine the reduced echelon form of each of the three given matrices. B and C are row equivalent, but the reduced echelon form of A differs from that of B and C in the $(1, 4)$ position.

Exercises 6.2

1. (i) Write a matrix Q with γ_i as column i for $i = 1, 2, 3$. The echelon form shows that Q has rank 3, so $\{\gamma_1, \gamma_2, \gamma_3\}$ is linearly independent and hence a basis.

(ii) $\gamma_1 = \varepsilon_1 + \varepsilon_2,$
$\gamma_2 = \varepsilon_1 \qquad + \varepsilon_3,$ Let $Q = \begin{pmatrix} 1 & 1 & 1 \\ 1 & 0 & -1 \\ 0 & 1 & 1 \end{pmatrix}.$
$\gamma_3 = \varepsilon_1 - \varepsilon_2 + \varepsilon_3.$

Then $Q^{-1} = \begin{pmatrix} 1 & 0 & -1 \\ -1 & 1 & 2 \\ 1 & -1 & -1 \end{pmatrix}.$ Hence $\begin{aligned} \varepsilon_1 &= \gamma_1 - \gamma_2 + \gamma_3, \\ \varepsilon_2 &= \qquad \gamma_2 - \gamma_3, \\ \varepsilon_3 &= -\gamma_1 + 2\gamma_2 - \gamma_3; \end{aligned}$

and $\xi = a\varepsilon_1 + b\varepsilon_2 + c\varepsilon_3 = (a - c)\gamma_1 + (-a + b + 2c)\gamma_2 + (a - b - c)\gamma_3.$

(iii) $M = Q^{-1}$, as in (ii).

(iv) $\xi = Q^{-1}\begin{pmatrix} a \\ b \\ c \end{pmatrix} = \begin{pmatrix} a - c \\ -a + b + 2c \\ a - b - c \end{pmatrix}$.

3. (i) ${}^{\beta}A_{\alpha} = \begin{pmatrix} 1 & 0 & -3 \\ 2 & 1 & 1 \end{pmatrix}$, $\begin{aligned} \beta_1 &= \tfrac{1}{3}(\delta_1 + 2\delta_2), \\ \beta_2 &= \tfrac{1}{3}(-\delta_1 + \delta_2). \end{aligned}$

$\mathbf{T}\,\gamma_1 = \mathbf{T}(\alpha_1 + \alpha_2 + \alpha_3) = -2\beta_1 + 4\beta_2 = 2\delta_1,$

$\mathbf{T}\,\gamma_2 = \mathbf{T}(\alpha_2 + \alpha_3) = -3\beta_1 + 2\beta_2 = \tfrac{1}{3}(-5\delta_1 - 4\delta_2),$

$\mathbf{T}\,\gamma_3 = \mathbf{T}(\alpha_1 + \alpha_3) = -2\beta_1 + 3\beta_2 = \tfrac{1}{3}(-5\delta_1 - \delta_2).$

Hence $B = \tfrac{1}{3}\begin{pmatrix} -6 & -5 & -5 \\ 0 & -4 & -1 \end{pmatrix}$.

4. Because $\gamma_i = \sum_{j=1}^{n} q_{ji}\alpha_j$ and $\alpha_j = \sum_{k=1}^{n} m_{kj}\gamma_k$, we have

$$\gamma_i = \sum_{j=1}^{n} q_{ji}\left(\sum_{k=1}^{n} m_{kj}\gamma_k \right) = \sum_{k=1}^{n} \left(\sum_{j=1}^{n} m_{kj}q_{ji} \right)\gamma_k.$$

The term in parentheses is the (k, i) entry of MQ, say c_{ki}. Then $\gamma_i = \sum_{k=1}^{n} c_{ki}\gamma_k$, so $c_{ki} = 1$ if $k = i$, and $c_{ki} = 0$ if $k \neq i$. Hence $MQ = I$.

5.

Let $\mathbf{R}\delta_i = \beta_i$ and let $\mathbf{S}\gamma_j = \alpha_j$. Then

$$\mathbf{T}_1\mathbf{S}\gamma_j = \mathbf{T}_1\alpha_j = \sum_{k=1}^{m} a_{kj}\beta_k,$$

and

$$\mathbf{R}\,\mathbf{T}_2\gamma_j = \mathbf{R}\sum_{k=1}^{m} a_{kj}\delta_k = \sum_{k=1}^{m} a_{kj}\mathbf{R}\delta_k = \sum_{k=1}^{m} a_{kj}\beta_k.$$

Hence $\mathbf{T}_1\mathbf{S} = \mathbf{R}\mathbf{T}_2$ for some nonsingular linear transformations $\mathbf{S}$, $\mathbf{R}$.

Exercises 6.3

1. The rank of each matrix is two, so A, B, and C are equivalent.

2. (i) $A^{-1} = QN = \frac{1}{8} \begin{pmatrix} -2 & 6 & 4 \\ 1 & -3 & 2 \\ 1 & 5 & 2 \end{pmatrix}$.

3. From Definition 6.2 B is equivalent to A if and only if B can be obtained by operating on A with suitable row operations and suitable column operations; the row operations multiply A on the left by elementary matrices and the column operations multiply A on the right. Because any product of elementary matrices is nonsingular, and conversely, the result follows.

5. Matrix equivalence partitions the set of all m-by-n matrices into disjoint equivalence classes. Equivalent matrices have the same rank because elementary row and column operations preserve rank. Furthermore, any matrix of rank k is equivalent to a matrix of the form described by Theorem 6.7. Because only one matrix of rank k has that form, the form is canonical for matrix equivalence.

6. Relative to matrix equivalence there are exactly $t + 1$ equivalence classes, where $t = \min (m, n)$, because the possible ranks of an m-by-n matrix are $0, 1, 2, \ldots, t$. Relative to row equivalence there are infinitely many equivalence classes, because if $n \neq 1$, there are infinitely many m-by-n matrices in reduced echelon form, no two of which are row equivalent.

7. (i) Equivalent. If $B = NAQ$, $B^t = Q^t A^t N^t$.

    (ii) Not necessarily equivalent. Let $A = \begin{pmatrix} 1 & 0 \\ 0 & 0 \end{pmatrix}$, $B = \begin{pmatrix} 0 & 1 \\ 0 & 0 \end{pmatrix}$.

8. Let $B = NAQ$, when N and Q are nonsingular. Let A represent a linear mapping $\mathbf{T}$ from $\mathcal{V}_n$ to $\mathcal{W}_m$ relative to bases $\{\alpha_1, \ldots, \alpha_n\}$ for $\mathcal{V}_n$ and $\{\beta_1, \ldots, \beta_m\}$ for $\mathcal{W}_m$. Let a new basis for $\mathcal{V}_n$ be defined by

$$\gamma_k = \sum_{j=1}^{n} q_{jk} \alpha_j, \ k = 1, \ldots, n.$$

Similarly let $N^{-1} = P = (p_{ij})$, and define a new basis for $\mathcal{W}_m$ by

$$\delta_j = \sum_{i=1}^{m} p_{ij} \beta_i, \ j = 1, \ldots, m.$$

Let B represent a linear mapping $\mathbf{S}$ from $\mathscr{V}_n$ to $\mathscr{W}_m$ relative to the γ- and δ-bases. Then $PB = AQ$, so

$$
\begin{aligned}
\mathbf{S}\gamma_k &= \sum_{j=1}^{m} b_{jk}\delta_j = \sum_{j=1}^{m} b_{jk}\left(\sum_{i=1}^{m} p_{ij}\beta_i\right) \\
&= \sum_{i=1}^{m}\left(\sum_{j=1}^{m} p_{ij}b_{jk}\right)\beta_i \\
&= \sum_{i=1}^{m}\left(\sum_{j=1}^{n} a_{ij}q_{jk}\right)\beta_i \\
&= \sum_{j=1}^{n} q_{jk}\left(\sum_{i=1}^{m} a_{ij}\beta_i\right) = \sum_{j=1}^{n} q_{jk}\mathbf{T}\alpha_j \\
&= \mathbf{T}\left(\sum_{j=1}^{n} q_{jk}\alpha_j\right) = \mathbf{T}\gamma_k.
\end{aligned}
$$

Hence $\mathbf{S} = \mathbf{T}$.

Exercises 6.4

1. Similarity is reflexive because $A = I^{-1}AI$ and symmetric because if $B = P^{-1}AP$, then $PBP^{-1} = A = Q^{-1}BQ$ where $Q = P^{-1}$. To show transitivity let $B = P^{-1}AP$ and $C = Q^{-1}BQ$. Then $C = Q^{-1}(P^{-1}AP)Q = (PQ)^{-1}A(PQ)$.

2. (i) $\mathbf{T}\alpha_1 = \begin{pmatrix} 1 & 1 \\ 1 & 1 \end{pmatrix}\begin{pmatrix} 1 \\ 1 \end{pmatrix} = \begin{pmatrix} 2 \\ 2 \end{pmatrix} = 2\alpha_1,$

$\mathbf{T}\alpha_2 = \begin{pmatrix} 1 & 1 \\ 1 & 1 \end{pmatrix}\begin{pmatrix} 1 \\ -1 \end{pmatrix} = \begin{pmatrix} 0 \\ 0 \end{pmatrix} = 0\alpha_2.$

Relative to the α basis $\mathbf{T}$ is represented by $B = \begin{pmatrix} 2 & 0 \\ 0 & 0 \end{pmatrix}$.

(ii) $P = \begin{pmatrix} 1 & 1 \\ 1 & -1 \end{pmatrix}$, $P^{-1} = \frac{1}{2}\begin{pmatrix} 1 & 1 \\ 1 & -1 \end{pmatrix}$,

$P^{-1}AP = \frac{1}{2}\begin{pmatrix} 1 & 1 \\ 1 & -1 \end{pmatrix}\begin{pmatrix} 2 & 0 \\ 2 & 0 \end{pmatrix} = \begin{pmatrix} 2 & 0 \\ 0 & 0 \end{pmatrix}.$

4. No. Let

$$
A = \begin{pmatrix} 1 & 0 \\ 0 & 0 \end{pmatrix}, B = \begin{pmatrix} 0 & 0 \\ 1 & 0 \end{pmatrix};
$$

$r(AB) \neq r(BA)$, so AB and BA are not similar. But if A is nonsingular, AB is similar to $A^{-1}(AB)A = BA$.

6. Let $B = P^{-1}AP$.

(i) $B^{-1} = P^{-1}A^{-1}P$, so A^{-1} and B^{-1} are similar.

(ii) $B^t = P^tA^t(P^{-1})^t = P^tA^t(P^t)^{-1} = Q^{-1}A^tQ$, where $Q^{-1} = P^t$, so A^t and B^t are similar.

7. Let $T^2 = T$. Then $\mathscr{R}(T^2) = \mathscr{R}(T)$ and $\mathscr{N}(T) = \mathscr{N}(T^2)$.

(i) By Theorem 3.6, $\mathscr{V} = \mathscr{R}(T) \oplus \mathscr{N}(T)$.

(ii) If $\eta \in \mathscr{R}(T)$, then $\eta = T\xi$ for some $\xi \in \mathscr{V}$; $T\eta = T^2\xi = T\xi = \eta$. By definition, if $v \in \mathscr{N}(T)$ then $Tv = \theta$.

(iii) Choose a basis for $\mathscr{R}(T)$ and a basis for $\mathscr{N}(T)$. Their union is a basis for $\mathscr{V}$, relative to which T is represented by a diagonal block matrix

$$\begin{pmatrix} I_k & Z \\ Z & Z \end{pmatrix}, \text{ where } k = r(T).$$

CHAPTER 7

Exercises 7.1

1. (i) $p(x) = \det(A - xI) = (x + 3)(x - 2)$, so $\lambda_1 = -3$ and $\lambda_2 = 2$. For $\lambda = 3$, $AX = \lambda X$ if and only if

$$X = a \begin{pmatrix} 2 \\ -3 \end{pmatrix} \text{ for some } a, \text{ so } \mathscr{C}(-3) = [X_1], \text{ where } X_1 = \begin{pmatrix} 2 \\ -3 \end{pmatrix}.$$

Similarly,

$$\mathscr{C}(2) = [X_2] \text{ where } X_2 = \begin{pmatrix} 1 \\ 1 \end{pmatrix}.$$

Hence $\{X_1, X_2\}$ is a maximal linearly independent set of characteristic vectors.

(iii) $p(x) = \det(C - xI) = (x - 1)^3(x - 3)$, so $\lambda_1 = 1 = \lambda_2 = \lambda_3$ and $\lambda_4 = 3$.

Then $CX_1 = \lambda_1 X_1$ if and only if $X_1 = a \begin{pmatrix} 1 \\ 0 \\ -2 \\ 0 \end{pmatrix}$ for some a, so $\mathscr{C}(1) = [X_1]$.

Similarly, $\mathscr{C}(3) = [X_4]$ where $X_4 = \begin{pmatrix} 1 \\ 0 \\ 0 \\ 0 \end{pmatrix}$.

Hence a maximal linearly independent set of characteristic vectors contains only two vectors; for example, $\{X_1, X_4\}$.

2. $p(x) = \det(A - xI) = -(x - 1)^2(x - 2)$, so $\lambda_1 = 1 = \lambda_2, \lambda_3 = 2$. $T\alpha_2$ is represented by

$$A \begin{pmatrix} 5 \\ 2 \\ -5 \end{pmatrix}, \text{ which by matrix computation is } \begin{pmatrix} 5 \\ 2 \\ -5 \end{pmatrix},$$

so $T\alpha_2 = \alpha_2$. Similarly, $T\alpha_3 = 2\alpha_3$. The second column of B shows that $T\alpha_2 = \alpha_2$. The third column shows that $T\alpha_3 = 2\alpha_3$.

4. The characteristic polynomial of A and of B is x^2. The characteristic subspace $\mathscr{C}(0)$ associated with B is $\mathbb{R}^2$; but $\mathscr{C}(0)$, associated with A, is

$$\left[\binom{0}{1} \right], \text{ having dimension one.}$$

A and B do not represent the same linear transformation because their null spaces do not coincide. Because similar matrices represent the same linear transformations, A and B are not similar.

6. (i) If $A^2 = A$, and $AX = \lambda X$, then $A^2 X = A(\lambda X) = \lambda^2 X$. Also $A^2 X = AX = \lambda X$, so $\lambda X = \lambda^2 X$ for some $X \neq Z$. Hence $\lambda = 0$ or 1.

(iii) If $AX = \lambda X$ and A is nonsingular, $AX \neq Z$ so $\lambda \neq 0$.

7. If A is nonsingular and $AX = \lambda X$, then $X = A^{-1}(\lambda X)$, so $A^{-1}X = \lambda^{-1}X$. Hence if λ is a characteristic value of A associated with X, then λ^{-1} is a characteristic value of A^{-1} associated with the same characteristic vectors as λ.

8. If $AX = \lambda X$, then $A^2 X = A(\lambda X) = \lambda(AX) = \lambda^2 X$. By induction assume that $A^k X = \lambda^k X$. Then $A(A^k X) = A(\lambda^k X)$, so $A^{k+1}X = \lambda^{k+1}X$.

10. Consider the terms involving x^{n-1} in the expansion of the given form of $p(x)$. It is the sum of terms obtained by multiplying x from $n-1$ factors with the constant term $(-\lambda_i)$ from the remaining factor. Hence it is

$$(-\lambda_1 - \lambda_2 - \cdots - \lambda_n)x^{n-1} = -(\text{tr } A)x^{n-1}.$$

11. (i) For $n = 1$, $p(x) = \det(c_1 - x) = -(x - c_1)$. Assume the result is valid for $n = k - 1$, and let $n = k$. Expand $\det(C - xI)$ by the entries of the first column to obtain

$$p(x) = -x \det \begin{pmatrix} -x & 0 & \cdots & 0 & c_2 \\ 1 & -x & \cdots & 0 & c_3 \\ \cdot & \cdot & & \cdot & \cdot \\ \cdot & \cdot & & \cdot & \cdot \\ \cdot & \cdot & & \cdot & \cdot \\ 0 & & \cdots & 1 & c_k - x \end{pmatrix} - \det \begin{pmatrix} 0 & 0 & \cdots & 0 & c_1 \\ 1 & -x & \cdots & 0 & c_3 \\ \cdot & \cdot & & \cdot & \cdot \\ \cdot & \cdot & & \cdot & \cdot \\ \cdot & \cdot & & \cdot & \cdot \\ 0 & 0 & \cdots & 1 & c_k - x \end{pmatrix}$$

$$= -x(-1)^{k-1}[x^{k-1} - c_k x^{k-2} - \cdots - c_3 x - c_2] - (-1)^k c_1$$
$$= (-1)^k[x^k - c_k x^{k-1} - \cdots - c_2 x - c_1].$$

(ii) Given a polynomial $q(x) = a_0 x^p + a_1 x^{p-1} + \cdots + a_p$ of degree p, let

$$c_j = -\frac{a_{p-j+1}}{a_0} \text{ for } j = 1, 2, \ldots, p.$$

Then

$$q(x) = (-1)^p a_0 [(-1)^p (x^p - c_p x^{p-1} - \cdots - c_2 x - c_1)].$$

The p-by-p companion matrix with $c_1, \ldots, c_p$ in its last column has a scalar multiple of $q(x)$ as its characteristic polynomial.

12. Relative to the basis described, **T** is represented by a matrix of the block form

$$\begin{pmatrix} \lambda I & B \\ Z & D \end{pmatrix},$$

where I is the identity matrix of size dim $\mathscr{C}(\lambda)$ and Z is a matrix of zeros.

13. Assume that $p \le k$ is an integer such that $\{\xi_1, \ldots, \xi_{p-1}\}$ is linearly independent but $\{\xi_1, \ldots, \xi_p\}$ is linearly dependent:

$$c_1 \xi_1 + c_2 \xi_2 + \cdots + c_p \xi_p = \theta,$$

where not all c_i are zero. Operating on each side by **T** and also multiplying each side by λ_p we have the two equations

$$c_1 \lambda_1 \xi_1 + c_2 \lambda_2 \xi_2 + \cdots + c_p \lambda_p \xi_p = \theta,$$
$$c_1 \lambda_p \xi_1 + c_2 \lambda_p \xi_2 + \cdots + c_p \lambda_p \xi_p = \theta.$$

Hence by subtraction

$$c_1(\lambda_1 - \lambda_p)\xi_1 + c_2(\lambda_2 - \lambda_p)\xi_2 + \cdots + c_{p-1}(\lambda_{p-1} - \lambda_p)\xi_{p-1} = \theta.$$

By our assumption and the fact that the λ's are distinct, we have $c_1 = c_2 = \cdots = c_{p-1} = 0$. Hence $c_p \xi_p = \theta$, so $c_p = 0$, a contradiction.

14. Let $\alpha \in \mathscr{C}(\lambda_k) \cap (\mathscr{C}(\lambda_1) + \cdots + \mathscr{C}(\lambda_{k-1}))$. If $\alpha \ne \theta$ then

$$\alpha = \gamma_1 + \cdots + \gamma_{k-1}, \text{ where } \gamma_i \in \mathscr{C}(\lambda_i),$$
$$\mathbf{T}\alpha = \lambda_1 \gamma_1 + \cdots + \lambda_{k-1} \gamma_{k-1},$$
$$= \lambda_k \alpha = \lambda_k \gamma_1 + \lambda_k \gamma_2 + \cdots + \lambda_k \gamma_{k-1}.$$

Hence $(\lambda_1 - \lambda_k)\gamma_1 + (\lambda_2 - \lambda_k)\gamma_2 + \cdots + (\lambda_{k-1} - \lambda_k)\gamma_{k-1} = \theta$. Because the λ's are distinct, the γ's are linearly dependent, which contradicts the result of Exercise 13. Hence $\alpha = \theta$, and by Definition 2.6

$$\mathscr{C}(\lambda_k) + (\mathscr{C}(\lambda_1) + \cdots + \mathscr{C}(\lambda_{k-1})) = \mathscr{C}(\lambda_k) \oplus (\mathscr{C}(\lambda_1) + \cdots + \mathscr{C}(\lambda_{k-1})).$$

An inductive argument then can be used to conclude that

$$\mathscr{C}(\lambda_1) + \mathscr{C}(\lambda_2) + \cdots + \mathscr{C}(\lambda_k) = \mathscr{C}(\lambda_1) \oplus \mathscr{C}(\lambda_2) \oplus \cdots \oplus \mathscr{C}(\lambda_k).$$

16. Let M be a Markov matrix, let $MX = \lambda X$, and let $a = \max |x_i| = |x_k|$. Then for each i, $\lambda x_i = \sum m_{ij} x_j$, so

$$|\lambda x_i| \le \sum |m_{ij} x_j| \le \sum |m_{ij}| a = a;$$

in particular, $|\lambda| |x_k| = |\lambda| a \le a$, so $|\lambda| \le 1$.

17. M has one as a characteristic value if and only if $M - I$ is singular. The sum of the column vectors of $M - I$ is θ because the sum of the entries in each row of M is 1. Hence the columns of $M - I$ are linearly dependent, and $M - I$ is singular.

Exercises 7.2

1. Because a diagonalizing matrix P is determined by the choice of a basis of characteristic vectors, P is not unique; solutions other than those listed below are also correct.

(i) $\lambda_1 = 1 = \lambda_2 = \lambda_3$; $X_1 = a\begin{pmatrix} 1 \\ 1 \\ 0 \end{pmatrix} + b\begin{pmatrix} -1 \\ 0 \\ 3 \end{pmatrix}$;

not diagonable because $2 = g_1 < s_1 = 3$.

(iv) $\lambda_1 = 1$, $\lambda_2 = -1$, $\lambda_3 = -2$; $P = \begin{pmatrix} 1 & 0 & 0 \\ 1 & 2 & 1 \\ 0 & 1 & 1 \end{pmatrix}$.

(v) $\lambda_1 = 1 = \lambda_2$, $\lambda_3 = -2$; $P = \begin{pmatrix} 1 & 0 & 1 \\ 0 & 0 & -1 \\ 0 & 1 & 2 \end{pmatrix}$.

2. Each of the matrices in (ii), (iv), (v), and (vi) has a basis of characteristic vectors and hence is diagonable. A diagonal form of each has the characteristic values λ_i along the diagonal, with λ_i appearing in s_i diagonal positions.

3. (i)

$$\lambda_1 = \lambda_2 = \lambda_3 = 1; \quad X_1 = a\begin{pmatrix} 1 \\ -2 \\ 1 \end{pmatrix};$$

not diagonable because $g_1 = 1 < 3 = s_1$.

(iii) $\lambda_1 = 0$, $\lambda_2 = 1$, $\lambda_3 = 2$, $\lambda_4 = -2$; diagonable because the four characteristic values are distinct.

5. By Theorem 7.4 A is diagonable if and only if the characteristic vectors of A span $\mathscr{V}_n$. But for each λ_i, the associated characteristic vectors span $\mathscr{C}(\lambda_i)$, so the set of all characteristic vectors of A span $\mathscr{C}(\lambda_1), + \cdots + \mathscr{C}(\lambda_k) \subseteq \mathscr{V}_n$. Equality holds if and only if A is diagonable.

6. (i) Let $\{\alpha_1, \ldots, \alpha_k\}$ be a basis for $\mathscr{C}(\lambda_1)$. Then $k = g_1$. Extend this to a basis $\{\alpha_1, \ldots, \alpha_k, \beta_{k+1}, \ldots, \beta_n\}$ for $\mathscr{V}_n$. Then $T\alpha_i = \lambda_i\alpha_i$ for $i = 1, \ldots, k$, so the matrix A that represents T relative to this basis has λ_1 in each of the first k diagonal positions, and has 0 in every nondiagonal position for the first k columns. The last $n - k$ columns of A depend entirely on the choice of the β_i.

(ii) $\det(A - xI) = \det((\lambda_1 - x)I(g_1)) \det(C - xI)$, so
$$p(x) = (\lambda_1 - x)^{g_1} q(x), \text{ where } q(x) = \det(C - xI).$$

(iii) From (ii) the algebraic multiplicity s_1 of λ_1 is at least g_1.

8. (i) $E_i = E_i I = E_i \sum_{j=1}^{t} E_j = \sum_{j=1}^{t} E_i E_j = E_i^2$.

(ii) $F_i^2 = (PE_i P^{-1})^2 = PE_i^2 P^{-1} = PE_i P^{-1} = F_i$.

$F_i F_j = (PE_i P^{-1})(PE_j P^{-1}) = PE_i E_j P^{-1} = Z$ if $i \neq j$.

$$\sum_{j=1}^{t} F_j = \sum_{j=1}^{t} PE_j P^{-1} = P\left(\sum_{j=1}^{t} E_i\right) P^{-1} = I.$$

9. Various correct spectral decompositions can be written; the following solutions are typical but not unique.

(i) Let $E_1 = \mathrm{diag}(1, 0, 0)$, $E_2 = \mathrm{diag}(0, 1, 0)$, $E_3 = \mathrm{diag}(0, 0, 1)$, and let

$$P = \begin{pmatrix} 1 & 0 & 0 \\ 1 & 2 & 1 \\ 0 & 1 & 1 \end{pmatrix}. \text{ Then compute } F_i = PE_i P^{-1} \text{ to obtain}$$

$$F_1 = \begin{pmatrix} 1 & 0 & 0 \\ 1 & 0 & 0 \\ 0 & 0 & 0 \end{pmatrix}, F_2 = \begin{pmatrix} 0 & 0 & 0 \\ -2 & 2 & -2 \\ -1 & 1 & -1 \end{pmatrix}, F_3 = \begin{pmatrix} 0 & 0 & 0 \\ 1 & -1 & 2 \\ 1 & -1 & 2 \end{pmatrix}.$$

Then $A = F_1 - F_2 - 2F_3$, where $F_i^2 = F_i$, $F_i F_j = Z$ when $i \neq j$, $\sum F_i = I$.

10. (i) Let $E_1 = \mathrm{diag}(1, 1, 0)$ and $E_2 = \mathrm{diag}(0, 0, 1)$. Let $F_i = PE_i P^{-1}$, obtaining a spectral decomposition for A:

$$F_1 = \begin{pmatrix} 1 & 1 & 0 \\ 0 & 0 & 0 \\ 0 & 2 & 1 \end{pmatrix}, F_2 = \begin{pmatrix} 0 & -1 & 0 \\ 0 & 1 & 0 \\ 0 & -2 & 0 \end{pmatrix}, A = F_1 - 2F_2.$$

(ii) Verify by computation that F_1 and F_2 are idempotent, orthogonal, and supplementary, and that $F_1 - 2F_2 = A$.

11. (i) P is the matrix in which column k is the n-tuple that represents X_k relative to the original basis.

(ii) If $n = 1$ or 2, $P^{-1}AP$ is upper triangular. Proceeding by induction we assume that the result is valid for any square matrix of dimension $n - 1$. Because C, as shown in (i) is of dimension $n - 1$ there exists a square matrix Q such that $Q^{-1}CQ$ is upper triangular. Let R be as defined in (ii), and let $S = PR$. Then

$$S^{-1}AS = (PR)^{-1}A(PR) = R^{-1}(P^{-1}AP)R$$

$$= \begin{pmatrix} 1 & Z \\ Z & Q^{-1} \end{pmatrix} \begin{pmatrix} \lambda_1 & B \\ Z & C \end{pmatrix} \begin{pmatrix} 1 & Z \\ Z & Q \end{pmatrix} = \begin{pmatrix} \lambda_1 & BQ \\ Z & Q^{-1}CQ \end{pmatrix}.$$

Thus $S^{-1}AS$ is upper triangular.

13. (i) Let $ABX = \lambda X$. Then $BABX = \lambda BX$. Hence if λ is a characteristic value of AB, associated with X, then λ also is a characteristic value of BA, associated with BX. The roles of A and B can be reversed in this argument so AB and BA have the same characteristic values.

(ii) To show that AB and BA have the same characteristic polynomial, first observe that for any n-by-n matrix C

$$C(AC - xI) = (CA - xI)C.$$

Let $C = B - yI$, and calculate the determinant of each side:

$$\det(B - yI)\det(A(B - yI) - xI) = \det((B - yI)A - xI)\det(B - yI).$$

Each side is a polynomial in two variables x and y, and each of those polynomials has the nonzero polynomial $\det(B - yI)$ as a factor. Hence

$$\det(A(B - yI) - xI) = \det((B - yI)A - xI).$$

Again the two sides are polynomials in x and y; for any numerical values of x and y, the values of these two polynomials are equal. Let x be arbitrary and let $y = 0$; then

$$\det(AB - xI) = \det(BA - xI),$$

as desired.

Exercises 7.3

1. $B = \begin{pmatrix} 0 & 0 & 0 \\ 0 & 2 & 1 \\ 0 & 0 & 2 \end{pmatrix}$.

The characteristic subspaces are $\mathscr{C}(0) = [\alpha_3]$ and $\mathscr{C}(2) = [\alpha_2]$, and for the given basis $\mathscr{W} = [\alpha_1]$.

Also $D = \begin{pmatrix} 0 & 0 \\ 0 & 2 \end{pmatrix}$, $R = \begin{pmatrix} 0 \\ 1 \end{pmatrix}$, and $S = (2)$.

3. (i) Because $\mathbf{T}$ is idempotent, $\mathbf{T}^2 = \mathbf{T}$. Then $\mathscr{V}_n = \mathscr{R}(\mathbf{T}) \oplus \mathscr{N}(\mathbf{T})$ by Theorems 3.6 and 7.8.

(ii) Let $\eta \in \mathscr{R}(\mathbf{T})$, $\eta = \mathbf{T}\xi$ for some $\xi \in \mathscr{V}_n$. Then $\mathbf{T}\eta = \mathbf{T}^2\xi = \mathbf{T}\xi = \eta$, so on $\mathscr{R}(\mathbf{T})$ the transformation $\mathbf{T}$ is the identity mapping. Now let $v \in \mathscr{N}(\mathbf{T})$. Then $\mathbf{T}v = \theta$, so on $\mathscr{N}(\mathbf{T})$ the transformation $\mathbf{T}$ is the zero mapping.

(iii) $\begin{pmatrix} I & Z \\ Z & Z \end{pmatrix}$, where I_r is the r-by-r identity matrix.

(iv) Let $\xi = \rho + v$ where $\rho \in \mathscr{R}(\mathbf{T})$, $v \in \mathscr{N}(\mathbf{T})$. The projection $\mathbf{T}_{\mathscr{R}}$ onto $\mathscr{R}(\mathbf{T})$ along $\mathscr{N}(\mathbf{T})$ is defined by $\mathbf{T}_{\mathscr{R}}\xi = \rho$, and $\mathbf{T}_{\mathscr{R}}$ coincides with $\mathbf{T}$ because $\mathbf{T}\xi = \mathbf{T}\rho + \mathbf{T}v = \rho + \theta$.

4. (i) Let **E** be a projection on $\mathcal{M}$ along $\mathcal{N}$ and let $\xi = \mu + v$ where $\mu \in \mathcal{M}$ and $v \in \mathcal{N}$. Then $\mathbf{E}\xi = \mathbf{E}\mu + \mathbf{E}v = \mu + \theta$, so $(\mathbf{I} - \mathbf{E})\xi = (\mu + v) - (\mu + \theta) = \theta + v$. Hence $\mathbf{I} - \mathbf{E}$ is a projection on $\mathcal{N}$ along $\mathcal{M}$. Similar calculations establish the converse statement.

(ii) $\mathcal{M}$ is **T**-invariant if and only if $\mathbf{T}\mu \in \mathcal{M}$ for all $\mu \in \mathcal{M}$. Let $\xi = \mu + v$. Then $\mathbf{TE}\xi = \mathbf{T}(\mu + \theta) = \mathbf{T}\mu$, and $\mathbf{ETE}\xi = \mathbf{ET}\mu$. If $\mathbf{T}\mu \in \mathcal{M}$, $\mathbf{ET}\mu = \mathbf{T}\mu$, so $\mathbf{TE} = \mathbf{ETE}$. Conversely, $\mathbf{ET}\mu \in \mathcal{M}$, so if $\mathbf{ETE} = \mathbf{TE}$, then $\mathbf{T}\mu \in \mathcal{M}$, and $\mathcal{M}$ is **T**-invariant.

(iii) Use the method of (ii) to show that $\mathcal{N}$ is **T**-invariant if and only if $\mathbf{ET} = \mathbf{ETE}$, and then combine this with (ii) to finish the proof.

5. Let $\xi = \mu_1 + v_1 = \mu_2 + v_2$ where $\mu_i \in \mathcal{M}_i$ and $v_i \in \mathcal{N}_i$. Then $\mathbf{E}_1\xi = \mu_1$ and $\mathbf{E}_2\xi = \mu_2$. Then $(\mathbf{E}_1 + \mathbf{E}_2)\xi = \mu_1 + \mu_2 \in \mathcal{M}_1 + \mathcal{M}_2$. But a linear transformation is a projection if and only if it is idempotent, so $\mathbf{E}_1 + \mathbf{E}_2$ is a projection if and only if $(\mathbf{E}_1 + \mathbf{E}_2)^2 = \mathbf{E}_1 + \mathbf{E}_2$. Because $\mathbf{E}_1$ and $\mathbf{E}_2$ are projections, this equation reduces to $\mathbf{E}_1\mathbf{E}_2 = -\mathbf{E}_2\mathbf{E}_1$; hence $\mathbf{E}_1\mathbf{E}_2 = \mathbf{E}_1^2\mathbf{E}_2 = -\mathbf{E}_1\mathbf{E}_2\mathbf{E}_1 = \mathbf{E}_2\mathbf{E}_1^2 = \mathbf{E}_2\mathbf{E}_1$. Thus if $\mathbf{E}_1 + \mathbf{E}_2$ is a projection, then $\mathbf{E}_1\mathbf{E}_2 = \mathbf{Z} = \mathbf{E}_2\mathbf{E}_1$. Conversely, if $\mathbf{E}_1\mathbf{E}_2 = \mathbf{Z} = \mathbf{E}_2\mathbf{E}_1$, then $(\mathbf{E}_1 + \mathbf{E}_2)^2 = (\mathbf{E}_1 + \mathbf{E}_2)$. If $\mathbf{E}_1 + \mathbf{E}_2$ is a projection on $\mathcal{M}$ along $\mathcal{N}$, then $\mathcal{N}_1 \cap \mathcal{N}_2 \subseteq \mathcal{N}$ because $(\mathbf{E}_1 + \mathbf{E}_2)v = \theta$ for each $v \in \mathcal{N}_1 \cap \mathcal{N}_2$. Conversely, if $(\mathbf{E}_1 + \mathbf{E}_2)\alpha = \theta$ then $\mathbf{E}_1\mathbf{E}_2 = \mathbf{Z}$, so $\mathbf{E}_1\alpha = -\mathbf{E}_2\alpha$, so $\mathbf{E}_1\alpha = \mathbf{E}_1^2\alpha = -\mathbf{E}_1\mathbf{E}_2\alpha = \theta$, and $\alpha \in \mathcal{N}_1$. Similarly $\alpha \in \mathcal{N}_2$ and $\mathcal{N}_1 \cap \mathcal{N}_2 = \mathcal{N}$. Then $\mathcal{M} \subseteq \mathcal{M}_1 + \mathcal{M}_2$; let $\alpha \in \mathcal{M}_1 + \mathcal{M}_2$. Then $\alpha = \mu_1 + \mu_2 = \mathbf{E}_1\mu_1$. Because $\mathbf{E}_1\mathbf{E}_2 = \mathbf{Z} = \mathbf{E}_2\mathbf{E}_1$, $(\mathbf{E}_1 + \mathbf{E}_2)\alpha = \mathbf{E}_1^2\mu_1 + \mathbf{E}_2^2\mu_2 = \alpha \in \mathcal{M} = \mathcal{R}(\mathbf{E}_1 + \mathbf{E}_2)$. Hence $\mathcal{M} = \mathcal{M}_1 + \mathcal{M}_2$.

7. (i) $AX = (3, 2, 1)$ and $A^2X = (8, 4, 4) = -4X + 4AX$. Hence $[(X)_\mathbf{T}] = \{X, \mathbf{T}X\}$, and $\dim[(X)_\mathbf{T}] = 2$.

(ii) To extend $[(X)_\mathbf{T}]$ to a **T**-cyclic basis for $\mathbb{R}^3$, adjoin any characteristic vector not in $[(X)_\mathbf{T}]$. The vector $Y = (1, -1, 1)$, for example, will do.

Exercises 7.4

1. Let $m(x)$ and $n(x)$ be two minimal polynomials of A. By Theorem 7.10 $n(x) = m(x)q(x)$ and $m(x) = n(x)t(x)$ for suitable polynomials $q(x)$ and $t(x)$. Then $m(x) = m(x)q(x)t(x)$ so $q(x)t(x) = 1$. Hence $q(x)$ and $t(x)$ are constant polynomials; because $m(x)$ and $n(x)$ are each monic, $q(x) = 1 = t(x)$, and $m(x) = n(x)$.

2.

(ii) $(A - I)^2 = \begin{pmatrix} 1 & 1 & 1 \\ -2 & -2 & -2 \\ 1 & 1 & 1 \end{pmatrix}$; $(A - I)^3 = Z$.

(iii) $(A - I)^2 = Z$; $(A - I)^3 = Z$.

(iv) $(A - I)(A - 5I) = Z$; $(A - I)^2(A - 5) = Z$.

3. (ii) $(A - I)^2 \neq Z$, so $m(x) = (x - 1)^3$. Not diagonable.

(iii) $(A - I) \neq Z$, but $m(x) = (x - 1)^2$. Not diagonable.

(iv) $(A - I) \neq Z \neq (A - 5I)$, but $m(x) = (x - 1)(x - 5)$. Diagonable.

7. If an n-by-n matrix has n distinct characteristic values, it is diagonable. Hence a necessary condition that the given matrix A is not diagonable is that its two characteristic values coincide: $(a - d)^2 + 4bc = 0$. Then the characteristic polynomial is

$$p(x) = \left(x - \frac{a + d}{2}\right)^2,$$

and the minimal polynomial is $p(x)$ provided that

$$\left(A - \frac{a + d}{2} I\right) = \begin{pmatrix} \dfrac{a - d}{2} & b \\ c & -\dfrac{a - d}{2} \end{pmatrix} \neq Z.$$

Hence A is not diagonable if and only if $(a - d)^2 + 4bc = 0$ and either b or c is nonzero.

8. (i) Let

$$\begin{pmatrix} a & b \\ c & d \end{pmatrix}^2 = -I.$$

Solve the resulting system of four equations in a, b, c, and d to obtain

$$A = \begin{pmatrix} a & b \\ c & -a \end{pmatrix}, \text{ where } bc = -1 - a^2.$$

(ii) If A is three-by-three, its characteristic polynomial is a real cubic, which has a real root. Hence A^2 has a real, nonnegative characteristic value, but the only real characteristic value of $-I$ is -1.

10. $\sum_{i=1}^{t} F_i = \sum_{i=1}^{t} q_i(A) = I$ because $r(A) = Z$.

$$F_i F_j = q_i(A)q_j(A) = \frac{p_i(A)p_j(A)}{p_i(\lambda_i)p_j(\lambda_j)}.$$

When $i \neq j$, $F_i F_j = m(A)s(A) = Z$, where $s(x)$ is a polynomial of degree $t - 2$. Thus when $i = j$ we have

$$F_i = F_i I = F_i(F_1 + F_2 + \cdots + F_t) = F_i F_i.$$

$$\sum_{i=1}^{t} \lambda_i F_i = \sum_{i=1}^{t} \lambda_i q_i(A)$$

$$= \sum_{i=1}^{t} \frac{\lambda_i p_i(A)}{p_i(\lambda_i)} = \sum_{i=1}^{t} \frac{A p_i(A)}{p_i(\lambda_i)}$$

$$= A \sum_{i=1}^{t} q_i(A),$$

where the next to last equation is obtained from $m(A) = Z = (A - \lambda_i I)p_i(A)$. The proof is completed by recalling that $\sum_{i=1}^{t} q_i(A) = I$.

Exercises 7.5

1. Each entry of N not on the subdiagonal of N is zero. The subdiagonal elements of N are these:

   (i) 1, 1, 1, 1, 0.

   (ii) 1, 1, 0, 1, 1.

2. The set C is a subset of $\mathscr{R}(\mathbf{T}) + \mathscr{N}(\mathbf{T})$ obtained by adjoining to the set A the vectors $\zeta_1, \ldots, \zeta_s$, because $(\zeta_i)_{\mathbf{T}} = \{\zeta_i\}$.

But
$$\begin{aligned}
\dim(\mathscr{R}(\mathbf{T}) + \mathscr{N}(\mathbf{T})) &= \dim \mathscr{R}(\mathbf{T}) + \dim \mathscr{N}(\mathbf{T}) - \dim(\mathscr{R}(\mathbf{T}) \cap \mathscr{N}(\mathbf{T})) \\
&= \dim \mathscr{R}(\mathbf{T}) + (t + s) - t \\
&= \dim \mathscr{R}(\mathbf{T}) + s.
\end{aligned}$$

Because A is a basis for $\mathscr{R}(\mathbf{T})$, the number of vectors in C coincides with the dimension of $\mathscr{R}(\mathbf{T}) + \mathscr{N}(\mathbf{T})$, and C is a basis for $\mathscr{R}(\mathbf{T}) + \mathscr{N}(\mathbf{T})$ if C is linearly independent. Suppose that

$$\sum_{k=1}^{q_1} a_{1k}\mathbf{T}^{k-1}\eta_1 + \cdots + \sum_{k=1}^{q_t} a_{tk}\mathbf{T}^{k-1}\eta_t + b_1\zeta_1 + \cdots + b_s\zeta_s = \theta.$$

If we operate on each side by $\mathbf{T}$ and recall that $\mathbf{T}^{q_i}\eta_i = \theta$ and $\mathbf{T}\zeta_j = \theta$, we obtain

$$\sum_{k=1}^{q_1-1} a_{1k}\mathbf{T}^k\eta_1 + \cdots + \sum_{k=1}^{q_t-1} a_{tk}\mathbf{T}^k\eta_t = \theta.$$

Because A is linearly independent each coefficient in this equation is zero, and the previous equation then becomes

$$a_{1q_1}\mathbf{T}^{q_1-1}\eta_1 + \cdots + a_{tq_t}\mathbf{T}^{q_t-1}\eta_t + b_1\zeta_1 + \cdots + b_s\zeta_s = \theta.$$

The linear independence of B then assures that each of these coefficients is zero, making C linearly independent.

5. (i) By matrix calculations $AN = NA$ if and only if $a_{i,\,j+1} = a_{i-1,\,j}$ for each $i > 1$ and each $j < n$, $a_{i-1,\,n} = 0$ if $1 < i \le n$ and $a_{1,\,j+1} = 0$ if $1 \le j < n$. Hence

$$A = \begin{pmatrix}
a_{11} & 0 & 0 & \cdots & 0 & 0 \\
a_{21} & a_{11} & 0 & \cdots & 0 & 0 \\
a_{31} & a_{21} & a_{11} & \cdots & 0 & 0 \\
. & . & & & . & . \\
. & . & & & . & . \\
. & . & & & . & . \\
a_{n1} & a_{n-1,\,1} & & \cdots & a_{21} & a_{11}
\end{pmatrix}.$$

(ii) The only characteristic value of A is a_{11}, and the corresponding characteristic subspace is $[(0, 0, \ldots, 0, 1)]$.

(iii) Under the given hypothesis, the system $(A - a_{11} I)X = Z$ has its first k equations of the form $0 = 0$, and the remaining $n - k$ equations are

$$
\begin{aligned}
a_{k+1,\,1} x_1 &= 0, \\
a_{k+2,\,1} x_1 + a_{k+1,\,1} x_2 &= 0, \\
&\ \ \vdots \\
\end{aligned}
$$

$$a_{n1} x_1 + a_{n-1,\,1} x_2 + \cdots + a_{k+1,\,1} x_{n-k} = 0.$$

Hence $x_1 = x_2 = \cdots = x_{n-k} = 0$, and the last k components of X are arbitrary. Hence k linearly independent characteristic vectors of A are associated with a_{11}.

6. (i) Each vector in the first column is mapped by $\mathbf{T}$ into θ and therefore is in $\mathscr{N}(\mathbf{T})$; every other vector of the array is mapped by $\mathbf{T}$ into the vector preceding it in the same row of the array. The fact that the array is linearly independent then shows that $\mathscr{N}(\mathbf{T})$ is spanned by the vectors of the first column. Hence $n(\mathbf{T}) = k$.

(ii) As in (i) $\mathbf{T}^2$ maps each vector in the first two columns into θ and maps every other vector of the array into the vector two positions to the left of it in the array. Hence the first two columns are a basis for $\mathscr{N}(\mathbf{T}^2)$. Hence the length of the second column is $n(\mathbf{T}^2) - n(\mathbf{T})$.

(iii) By induction, arguing as above, the first k columns of the array form a basis for $\mathscr{N}(\mathbf{T}^k)$, and the length of column k is $n(\mathbf{T}^k) - n(\mathbf{T}^{k-1})$, which is determined by $\mathbf{T}$.

(iv) The length of the first row is the index of nilpotency of $\mathbf{T}$.

(v) The array forms a basis for $\mathscr{V}_n$.

(vi) If n numbers are arranged in a rectangular array having p_1 columns, with the length of each column specified, then the length p_i of each row and number k of rows are uniquely determined.

7. If $\eta \in \mathscr{R}(\mathbf{T}^{p-k})$ then $\eta = \mathbf{T}^{p-k}\xi$ for some ξ, and $\mathbf{T}^k\eta = \theta$, so $\eta \in \mathscr{N}(\mathbf{T}^k)$. As in Theorem 3.6 we then have

$$[\theta] \subseteq \mathscr{R}(\mathbf{T}^{p-1}) \subseteq \cdots \subseteq \mathscr{R}(\mathbf{T}^{p-k}) \subseteq \cdots \subseteq \mathscr{R}(\mathbf{T}) \subseteq \mathscr{V},$$
$$[\theta] \subseteq \mathscr{N}(\mathbf{T}) \subseteq \cdots \subseteq \mathscr{N}(\mathbf{T}^k) \subseteq \cdots \subseteq \mathscr{N}(\mathbf{T}^{p-1}) \subseteq \mathscr{V},$$

where if equality holds at any position in either chain it holds at the same position and at each subsequent position in both chains. But any equality in the second chain implies that $\mathscr{N}(\mathbf{T}^{p-1}) = \mathscr{V}$, contrary to the definition of p as the index of nilpotency of $\mathbf{T}$. Hence strict inequality holds throughout both chains. Now suppose that for some k, $\mathscr{R}(\mathbf{T}^{p-k}) = \mathscr{N}(\mathbf{T}^k)$. Show that $\mathscr{R}(\mathbf{T}^{p-k-1}) = \mathscr{N}(\mathbf{T}^{k+1})$, and then use Theorem 3.5 to show that if $\mathscr{R}(\mathbf{T}^{p-1}) \neq \mathscr{N}(\mathbf{T})$, then $\mathscr{R}(\mathbf{T}) \neq \mathscr{N}(\mathbf{T}^{p-1})$.

Exercises 7.6

1. (i) $p(x) = -(x-3)^3$.

    (ii) $\lambda = 3$ with multiplicity 3.

    (iii) $m(x) = (x-3)^3$.

    (iv) $J = \begin{pmatrix} 3 & 0 & 0 \\ 1 & 3 & 0 \\ 0 & 1 & 3 \end{pmatrix}$.

    (v) $\{(3)\}$.

3. For Exercise 7.4-2(ii), $p(x) = -(x-1)^3 = -m(x)$, and $J = \begin{pmatrix} 1 & 0 & 0 \\ 1 & 1 & 0 \\ 0 & 1 & 1 \end{pmatrix}$.

   For Exercise 7.4-2(iii), $p(x) = -(x-1)^3$, but $m(x) = (x-1)^2$, and

   $$J = \begin{pmatrix} 1 & 0 & 0 \\ 1 & 1 & 0 \\ 0 & 0 & 1 \end{pmatrix}.$$

4. (i) There are two possible Jordan forms. If $p(x) = (x-1)^2(x-2)^2$ then

    $$J_1 = \begin{pmatrix} 1 & 0 & 0 & 0 \\ 0 & 1 & 0 & 0 \\ 0 & 0 & 2 & 0 \\ 0 & 0 & 1 & 2 \end{pmatrix}. \text{ If } p(x) = (x-1)(x-2)^3, \text{ then } J_2 = \begin{pmatrix} 1 & 0 & 0 & 0 \\ 0 & 2 & 0 & 0 \\ 0 & 1 & 2 & 0 \\ 0 & 0 & 0 & 2 \end{pmatrix}.$$

    (ii) The characteristic polynomial determines which form J has.

    (iii) J_1 and J_2 are not similar but they have the same minimal polynomial.

6. The minimal polynomial is $(x-3)^{m_1}(x-2)^{m_2}$, where $1 \le m_1 \le 3$ and $1 \le m_2 \le 2$. Hence there are six corresponding cases:

m_1	m_2	Subdiagonal
3	2	1, 1, 0, 1
2	2	1, 0, 0, 1
1	2	0, 0, 0, 1
3	1	1, 1, 0, 0
2	1	1, 0, 0, 0
1	1	0, 0, 0, 0

8. If $m_1 = 3$, three linearly independent characteristic vectors are associated with $\lambda_1 = 0$, so there are 2 zeros (and hence 4 ones) on the subdiagonal. The first two subdiagonal entries must be 1, 1; and the next two must be 0, 1. But the last two could be either 1, 0 or 0, 1.

10. (i) J has three major blocks, of sizes 6, 4, and 2. The first block has two sub-blocks, of sizes four and two, and has 2 in each diagonal position. The second block has one sub-block of size four and has 0 in each diagonal position. The last block has two sub-blocks each of size one and has 1 in each diagonal position.

12. There are eleven such matrices, one for each-of the following subdiagonals.

$$
\begin{array}{l}
1\ 0\ 0\ 0\ 0, \qquad 1\ 1\ 1\ 1\ 0, \\
1\ 1\ 0\ 0\ 0, \qquad 1\ 1\ 1\ 0\ 1, \\
1\ 0\ 1\ 0\ 0, \qquad 1\ 1\ 0\ 1\ 1, \\
1\ 1\ 1\ 0\ 0, \qquad 1\ 1\ 1\ 1\ 1, \\
1\ 1\ 0\ 1\ 0, \qquad 0\ 0\ 0\ 0\ 0. \\
1\ 0\ 1\ 0\ 1,
\end{array}
$$

13. $p(x) = (x - 2)^2(x - 1)(x + 1)$. The characteristic subspace $\mathscr{C}(2)$ is one-dimensional, $\mathscr{C}(2) = [(2, -1, -2, 1)]$. Hence J has 2, 2, 1, -1 along the diagonal and 1, 0, 0 along the subdiagonal. Vectors $X_4 = (-4, 8, -5, 1)$, $X_3 = (4, 0, -3, 1)$, and $X_2 = (2, -1, -2, 1)$ form a maximal linearly independent set of characteristic vectors. Choose X_1 so that $AX_1 = 2X_1 + X_2$; then $X_1 = (-1, 0, 1, 0)$. Use these vectors to form the columns (corresponding to their subscripts) of a matrix P, and then verify by matrix multiplication that $P^{-1}AP = J$.

Exercises 7.7

1. (i) $p(x) = (1 - x)(2 - x)^2$. Let $\lambda_1 = 2 = \lambda_2$. Then $g_1 = \dim \mathscr{C}(2) = 2$. Choose $X_1 = (2, 0, 1)$ and $X_2 = (2, 1, 0)$ as a basis for $\mathscr{C}(2)$. For $\lambda_3 = 1$ let $X_3 = (-3, 1, -3)$. Then $P^{-1}AP = J = \operatorname{diag}(2, 2, 1)$ where

$$
P = \begin{pmatrix} 2 & 2 & -3 \\ 0 & 1 & 1 \\ 1 & 0 & -3 \end{pmatrix} \quad \text{and} \quad P^{-1} = \begin{pmatrix} 3 & -6 & -5 \\ -1 & 3 & 2 \\ 1 & -2 & -2 \end{pmatrix}.
$$

(iii) $p(x) = (4 - x)^3$. Reduce the system $(A - 4I)X = Y$ as in (7.2) to obtain the reduced echelon form,

$$
\begin{pmatrix} 1 & -1 & -1 & -y_1 \\ 0 & 0 & 0 & y_2 \\ 0 & 0 & 0 & y_3 - y_1 \end{pmatrix}.
$$

The consistency condition is $y_3 = y_1$ and $y_2 = 0$. To determine characteristic vectors let $Y = Z$. Then $X = (x_2 + x_3, x_2, x_3)$. Let $X_3 = (1, 1, 0)$ and $X_2 = (1, 0, 1)$. Then choose X_1 to satisfy $(A - I)X_1 = X_2$, obtaining $X_1 = (1, 1, 1)$ as a solution. Then

$$
P^{-1}AP = \begin{pmatrix} 4 & 0 & 0 \\ 1 & 4 & 0 \\ 0 & 0 & 4 \end{pmatrix}, \quad \text{where } P = \begin{pmatrix} 1 & 1 & 1 \\ 1 & 0 & 1 \\ 1 & 1 & 0 \end{pmatrix} \text{ and } P^{-1} = \begin{pmatrix} -1 & 1 & 1 \\ 1 & -1 & 0 \\ 1 & 0 & -1 \end{pmatrix}.
$$

(iv) To find $p(x)$ expand $\det(A - xI)$ by the entries of the fourth row to obtain $(2 - x)q(x)$, and then by the entries of row 2: $p(x) = (x - 2)^2(x - 3)^2$. Then show that dim $\mathscr{C}(2) = 2$, dim $\mathscr{C}(3) = 1$, that $X_3 = (-5, 0, 1, 2)$ and $X_4 = (-1, 0, 2, 1)$ are characteristic vectors associated with $\lambda = 2$, and that $X_2 = (1, 0, 1, 0)$ is a characteristic vector associated with $\lambda = 3$. Solve $(A - 3I)X_1 = X_2$ to obtain a fourth basis vector, say $X_1 = (2, 0, 1, 0)$. Verify that if

$$P = \begin{pmatrix} 2 & 1 & -5 & -1 \\ 0 & 0 & 1 & 0 \\ 1 & 1 & 0 & 2 \\ 0 & 0 & 2 & 1 \end{pmatrix},$$

$$P^{-1} = \begin{pmatrix} 1 & -1 & -1 & 3 \\ -1 & 5 & 2 & -5 \\ 0 & 1 & 0 & 0 \\ 0 & -2 & 0 & 1 \end{pmatrix} \quad \text{and} \quad P^{-1}AP = \begin{pmatrix} 3 & 0 & 0 & 0 \\ 1 & 3 & 0 & 0 \\ 0 & 0 & 2 & 0 \\ 0 & 0 & 0 & 2 \end{pmatrix}.$$

2. Each major block B_i has dimension s_i and has $s_i - 1$ entries on the subdiagonal (in columns 1, 2, ..., $s_i - 1$). A given subdiagonal entry is 0 if and only if the basis vector corresponding to that column is a characteristic vector, and there are g_i such vectors in the chosen basis because $g_i = \dim \mathscr{C}(\lambda_i)$. One of these vectors corresponds to column s_i, so there are exactly $g_i - 1$ zero entries on the subdiagonal. Hence the number of ones on the subdiagonal is $(s_i - 1) - (g_i - 1) = s_i - g_i$.

3. (i) The given values of $N(m)$ for $m = 1$ and $m = 2$ are obvious. The values of $N(m, p)$ can be calculated as follows:

$N(m, m) = 1$ because the canonical form has one in each subdiagonal position. $N(m, m - 1) = 1$ because the subdiagonal can have 0 only in the last position. In general $N(m, p)$ counts the number of m-by-m matrices that are nilpotent of index p and in canonical form. Any such matrix is of diagonal block form

$$\begin{pmatrix} A & Z \\ Z & B \end{pmatrix},$$

where A is p-by-p with 1 in each subdiagonal position. B is $(m - p)$-by-$(m - p)$ and can be either zero or nilpotent of any index from two up to the smaller of the two numbers p and $m - p$. Therefore,

$$N(m, p) = 1 + N(m - p, 2) + \cdots + N(m - p, M)$$
$$= 1 + \sum_{k=2}^{M} N(m - p, k), \text{ where } M = \min(m - p, p).$$

This recursion relation and the formula for $N(m)$ yield the values stated for $1 \le m \le 9$.

Exercises 7.8

1. Calculate A^2 and A^3; you should obtain $A^3 = \begin{pmatrix} 2 & -6 & 2 & -4 \\ 0 & 2 & 0 & 2 \\ 2 & -2 & -2 & 0 \\ 0 & 2 & 0 & -2 \end{pmatrix}$.

Then tr $A = 0$, tr $A^2 = 8$, tr $A^3 = 0$, tr $A^4 = 16$,
   $c_1 = 0$, $c_2 = -4$, $c_3 = 0$, $c_4 = 4$.
Hence $p(x) = x^4 - 4x^2 + 4$, and $A^{-1} = -4^{-1}[A^3 - 4A]$,

$$A^{-1} = \tfrac{1}{2}\begin{pmatrix} 1 & 1 & 1 & 0 \\ 0 & 1 & 0 & 1 \\ 1 & 1 & -1 & 0 \\ 0 & 1 & 0 & -1 \end{pmatrix}.$$

3. The coefficient of x^{n-2} in $p(x)$ is $(-1)^n c_2$. From the factored form, $p(x) = (-1)^n (x - \lambda_1) \cdots (x - \lambda_n)$ the coefficient of x^2 is $(-1)^n \sum_{i \neq j} \lambda_i \lambda_j$. But

$$(\operatorname{tr} A)^2 = (\lambda_1 + \cdots + \lambda_n)^2 = (\lambda_1^2 + \cdots + \lambda_n^2) + 2\sum_{i \neq j} \lambda_i \lambda_j,$$

so

$$c_2 = \sum_{i \neq j} \lambda_i \lambda_j = \tfrac{1}{2}[(\operatorname{tr} A)^2 - \operatorname{tr}(A^2)] = -\tfrac{1}{2}[c_1 \operatorname{tr} A + \operatorname{tr}(A^2)].$$

4. (i) If A is nilpotent, every characteristic value is zero.

(ii) From Example C of Section 3.4 or Exercise 6.4-7(iii), the Jordan form of an idempotent matrix has 1 in the first $r(A)$ diagonal positions and 0 elsewhere. Hence tr $A = r(A)$.

5. (i) The characteristic polynomial of a two-by-two matrix A is

$$p(x) = x^2 - (\operatorname{tr} A)x + \det A.$$

Hence if A and B have the same trace and the same determinant, they have the same characteristic polynomial and the same characteristic values.

7. (i) H_p is the matrix with 1 in position $(k, p - k + 1)$ for $k = 1, \ldots, p$, and 0 elsewhere; that is, each entry on the "reverse diagonal" is 1 and every other entry is 0. Thus $h_{ij} = \delta_{i, p+1-j}$. The (i, j) entry of $H_p A$ is

$$\sum_{k=1}^{p} h_{ik} a_{kj} = \sum_{k=1}^{p} \delta_{i, p+1-k} a_{kj} = a_{p+1-i, j},$$

as described.

(ii) The (i, j) entry of $A H_p$ is

$$\sum_{k=1}^{p} a_{ik} h_{kj} = \sum_{k=1}^{p} a_{ik} \delta_{k, p+1-j} = a_{i, p+1-j}.$$

Hence column j of A becomes column $p + 1 - j$ of AH_p, as described.

(iii) $H_p^{-1}AH_p = H_pAH_p$ is obtained by reflecting the entries of A through the center point of the matrix (a rotation of 180° around the center point).

8. Let $J = P^{-1}AP$ be the Jordan form of A. Then J^t is similar to A^t. J^t has the same diagonal sub-block decomposition as does J, with the same diagonal entries in corresponding sub-blocks but with 1 in each *superdiagonal* (rather than subdiagonal) position within each sub-block. If each sub-block of J^t is rotated through 180° about its center, the resulting matrix is J. To see that such a rotation can be performed by a similarity transformation let H_n be a block diagonal matrix with the sizes of the diagonal blocks coinciding with those of the sub-blocks of J and with each block of H_n being of the form H_p described in Exercise 7(i). Then $H_n^{-1}J^tH = J$, so A^t is similar to A.

CHAPTER **8**

Exercises 8.1

1. (i) Let $\alpha = (a_1, a_2)$ and $\beta = (b_1, b_2)$. Then

$$c\alpha + d\beta = (ca_1 + db_1, ca_2 + db_2),$$
$$p(c\alpha + d\beta, \eta) = (ca_1 + db_1)y_1 + (ca_1 + db_1)y_2 + (ca_2 + db_2)y_1$$
$$+ k(ca_2 + db_2)y_2$$
$$= cp(\alpha, \eta) + dp(\beta, \eta).$$

Also $p(\eta, \xi) = p(\xi, \eta)$. Because p is symmetric and linear in the first component, it is also linear in the second component.

(ii) From (i) p is an inner product if and only if

$$p(\xi, \xi) > 0 \text{ for } \xi \neq \theta.$$

But $p(\xi, \xi) = x_1^2 + 2x_1x_2 + kx_2^2 = (x_1 + x_2)^2 + (k - 1)x_2^2$, which is positive whenever $\xi \neq \theta$ if and only if $k > 1$.

3. (i) $p(c\xi, d\eta) = cp(\xi, d\eta) = c\bar{d}p(\xi, \eta)$.

(ii) $p(\theta, \eta) = p(0\xi, \eta) = 0p(\xi, \eta) = 0$.

$p(\xi, \theta) = p(\xi, 0\eta) = \bar{0}p(\xi, \eta) = 0$.

4. If $p(\xi, \eta_1) = p(\xi, \eta_2)$ for all ξ, then $p(\xi, \eta_1 - \eta_2) = 0$ for all ξ. Hence for $\xi = \eta_1 - \eta_2$ we have $p(\eta_1 - \eta_2, \eta_1 - \eta_2) = 0$, and $\eta_1 - \eta_2 = \theta$ because p is positive-definite.

6. If $\{\xi, \eta\}$ is linearly dependent, either $\eta = k\xi$ for some $k \neq 0$, or $\xi = \theta$ or $\eta = \theta$. If $\xi = \theta$ or $\eta = \theta$, each side of the Schwarz inequality reduces to zero. If $\eta = k\xi$, we have

$$|\langle \xi, \eta \rangle|^2 = |\langle \xi, k\xi \rangle|^2 = |\bar{k}|^2 |\langle \xi, \xi \rangle|^2 = |\langle \xi, \xi \rangle| \, |\langle k\xi, k\xi \rangle|$$
$$= \langle \xi, \xi \rangle \langle \eta, \eta \rangle.$$

Conversely, if equality holds in the Schwarz inequality, use the calculations as in Exercise 8.1-5(i) for any real k to obtain

$$\langle \xi + k\eta, \xi + k\eta \rangle = \langle \xi, \xi \rangle + 2\mathrm{Rl}\langle \xi, \eta \rangle k + \langle \eta, \eta \rangle k^2$$
$$\leq \langle \xi, \xi \rangle + 2\langle \xi, \xi \rangle^{1/2}\langle \eta, \eta \rangle^{1/2}k + \langle \eta, \eta \rangle k^2$$
$$= [\langle \xi, \xi \rangle^{1/2} + k\langle \eta, \eta \rangle^{1/2}]^2$$
$$= 0 \text{ when } k = -\langle \xi, \xi \rangle^{1/2}\langle \eta, \eta \rangle^{-1/2}.$$

Then $\xi + k\eta = \theta$ for that value of k.

7. (i) This is the Schwarz inequality for the dot product in $\mathscr{E}^n$.

(ii) This is the Schwarz inequality for the inner product as defined in Example D.

(iii) This is the Schwarz inequality for the inner product as defined in Example C with the interval of integration generalized from $[0, 1]$ to $[a, b]$.

9. Let $f(A, B) = \mathrm{tr}(A\bar{B}^t)$;

$$f(aA + cC, B) = \mathrm{tr}((aA + cC)\bar{B}^t) = \mathrm{tr}(aA\bar{B}^t + cC\bar{B}^t) = a\,\mathrm{tr}(AB^t) + c\,\mathrm{tr}(CB^t),$$

so f is linear in its first component. Next, $f(B, A) = \mathrm{tr}(B\bar{A}^t)$. But $B\bar{A}^t = \overline{(A\bar{B}^t)^t}$. But $\mathrm{tr}(X^t) = \mathrm{tr}\,X$ and $\mathrm{tr}\,\bar{X} = \overline{\mathrm{tr}\,X}$, so $f(B, A) = \overline{\mathrm{tr}(A, B)}$. Finally $f(A, A) = \mathrm{tr}(A\bar{A}^t) = \sum_{i=1}^n \left(\sum_{k=1}^n a_{ik}\bar{a}_{ik} \right) > 0$ unless each $a_{ik} = 0$.

11. Let $N(f) = \max |f(t)|, 0 \leq t \leq 1$. Then $N(cf) = \max|cf(t)| = |c|N(f)$. If f is not zero on $[0, 1]$, $|f(t)| > 0$ for some t in $[0, 1]$, so $N(f) > 0$. The triangle inequality follows from the observation that

$$\max |f(x) + g(x)| \leq \max |f(x)| + \max |g(x)|.$$

Exercises 8.2

2. (a) If $\xi = \theta$ then $\langle \xi, \eta \rangle = \langle 0\eta, \eta \rangle = 0\langle \eta, \eta \rangle = 0$ for each η in $\mathscr{V}$. If $\langle \xi, \eta \rangle = 0$ for every η, then $\langle \xi, \xi \rangle = 0$, so $\xi = \theta$.

(b) Let $\langle \xi, \sigma \rangle = 0$ for every $\sigma \in S$. If $\eta \in [S]$ then $\eta = \sum_{i=1}^k a_i\sigma_i$ for some set of scalars a_i and vectors $\sigma_i \in S$. Then

$$\langle \xi, \eta \rangle = \langle \xi, \sum a_i\sigma_i \rangle = \sum \bar{a}_i\langle \xi, \sigma_i \rangle = \left(\sum \bar{a}_i \right)0 = 0.$$

(c) Let S be any set of vectors such that $\langle \sigma_i, \sigma_j \rangle = 0$ for all $\sigma_i \in S$, $\sigma_j \in S$, $i \neq j$. Suppose $\sum_{i=1}^n a_i \sigma_i = \theta$ for distinct vectors $\sigma_i \in S$. For each j $\langle \sum_{i=1}^n a_i \sigma_i, \sigma_j \rangle = \langle \theta, \sigma_j \rangle = 0 = \sum_{i=1}^n a_i \langle \sigma_i, \sigma_j \rangle = a_j \langle \sigma_j, \sigma_j \rangle$, so $a_j = 0$. Hence S is linearly independent.

4. $\langle \xi + \eta, \xi - \eta \rangle = \langle \xi, \xi \rangle - \langle \eta, \eta \rangle$ in $\mathscr{E}^n$, because $\langle \eta, \xi \rangle = \langle \xi, \eta \rangle$. This is not valid in unitary space because $\langle \xi, \eta \rangle \neq \langle \eta, \xi \rangle$ there. In $\mathscr{E}^n$ this result expresses the theorem that states that the diagonals of a parallelogram are orthogonal if and only if the parallelogram is equilateral (a rhombus).

5. (i) $\|\alpha + \beta\|^2 = \langle \alpha + \beta, \alpha + \beta \rangle = \langle \alpha, \alpha \rangle + \langle \alpha, \beta \rangle + \langle \beta, \alpha \rangle + \langle \beta, \beta \rangle$
$$= \langle \alpha, \alpha \rangle + (\langle \alpha, \beta \rangle + \overline{\langle \alpha, \beta \rangle}) + \langle \beta, \beta \rangle$$
$$= \|\alpha\|^2 + 2\mathrm{Rl}\langle \alpha, \beta \rangle + \|\beta\|^2$$

(ii) As in (i), $\|\alpha - \beta\|^2 = \|\alpha\|^2 - 2\mathrm{Rl}\langle \alpha, \beta \rangle + \|\beta\|^2$.
Hence $\|\alpha + \beta\|^2 - \|\alpha - \beta\|^2 = 4\mathrm{Rl}\langle \alpha, \beta \rangle$.

(iii) If $\|\xi\|$ is known for each vector ξ, then $\langle \alpha, \beta \rangle$ is determined by
$$\mathrm{Rl}\langle \alpha, \beta \rangle = \tfrac{1}{4}[\|\alpha + \beta\|^2 - \|\alpha - \beta\|^2],$$
$$\mathrm{Im}\langle \alpha, \beta \rangle = \mathrm{Rl}\langle \alpha, i\beta \rangle = \tfrac{1}{4}[\|\alpha + i\beta\|^2 - \|\alpha - i\beta\|^2].$$

The second equation uses the observation that if $z = x + iy$, then
$$y = \mathrm{Im}\, z = \mathrm{Rl}(-iz).$$

8. Let $\langle p, q \rangle = \int_0^1 p(x)q(x)\, dx$, and use the Gram-Schmidt process on the usual basis $\{1, x, x^2\}$ for $\mathscr{P}_2$.

(i) $\{1, x, x^2\}$ is not orthogonal, because $\langle 1, x \rangle = \int_0^1 x\, dx = \tfrac{1}{2} \neq 0$.

(ii) Let $q_1(x) = 1 = p_1(x)$, $q_2(x) = x$, and
$$p_2(x) = q_2(x) - \frac{\langle q_2, p_1 \rangle}{\langle p_1, p_1 \rangle} p_1(x) = x - \frac{1}{2}.$$

Then let $q_3(x) = x^2$ and $p_3(x) = q_3(x) - \dfrac{\langle q_3, p_2 \rangle}{\langle p_2, p_2 \rangle} p_2(x) - \dfrac{\langle q_3, p_1 \rangle}{\langle p_1, p_1 \rangle} p_1(x).$

That is,
$$p_3(x) = x^2 - \frac{\int_0^1 x^2 \left(x - \frac{1}{2}\right) dx}{\int_0^1 \left(x - \frac{1}{2}\right)^2 dx}\left(x - \frac{1}{2}\right) - \frac{\int_0^1 x^2\, dx}{1}\,1$$
$$= x^2 - \left(x - \frac{1}{2}\right) - \frac{1}{3}.$$

Then $\|p_1(x)\| = 1$, $\|p_2(x)\| = \dfrac{1}{2\sqrt{3}}$, and $\|p_3(x)\| = \dfrac{1}{6\sqrt{5}}$.

Let $r_i(x) = \dfrac{p_i(x)}{\|p_i(x)\|}$, $i = 1, 2, 3$ to obtain the normal orthogonal basis N.

(iii) We wish to determine a, b, c, such that

$$p(x) = x^2 = a(1) + b[\sqrt{3}(2x - 1)] + c[\sqrt{5}(6x^2 - 6x + 1)].$$

By equating coefficients of like powers of x we obtain

$$1 = 6c\sqrt{5}$$
$$0 = 2b\sqrt{3} - 6c\sqrt{5}$$
$$0 = a - \sqrt{3}b + \sqrt{5}c,$$

so $(a, b, c) = (1/3, 1/2\sqrt{3}, 1/6\sqrt{5})$. Similarly determine d, e, f such that

$$q(x) = 2x - 6x^2 = d(1) + e[\sqrt{3}(2x - 1)] + f[\sqrt{5}(6x^2 - 6x + 1)],$$

obtaining $(d, e, f) = (-1, -2/\sqrt{3}, -1/\sqrt{5})$. Then

$$\langle p, q \rangle = \int_0^1 x^2(2x - 6x^2)\, dx = -\frac{7}{10},$$

$$(a, b, c) \cdot (d, e, f) = \frac{1}{3}(-1) + \frac{1}{6\sqrt{5}}\left(-\frac{1}{\sqrt{5}}\right) = -\frac{7}{10}.$$

9. (i) Let $\xi = \sum x_j \alpha_j$. For each i, $\langle \xi, \alpha_i \rangle^2 = x_i^2$, and $\|\xi\|^2 = x_1^2 + \cdots + x_n^2$, because the inner product has the form of the dot product. Clearly

$$x_1^2 + \cdots + x_k^2 \le x_1^2 + \cdots + x_n^2 \text{ whenever } k \le n.$$

(ii) $\langle \xi, \eta \rangle = \sum x_k y_k = \sum \langle \xi, \alpha_k \rangle y_k = \sum \langle \xi, \alpha_k \rangle \langle \alpha_k, \eta \rangle.$

11. (i) Let $\xi, \eta \in \mathscr{S}^\perp$. Then $\langle a\xi + b\eta, \sigma \rangle = a\langle \xi, \sigma \rangle + b\langle \eta, \sigma \rangle = 0$ so $\mathscr{S}^\perp$ is a subspace of $\mathscr{S}$.

(ii) By Exercise 9 each vector ξ of $\mathscr{E}$ has a unique decomposition $\xi = \sigma + \tau$ where $\sigma \in \mathscr{S}$ and $\tau \in \mathscr{S}^\perp$. Hence $\mathscr{E} = \mathscr{S} \oplus \mathscr{S}^\perp$.

(iii) $(\mathscr{S}^\perp)^\perp$ is the set of all vectors orthogonal to $\mathscr{S}^\perp$. This certainly includes every vector in $\mathscr{S}$, so $\mathscr{S} \subseteq (\mathscr{S}^\perp)^\perp$. If $\mathscr{E}$ is finite dimensional, $\dim \mathscr{S} = \dim(\mathscr{S}^\perp)^\perp$ because

$$\mathscr{E} = \mathscr{S} \oplus \mathscr{S}^\perp = \mathscr{S}^\perp \oplus (\mathscr{S}^\perp)^\perp.$$

(iv) If $\alpha \in (\mathscr{S} + \mathscr{T})$ then α is orthogonal to $\sigma + 0$ and to $0 + \tau$ for every $\sigma \in \mathscr{S}$ and $\tau \in \mathscr{T}$. Thus $(\mathscr{S} + \mathscr{T})^\perp \subseteq \mathscr{S}^\perp \cap \mathscr{T}^\perp$. Conversely, if α is orthogonal to σ and to τ, α is orthogonal to $\sigma + \tau$, so equality holds.

(v) Let $\alpha \in \mathscr{S}^\perp + \mathscr{T}^\perp$. Then $\alpha = \beta + \gamma$ where $\beta \in \mathscr{S}^\perp$ and $\gamma \in \mathscr{T}^\perp$. Hence β and γ are both orthogonal to $\mathscr{S} \cap \mathscr{T}$, so $\alpha \in (\mathscr{S} \cap \mathscr{T})^\perp$. If $\mathscr{E}$ is n-dimensional, $(\mathscr{S}^\perp)^\perp = \mathscr{S}$ by (iii). By (iv) we have $(\mathscr{S}^\perp + \mathscr{T}^\perp)^\perp = \mathscr{S} \cap \mathscr{T}$, so $(\mathscr{S}^\perp + \mathscr{T}^\perp) = (\mathscr{S} \cap \mathscr{T})^\perp.$

Exercises 8.3

2. (i) Relative to the standard basis, $R\varepsilon_1$ and $R\varepsilon_2$ have coordinates $(\cos\Psi, \sin\Psi)$ and $(\cos(\Psi + \pi/2), \sin(\Psi + \pi/2))$, so $R(\Psi)$ is represented by

$$A = \begin{pmatrix} \cos\Psi & -\sin\Psi \\ \sin\Psi & \cos\Psi \end{pmatrix}.$$

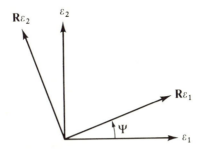

(ii) By computation, the column vectors are normal and orthogonal. Also by matrix multiplication $AA^1 = 1$.

3. (i) Consider the effect of the motion on the vectors of the standard basis to obtain

$$B = \begin{pmatrix} 0 & 1 & 0 \\ 1 & 0 & 0 \\ 0 & 0 & -1 \end{pmatrix}.$$

(ii) $B^t = B$, and $B^t B = I$ by matrix multiplication.

(iii) $B^{-1} = B$, so B^{-1} represents the same transformation as B.

6. If A is unitary, $A^*A = I$, so $\det(A^*A) = 1 = (\det A^*)(\det A)$
$= (\det \bar{A}^t)(\det A) = (\overline{\det A^t})\det A = (\overline{\det A})\det A = |\det A|^2.$

The same argument is valid when A is orthogonal.

7. (i) The condition $A^{-1} = A^t$ implies that $\det^2 A = 1$; hence necessary and sufficient conditions are

$$d = a(ad - bc),$$
$$-b = c(ad - bc),$$
$$-c = b(ad - bc),$$
$$a = d(ad - bc),$$
$$\pm 1 = ad - bc.$$

(ii) From these equations it follows that $d = \pm a, c = \pm b$; hence the possible matrices are

$$\begin{pmatrix} a & b \\ b & a \end{pmatrix}, \begin{pmatrix} a & b \\ -b & -a \end{pmatrix}, \begin{pmatrix} a & b \\ -b & a \end{pmatrix}, \begin{pmatrix} a & b \\ b & -a \end{pmatrix}.$$

Because the column vectors must be normal and orthogonal, $a^2 + b^2 = 1$, and for the first two of these matrices, $ab = 0$. Hence, the first of these four matrices yields

$$\pm I, \ \pm \begin{pmatrix} 0 & 1 \\ 1 & 0 \end{pmatrix}; \text{ and the second yields } \pm \begin{pmatrix} 1 & 0 \\ 0 & -1 \end{pmatrix}, \ \pm \begin{pmatrix} 0 & 1 \\ -1 & 0 \end{pmatrix}.$$

Each of these is a special case of the third or fourth of these matrices.

(iii) If $\cos \Psi = a$, the matrix $\begin{pmatrix} a & b \\ -b & a \end{pmatrix}$ represents a counterclockwise rotation of axis through the angle Ψ. Also

$$\begin{pmatrix} a & b \\ b & -a \end{pmatrix} = \begin{pmatrix} 1 & 0 \\ 0 & -1 \end{pmatrix} \begin{pmatrix} a & b \\ -b & a \end{pmatrix}, \quad \text{so} \quad \begin{pmatrix} a & b \\ b & -a \end{pmatrix}$$

represents a rotation of axes through the angle Ψ followed by a reflection across the new first axis.

9. Let $K^t = -K$ with $I + K$ nonsingular. Because $(I - K)$ and $(I + K)$ commute, $(I - K)^{-1}$ and $(I + K)^{-1}$ also commute. Let $A = (I - K)(I + K)^{-1}$. We want to prove that $A^t = A^{-1}$. But $A^{-1} = (I + K)(I - K)^{-1} = (I - K)^{-1}(I + K)$. We have $A(I + K) = (I - K)$, so $(I + K)^t A^t = (I - K)^t$; thus

$$(I - K)A^t = (I + K), \text{ so } A^t = (I - K)^{-1}(I + K) = A^{-1}.$$

10. (i) $c_{ij} = \sum_k a_{ik} a_{jk} = p(\alpha_i, \alpha_j)$, where p is the dot product.

(ii) If $\{\alpha_1, \ldots, \alpha_n\}$ is an orthogonal set, $c_{ij} = 0$ if $i \neq j$ and $c_{ii} = \sum_k a_{ik}^2 = \|\alpha_i\|^2$. Then C is diagonal, and

$$\det C = c_{11} \ldots c_{nn} = (\|\alpha_1\| \cdots \|\alpha_n\|)^2.$$

(iii) If each a_{ij} satisfies $|a_{ij}| \leq k$, then $\|\alpha_i\|^2 = \sum_j a_{ij}^2 \leq nk^2$.

(iv) Under the assumptions of (ii) and (iii) we have

$$\det C = \|\alpha_1\|^2 \cdots \|\alpha_n\|^2 \leq (nk^2) \cdots (nk^2) = n^n k^{2n}$$
$$= \det(AA^t) = \det^2 A;$$
$$|\det A| \leq n^{n/2} k^n.$$

CHAPTER 9

Exercises 9.1

1. If f is nonzero, there exists a vector $\alpha \in \mathscr{V}$ such that $f(\alpha) \neq 0$. If $c \neq 0$ in $\mathbb{R}$, let $\xi = k\alpha$, where $k = c/f(\alpha)$. Then $f(\xi) = kf(\alpha) = c$.

3. We have $\mathbf{f}_1\alpha_1 = \mathbf{f}_1\varepsilon_2 + \mathbf{f}_1\varepsilon_3 = 1,$
$\qquad \mathbf{f}_1\alpha_2 = \mathbf{f}_1\varepsilon_1 + \mathbf{f}_1\varepsilon_3 = 0,$ so $\begin{cases} \mathbf{f}_1\varepsilon_1 = -\frac{1}{2}, \\ \mathbf{f}_1\varepsilon_2 = \frac{1}{2}, \\ \mathbf{f}_1\varepsilon_3 = \frac{1}{2}. \end{cases}$
$\qquad \mathbf{f}_1\alpha_3 = \mathbf{f}_1\varepsilon_1 + \mathbf{f}_1\varepsilon_2 = 0,$

Similarly, $\begin{matrix} \mathbf{f}_2\varepsilon_1 = \frac{1}{2}, \\ \mathbf{f}_2\varepsilon_2 = -\frac{1}{2}, \\ \mathbf{f}_2\varepsilon_3 = \frac{1}{2}, \end{matrix}$ and $\begin{cases} \mathbf{f}_3\varepsilon_1 = \frac{1}{2}, \\ \mathbf{f}_3\varepsilon_2 = \frac{1}{2}, \\ \mathbf{f}_3\varepsilon_3 = -\frac{1}{2}. \end{cases}$

Hence $\mathbf{f}_1(a, b, c) = \frac{1}{2}(-a + b + c),$
$\qquad \mathbf{f}_2(a, b, c) = \frac{1}{2}(a - b + c),$
$\qquad \mathbf{f}_3(a, b, c) = \frac{1}{2}(a + b - c).$

6. By definition $\mathbf{f}_\gamma(\xi) = \xi \cdot \gamma$. If $\mathbf{f}_\beta = \mathbf{f}_\gamma$ then $\xi \cdot \beta = \xi \cdot \gamma$ for all ξ, so $\beta = \gamma$, and the mapping is one-to-one. To see that it maps $\mathscr{V}_n$ onto $\mathscr{V}_n'$, let $g \in \mathscr{V}_n'$, let $\{\alpha_1, \ldots, \alpha_n\}$ be a normal orthogonal basis for $\mathscr{V}_n$, let $\gamma = \sum \mathbf{g}(\alpha_i)\alpha_i$, and let $\xi = \sum x_i \alpha_i$. Then $\mathbf{g}(\xi) = \sum x_i \mathbf{g}(\alpha_i) = \xi \cdot \gamma$ so $\mathbf{g} = \mathbf{f}_\gamma$.

7. (a) Let $S \subseteq T$ and $\mathbf{f} \in T^0$. Then $\mathbf{f}\tau = 0$ for each $\tau \in T$, so $\mathbf{f}\tau = 0$ for each $\tau \in S$. Hence $\mathbf{f} \in S^0$, and $T^0 \subseteq S^0$.

(b) $\mathscr{M} \cap \mathscr{N} \subseteq \mathscr{M} \subseteq \mathscr{M} + \mathscr{N}$, so $(\mathscr{M} + \mathscr{N})^0 \subseteq \mathscr{M}^0 \subseteq (\mathscr{M} \cap \mathscr{N})^0$, and similarly for $\mathscr{N}$. Hence $(\mathscr{M} + \mathscr{N})^0 \subseteq \mathscr{M}^0 \cap \mathscr{N}^0 \subseteq \mathscr{M}^0 + \mathscr{N}^0 \subseteq (\mathscr{M} \cap \mathscr{N})^0$. Let $\mathbf{f} \in \mathscr{M}^0 \cap \mathscr{N}^0$. Then $\mathbf{f}\alpha = 0$ for all $\alpha \in \mathscr{M}$ and $\mathbf{f}\beta = 0$ for all $\beta \in \mathscr{N}$. Then for $\xi \in \mathscr{M} + \mathscr{N}$, $\mathbf{f}(\xi) = \mathbf{f}(\alpha + \beta) = \mathbf{f}\alpha + \mathbf{f}\beta = 0$, where $\alpha \in \mathscr{M}$ and $\beta \in \mathscr{N}$. Hence $\mathbf{f} \in (\mathscr{M} + \mathscr{N})^0$ and so $(\mathscr{M} + \mathscr{N})^0 = \mathscr{M}^0 \cap \mathscr{N}^0$. To show that $\mathscr{M}^0 + \mathscr{N}^0 = (\mathscr{M} \cap \mathscr{N})^0$ we need to use Theorem 9.2(c) and the dimension of $\mathscr{V}$. Let dim $\mathscr{V} = n$;

$$\dim(\mathscr{M} \cap \mathscr{N})^0 = n - \dim(\mathscr{M} \cap \mathscr{N})$$
$$= n - \dim \mathscr{M} - \dim \mathscr{N} + \dim(\mathscr{M} + \mathscr{N})$$
$$= \dim \mathscr{M}^0 + \dim \mathscr{N}^0 - \dim(\mathscr{M} + \mathscr{N})^0$$
$$= \dim \mathscr{M}^0 + \dim \mathscr{N}^0 - \dim(\mathscr{M}^0 \cap \mathscr{N}^0) = \dim(\mathscr{M}^0 + \mathscr{N}^0).$$

Because $\mathscr{M}^0 + \mathscr{N}^0$ is a subspace of $(\mathscr{M} \cap \mathscr{N})^0$ and they have the same dimension, equality holds.

8. $\mathscr{H} = \{\xi \in \mathscr{V}_n | \mathbf{f}\xi = 0\}$. Let α, $\beta \in \mathscr{H}$ and a, $b \in \mathscr{F}$. Then $\mathbf{f}(a\alpha + b\beta) = a\mathbf{f}\alpha + b\mathbf{f}\beta = a0 + b0 = 0$. $\mathscr{H}$ is a subspace of $\mathscr{V}_n$, the null space of the mapping $\mathbf{f}$ from $\mathscr{V}_n$ to $\mathscr{F}$. Because $\mathbf{f} \neq \mathbf{Z}$, the range space of $\mathbf{f}$ is $\mathscr{F}$. Hence $r(\mathbf{f}) = 1$ and $n(\mathbf{f}) = n - 1 = \dim \mathscr{H}$.

9. (i) Let $\sigma = \{c_i \,|\, i = 0, 1, \ldots\}$ and let $\mathbf{f}_\sigma$ be the linear mapping on $\mathscr{P}$ defined by $\mathbf{f}_\sigma(x^k) = c_k$. Then

$$\mathbf{f}_\sigma(a_0 + a_1 x + a_2 x^2) = a_0 c_0 + a_1 c_1 + a_2 c_2.$$

(ii) $\mathbf{f}_\sigma$ is a linear mapping from $\mathscr{P}$ to $\mathbb{R}$, so $\mathbf{f}_\sigma \in \mathscr{P}'$.

(iii) Let $\mathbf{h}$ denote the mapping from $\mathscr{S}$ to $\mathscr{P}'$ defined by $\mathbf{h}(\sigma) = \mathbf{f}_\sigma$. If $\mathbf{h}(\sigma) = \mathbf{h}(\tau)$ then $\mathbf{f}_\sigma = \mathbf{f}_\tau$, so $\mathbf{f}_\sigma(x^k) = \mathbf{f}_\tau(x^k)$ so $c_k = d_k$ for all k where $\sigma = \{c_i\}$ and $\tau = \{d_i\}$. Thus $\sigma = \tau$ whenever $\mathbf{h}(\sigma) = \mathbf{h}(\tau)$, so $\mathbf{h}$ is one-to-one. Also, given any $\mathbf{f} \in \mathscr{P}'$ let $c_k = \mathbf{f}(x^k)$ for $k = 0, 1, \ldots$; and let $\sigma = \{c_k\}$. Then $\mathbf{f} = \mathbf{f}_\sigma$, so $\mathbf{h}$ is onto $\mathscr{P}'$.

(iv) Let $\alpha = \{a_i\}$ and $\beta = \{b_i\}$ be in $\mathscr{S}$; then $\alpha + \beta = \{a_i + b_i\}$ so $\mathbf{f}_{\alpha+\beta}(x^k) = a_k + b_k = \mathbf{f}_\alpha(x^k) + \mathbf{f}_\beta(x^k)$ for each k. Because $\mathbf{f}_{\alpha+\beta}$ and $\mathbf{f}_\alpha + \mathbf{f}_\beta$ agree on a basis for $\mathscr{P}$ and are linear, they are equal. Similarly, $\mathbf{f}_{c\alpha}(x^k) = ca_k = c\mathbf{f}_\alpha(x^k)$ for each k, so $\mathbf{f}_{c\alpha} = c\mathbf{f}_\alpha$. Thus the one-to-one mapping $\sigma \to \mathbf{f}_\sigma$ of $\mathscr{S}$ onto $\mathscr{P}'$ preserves the vector space operations of $\mathscr{S}$; hence $\mathscr{S}$ and $\mathscr{P}'$ are isomorphic.

Exercises 9.2

2. (i) $\mathbf{I}^t = \mathbf{I}$ means that $\mathbf{I}^t$ is the identity linear transformation on $\mathscr{V}'$. For each $f \in \mathscr{V}'\,\mathbf{I}^t f = f\mathbf{I} = f$, so $\mathbf{I}^t = \mathbf{I}$.

(ii) $(c\mathbf{T})^t g = g(c\mathbf{T}) = cg\mathbf{T} = c\mathbf{T}^t g$ for all $g \in \mathscr{V}'$, so $(c\mathbf{T})^t = c\mathbf{T}^t$.

3. To show that $\mathbf{S}^t$ is nonsingular, suppose that $\mathbf{S}^t$ maps g in $\mathscr{W}'$ onto the zero linear functional z in $\mathscr{V}'$. Then $\mathbf{S}^t g = z = g\mathbf{S}$, so for each $\xi \in \mathscr{V}$, $(g\mathbf{S})\xi = 0 = g(\mathbf{S}\xi)$. But $\{\mathbf{S}\xi \mid \xi \in \mathscr{V}\} = \mathscr{W}$ because $\mathbf{S}$ is nonsingular from $\mathscr{V}$ onto $\mathscr{W}$. Because g maps each vector in $\mathscr{W}$ onto 0, $g = z$ in $\mathscr{W}'$. Hence $\mathbf{S}^t$ is nonsingular. By Exercise 2

$$(\mathbf{S}^{-1})^t \mathbf{S}^t = (\mathbf{S}\mathbf{S}^{-1})^t = \mathbf{I}^t,$$

the identity linear transformation on $\mathscr{W}'$. Hence $(\mathbf{S}^{-1})^t$ is the inverse of $\mathbf{S}^t$.

4. $(\mathbf{T}^t)^t = \mathbf{T}$ is interpreted in terms of the isomorphism between a vector space $\mathscr{V}$ and its bidual $\mathscr{V}''$; each vector $\xi \in \mathscr{V}$ corresponds to a linear functional $\xi \in \mathscr{V}''$ such that $\xi f = f\xi$ for each $f \in \mathscr{V}'$. $\mathscr{V}$ is isomorphic to a subspace $\tilde{\mathscr{V}}$ of $\mathscr{V}''$. Thus $(\mathbf{T}^t)^t$ maps $\tilde{\mathscr{V}}$ into $\tilde{\mathscr{W}}$ by the rule $(\mathbf{T}^t)^t \xi = \xi \mathbf{T}^t = \mathbf{T}\xi$ for each $\xi \in \tilde{\mathscr{V}}$.

Exercises 9.3

1. (i) The table of Section 4.3 can be extended readily as follows.

Elementary Matrix: E	Conjugate: $\bar{E}$	Adjoint: E^*
$M_i(c)$	$M_i(\bar{c})$	$M_i(\bar{c})$
$R_{i,\,i+cj}$	$R_{i,\,i+\bar{c}j}$	$R_{j,\,j+\bar{a}}$
P_{ij}	P_{ij}	P_{ij}

(ii) As row operations

$\bar{M}_i(c)$ and M_i^* multiply row i by $\bar{c}$,

$\bar{R}_{i+cj}$ replaces row i by $R_i + \bar{c}R_j$,

R_{i+cj}^* replaces row j by $R_j + \bar{c}R_i$,

$\bar{P}_{ij}$ and P_{ij}^* permute rows i and j.

(iii) As column operations

$\bar{M}_i(c)$ and $M_i^*(c)$ multiply column i by $\bar{c}$,

$\bar{R}_{i+cj}$ replaces column j by $C_j + \bar{c}C_i$,

R_{i+cj}^* replaces column i by $C_i + \bar{c}C_j$,

$\bar{P}_{ij}$ and P_{ij}^* permute columns i and j.

3. (i) $A = \begin{pmatrix} 4 & 2 & 0 \\ 4 & 4 & -1 \end{pmatrix}$.

4. $f(\alpha_2, \beta_1) = a_{12} = 3,$

$f(\alpha_1, \beta_2) = a_{21} = 2,$

$f(\alpha_2, \beta_2) = a_{22} = -1.$

5. (i) $(A^*)^* = (\bar{A}^t)^* = ((\bar{A})^t)^t = (A^t)^t = A.$

(ii) $r(\bar{A}) = r(A) = r(A^t)$, so $r(A^*) = r(A).$

6. In Theorem 9.5 let $m = n$, let $\mathscr{W}_m(\mathbb{C}) = \mathscr{V}_n(\mathbb{C})$, and let $\beta_i = \alpha_i$ for $i = 1, \ldots, n$. Then $a_{ij} = f(\alpha_j, \alpha_i)$. Also let $\delta_i = \gamma_i$ for $i = 1, \ldots, n$ so that $P^* = Q$. With these specializations the statement of Theorem 9.5 becomes Theorem 9.8.

7. A real bilinear function f on $\mathscr{V}_n(\mathbb{R})$ is represented uniquely relative to a basis $\{\alpha_1, \ldots, \alpha_n\}$ for $\mathscr{V}_n(\mathbb{R})$ by the n-by-n matrix

$$A = (a_{ij}), \text{ where } a_{ij} = f(\alpha_j, \alpha_i).$$

If ξ and η in $\mathscr{V}_n(\mathbb{R})$ are represented by X and Y, respectively, then

$$f(\xi, \eta) = Y^t A X.$$

An n-by-n matrix C also represents f relative to a suitable basis if and only if

$$C = Q^t A Q$$

for some nonsingular matrix Q.

9. Let A and C be m-by-n complex matrices such that $C = P^* A Q$ for nonsingular matrices P and Q. Let f be the function defined from $\mathscr{V}_n(\mathbb{C}) \times \mathscr{W}_m(\mathbb{C})$ to $\mathbb{C}$ by the following rule: if $\xi = \sum_{s=1}^n x_s \alpha_s \in \mathscr{V}_n$ and $\eta = \sum_{r=1}^m y_r \beta_r \in \mathscr{W}_m$, then

$$f(\xi, \eta) = \sum_{r=1}^m \sum_{s=1}^n \bar{y}_r a_{rs} x_s = Y^* A X.$$

Then, in particular, for each vector in the basis $\{\alpha_1, \ldots, \alpha_n\}$ for $\mathscr{V}_n$ and each vector in the basis $\{\beta_1, \ldots, \beta_m\}$ for $\mathscr{W}_m$,

$$f(\alpha_s, \beta_r) = a_{rs}.$$

Because Q is a nonsingular n-by-n matrix, if we let

$$\gamma_j = \sum_{s=1}^n q_{sj} \alpha_s, j = 1, \ldots, n,$$

then $\{\gamma_1, \ldots, \gamma_n\}$ is a basis for $\mathscr{V}_n$. Similarly, because P is a nonsingular m-by-m matrix, $\{\delta_1, \ldots, \delta_m\}$ is a bsis for $\mathscr{W}_m$, where

$$\delta_i = \sum_{r=1}^{m} p_{ri}\beta_r.$$

Then for each i and each j

$$f(\gamma_j, \delta_i) = f\left(\sum_{s=1}^{n} q_{sj}\alpha_s, \sum_{r=1}^{n} p_{ri}\beta_r\right)$$

$$= \sum_{r=1}^{m}\sum_{s=1}^{n} \bar{p}_{ri} f(\alpha_s, \beta_r)q_{sj}$$

$$= \sum_{r=1}^{m} p_{ir}^{*}\left(\sum_{s=1}^{n} a_{rs} q_{sj}\right) = c_{ij}.$$

Exercises 9.4

1. By Theorem 9.6 the two forms represent the same function if and only if their matrix representations are equivalent. But both matrices have the same rank and hence are equivalent.

4. C is conjunctive to A if and only if $C = Q^*AQ$ for some nonsingular Q. Express Q as a product of elementary matrices;

$$Q = E_1 E_2 \cdots E_k,$$
$$C = E_k^* \cdots E_2^*(E_1^*AE_1)E_2 \cdots E_k.$$

But E_1^* is an elementary matrix, which performs an elementary row operation on A, and $E_1 = (E_1^*)^*$ performs the corresponding conjugate column operation on A. Similarly for E_2^*, and so on.

6. (i) $(AA^*)^* = A^{**}A^* = AA^*$, so AA^* is Hermitian; similarly for A^*A.

(ii) If $A^* = A$, then $\bar{a}_{ii} = a_{ii}$, so a_{ii} is real. If $A^t = A$, then $A^* = \bar{A} = A$ if A is real.

8. (i) If $H^* = H$, then $(iH)^* = \bar{i}H^* = -iH$, and conversely.

(ii) By (i) K is skew-Hermitian if and only if $-iK$ is Hermitian. A canonical form for $-iK$, therefore, is the block diagonal matrix diag $(I_p, -I_{r-p}, Z)$ where r is the rank of $-iK$ and $s = 2p - r$ is its signature. Thus K is conjunctive to one and only one matrix of the block diagonal form diag $(iI_p, -iI_{r-p}, Z)$.

9. If A is skew-symmetric, so is any matrix congruent to A. Let $a_{ij} \neq 0$. Then $B = P_{i1}AP_{i1}^t$ is congruent to A and has a_{ij} in position $(1, j)$. Then $C = P_{2j}BP_{2j}^t$ has a_{ij} in position $(1, 2)$, $-a_{ij}$ in position $(2, 1)$, and zero in each diagonal position. Next

$$D = M_1(a_{ij}^{-1})CM_1(a_{ij}^{-1})^t \text{ has } \begin{pmatrix} 0 & 1 \\ -1 & 0 \end{pmatrix} \text{ in the upper left-hand corner.}$$

Additional row operations can then be used to produce zero in rows three to n of the first two columns, and the corresponding column operations produce zero in columns three to n of the first two rows. Thus A is congruent to

$$E = \left(\begin{array}{cc|c} 0 & 1 & \\ & & Z \\ -1 & 0 & \\ \hline & Z & F \end{array} \right).$$

The argument can be repeated on F unless $F = Z$. Hence A is congruent to diag $(A_1, \ldots, A_t, Z)$, and the rank of A is $2t$. Because the rank of A determines this form, any two skew matrices of the same (even) rank are congruent to the same matrix of this form and hence to each other.

11. (i) One correct answer is $\frac{1}{2} \begin{pmatrix} 5 & 0 & 0 \\ 0 & -20 & 0 \\ 0 & 0 & 0 \end{pmatrix}$.

(ii) $\begin{pmatrix} 1 & 0 & 0 \\ 0 & -1 & 0 \\ 0 & 0 & 0 \end{pmatrix}$; $r(A) = 2$, $s(A) = 0$.

(iii) $\begin{pmatrix} 1 & 0 & 0 \\ 0 & 1 & 0 \\ 0 & 0 & 0 \end{pmatrix}$.

(iv) The answers to parts (i–iii) are in the form of Theorem 9.10 but all are different because of different properties of the scalar fields.

13. By Theorem 9.10 a complex symmetric matrix of rank r is congruent over $\mathbb{C}$ to diag$(d_1, \ldots, d_r, 0, \ldots, 0)$ where each d_i is nonzero in $\mathbb{C}$. Let $d_i^{1/2}$ be a complex square root of d_i, and multiply row i and column i by $d_i^{1/2}$ for $i = 1, \ldots, r$. The resulting matrix is of the form stated in Theorem 9.15 and it is congruent to the given matrix. Any two complex symmetric matrices of rank r are congruent to diag(I_r, Z) and hence to each other. Furthermore (I_r, Z) is congruent to (I_s, Z) only if $r = s$.

Exercises 9.5

1. (i) If $a = 0$, the form is not positive definite; if $a \neq 0$, then

$$ax_1^2 + 2bx_1x_2 + cx_2^2 = a\left(x_1^2 + \frac{2b}{a}x_1x_2 + \frac{b^2}{a^2}x_2^2 - \frac{b^2}{a^2}x_2^2 + \frac{c}{a}x_2^2\right)$$
$$= a\left(x_1 + \frac{b}{a}x_2\right)^2 + \left(\frac{ca - b^2}{a^2}\right)x_2^2.$$

Each term is nonnegative only if the two coefficients a and $ca - b^2$ are positive. And then both terms are 0 if and only if $x_2 = 0 = x_1$.

(ii) If $a > 0$ and $ca - b^2 > 0$, then the form is positive definite so $r = s = 2$. Then the conic has an equation of the form $AX^2 + BY^2 = 1$ with A, B positive—an ellipse. If a and $ca - b^2$ have opposite sign, then $r = 2$ and $s = 0$, and the conic has an equation of the form $AX^2 - BY^2 = 1$, where A and B have the same sign—a hyperbola.

2. (i)
$$A = \begin{pmatrix} 1 & 0 & -1 \\ 0 & 2 & 2 \\ -1 & 2 & 6 \end{pmatrix}.$$

Use row operations and the corresponding column operations to show that A is congruent to diag$(1, 1, 1)$. Hence $r = 3 = s$.

(ii)
$$A = \begin{pmatrix} 0 & 8 & 0 \\ 8 & 0 & 0 \\ 0 & 0 & -1 \end{pmatrix}.$$

The characteristic polynomial of A is $p(x) = -(x + 1)(x + 8)(x - 8)$, so A is similar and congruent to diag$(8, -8, -1)$. Hence $r = 3$, and $s = -1$.

3. (i) The forms in (ii), (iii), (v) have $r = 3$, $s = -1$, and hence represent the same quadratic function. Similarly (iv) and (vi) have $r = 3$, $s = 1$.

(ii) Only (i) is positive definite, a necessary condition for an inner product.

4. If $A = H + K$ and X^*AX is real for each X, then X^*KX is also real for each X, where K is skew-Hermitian. Thus $X^*KX = (X^*KX)^* = X^*K^*X = -X^*KX$, so $X^*KX = 0$, and $X^*AX = X^*HX$.

5. A real quadratic function of rank r and signature s on $\mathscr{V}_n$ is represented by a matrix that is congruent to diag $(I_p, -I_{r-p}, Z)$ where $2p - r = s$. Hence in some coordinate system

$$q(x_1, \ldots, x_n) = x_1^2 + \cdots + x_p^2 - x_{p+1}^2 - \cdots - x_r^2.$$

For some nonzero vector X, $q(X) = 0$ if $r < n$, and for some nonzero vector Y, $q(Y) < 0$ if $p < r$. Hence $r = n = p = s$ is a necessary and sufficient condition that q be positive definite.

9. (i) If $r = 2$ and $s = -2$, the quadratic form is negative definite. Hence $f(a + h, b + k) < f(a, b)$ for all sufficiently small h and k, and f has a relative maximum at (a, b).

(ii) If $r = 2$ and $s = 2$, the quadratic form is positive definite. Hence $f(a + h, b + k) > f(a, b)$ for all sufficiently small h and k, and f has a relative minimum at (a, b).

(iii) If $r = 2$ and $s = 0$, the value of the quadratic form is positive for some small h and k, but negative for others. Then f has a saddle point at (a, b).

11. (i) $(A^tA)^t = A^tA$, so B_0 is symmetric.

(ii) If A is nonsingular, then $r(A) = n = r(A^t) = r(B_0)$. Also $X^t B_0 X = X^t(A^t A)X = (AX)^t(AX) > 0$ if $X = Z$.

(iii) If $D = \text{diag}(x_1, \ldots, x_n)$, then $C = DBD = D^t BD$. Thus C is symmetric if B is. Also $X^t CX = (DX)^t B(DX) > 0$ for $X \neq Z$ because D is nonsingular. Hence C is positive definite.

(iv) Because the characteristic values of C are real and positive,

$$\det C = \lambda_1 \lambda_2 \cdots \lambda_n \leq \left[\frac{\lambda_1 + \cdots + \lambda_n}{n}\right]^n = \left[\frac{\text{tr } C}{n}\right]^n,$$

where the inequality is valid because the geometric mean of positive real numbers never exceeds their arithmetic mean.

(v) Hadamard's inequality is trivial if A is singular. If A is nonsingular, $A^t A$ is real and positive definite by (ii), so we can apply the conclusion of (iv) with $C = A^t A$.

$$\det A^t A = (\det A)^2 \leq \left[\frac{\text{tr}(A^t A)}{n}\right]^n$$

But $\text{tr}(A^t A)$ is the sum of the n diagonal entries $\sum_{j=1}^{n} a_{ji} a_{ji}$ of $A^t A$, each of which is less than or equal to nk^2. Hence

$$(\det A)^2 \leq \left[\frac{n(nk^2)}{n}\right]^n,$$

which yields Hadamard's inequality.

12. The number c_j is the coefficient of λ^{n-j} in the characteristic polynomial of A. From the expression of $\det(A - \lambda I)$ as a sum of signed products of elements of $A - \lambda I$, one from each row and each column, it follows that c_j is a sum of signed products of n terms, $n - j$ of which are $(-\lambda)$ obtained from $n - j$ diagonal entries and j of which are entries of A. For a given choice of $n - j$ diagonal entries, the j entries of A in any product are the terms of the determinant of the j-by-j matrix D_j that remains after each row and column of the chosen diagonal entries are deleted. By Hadamard's inequality $|\det D_j| \leq k^j j^{j/2}$. Because $n - j$ diagonal entries can be chosen in

$$\binom{n}{j} \text{ ways, we obtain } |c_j| \leq \binom{n}{j} j^{j/2} k^j, \text{ for } j = 1, \ldots, n.$$

Exercises 9.6

2. When each entry and each characteristic value of A is real, the proof of Theorem 9.28 can be carried out in $\mathbb{R}$ to yield the conclusion that $Q^t AQ$ is real and upper triangular for some orthogonal Q. Then the proof of Theorem 9.29 can be specialized to $\mathbb{R}$, to conclude that A and A^t commute (that is, A is normal) if and only if $P^t AP$ is diagonal for some orthogonal P.

3. By Theorem 9.29 A is normal if and only if A is isometrically diagonable; $U^{-1}AU = D = \text{diag}(\lambda_1, \ldots, \lambda_n) =$ where $U^{-1} = U^*$. For each i the ith unit n-tuple $X_i = (0, \ldots, 0, 1, 0, \ldots, 0)$ is a characteristic vector of D, associated with λ_i, and the set $\{X_1, \ldots, X_n\}$ is mutually orthogonal. Because D and A represent the same linear transformation, A also has a set of n mutually orthogonal characteristic vectors.

4. The characteristic polynomial of A is $-(x-1)(x-2)^2$, and the minimal polynomial of A is $(x-1)(x-2)$, so A is diagonable. But $AA^t \neq A^tA$ by matrix calculations, so A is not normal and hence not isometrically diagonable. Any characteristic vector associated with 1 is of the form $(0, b, -2b)$, and any characteristic vector associated with 2 is of the form $(a, 0, c)$. It follows that no three characteristic vectors are mutually orthogonal.

6. Let $\{\alpha_1, \ldots, \alpha_n\}$ be a normal orthogonal basis and let $\mathbf{T}\alpha_j = \sum_{k=1}^n a_{kj}\alpha_k$. Then

$$\langle \mathbf{T}\alpha_j, \alpha_i \rangle = \left\langle \sum_{k=1}^n a_{kj}\alpha_k, \alpha_i \right\rangle = \sum_{k=1}^n a_{kj}\langle \alpha_k, \alpha_i \rangle = a_{ij}.$$

Hence the (i, j) entry of A is $\langle \mathbf{T}\alpha_j, \alpha_i \rangle$, so the (i, j) entry of the matrix B that represents $\mathbf{T}^*$ relative to this basis is

$$\langle \mathbf{T}^*\alpha_j, \alpha_i \rangle = \langle \alpha_i, \mathbf{T}^*\alpha_j \rangle = \overline{\langle \mathbf{T}\alpha_i, \alpha_j \rangle} = \overline{a_{ji}}.$$

Thus $B = A^*$.

7. (i) $\|\mathbf{T}\xi\|^2 = \langle \mathbf{T}\xi, \mathbf{T}\xi \rangle = \langle \xi, \mathbf{T}^*\mathbf{T}\xi \rangle$

$= \langle \xi, \mathbf{T}\mathbf{T}^*\xi \rangle = \langle \mathbf{T}^*\xi, \mathbf{T}^*\xi \rangle = \|\mathbf{T}^*\xi\|^2.$

(ii) $\mathbf{T}$ is normal if and only if $\mathbf{T} - \lambda\mathbf{I}$ is normal because $\mathbf{T}$ commutes with a transformation $\mathbf{S}$ if and only if $(\mathbf{T} - c\mathbf{I})$ commutes with $(\mathbf{S} - c\mathbf{I})$ for any scalar c. Then from (i)

$$\|(\mathbf{T} - \lambda\mathbf{I})\xi\|^2 = \|(\mathbf{T} - \lambda\mathbf{I})^*\xi\|^2 = \|(\mathbf{T}^* - \bar{\lambda}\mathbf{I})\xi\|^2.$$

Hence ξ is a characteristic vector of $\mathbf{T}$ associated with λ if and only if ξ is a characteristic vector of $\mathbf{T}^*$ associated with $\bar{\lambda}$.

8. (i) A is not normal because $AA^* \neq A^*A$.

9. (i) If A is normal, $D = U^*AU = \text{diag}(\lambda_1, \ldots, \lambda_n)$, where $U^* = U^{-1}$. If $A^p = Z$, then $D^p = \text{diag}(\lambda_1^p, \ldots, \lambda_n^p) = U^*A^pU = Z$, so each $\lambda_i = 0$ and $D = Z$. Because U is nonsingular, $A = Z$.

10. (i) $(c_1\mathbf{T} + c_2\mathbf{T}^*)(c_1\mathbf{T} + c_2\mathbf{T}^*)^* = (c_1\mathbf{T} + c_2\mathbf{T}^*)(\bar{c}_1\mathbf{T}^* + \bar{c}_2\mathbf{T})$

$= c_1\bar{c}_1\mathbf{T}\mathbf{T}^* + c_2\bar{c}_2\mathbf{T}^*\mathbf{T} + c_1\bar{c}_2\mathbf{T}^2 + c_2\bar{c}_1(\mathbf{T}^*)^2.$

Expand $(c_1\mathbf{T} + c_2\mathbf{T}^*)^*(c_1\mathbf{T} + c_2\mathbf{T}^*)$ and use $c_1\bar{c}_1 = c_2\bar{c}_2$ to obtain the same result, showing that $c_1\mathbf{T} + c_2\mathbf{T}^*$ is normal.

(ii) $\langle \mathbf{T}\xi,\ \xi \rangle = \langle \xi,\ \mathbf{T}^*\xi \rangle = \overline{\langle \mathbf{T}^*\xi,\ \xi \rangle} = \langle \mathbf{T}^*\xi,\ \xi \rangle$ if $\langle \mathbf{T}\xi,\ \xi \rangle$ is real. Then $\mathbf{T}\xi = \mathbf{T}^*\xi$ for each ξ, and $\mathbf{T} = \mathbf{T}^*$.

(iii) If $\langle \mathbf{T}^*\mathbf{T}\xi,\ \xi \rangle = \langle \mathbf{T}\xi,\ \mathbf{T}\xi \rangle = \langle \mathbf{T}^*\xi,\ \mathbf{T}^*\xi \rangle = \langle \mathbf{T}\mathbf{T}^*\xi,\ \xi \rangle$ for all ξ, then $\mathbf{T}^*\mathbf{T}\xi = \mathbf{T}\mathbf{T}^*\xi$ for all ξ, and $\mathbf{T}$ is normal.

12. To see that p is linear in the first component, compute

$$p(af_1 + bf_2, g) = \langle \varphi(g), \varphi(af_1 + bf_2) \rangle = \langle \varphi(g), \bar{a}\varphi(f_1) + \bar{b}\varphi(f_2) \rangle$$
$$= a\langle \varphi(g), \varphi(f_1) \rangle + b\langle \varphi(g), \varphi(f_2) \rangle$$
$$= ap(f_1, g) + bp(f_2, g).$$

To see that p is conjugate symmetric, compute

$$p(g, f) = \langle \varphi(f), \varphi(g) \rangle = \overline{\langle \varphi(g), \varphi(f) \rangle} = \overline{p(f, g)}.$$

Finally to see that p is positive definite, observe that

$$p(f, f) = \langle \varphi(f), \varphi(f) \rangle > 0 \text{ if } \varphi(f) \neq \theta.$$

Because φ is one-to-one, it follows that $p(f, f) > 0$ if $f \neq z$ in $\mathscr{V}'$. Hence p is an inner product on $\mathscr{V}'$.

13. (i) Exercise 9.4-6(i). It follows also that A^*A is normal.

(ii) Theorem 9.29 shows that A^*A is isometrically diagonable, and the diagonal entries of D are the characteristic values λ of A^*A, which are real and nonzero because A^*A is Hermitian and nonsingular. Let $A^*AX = \lambda X$, $X \neq Z$. Then $(XA)^*(XA) = \lambda X^*X$. Because Y^*Y is a positive number for each $Y \neq Z$, λ is the ratio of two positive real numbers, and thus is positive.

(iii) Let $e_{ii} = \sqrt{d_{ii}}$ for each i.

(iv) Because E is real and diagonal, $H^* = H$. Also H is conjunctive to E and hence has the same rank and signature as E; H is positive definite because E is.

(v) Let $Q = AH^{-1} = A(PEP^*)^{-1} = APE^{-1}P^*$. Then
$$Q^* = PE^{-1}P^*A^*,$$
$$Q^*Q = (PE^{-1}P^*A^*)(APE^{-1}P^*) = PE^{-1}(P^*A^*AP)E^{-1}P^*$$
$$= P(E^{-1}DE^{-1})P^* = PP^* = I.$$

(vi) By the definition of Q in (v), A is the product QH of a unitary matrix and a positive definite Hermitian matrix.

CHAPTER 10

Exercises 10.1

1. (i)

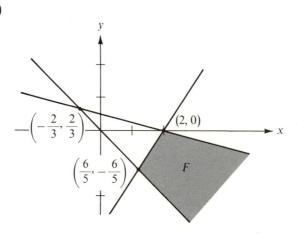

(ii) The corner points are $(2, 0)$ and $(6/5, -6/5)$. The value of $5x - 3y$ is 10 at $(2, 0)$ and is 9.6 at $(6/5, -6/5)$. Because the value of that function is positive at each point of F and unbounded on F, it has no maximum on F and its minimum on F is 9.6, attained at $(6/5, -6/5)$.

3. (i)

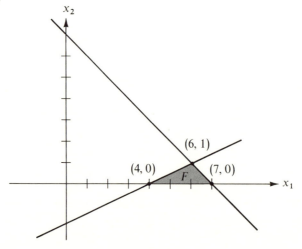

4. (i) The maximum value of $3x_1 + 2x_2$ on F is 21 at $(7, 0)$. The minimum value is 12 at $(4, 0)$.

5. (i) By vector geometry a point on the line segment from U to V is given by $Y = U + t(V - U) = (1 - t)U + tV$ for some t in $[0, 1]$.

(ii) We have $\sum_{i=1}^{n} a_i u_i < c$ and $\sum_{i=1}^{n} a_i v_i < c$, so

$$\sum_{i=1}^{n} a_i y_i = \sum_{i=1}^{n} a_i[(1 - t)u_i + tv_i] = (1 - t)\sum_{i=1}^{n} a_i u_i + t\sum_{i=1}^{n} a_i v_i$$
$$< (1 - t)c + tc = c.$$

7. All except (iii) and (v) are convex.

8. The daily picking output and the contractual agreements can be expressed as follows:

Grade I: $6y_1 + 2y_2 \geq 12$,

Grade II: $2y_1 + 2y_2 \geq 8$,

Grade III: $4y_1 + 12y_2 \geq 24$.

The daily picking cost for labor and equipment is

$$C(y_1, y_2) = 20y_1 + 16y_2.$$

10. (i) Let $X, Y \in \mathscr{E}^n$. Any point V on the line segment from X to Y is of the form $V = (1 - t)X + tY$ for some t in $[0, 1]$.
Because f is linear, $f(V) = (1 - t)f(X) + tf(Y)$. If $f(X) = f(Y)$ then

$$f(V) = f(X) = f(Y)$$

for all V. If $f(X) < f(Y)$, then

$$f(X) = tf(X) + (1 - t)f(X) \leq tf(X) + (1 - t)f(Y) = f(V)$$
$$\leq tf(Y) + (1 - t)f(Y) = f(Y).$$

(ii) Let V be any interior point of C; any line through V intersects the boundary of C in two points X and Y, and we can assume that $f(X) \leq f(Y)$. Then by (i) $f(X) \leq f(V) \leq f(Y)$. Hence f assumes on its boundary a value $f(X)$ at least as small as $f(V)$ and a value $f(Y)$ at least as large as $f(V)$.

Exercises 10.2

2. First rewrite the five inequalities of Example B in the form (10.3) with the variables $y_1, y_2, \ldots, y_6$ in place of $x_{11}, x_{12}, \ldots, x_{23}$. Use new variables $x_1, \ldots, x_5$ for the dual problem:

$$-x_1 \qquad\quad + x_3 \qquad\qquad\qquad \leq 1,$$
$$-x_1 \qquad\qquad\qquad + x_4 \qquad\quad \leq 2,$$
$$-x_1 \qquad\qquad\qquad\qquad\quad + x_5 \leq 3,$$
$$-x_2 \; + x_3 \qquad\qquad\qquad \leq 2,$$
$$-x_2 \qquad\qquad + x_4 \qquad\quad \leq 4,$$
$$-x_2 \qquad\qquad\qquad\quad + x_5 \leq 6,$$
$$x_i \geq 0, \qquad i = 1, \ldots, 5.$$

Maximize $-4000x_1 - 7000x_2 + 2000x_3 + 3000x_4 + 5000x_5$.

3.
$$6x_1 + 2x_2 + \; 4x_3 \leq 20,$$
$$2x_1 + 2x_2 + 12x_3 \leq 16,$$
$$x_i \geq 0, \qquad i = 1, 2, 3.$$

Maximize $12x_1 + 8x_2 + 24x_3$.

There are many feasible vectors; for example $X_0 = (2, 1, 1/2)$. The value $f(X_0)$ of the dual objective function at X_0 is 44. There also are many feasible vectors for the orchard problem; for example $Y_0 = (1, 4)$. Then $g(Y_0) = 84 \geq 44 = f(X_0)$.

5. (i)
$$40x_1 + 60x_2 \leq 240,$$
$$60x_1 + 40x_2 \leq 180,$$
$$80x_1 + 20x_2 \leq 160,$$
$$70x_1 + 30x_2 \leq 210,$$
$$x_1, x_2 \geq 0.$$

Maximize $60x_1 + 35x_2$.

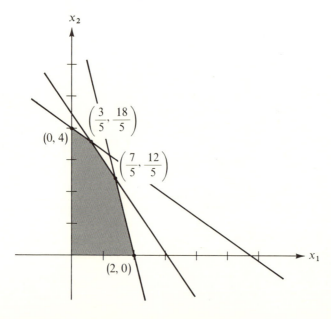

(ii) The corner points of the feasible region are $(0, 4)$, $(3/5, 12/5)$, $(2, 0)$. The values of the objective function at those points are 140, 162, 168, 120. Maximum value is 168 at $(7/5, 12/5)$.

(iii) At $(7/5, 12/5)$ the value of $65x_1 + 35x_2$ is 175.

(iv) Let y_j be the proportion of alloy A_j used, $j = 1, \ldots, 4$. If $y_j = 1/4$ for each j, all constraints are satisfied at cost 197.5. To reduce this cost, the processor should reduce the proportion of A_1. Suppose y_1 is reduced to zero. Then for the percentage of tin to be at least 35, the proportion of A_2 must be at least one-half. The proportions of $y_1 = 0$, $y_2 = 1/2$, $y_3 = 0$, $y_4 = 1/2$ satisfies the constraints at a cost of 195. To reduce costs further, the processor can increase the proportion of A_2 and A_3 and reduce that of A_4. The values $y_1 = 0$, $y_2 = 3/4$, $y_3 = 1/4$, $y_4 = 0$ produce an alloy of 65 percent lead and 35 percent tin at a cost of 175.

(v) Because the objective function (cost) of the alloy problem always is as great as the objective function of the dual problem (for feasible vectors in each case), the minimum cost of the alloy must be at least 175, the maximum value of the dual objective function. Hence $(0, 3/4, 1/4, 0)$ is optimal.

6. (i)

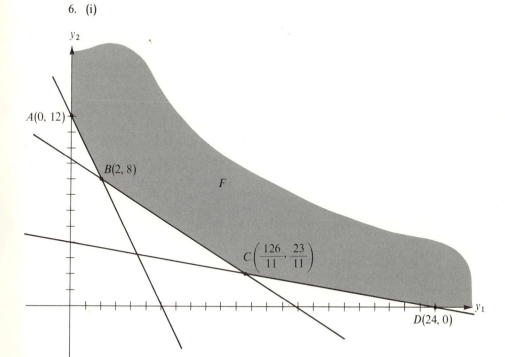

(ii) $\min(y_1 + y_2) = 10$ at B.

$\min(3y_1 + y_2) = 12$ at A.

$\min(y_1 + 3y_2) = 195/11$ at C.

$\min(3y_1 - y_2)$ does not exist on F.

Exercises 10.3

1. (i)

	x_1	x_2	x_3	-1	$=$
y_1	6	2	4	20	$-v_1$
y_2	2	2	12	16	$-v_2$
-1	12	8	24		
$=$	u_1	u_2	u_3		

(iii)

	x_1	x_2	x_3	x_4	x_5	-1	$=$
y_{11}	-1	0	1	0	0	1	$-v_1$
y_{12}	-1	0	0	1	0	2	$-v_2$
y_{13}	-1	0	0	0	1	3	$-v_3$
y_{21}	0	-1	1	0	0	2	$-v_4$
y_{22}	0	-1	0	1	0	4	$-v_5$
y_{23}	0	-1	0	0	1	6	$-v_6$
-1	-4	-7	2	3	5		
$=$	u_1	u_2	u_3	u_4	u_5		

2.

	x_1	x_2	v_2	-1	$=$
y_1	1/4	1/2	$-1/4$	95/2	$-v_1$
u_3	3/4	1/2	1/4	105/2	$-x_3$
y_3	3	2	0	150	$-v_3$
-1	1/2	1	$-3/2$	-315	
$=$	u_1	u_2	y_2		

3. (ii)

	v_3	v_2	x_3	-1	$=$
y_1	0	$-1/3$	5/3	1/3	$-v_1$
u_2	$-1/9$	1/9	$-1/3$	2/9	$-x_2$
u_1	2/9	1/9	0	11/9	$-x_1$
-1	$-40/9$	$-35/9$	55/3		
$=$	y_3	y_2	u_3		

(iv) The objective function of the maximum problem has the value

$$25(11/9) + 10(13/45) + 15(1/5) = 328/9,$$

and the objective function of the minimal problem has the value

$$2(11) + 5(2/9) + 3(40/9) = 328/9.$$

The given vector X is a solution of the linear programming problem, and Y is a solution of the dual problem.

4. (iii)

	x_1	x_2	x_3	x_4	-1	$=$
y_1	5	6	0	0	12000	$-v_1$
y_2	3	1	0	0	15000	$-v_2$
y_3	0	0	5	6	12000	$-v_3$
y_4	0	0	3	1	15000	$-v_4$
y_5	1	0	1	0	3000	$-v_5$
y_6	0	1	0	1	3000	$-v_6$
-1	20	18	40	35		
$=$	u_1	u_2	u_3	u_4		

Exercises 10.4

1. (i)

	x_1	x_2	x_3	-1	$=$
y_1	6	2	4	20	$-v_1$
y_2	2	2	12*	16	$-v_2$
-1	12	8	24	0	
$=$	u_1	u_2	u_3		

	x_1	x_2	v_2	-1	$=$
y_1	16/3*	4/3	$-1/3$	44/3	$-v_1$
u_3	1/6	1/6	1/12	4/3	$-x_3$
-1	8	4	-2	-32	
$=$	u_1	u_2	y_2		

	v_1	x_2	v_2	-1	$=$
u_1	3/16	1/4	$-1/16$	11/4	$-x_1$
u_3	$-1/32$	1/8*	3/32	7/8	$-x_3$
-1	$-3/2$	2	$-3/2$	-54	
$=$	y_1	u_2	y_2		

	v_1	x_2	v_2	-1	$=$
u_1				1	$-x_1$
u_2				7	$-x_2$
-1	-1	-16	-3	-68	
$=$	y_2	u_3	y_2		

$$(X; V) = (1, 7, 0; 0, 0).$$
$$(U; Y) = (0, 0, 16; 1, 3).$$
$$\text{Max } BX = 12(1) + 8(7) + 24(0) = 68.$$
$$\text{Min } YC = 20(1) + 16(3) = 68.$$

(iii) Beginning with the tableau of Exercise 10.3-1(iii), successive pivots on the entries in positions (3, 5), (6, 1), (5, 4) and (4, 3) produce this solution:

$$(X; V) = (3, 0, 2, 4, 6; 2, 1, 0, 0, 0, 0),$$
$$(U; Y) = (0, 1, 0, 0, 0; 0, 0, 4, 2, 3, 1),$$
$$\text{Min } YC = (0, 0, 4, 2, 3, 1) \cdot (0, 0, 3, 2, 4, 6) = 34,$$
$$\text{Max } BX = (-4, -7, 2, 3, 5) \cdot (3, 0, 2, 4, 6) = 34.$$

(v) After making the pivot described in Exercise 10.3-2, pivot on the (3, 2) entry of the resulting tableau to obtain this tableau.

	x_1	v_3	v_2	-1	$=$
y_1				10	$-v_1$
u_3				15	$-x_3$
u_2				75	$-x_2$
-1	-1	$-1/2$	$-3/2$	-390	
$=$	u_1	y_3	y_2		

Then $(X; V) = (0, 75, 15; 10, 0, 0),$

$(U, Y) = (-1, 0, 0; 0, 3/2, 1/2),$

$BX = (5, 4, 6) \cdot (0, 75, 15) = 390,$

$YC = (0, 3/2, 1/2) \cdot (100, 210, 150) = 390.$

2. Let x_{iL} be the number of items of product L produced by machine i, where $i = 1, 2, 3$ and L = A, B, C, D. Then

$$10x_{1A} + 5x_{1B} + 2x_{1C} + x_{1D} \le 50,$$
$$6x_{2A} + 6x_{2B} + 2x_{2C} + 2x_{2D} \le 36,$$
$$9/2x_{3A} + 18x_{3B} + 3/2x_{3C} + 6x_{3D} \le 81,$$

Profit $= 9(x_{1A} + x_{2A} + x_{3A}) + 7(x_{1B} + x_{2B} + x_{3B})$
$+ 6(x_{1C} + x_{2C} + x_{3C}) + 4(x_{1D} + x_{2D} + x_{3D}).$

	x_{1A}	x_{1B}	x_{1C}	x_{1D}	x_{2A}	x_{2B}	x_{2C}	x_{2D}	x_{3A}	x_{3B}	x_{3C}	x_{3D}	-1	$=$
y_1	10	5	2	1	0	0	0	0	0	0	0	0	50	$-v_1$
y_2	0	0	0	0	6	6	2	2	0	0	0	0	36	$-v_2$
y_3	0	0	0	0	0	0	0	0	9/2	18	3/2	6	81	$-v_3$
-1	9	7	6	4	9	7	6	4	9	7	6	4	0	
$=$	u_1	u_2	u_3	u_4	u_5	u_6	u_7	u_8	u_9	u_{10}	u_{11}	u_{12}		

Successive pivots on the entries in positions $(3, 11)$, $(2, 7)$, and $(1, 3)$ produce this solution:

$$(X; V) = (0, 0, 25, 0, 0, 0, 18, 0, 0, 0, 54, 0; 0, 0, 0),$$
$$(U; Y) = (-11, -3, 0, 0, -9, -11, 0, -2, -9, -65, 0, -20; 2, 3, 4),$$
$$BX = 6(25) + 6(18) + 6(54) = 582,$$
$$YC = 3(50) + 3(36) + 4(81) = 582.$$

For maximum profit each machine should produce only product C, and all machine time will be used (because v_i represents the slack time on machine i).

4. (i) Begin by pivoting on the $(2, 2)$ entry, producing positive entries throughout the final column (except for the value -20 in the lower right-hand corner. After one more pivot the solution can be obtained:

$$(X; V) = (5, 15, 0; 5, 0, 0),$$
$$(U; Y) = (0, 0, 9; 0, 3, 5),$$
$$BX = 45 = YC.$$

5. Let x_{Lj} denote the proportion of funds invested in bonds of type L_j, where $L = A, B, C$ and $j = 1, 2$. Then

$$x_{A1} + x_{A2} \geq 0.40,$$
$$x_{B1} + x_{B2} \leq 0.35,$$
$$x_{C1} + x_{C2} \leq 0.35,$$

$$x_{A1} + x_{A2} + x_{B1} + x_{B2} + x_{C1} + x_{C2} = 1.$$

Return $= 0.06x_{A1} + 0.05x_{A2} + 0.07x_{B1} + 0.08x_{B2} + 0.10x_{C1} + 0.09x_{C2}$.

Any equation $y = c$ can be written as a pair of inequalities, $y \leq c$ and $y \geq c$, or equivalently, $y \leq c$ and $-y \leq -c$. Using the latter formulation, we obtain the following tableau.

	x_{A1}	x_{A2}	x_{B1}	x_{B2}	x_{C1}	x_{C2}	-1	$=$
y_1	-1	-1	0	0	0	0	$-.40$	$-v_1$
y_2	0	0	1	1	0	0	$.35$	$-v_2$
y_3	0	0	0	0	1	1	$.35$	$-v_3$
y_4	1	1	1	1	1	1	1.00	$-v_4$
y_5	-1	-1	-1	-1	-1	-1	-1.00	$-v_5$
-1	0.06	0.05	0.07	0.08	0.10	0.09		
$=$	u_1	u_2	u_3	u_4	u_5	u_6		

The initial pivot should be chosen as described in the text to decrease the number of negative entries in the last column. After five or six pivots the solution is obtained:

the maximum yield is 7.9 percent, obtained with 40 percent A_1 bonds, 25 percent B_2 bonds, and 35 percent C_1 bonds.

$$(X; V) = (0.40, 0, 0, 0.25, 0.35, 0; 0, 0.10, 0, 0, 0),$$
$$(U; Y) = (0, 0.01, 0.01, 0, 0, 0.01; 0.02, 0, 0.02, 0.08, 0),$$
$$BX = YC = 0.079.$$

CHAPTER 11

Exercises 11.1

1. (i) $\{A^{(k)}\}$ converges to $\begin{pmatrix} 0 & 0 \\ 1 & -3 \end{pmatrix}$.

    (ii) $\{A^{(k)}\}$ fails to converge because $\{\sin \pi k\}$ does not converge.

2. $\sum_{k=0}^{\infty} A^{(k)}$ converges if and only if $\sum_{k=0}^{\infty} a_{ij}^{(k)}$ converges for each i and j. The latter converges if and only if the sequence of partial sums $\{s_{ij}^{(m)}\}$ converges, where $s_{ij}^{(m)} = \sum_{k=0}^{m} a_{ij}^{(k)}$. This is valid for each i and j if and only if $\{S^{(m)}\}$ converges, where

$$S^{(m)} = A^{(0)} + A^{(1)} + \cdots + A^{(m)}.$$

6. Let $S^{(t)} = \sum_{k=0}^{t} P A^{(k)} Q$. Fix i and j. Then

$$s_{ij}^{(t)} = \sum_{k=0}^{t} \left(\sum_{r=1}^{m} \sum_{s=1}^{n} p_{ir} a_{rs}^{(k)} q_{sj} \right) = \sum_{r=1}^{m} \left(p_{ir} \sum_{s=1}^{n} \left(\sum_{k=0}^{t} a_{rs}^{(k)} q_{sj} \right) \right)$$

Because

$$\lim_{t \to \infty} \sum_{k=0}^{t} a_{rs}^{(k)} = a_{rs}, \quad \lim_{t \to \infty} s_{ij}^{(t)} = \sum_{r=1}^{m} p_{ir} \left(\sum_{s=1}^{n} a_{rs} q_{sj} \right),$$

which is the (i, j) entry of PAQ.

Exercises 11.2

1. (iii) The characteristic values of A are 2 and -2, so A is diagonable: $P^{-1} A P = \text{diag}(2, -2) = J$, where

$$P = \begin{pmatrix} 1 & -3 \\ 1 & 1 \end{pmatrix} \text{ and } P^{-1} = \tfrac{1}{4} \begin{pmatrix} 1 & 3 \\ -1 & 1 \end{pmatrix}.$$

Hence $e^A = P e^J P^{-1} = \tfrac{1}{4} \begin{pmatrix} e^2 + 3e^{-2} & 3e^2 - 3e^{-2} \\ e^2 - e^{-2} & 3e^2 + e^{-2} \end{pmatrix}.$

(iv) $e^A = \tfrac{1}{4} \begin{pmatrix} 4e^2 & 0 & 0 \\ 0 & e^2 + 3e^{-2} & 3e^2 - 3e^{-2} \\ 0 & e^2 - e^{-2} & 3e^2 + e^{-2} \end{pmatrix}.$

2. $P^{-1}AP = \text{diag}(2, 2, 1) = J$. Hence

$$e^A = Pe^J P^{-1} = \begin{pmatrix} 2 & 2 & -3 \\ 0 & 1 & 1 \\ 1 & 0 & -3 \end{pmatrix} \begin{pmatrix} e^2 & 0 & 0 \\ 0 & e^2 & 0 \\ 0 & 0 & e \end{pmatrix} \begin{pmatrix} 3 & -6 & -5 \\ -1 & 3 & 2 \\ 1 & -2 & -2 \end{pmatrix}$$

$$= \begin{pmatrix} 4e^2 - 3e & 6e & -6e^2 + 6e \\ -e^2 + e & -e^2 + e & 2e^2 - 2e \\ 3e^2 - 3e & 3e^2 - 3e & -5e^2 + 6e \end{pmatrix}.$$

4. (i) Let $f(X) = \sum_{k=0}^{\infty} a_k X^k$. If $Y = \begin{pmatrix} Y_1 & Z \\ Z & Y_2 \end{pmatrix}$, then $Y^k = \begin{pmatrix} Y_1^k & Z \\ Z & Y_2^k \end{pmatrix}$,

so $f(Y) = \sum_{k=0}^{\infty} a_k Y^k$. The partial sum is

$$\sum_{k=0}^{m} a_k Y^k = \begin{pmatrix} \sum_{k=0}^{m} a_k Y_1^k & Z \\ Z & \sum_{k=0}^{m} a_k Y_2^k \end{pmatrix},$$

so $f(Y)$ converges if and only if $f(Y_1)$ and $f(Y_2)$ both converge. In that case $f(Y)$ converges to

$$\begin{pmatrix} f(Y_1) & Z \\ Z & f(Y_2) \end{pmatrix}.$$

(ii) When Y is nilpotent any power series in Y reduces to a finite sum because $Y^m = Z$ for all large m.

(iii) From Exercise 7.1-9 det A is the product of the n characteristic values. By Theorem 11.6 the characteristic values of e^X are e^λ where λ is a characteristic value of X. Hence

$$\det(e^X) = e^{\lambda_1} e^{\lambda_2} \cdots e^{\lambda_n} = e^{(\lambda_1 + \cdots + \lambda_n)} = e^{\text{tr } X}.$$

Because $e^{\text{tr } X} \neq 0$, e^X is nonsingular.

Because X and $-X$ commute, $e^{-X} e^X = e^X e^{-X} = e^Z = I$. Hence e^{-X} is the inverse of e^X.

5. (i) The characteristic values of A are $\lambda_1 = 1$, $\lambda_2 = 2$, $\lambda_3 = 3$. Let $f(\lambda_1) = a$, $f(\lambda_2) = b$, $f(\lambda_3) = c$. Then $D = 2$, $D_0 = 6a - 6b + 2c$, $D_1 = -5a + b - 3c$, and $D_2 = a - 2b + c$;

$$f(A) = \begin{pmatrix} -a + 2b & -b + c & a - c \\ -2a + 2b & -b + 2c & 2a - 2c \\ -2a + 2b & -b + c & 2a - c \end{pmatrix}.$$

7. $e^A e^B = (I + A + A^2/2! + \cdots + A^k/k! + \cdots)(I + B + \cdots + B^m/m! + \cdots)$. The term of order n in the product is

$$\frac{A^n}{n!} + \frac{A^{n-1}B}{(n-1)!} + \frac{A^{n-2}B^2}{(n-2)!2!} + \cdots + \frac{A^{n-k}B^k}{(n-k)!k!} + \cdots + \frac{B^n}{n!},$$

or

$$\frac{1}{n!} \sum_{k=0}^{n} C(n, k) A^{n-k} B^{k}.$$

By the binomial theorem, this is precisely the expression for

$$\frac{1}{n!} (A + B)^n$$

when A and B commute. In that case $e^A e^B = e^{A+B}$. But Exercise 11.2-6 demonstrates an example in which $e^A e^B$, $e^B e^A$, and e^{A+B} are all distinct and $AB \neq BA$.

Exercises 11.3

2. (a) The (i, j) entry of YX is $\sum_{k=1}^{n} y_{ik}(t) x_{kj}(t)$. Hence the (i, j) entry of $\mathscr{D}(YX)$ is $\sum_{k=1}^{n} (y_{ik}'(t) x_{kj}(t) + y_{ik}(t) x_{kj}'(t))$, so $\mathscr{D}(YX) = (\mathscr{D}Y)X + Y(\mathscr{D}X)$.

(b) By writing $X^r = X^{r-1} X$ and using Theorem 11.8(a) and finite induction, we obtain the statement of Theorem 11.8(b).

(c) If X is nonsingular, $Z = \mathscr{D}(X^{-1}X) = (\mathscr{D}X^{-1})X + X^{-1}\mathscr{D}X$. Hence $(\mathscr{D}X^{-1})X = -X^{-1}\mathscr{D}X$, and $\mathscr{D}(X^{-1}) = -X^{-1}(\mathscr{D}X)X^{-1}$.

3.

$$\int_a^b X(t)\, dt = \left(\int_a^b x_{ij}(t)\, dt \right) = (y_{ij}(b) - y_{ij}(a)),$$

where $y_{ij}'(t) = x_{ij}(t)$. Hence if $Y(t)$ is any matrix such that $\mathscr{D}Y(t) = X(t)$, then

$$\int_a^b X(t)\, dt = Y(b) - Y(a).$$

5. Use induction on m and the Laplace expansion by elements of the first column.

Exercises 11.4

1. (i) The Jordan form of A is $J = \begin{pmatrix} 4 & 0 & 0 \\ 1 & 4 & 0 \\ 0 & 0 & 4 \end{pmatrix} = P^{-1}AP$

where $P = \begin{pmatrix} 1 & 1 & 1 \\ 1 & 0 & 1 \\ 1 & 1 & 0 \end{pmatrix}$ and $P^{-1} = \begin{pmatrix} -1 & 1 & 1 \\ 1 & -1 & 0 \\ 1 & 0 & -1 \end{pmatrix}$.

Hence

$$Y(0) = P^{-1}X(0) = \begin{pmatrix} -1 \\ 1 \\ 2 \end{pmatrix}.$$

To calculate e^{tJ} we first let $f(x) = e^{tx}$; then $f'(x) = te^{tx}$. Because $\lambda = 2$,

$$e^{tJ} = \begin{pmatrix} f(4) & 0 & 0 \\ f'(4) & f(4) & 0 \\ 0 & 0 & f(4) \end{pmatrix} = e^{4t} \begin{pmatrix} 1 & 0 & 0 \\ t & 1 & 0 \\ 0 & 0 & 1 \end{pmatrix}.$$

Then

$$Y(t) = e^{tJ} Y(0) = e^{4t} \begin{pmatrix} -1 \\ 1-t \\ 2 \end{pmatrix}, \text{ and } X(t) = PY(t) = e^{4t} \begin{pmatrix} 2-t \\ 1 \\ -t \end{pmatrix}.$$

2. A Jordan form of A is $J = \begin{pmatrix} 1 & 0 & 0 \\ 1 & 1 & 0 \\ 0 & 0 & 2 \end{pmatrix} = P^{-1}AP,$

where $P = \begin{pmatrix} 0 & 5 & 1 \\ 2 & 2 & 1 \\ -3 & -5 & -2 \end{pmatrix}$ and $P^{-1} = \begin{pmatrix} 1 & 5 & 3 \\ 1 & 3 & 2 \\ -4 & -15 & -10 \end{pmatrix}.$

Then $e^{tJ} = \begin{pmatrix} e^t & 0 & 0 \\ te^t & e^t & 0 \\ 0 & 0 & e^{2t} \end{pmatrix}$ and $X(t) = e^{tA}X(0)$, where

$$e^{tA} = Pe^{tJ}P^{-1} = \begin{pmatrix} 5e^t + 5te^t - 4e^{2t} & 15e^t + 25te^t - 15e^{2t} & 10e^t + 15te^t - 10e^{2t} \\ 4e^t + 2te^t - 4e^{2t} & 16e^t + 10te^t - 15e^{2t} & 10e^t + 6te^t - 10e^{2t} \\ -8e^t + 5te^t + 8e^{2t} & -30e^t - 25te^t + 30e^{2t} & -19e^t - 15te^t + 20e^{2t} \end{pmatrix}.$$

(i) If $X(0) = \begin{pmatrix} 2 \\ 0 \\ 1 \end{pmatrix}$, then $X(t) = \begin{pmatrix} -5te^t + 2e^{2t} \\ -2e^t - 2te^t + 2e^{2t} \\ 3e^t + 5te^t - 4e^{2t} \end{pmatrix}.$

(ii) If $X(0) = \begin{pmatrix} -1 \\ 2 \\ 3 \end{pmatrix}$, then $X(t) = \begin{pmatrix} -5e^t + 4e^{2t} \\ -2e^t + 4e^{2t} \\ 5e^t - 8e^{2t} \end{pmatrix}.$

(iii) If $X(0) = \begin{pmatrix} 5 \\ -2 \\ 1 \end{pmatrix}$, then $X(t) = \begin{pmatrix} 5e^t - 10te^t \\ -2e^t - 4te^t \\ 25e^t + 10te^t \end{pmatrix}.$

3. (i) Let $y_1 = x(t)$, $y_2 = x'(t)$, $y_3 = x''(t)$. Then

$$\begin{aligned} y_1' &= & y_2, \\ y_2' &= & y_3, \\ y_3' &= y_1 - 3y_2 + 3y_3. \end{aligned}$$

In matrix form $Y' = AY$, where

$$A = \begin{pmatrix} 0 & 1 & 0 \\ 0 & 0 & 1 \\ 1 & -3 & 3 \end{pmatrix} \text{ and } Y = \begin{pmatrix} y_1 \\ y_2 \\ y_3 \end{pmatrix}.$$

The solution is $Y(t) = e^{tA}Y(0)$. The characteristic polynomial of A is $p(x) = -(x^3 - 3x^2 + 3x - 1) = (1 - x)^3$, and the Jordan form of A can be determined to be

$$J = P^{-1}AP = \begin{pmatrix} 1 & 0 & 0 \\ 1 & 1 & 0 \\ 0 & 1 & 1 \end{pmatrix}, \text{ where } P = \begin{pmatrix} 3 & -2 & 1 \\ 1 & -1 & 1 \\ 0 & 0 & 1 \end{pmatrix}, P^{-1} = \begin{pmatrix} 1 & -2 & 1 \\ 1 & -3 & 2 \\ 0 & 0 & 1 \end{pmatrix}.$$

$$e^{tJ} = \begin{pmatrix} e^t & 0 & 0 \\ te^t & e^t & 0 \\ \frac{t^2}{2}e^t & te^t & e^t \end{pmatrix} \text{ and } e^{tA} = e^t \begin{pmatrix} 1 - t + \frac{t^2}{2} & t - t^2 & \frac{t^2}{2} \\ \frac{t^2}{2} & 1 - t - t^2 & t + \frac{t^2}{2} \\ t + \frac{t^2}{2} & -3t - t^2 & 1 + 2t + \frac{t^2}{2} \end{pmatrix}.$$

Because $Y(0) = \begin{pmatrix} 2 \\ -1 \\ 0 \end{pmatrix}$, $Y(t) = e^t \begin{pmatrix} 2 - 3t + 2t^2 \\ -1 + t + 3t^2/2 \\ 5t + 2t^2 \end{pmatrix}.$

The desired solution is $y_1(t) = x(t) = (2 - 3t + 2t^2)e^t$.

  4. (ii) The characteristic values are -2 and 2, with the latter having multiplicity two. Hence $\{e^{-2t}, e^{2t}, te^{2t}\}$ is a basis of the solution space. Let

$$x(t) = ae^{-2t} + (b + ct)e^{2t},$$

where a, b, and c are chosen to meet the initial conditions:

$$x(0) = \quad 1 = \quad a + \ b,$$
$$x'(0) = \quad 4 = -2a + 2b + \ c,$$
$$x''(0) = -4 = \quad 4a + 4b + 4c.$$

The only solution of this linear system is $a = -1$, $b = 2$, $c = -2$, and the unique solution of the differential equation is

$$x(t) = -e^{-2t} + 2(1 - t)e^{2t}.$$

  5. Let $y_1 = t^r e^{at}$ and $y_2 = t^s e^{at}$, where $r \neq s$. Then $y_1' = e^{at}(at^r + rt^{r-1})$, $y_2' = e^{at}(at^s + st^{s-1})$, and

$$W(y_1, y_2) = \det \begin{pmatrix} y_1 & y_2 \\ y_1' & y_2' \end{pmatrix} = e^{2at} \det \begin{pmatrix} t^r & t^s \\ at^r + rt^{r-1} & at^s + st^{s-1} \end{pmatrix}$$

$$= e^{2at}t^{r+s-1}[(at + s) - (at + r)] = (s - r)e^{2at}t^{r+s-1}.$$

Because $W(y_1, y_2) = 0$ only at $t = 0$, the given set is linearly independent.

INDEX